W9-BSA-996

Chemical Process Safety

PRENTICE HALL INTERNATIONAL SERIES IN THE PHYSICAL AND CHEMICAL ENGINEERING SCIENCES

AMUNDSON *Mathematical Methods in Chemical Engineering: Matrices and Their Applications*

BALZHIZER, SAMUELS, AND ELLIASSEN *Chemical Engineering Thermodynamics*

BRIAN *Staged Cascades in Chemical Processing*

BUTT *Reaction Kinetics and Reactor Design*

CROWL AND LOUVAR, *Chemical Process Safety: Fundamentals with Applications*

DENN *Process Fluid Mechanics*

FOGLER *Elements of Chemical Reaction Engineering*

FOGLER *Elements of Chemical Reaction Engineering, 2nd edition*

HIMMELBLAU *Basic Principles and Calculations in Chemical Engineering, 5th edition*

HOLLAND *Fundamentals and Modeling of Separation Processes: Absorption, Distillation, Evaporation, and Extraction*

HOLLAND AND ANTHONY *Fundamentals of Chemical Reaction Engineering*

KUBICEK AND HAVACEK *Numerical Solution of Nonlinear Boundary Value Problems with Applications*

LEVICH *Physiochemical Hydrodynamics*

MODELL AND REID *Thermodynamics and its Applications, 2nd edition*

MYERS AND SEIDER *Introduction to Chemical Engineering and Computer Calculations*

NEWMAN *Electrochemical Systems*

PRAUSNITZ, LICHTENTHALER, AND DE AZEVEDO *Molecular Thermodynamics of Fluid-Phase Equilibria, 2nd edition*

PRAUSNITZ ET AL. *Computer Calculations for Multicomponent Vapor-Liquid and Liquid-Liquid Equilibria*

RHEE ET AL. *First-Order Partial Differential Equations: Theory and Applications of Single Equations. Volumes I and II*

RUDD ET AL. *Process Synthesis*

SCHULTZ *Diffraction for Materials Scientists*

SCHULTZ *Polymer Materials Science*

SCHULTZ AND FAKIROV, eds. *Solid State Behavior of Linear Polyesters and Polyamides*

VILLADSEN AND MICHELSEN *Solution of Differential Equation by Polynomial Approximation*

WHITE *Heterogeneous Catalysis*

WILLIAMS *Polymer Science Engineering*

Chemical Process Safety
Fundamentals with Applications

Daniel A. Crowl
Michigan Technological University

Joseph F. Louvar
BASF Corporation
Wyandotte, Michigan

PRENTICE HALL P T R, Englewood Cliffs, New Jersey 07632

Library of Congress Cataloging-in-Publication Data

CROWL, DANIEL A.

Chemical process safety: fundamentals with applications/Daniel A. Crowl, Joseph F. Louvar.

p. cm.

Bibliography: p.

Includes index.

ISBN 0-13-129701-5

1. Chemical plants—Safety measures. I. Louvar, Joseph F.

II. Title.

TP155.5.C76 1990

660'.2804—dc20

89-8766

CIP

Editorial/production supervision
and interior design: *Jean Lapidus*

Cover design: *Karen Stephens*

Manufacturing buyer: *Ray Sintel*

Prentice-Hall, Inc.

A Simon & Schuster Company

Englewood Cliffs, New Jersey 07632

Printed in the United States of America

20 19 18 17 16 15 14 13 12 11

ISBN 0-13-129701-5

PRENTICE-HALL INTERNATIONAL (UK) LIMITED, *London*

PRENTICE-HALL OF AUSTRALIA PTY. LIMITED, *Sydney*

PRENTICE-HALL CANADA INC., *Toronto*

PRENTICE-HALL HISPANOAMERICANA, S.A., *Mexico*

PRENTICE-HALL OF INDIA PRIVATE LIMITED, *New Delhi*

PRENTICE-HALL OF JAPAN, INC., *Tokyo*

SIMON & SCHUSTER ASIA PTE. LTD., *Singapore*

EDITORA PRENTICE-HALL DO BRASIL, LTDA., *Rio de Janeiro*

Contents

Preface

This textbook is designed to teach and apply the fundamentals of chemical process safety. It is appropriate for an industrial reference, a senior level undergraduate course or a graduate course in chemical process safety. It can be used by anyone interested in improving chemical process safety, including chemical and mechanical engineers and chemists. More material is presented than can be accommodated in a 3 credit semester course.

The primary objective of this textbook is to encapsulate the important technical fundamentals of chemical process safety. The emphasis on the fundamentals will hopefully help the student and practicing scientist *understand* the concepts and apply them accordingly. This application requires a significant quantity of fundamental knowledge and technology.

Since a textbook fundamentally oriented on safety has never been undertaken, we experienced several difficulties. First, an outline that presented the material in a logical and expanding fashion was very difficult to develop. Second, since chemical process safety involves a very large quantity of material, the motivation was always present to include more and more. We have tried to limit the material to those items deemed essential for industrial and university practice. Third, the fundamental basis for much of the material was long lost and had to be redeveloped.

We firmly believe that a textbook on safety could only have been possible with both industrial and academic inputs. The industrial input insured that the material was industrially relevant. The academic input insured that the material was presented on a fundamental basis to help professors and students understand the concepts.

When this textbook is completed, few universities will have required courses in chemical process safety. While the need is recognized, little flexibility is presently available to add additional courses to the undergraduate chemical engineering curriculum. Also, faculty have little background in chemical process safety, and, finally, inadequate instructional materials are available to teach such a course. We hope this textbook solves the instructional material problem and that the remaining problems will be solved in time.

This textbook is designed for a dedicated course in chemical process safety. However, we feel that chemical process safety should be part of every undergraduate and graduate course in chemistry and chemical and mechanical engineering, just as it is a part of all the industrial experiences. This text can be used as a reference resource for these courses.

We will inevitably be criticized by some who believe our presentation is not complete or that some details are missing. The purpose of this book is not completeness but to provide a starting point for those who wish to learn about this important area.

We would like to thank Harold Fisher of Union Carbide and Joseph Leung of Fauske and Associates for reviewing the sections on two-phase flow relief and providing a number of important corrections and modifications. Edward J. Kerfoot, Michael Capraro and Jim Strickland of BASF Corporation reviewed specific chapters. Dan Goerke helped with the homework problems and solutions and suggested a number of important changes. Bruce Barna, Dave Leddy, and Nam Kim of Michigan Technological University helped develop a number of instructional modules that contributed towards the initial basis for this text. Michigan Technological University also provided a sabbatical experience to one of us (D.A.C.) which contributed to the initial textbook development. Professor Edward R. Fisher of Michigan Tech is thanked for this opportunity.

We thank our coworkers at BASF Corporation for giving us the knowledge and deep appreciation for the importance of process safety in laboratories, pilot plants and plants. This especially includes M. Capraro, J. Strickland, and K. Kilponen.

We also acknowledge and thank the members of the Curriculum Committee of the Center for Chemical Process Safety (CCPS) and the Safety and Loss Prevention Committee of the American Institute of Chemical Engineers. We are honored to be members of both committees. The members of these committees are the experts in safety; their enthusiasm and knowledge have been truly educational and a key inspiration to the development of this text.

We thank John Davenport of Industrial Risk Insurers for providing information concerning several case histories which he particularly chose because of their significant relevance to process safety. John also reviewed Chapters 6 and 7 and provided very valuable information on sprinkler system design. Walter Howard, presently a Process Safety Consultant, has given us important safety information and facts and continued encouragement to complete this project.

Finally, we acknowledge our families, who provided patience, understanding and encouragement.

We hope that this textbook helps prevent chemical plant and university accidents and contributes to a much safer future.

Detroit, Michigan

Daniel A. Crowl

Joseph F. Louvar

Nomenclature

a	velocity of sound (length/time)
A	area (length^2) or Helmholtz free energy (energy); or process component availability
A_t	tank area (length^2)
ΔA	change in Helmoltz free energy (energy/mole)
C	mass concentration (mass/volume) or capacitance (Farads)
C_0, C_1	discharge coefficients (unitless) or concentration at a specified time (mass/volume)
C_p	heat capacity at constant pressure (energy/mass deg)
C_V	heat capacity at constant volume (energy/mass deg)
C_{ppm}	concentration in parts per million by volume
C_{vent}	deflagration vent constant (pressure$^{1/2}$)
$\langle C \rangle$	average or mean mass concentration (mass/volume)
d	diameter (length)
d_p	particle diameter (length)
d_f	diameter of flare stack (length)
D	diffusion coefficient (area/time)
D_0	reference diffusion coefficient (area/time)
D_m	molecular diffusivity (area/time)
D_{tid}	total integrated dose due to a passing puff of vapor (mass time/volume)

E_a	activation energy (energy/mole)
f	Fanning friction factor (unitless)
$f(t)$	failure density function
f_v	mass fraction of vapor (unitless)
F	frictional fluid flow loss term (energy/mass)
g	gravitational acceleration (length/time^2)
g_c	gravitational constant
G	Gibbs free energy (energy/mole) or mass flux (mass/area time)
G_T	mass flux during relief (mass/area time)
ΔG	change in Gibbs free energy (energy/mole)
h	specific enthalpy (energy/mass)
h_L	fluid level above leak in tank (length)
h_L^0	initial fluid level above leak in tank (length)
h_s	leak height above ground level (length)
H	enthalpy (energy/mole) or height (length)
H_f	flare height (length)
H_r	effective release height in plume model (length)
ΔH	change in enthalpy (energy/mole)
ΔH_c	heat of combustion (energy/mass)
ΔH_v	enthalpy of vaporization (energy/mass)
I	sound intensity (decibels)
I_0	reference sound intensity (decibels)
I_s	streaming current (amps)
j	number of inerting purge cycles (unitless)
J	electrical work (energy)
k	non-ideal mixing factor for ventilation (unitless)
k_1, k_2	constants in probit equations
K	mass transfer coefficient (length/time)
K_b	backpressure correction for relief sizing (unitless)
K_g	explosion constant for vapors (length pressure/time)
K_j	eddy diffusivity in x, y or z direction (area/time)
K_P	overpressure correction for relief sizing (unitless)
K_{St}	explosion constant for dusts (length pressure/time)
K_V	viscosity correction for relief sizing (unitless)
K_0	reference mass transfer coefficient (length/time)
K^*	constant eddy diffusivity (area/time)
L	length
LEL	lower explosion limit (volume %)
$LFL = LEL$	lower flammability limit (volume %)

m	mass
m_0	total mass contained in reactor vessel (mass)
m_{TNT}	mass of TNT
m_v	mass of vapor
M	molecular weight (mass/mole)
M_0	reference molecular weight (mass/mole)
Ma	Mach number (unitless)
MTBC	mean time between coincidence (time)
MTBF	mean time between failure (time)
n	number of moles
p	partial pressure (force/area)
p_d	number of dangerous process episodes
P	total pressure or probability
P_b	backpressure for relief sizing (psig)
P_g	gauge pressure (force/area)
P_{max}	maximum pressure for relief sizing (psig)
P_s	set pressure for relief sizing (psig)
P^{sat}	saturation vapor pressure
q	heat (energy/mass)
q_f	heat intensity of flare (energy/time area)
q_s	specific energy release rate at set pressure during reactor relief (energy/mass)
Q	heat (energy) or electrical charge (coulombs)
Q_m	mass discharge rate (mass/time)
Q_m^*	instantaneous mass release (mass)
Q_v	ventilation rate (volume/time)
r	radius (length)
R	electrical resistance (ohms) or reliability
RHI	reaction hazard index defined by Equation 13-1
r_f	vessel filling rate ($time^{-1}$)
R_g	ideal gas constant (pressure volume/mole deg)
Re	Reynolds number (unitless)
S	entropy (energy/mole deg)
S_m	material strength (force/area)
t	time
t_e	emptying time
t_p	time to form a puff of vapor
t_v	vessel wall thickness (length)
t_w	worker shift time

Δt_v	venting time for reactor relief
T	temperature (deg)
T_d	material decomposition temperature (deg)
T_i	time interval
T_m	maximum temperature during reactor relief (deg)
T_s	saturation temperature at set pressure during reactor relief (deg)
u	velocity (length/time)
u_d	dropout velocity of a particle (length/time)
$\bar{u}$	average velocity (length/time)
$\langle u \rangle$	mean or average velocity (length/time)
U	internal energy (energy/mole) or overall heat transfer coefficient (energy/area time) or process component unavailability
UEL	upper explosion limit (volume %)
$UFL = UEL$	upper flammability limit (volume %)
v	specific volume (volume/mass)
v_f	specific volume of liquid (volume/mass)
v_g	specific volume of vapor (volume/mass)
v_{fg}	specific volume change with liquid vaporization (volume/mass)
V	total volume or electrical potential (volts)
V_c	container volume
W	width (length)
W_e	expansion work (energy)
W_s	shaft work (energy)
x	mole fraction or Cartesian coordinant (length)
X_f	distance from flare at grade (length)
y	mole fraction of vapor (unitless) or Cartesian coordinate (length)
Y	probit variable (unitless)
z	height above datum (length) or Cartesian coordinate (length) or compressibility (unitless)
z_e	scaled distance for explosions (length/mass$^{1/3}$)

GREEK LETTERS

α	velocity correction factor (unitless)
β	thermal expansion coefficient (deg^{-1})
δ	double layer thickness (length)
ε	pipe roughness (length) or emissivity (unitless)

ε_r	relative dielectric constant (unitless)
ε_0	permittivity constant for free space (charge^2/force length^2)
Φ	nonideal filling factor (unitless)
γ	heat capacity ratio (unitless)
γ_c	conductivity (mho/cm)
χ	function defined by Equation 9-6
λ	frequency of dangerous episodes
λ_d	average frequency of dangerous episodes
μ	viscosity (mass/length/time) or mean value or failure rate (faults/time)
μ_V	vapor viscosity (mass/length/time)
Ψ	overall discharge coefficient used in Equation 9-14 (unitless)
ρ	density (mass/volume)
ρ_L	liquid density (mass/volume)
ρ_{ref}	reference density for specific gravity (mass/volume)
ρ_V	vapor density (mass/volume)
σ	standard deviation (unitless)
σ_x, σ_y, σ_z	dispersion coefficient (length)
τ	relaxation time
τ_i	inspection period for unrevealed failures
τ_0	operation period for a process component
τ_r	period required to repair a component
τ_u	period of unavailability for unrevealed failures
ζ	zeta potential (volts)

SUBSCRIPTS

c	combustion
f	formation or liquid
g	vapor or gas
m	maximum
s	set pressure
o	initial or reference

SUPERSCRIPTS

°	standard
′	stochastic or random variable

Chemical Process Safety

Introduction 1

In 1987, Robert M. Solow, an economist at the Massachusetts Institute of Technology, received the Nobel Prize in economics for his work in determining the sources of economic growth. Professor Solow concluded that the bulk of an economy's growth is the result of technological advances.

It is reasonable to conclude that the growth of an industry is also dependent on technological advances. This is especially true in the chemical industry, which is entering an era of more complex processes: higher pressure, more reactive chemicals, and exotic chemistry.

More complex processes require more complex safety technology. Many industrialists even believe that the development and application of safety technology is actually a constraint on the growth of the chemical industry.

As chemical process technology becomes more complex, chemical engineers will need a more detailed and fundamental understanding of safety. H. H. Fawcett said, "To know is to survive and to ignore fundamentals is to court disaster."[1] This book sets out the fundamentals of chemical process safety.

Since 1950, significant technological advances have been made in chemical process safety. Today, safety is equal in importance to production and has developed into a scientific discipline which includes many highly technical and complex theories and practices. Examples of the technology of safety include

- Hydrodynamic models representing two-phase flow through a vessel relief.
- Dispersion models representing the spread of toxic vapor through a plant after a release.
- Mathematical techniques to determine the various ways that processes can fail, and the probability of failure.

[1]H. H. Fawcett and W. S. Wood, *Safety and Accident Prevention in Chemical Operations,* 2nd ed. (New York: John Wiley and Sons, 1982), p. 1.

Recent advances in chemical plant safety emphasize the use of appropriate technological tools to provide information for making safety decisions with respect to plant design and operation.

The word "safety" used to mean the older strategy of accident prevention through the use of hard hats, safety shoes, and a variety of rules and regulations. The main emphasis was on worker safety. Much more recently, "safety" has been replaced by "loss prevention." This term includes hazard identification, technical evaluation, and the design of new engineering features to prevent loss. The subject of this text is loss prevention, but for convenience, the words "safety" and "loss prevention" will be used synonymously throughout.

Safety, hazard, and *risk* are frequently-used terms in chemical process safety. Their definitions follow.

- **Safety** or **loss prevention** is the prevention of accidents by the use of appropriate technologies to identify the hazards of a chemical plant and to eliminate them before an accident occurs.
- A **hazard** is anything with the potential for producing an accident.
- **Risk** is the probability of a hazard resulting in an accident.

Chemical plants contain a large variety of hazards. First, there are the usual mechanical hazards that cause worker injuries from tripping, falling, or moving equipment. Second, there are chemical hazards. These include fire and explosion hazards, reactivity hazards, and toxic hazards.

As will be shown later, chemical plants are the safest of all manufacturing facilities. However, the potential always exists for an accident of catastrophic proportions. Despite substantial safety programs by the chemical industry, headlines of the type shown in Figure 1-1 continue to appear in the newspapers.

1-1 SAFETY PROGRAMS

A successful safety program requires several ingredients, as shown in Figure 1-2. These ingredients are

- Safety knowledge
- Safety experience
- Technical competence
- Safety management support
- Commitment

First, the participants must have the technical knowledge of safety gained through classroom or independent study. This includes knowledge of the fundamental laws of nature, the dangerous properties of chemicals, and how these laws and properties apply to safety. Second, the participants must have experience with the process and the safety procedures used to prevent accidents. Third, they must have technical competence. This includes the ability to make proper safety decisions based on a technical evaluation. Fourth, a safety management support system

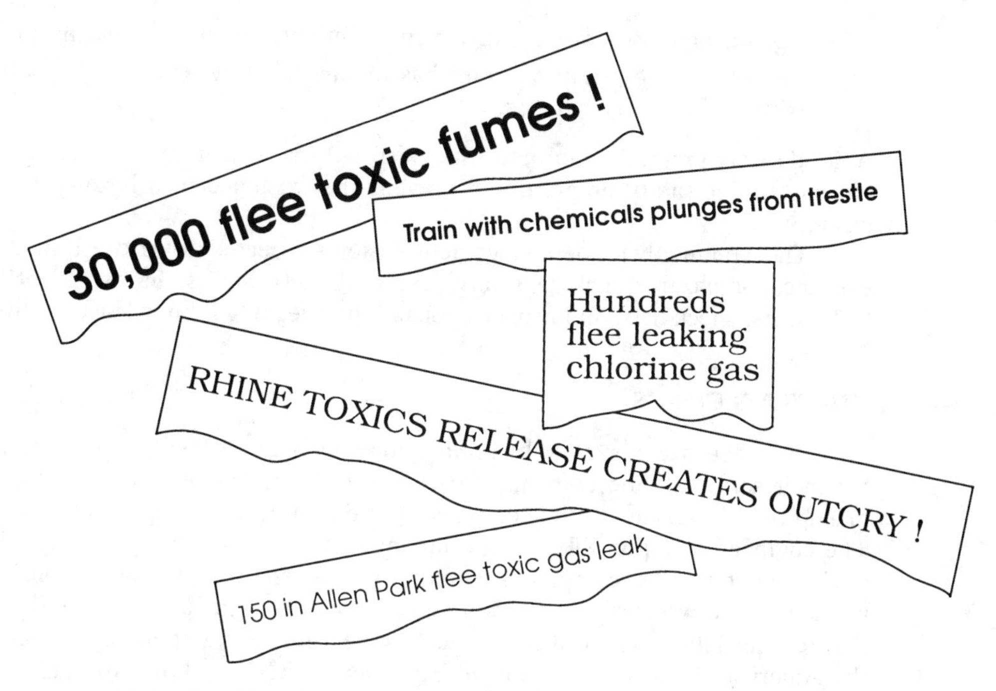

Figure 1-1 Headlines are indicative of the public's concern over chemical safety.

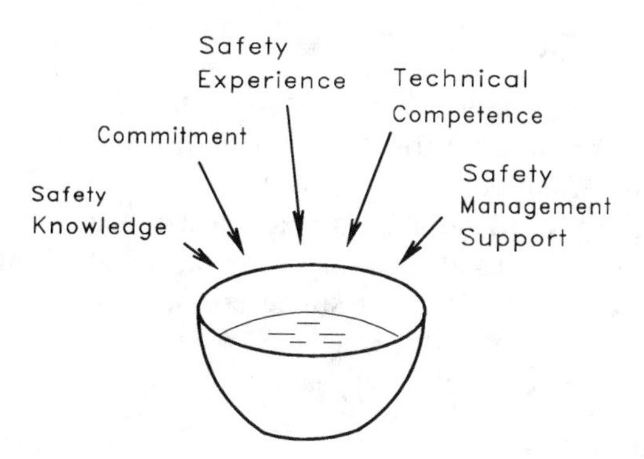

Figure 1-2 The ingredients of a successful safety program.

must be in place to monitor losses and "near misses" and to make the necessary changes to prevent future losses from occurring. Finally, and perhaps most importantly, a safety program must have the commitment from the entire company to work toward improving and practicing safety. Safety must be given importance equal to production.

The most effective means of implementing a safety program is to make it everyone's responsibility in a chemical process plant. The older concept of identifying a few employees to be responsible for safety is inadequate by today's standards. All employees have the responsibility to be knowledgeable about safety and to practice safety.

It is important to recognize the distinction between a good and an outstanding safety program.

- A *good* safety program identifies and eliminates existing safety hazards.
- An *outstanding* safety program has management systems which prevent the existence of safety hazards.

A good safety program eliminates the existing hazards as they are identified while an outstanding safety program will prevent the existence of a hazard in the first place.

The commonly-used management systems directed toward eliminating the existence of hazards include safety reviews, safety audits, hazard identification techniques, check lists, and proper application of technical knowledge, to list a few.

1-2 ENGINEERING ETHICS

Most engineers are employed by private companies that provide wages and benefits for their services. The company earns profits for its shareholders, and engineers must provide a service to the company by maintaining and improving these profits. The engineer is responsible for minimizing losses and providing a safe and secure environment for the company's employees. The engineer has a responsibility to himself, fellow workers, family, community and the engineering profession. Part of this responsibility is described in the Engineering Ethics statement developed by the American Institute of Chemical Engineers (AICHE), shown in Table 1-1.

1-3 ACCIDENT AND LOSS STATISTICS

Accident and loss statistics are important measures of the effectiveness of safety programs. These statistics are valuable for determining if a process is safe or if a safety procedure is working effectively.

Many statistical methods are available to characterize accident and loss performance. These statistics must be used carefully. Like most statistics they are only averages and do not reflect the potential for single episodes involving substantial losses. Unfortunately, no single method is capable of measuring all required aspects. The three systems to be considered here are

- OSHA incidence rate,
- Fatal accident rate (FAR), and
- Fatality rate or deaths per person per year.

All three of these methods report the number of accidents and/or fatalities for a fixed number of workers during a specified period.

OSHA stands for the Occupational Safety and Health Administration of the United States Government. OSHA is responsible for ensuring that workers are provided with a safe working environment. Table 1-2 contains several OSHA definitions applicable to accident statistics.

The OSHA incidence rate is based on cases per 100 worker years. A worker year is assumed to contain 2000 hours (50 work weeks/year $\times$ 40 hours/week). The OSHA incidence rate is therefore based on 200,000 hours of worker exposure to a

TABLE 1-1 AMERICAN INSTITUTE OF CHEMICAL ENGINEERS CODE OF PROFESSIONAL ETHICS

Fundamental principles

Engineers shall uphold and advance the integrity, honor and dignity of the engineering profession by

1. using their knowledge and skill for the enhancement of human welfare;
2. being honest and impartial and serving with fidelity the public, their employers, and clients;
3. striving to increase the competence and prestige of the engineering profession.

Fundamental canons

1. Engineers shall hold paramount the safety, health, and welfare of the public in the performance of their professional duties.
2. Engineers shall perform services only in areas of their competence.
3. Engineers shall issue public statements only in an objective and truthful manner.
4. Engineers shall act in professional matters for each employer or client as faithful agents or trustees, and shall avoid conflicts of interest.
5. Engineers shall build their professional reputations on the merits of their services.
6. Engineers shall act in such a manner as to uphold and enhance the honor, integrity, and dignity of the engineering profession.
7. Engineers shall continue their professional development throughout their careers and shall provide opportunities for the professional development of those engineers under their supervision.

TABLE 1-2 A GLOSSARY OF TERMS USED BY OSHA TO REPRESENT WORK RELATED LOSSES[1].

OCCUPATIONAL INJURY is any injury such as a cut, fracture, sprain, amputation, etc., which results from a work accident or from an exposure involving a single incident in the work environment.

OCCUPATIONAL ILLNESS of an employee is any abnormal condition or disorder, other than one resulting from an occupational injury caused by exposure to environmental factors associated with employment. It includes acute and chronic illnesses or diseases which may be caused by inhalation, absorption, ingestion, or direct contact.

LOST WORKDAYS are those days which the employee would have worked but could not because of occupational injury or illness. The number of lost workdays should not include the day of injury or onset of illness. The number of days includes all days (consecutive or not) on which, because of injury or illness (1) the employee would have worked but could not or (2) the employee was assigned to a temporary job, or (3) the employee worked at a permanent job less than full time, or (4) the employee worked at a permanently assigned job but could not perform all duties normally connected with it.

RECORDABLE CASES are those involving an occupational injury or occupational illness, including deaths. Not recordable are first aid cases which involve onetime treatment and subsequent observation of minor scratches, cuts, burns, splinters, etc., which do not ordinarily require medical care, even though such treatment is provided by a physician or registered professional personnel.

NONFATAL CASES WITHOUT LOST WORKDAYS are cases of occupational injury or illness which did not involve fatalities or lost workdays but did result in (1) transfer to another job or termination of employment, or (2) medical treatment other than first aid, or (3) diagnosis of occupational illness, or (4) loss of consciousness, or (5) restriction of work or motion.

[1]*Accident Facts,* 1985 Edition (Chicago: National Safety Council, 1985), p. 30.

hazard. The OSHA incidence rate is calculated from the number of occupational injuries and illnesses and the total number of employee hours worked during the applicable period. The following equation is used

$$\text{OSHA Incidence Rate (based on injuries and illness)} = (\text{Number of injuries \& illnesses} \times 200{,}000) \Big/ \text{Total hours worked by all employees during period covered} \tag{1-1}$$

An incidence rate can also be based on lost workdays instead of injuries and illnesses. For this case

$$\text{OSHA Incidence Rate (based on lost workdays)} = (\text{Number of lost workdays} \times 200{,}000) \Big/ \text{Total hours worked by all employees during period covered.} \tag{1-2}$$

The definition of a "lost workday" is given in Table 1-2.

The OSHA incidence rate provides information on all types of work related injuries and illnesses, including fatalities. This provides a better representation of worker accidents than systems based on fatalities alone. For instance, a plant might experience many small accidents with resulting injuries, but no fatalities. On the other hand, fatality data cannot be extracted from the OSHA incidence rate without additional information.

The Fatal Accident Rate (FAR) is used mostly by the British chemical industry. This statistic is used here because there are some useful and interesting FAR data available in the open literature. The FAR reports the number of fatalities based on 1000 employees working their entire lifetime. The employees are assumed to work a total of 50 years. Thus, the FAR is based on 10^8 working hours. The resulting equation is

$$\text{FAR} = (\text{Number of fatalities} \times 10^8) \Big/ \text{Total hours worked by all employees during period covered} \tag{1-3}$$

The last method considered is the fatality rate or deaths per person per year. This system is independent of the number of hours actually worked and reports only the number of fatalities expected per person per year. This approach is useful for performing calculations on the general population where the number of exposed hours is poorly defined. The applicable equation is

$$\text{Fatality rate} = (\text{Number of fatalities per year}) \Big/ \text{Total number of people in applicable population} \tag{1-4}$$

Both the OSHA incidence rate and the FAR are dependent on the number of exposed hours. An employee working a ten-hour shift is at greater total risk than one working an eight-hour shift. A FAR can be converted to a fatality rate (or vice versa) if the number of exposed hours is known. The OSHA incidence rate cannot be readily converted to a FAR or fatality rate because it contains both injury and fatality information.

Example 1-1

A process has a reported FAR of 2. If an employee works a standard eight-hour shift 300 days per year, compute the deaths per person per year.

Solution

$$\begin{aligned}\text{Deaths per person per year} &= (8 \text{ hours/day}) \times (300 \text{ days/year}) \\ &\quad \times (2 \text{ deaths}/10^8 \text{ hours}) \\ &= 4.8 \times 10^{-5}\end{aligned}$$

Typical accident statistics for various industries are shown in Table 1-3. A FAR of 4.0 is reported in Table 1-3 for the chemical industry. Approximately half of these deaths are due to ordinary industrial accidents (falling downstairs, being run over), the other half to chemical exposures.[2]

The FAR figures show that if a thousand workers begin employment in the chemical industry, four of the workers will die as a result of their employment throughout all of their working lifetimes. Two of these deaths will be due to direct chemical exposure. However, 20 of these same 1000 people will die as a result of non-industrial accidents (mostly at home or on the road) and 370 will die from disease. Of those that perish from disease, 40 will die as a direct result of smoking.[3]

Table 1-4 lists the FAR's for various common activities. The table is divided into voluntary and involuntary risks. Based on these data, it appears individuals are willing to take a substantially greater risk if it is voluntary. It is also evident that many common everyday activities are substantially more dangerous than working in a chemical plant.

For example, Table 1-4 indicates that canoeing is much more dangerous than traveling by motorcycle, despite general perceptions otherwise. This phenomenon is

TABLE 1-3 ACCIDENT STATISTICS FOR VARIOUS SELECTED INDUSTRIES

Industry	OSHA incidence rate[1] (Cases involving days away from work and deaths)	FAR[2] (deaths/10^8 hours)
Chemical	0.49	4.0
Vehicle	1.08	1.3
Steel	1.54	8
Paper	2.06	
Coal mining	2.22	40
Food	3.28	
Construction	3.88	67
Agricultural	4.53	10
Meat products	5.27	
Trucking	7.28	

[1] *Accident Facts,* 1985 Edition (Chicago: National Safety Council, 1985) p. 30.

[2] Frank P. Lees, *Loss Prevention in the Process Industries* (London: Butterworths, 1986), p. 177.

[2] T. A. Kletz, "Eliminating Potential Process Hazards," *Chemical Engineering,* April 1, 1985.

[3] Kletz, "Eliminating Potential Process Hazards."

TABLE 1-4 FATALITY STATISTICS FOR COMMON NONINDUSTRIAL ACTIVITIES[1]

Activity Voluntary activity	FAR (Deaths/10^8 hours)	Fatality Rate (Deaths per person per year)
Staying at Home	3	
Travelling by		
car	57	17×10^{-5}
bicycle	96	
air	240	
motorcycle	660	
Canoeing	1000	
Rock Climbing	4000	4×10^{-5}
Smoking (20 cigarettes/day)		500×10^{-5}
Involuntary activity		
Struck by meteorite		6×10^{-11}
Struck by lightning (U.K.)		1×10^{-7}
Fire (U.K.)		150×10^{-7}
Run over by vehicle		600×10^{-7}

[1]Frank P. Lees, *Loss Prevention in the Process Industries* (London: Butterworths, 1986), p. 178.

due to the number of exposed hours. Canoeing produces more fatalities per hour of activity than traveling by motorcycle. The total number of motorcycle fatalities is larger because more people travel by motorcycle than canoe.

Example 1-2

If twice as many people used motorcycles for the same average amount of time each, what will happen to (1) the OSHA incidence rate, (2) FAR, (3) the fatality rate, and (4) the total number of fatalities?

Solution

1. The OSHA incidence rate will remain the same. The number of injuries and deaths will double, but the total number of hours exposed will double as well.
2. The FAR will remain unchanged for the same reason as part 1.
3. The fatality rate, or deaths per person per year, will double. The fatality rate does not depend on exposed hours.
4. The total number of fatalities will double.

Example 1-3

If all riders used their motorcycles twice as much, what will happen to (1) the OSHA incidence rate, (2) FAR, (3) the fatality rate, and (4) the total number of fatalities?

Solution

1. The OSHA incidence rate will remain the same. The same reasoning applies as for Example 1-2, Part 1.
2. The FAR will remain unchanged for the same reason as Part 1.
3. The fatality rate will double. Twice as many fatalities will occur within this group.
4. The number of fatalities will double.

Example 1-4

A friend states that more rock climbers are killed traveling by automobile than are killed rock climbing. Is this statement supported by the accident statistics?

Solution The data from Table 1-4 shows that travelling by car (FAR = 57) is safer than rock climbing (FAR = 4000). Rock climbing produces many more fatalities per exposed hour than traveling by car. However, the rock climbers probably spend more time traveling by car than rock climbing. As a result, the statement might be correct but more data is required.

Recognizing that the chemical industry is very safe, why is there so much concern about chemical plant safety? The concern has to do with the industry's potential for many deaths, as, for example, in the Bhopal, India, tragedy. Accident statistics do not include information on the total number of deaths from a single incident. Accident statistics can be somewhat misleading in this respect. For example, consider two separate chemical plants. Both plants have a probability of explosion and complete devastation once every 1000 years. The first plant employs a single operator. When the plant explodes the operator is the sole fatality. The second plant employs ten operators. When this plant explodes all ten operators succumb. In both cases, the FAR and OSHA incidence rate are the same; the second accident kills more people but there are a correspondingly larger number of exposed hours. In both cases the risk taken by an individual operator is the same.[4]

It is human nature to perceive the accident with the greater loss of life as the greater tragedy. The potential for large loss of life gives the perception that the chemical industry is unsafe.

Loss data[5] published for losses after 1957 indicate that the total number of losses, the total dollar amount lost, and the average amount lost per incident, has steadily increased. The total loss figure has doubled every ten years despite increased efforts by the chemical process industry to improve safety. The increases are mostly due to an expansion in the number of chemical plants, an increase in chemical plant size, and an increase in the use of more complicated and dangerous chemicals. Furthermore, while the number of disabling injuries has declined, the cost per injury has risen significantly.[6]

Property damage and loss of production must also be considered in loss prevention. These losses can be substantial. Accidents of this type are much more common than fatalities. This is demonstrated in the accident pyramid shown in Figure 1-3. The numbers provided are only approximate. The exact numbers vary by industry, location, and time. "No Damage" accidents are frequently called "near misses" and provide a good opportunity for companies to determine that a problem exists and to correct it before a more serious accident occurs. It is frequently said "the cause of an accident is visible the day before it occurs." Inspections, safety re-

[4]Kletz, "Eliminating Potential Process Hazards."

[5]*One Hundred Largest Losses, A Thirty Year Review of Property Damage Losses in the Hydrocarbon-Chemical Industries* (Chicago: M & M Protection Consultants, 1986).

[6]"Industrial Injuries Fall 20 Percent, Costs More Than Triple Since 1973," *Plant/Operations Progress,* Vol. 6, No. 3 (1987), p. J12.

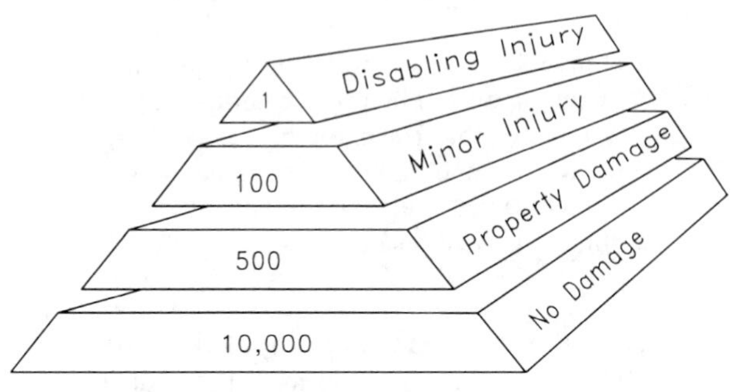

Figure 1-3 The accident pyramid.

views, and careful evaluation of near misses will identify hazardous conditions which can be corrected before real accidents occur.

Safety is good business and, like most business situations, has an optimal level of activity beyond which there are diminishing returns. As shown by Kletz,[7] if initial expenditures are made on safety, plants are prevented from blowing up and experienced workers are spared. This results in increased return due to reduced loss expenditures. If safety expenditures increase then the return increases more, but it may not be as much as before and not as much as achieved by spending money elsewhere. If safety expenditures increase further, the price of the product increases and sales diminish. Indeed, people are spared from injury (good humanity) but the cost is decreased sales. Finally, even higher safety expenditures will result in uncompetitive product pricing: The company will go out of business. Each company needs to determine an appropriate level for safety expenditures. This is part of risk management.

From a technical viewpoint, excessive expenditures for safety equipment to solve single safety problems may make the system unduly complex and consequently cause new safety problems due to this complexity. This excessive expense could have a higher safety return if assigned to a different safety problem. Engineers need to also consider other alternatives when designing safety improvements.

1-4 ACCEPTABLE RISK

We cannot eliminate risk entirely. Every chemical process has a certain amount of risk associated with it. At some point in the design stage someone needs to decide if the risks are "acceptable." That is, are the risks greater than the normal day-to-day risks taken by individuals in their nonindustrial environment? Certainly it would require a substantial effort and considerable expense to design a process with a risk comparable to being struck by lightning (see Table 1-4). Is it satisfactory to design a process with a risk comparable to the risk of sitting at home? For a single chemical process in a plant comprised of several processes, this risk may be too high since the risks due to multiple exposures are additive.[8]

[7] T. A. Kletz, "Eliminating Potential Process Hazards."

[8] Modern site layouts require sufficient separation of plants within the site to minimize risks of multiple exposures.

The engineer must make every effort to minimize risks within the economic constraints of the process. No engineer should ever design a process that will knowingly result in certain human loss or injury, despite any statistics.

1-5 PUBLIC PERCEPTIONS

The general public has great difficulty with the concept of acceptable risk. The major objection is due to the involuntary nature of acceptable risk. Chemical plant designers who specify the acceptable risk are assuming that these risks are satisfactory to the civilians living near the plant. Frequently these civilians are not aware there is any risk at all.

The results of a public opinion survey on the hazards of chemicals are shown in Figure 1-4. This survey asked the participants if they would say chemicals do more good than harm, more harm than good, or about the same amount of each. The results show an almost even three-way split, with a small margin to those who considered the good and harm to be equal.

Some naturalists suggest eliminating chemical plant hazards by "returning to nature." One alternative, for example, is to eliminate synthetic fibers produced by chemicals and use natural fibers such as cotton. As suggested by Kletz,[9] accident statistics demonstrate that this will result in a greater number of fatalities because the FAR for agriculture is higher.

Example 1-5

List six different products produced by chemical engineers that are of significant benefit to mankind.

Solution Penicillin, gasoline, synthetic rubber, paper, plastic, concrete.

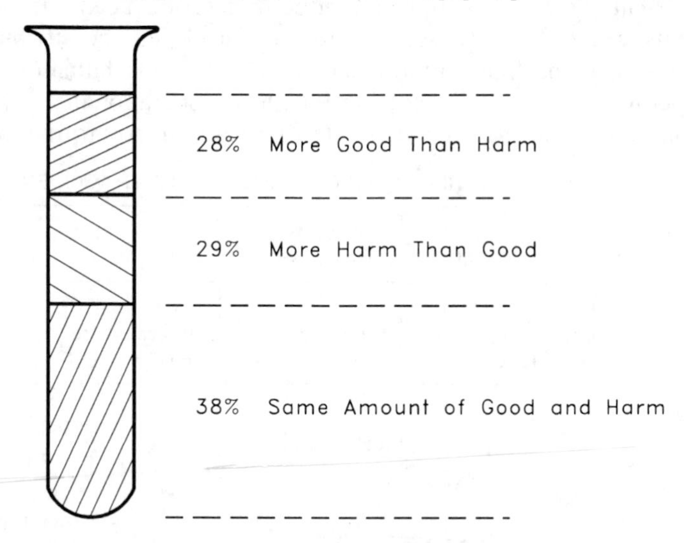

Figure 1-4 Results from a public opinion survey asking the question "Would you say chemicals do more good than harm, more harm than good, or about the same amount of each?" (Source: *The Detroit News.*)

[9]T. A. Kletz, "Eliminating Potential Process Hazards."

1-6 THE NATURE OF THE ACCIDENT PROCESS

Chemical plant accidents follow typical patterns. It is important to study these patterns in order to anticipate the types of accidents that will occur. As shown in Table 1-5, fires are the most common, followed by explosion and toxic release. With respect to fatalities, the order reverses, with toxic release having the greatest potential for fatalities.

Economic loss is consistently high for accidents involving explosions. The most damaging type of explosion is an unconfined vapor cloud explosion where a large cloud of volatile and flammable vapor is released and dispersed throughout the plant site followed by ignition and explosion of the cloud. An analysis of the largest chemical plant accidents (largest based on trended economic loss) is provided in Figure 1-5. Vapor cloud explosions account for the largest percentage of these large losses.

Toxic release typically results in little damage to capital equipment. Personnel injuries, employee losses, legal compensation, and cleanup liabilities can be significant.

Figure 1-6 presents the causes for loss for the largest chemical accidents. By far the largest cause of loss in a chemical plant is due to mechanical failure. Failures of this type are usually due to a problem with maintenance. Pumps, valves, and control equipment will fail if not properly maintained. The second largest cause is operational error. For example, valves are not opened or closed in the proper sequence, or reactants are not charged to a reactor in the correct order. Process upsets caused by, for example, power or cooling water failures account for 10% of the losses.

"Human error" is frequently used to describe a cause for losses. Almost all accidents, except those caused by natural hazards, can be attributed to human error. For instance, mechanical failures could all be due to human error as a result of improper maintenance or inspection. The term "operational error," used in Figure 1-6, includes human errors made on-site that lead directly to the loss.

TABLE 1-5 THREE TYPES OF CHEMICAL PLANT ACCIDENTS

Type of accident	Probability of occurrence	Potential for fatalities	Potential for economic loss
Fire	High	Low	Intermediate
Explosion	Intermediate	Intermediate	High
Toxic Release	Low	High	Low

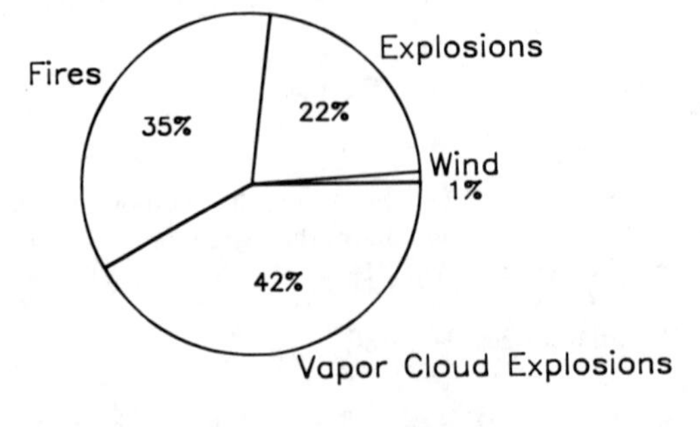

Figure 1-5 Types of loss for large chemical plant accidents. (*A Thirty Year Review of One Hundred of the Largest Property Damage Losses in the Hydrocarbon-Chemical Industries,* 1987. Used by permission of M&M Protection Consultants, Chicago.)

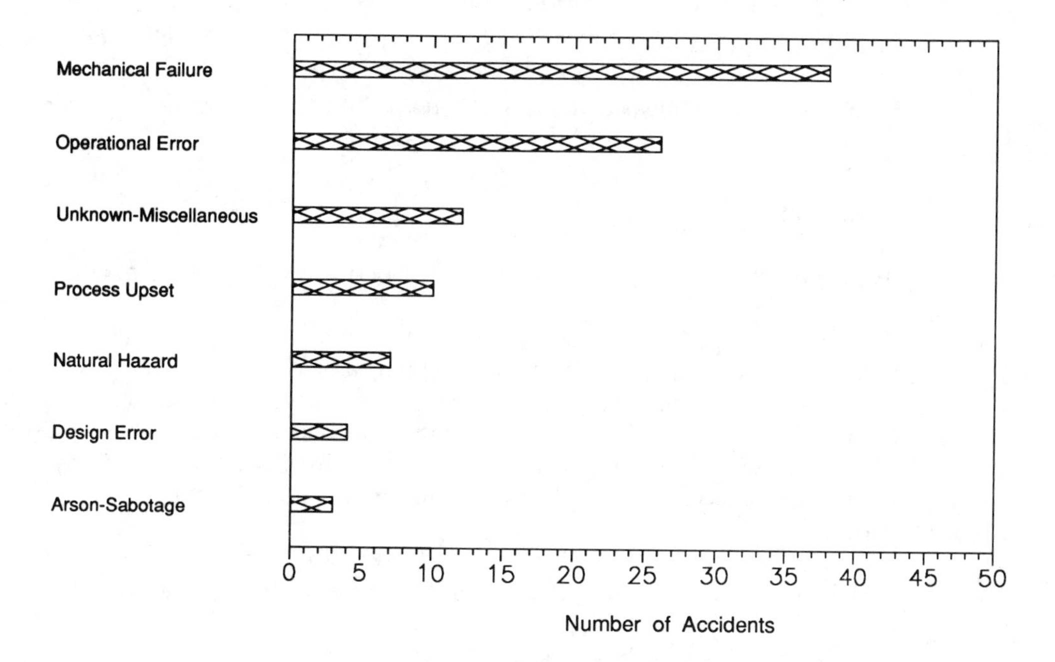

Figure 1-6 Causes for loss in the largest chemical accidents. (*A Thirty Year Review of One Hundred of the Largest Property Damage Losses in the Hydrocarbon-Chemical Industries,* 1987. Used by permission of M&M Protection Consultants, Chicago.)

Figure 1-7 presents a survey of the type of hardware associated with large accidents. Piping system failure represents the bulk of the accidents, followed by storage tanks and reactors. An interesting result of this study is that the most

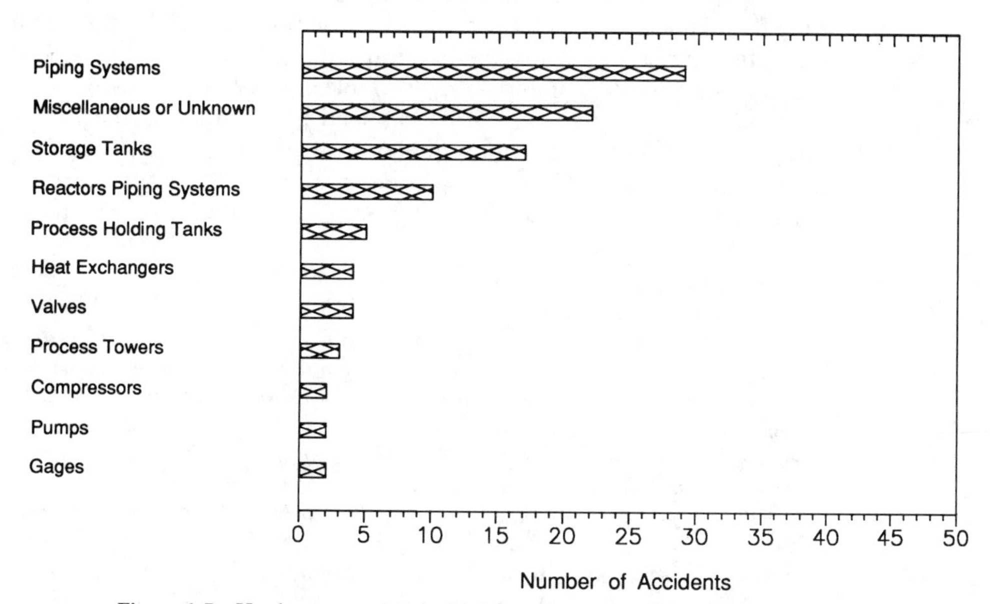

Figure 1-7 Hardware associated with largest losses. (*A Thirty Year Review of One Hundred of the Largest Property Damage Losses in the Hydrocarbon-Chemical Industries,* 1987. Used by permission of M&M Protection Consultants, Chicago.)

complicated mechanical components (pumps and compressors) are minimally responsible for large losses.

1-7 THE ACCIDENT PROCESS

Accidents follow a three step process. The following chemical plant accident illustrates these steps.

A worker walking across a high walkway in a process plant stumbles and falls towards the edge. To prevent the fall he grabs a nearby valve stem. Unfortunately, the valve stem shears off and flammable liquid begins to spew out. A cloud of flammable vapor rapidly forms and is ignited by a nearby truck. The explosion and fire quickly spread to nearby equipment. The resulting fire lasts for six days until all flammable materials in the plant are consummed, and the plant is completely destroyed.

This disaster occurred in 1969[10] and led to an economic loss of $4,161,000. It demonstrates an important point: Even the simplest accident can result in a major catastrophe.

Most accidents follow a three-step sequence.

- Initiation: the event that starts the accident.
- Propagation: the event or events that maintain or expand the accident.
- Termination: the event or events that stop the accident or diminish it in size.

In the example cited above, the worker tripped to initiate the accident. The accident was propagated by the shearing of the valve and the resulting explosion and growing fire. The event was terminated by consumption of all flammable materials.

Safety engineering involves eliminating the initiating step and replacing the propagation steps by termination events. Table 1-6 presents a few ways to accomplish this. In theory, accidents can be stopped by eliminating the initiating step. In practice this is not very effective: It is unrealistic to expect elimination of all initiations. A much more effective approach is to work on all three areas to insure that accidents, once initiated, do not propagate and will terminate as quickly as possible.

Example 1-6

The following accident report has been filed.[11]

Failure of a threaded 1-1/2" drain connection on a rich oil line at the base of an absorber tower in a large (1.35 MCF/D) gas producing plant allowed the release of rich oil and gas at 850 psi and −40°F. The resulting vapor cloud probably ignited from the ignition system of engine driven recompressors. The 75′ high × 10′ diameter absorber tower eventually collapsed across the pipe rack and on two exchanger trains. Breaking pipelines added more fuel to the fire. Severe flame impingment on an 11,000 horsepower gas turbine driven compressor-waste heat recovery and super-heater train resulted in its near total destruction.

Identify the initiation, propagation, and termination steps for this accident.

[10] *One Hundred Largest Losses,* p. 3.

[11] *One Hundred Largest Losses,* p. 10.

Solution

Initiation: Failure of threaded 1-1/2″ drain connection

Propagation: Release of rich oil and gas
Formation of vapor cloud
Ignition of vapor cloud by recompressors
Collapse of absorber tower across pipe rack

Termination: Consumption of combustible materials in process

1-8 THREE SIGNIFICANT DISASTERS

The study of case histories provides valuable information to the chemical engineer involved with safety. This information is used to improve procedures to prevent similiar accidents in the future.

The three most cited accidents (Flixborough, England; Bhopal, India; and Seveso, Italy) are presented here. All of these accidents had a significant impact on public perceptions and the chemical engineering profession which added new emphasis and standards in the practice of safety. Chapter 13 will present case histories in considerably more detail.

The Flixborough accident is perhaps the most documented chemical plant disaster. The British government insisted on an extensive investigation.

Flixborough, England

The accident at Flixborough, England occurred on a Saturday in June of 1974. While it was not reported to any great extent in the United States, it had a major

TABLE 1-6 DEFEATING THE ACCIDENT PROCESS

Step	Desired effect	Procedure
Initiation	Diminish	Grounding and Bonding Inerting Explosion proof electrical Guardrails and guards Maintenance procedures Hot-work permits Human factors design Process design Awareness of dangerous properties of chemicals
Propagation	Diminish	Emergency material transfer Reduce inventories of flammables Equipment spacing and layout Nonflammable construction materials Installation of check and emergency shut-off valves
Termination	Increase	Firefighting equipment and procedures Relief systems Sprinkler systems Installation of check and emergency shut-off valves

impact on chemical engineering in the United Kingdom. As a result of the accident, safety achieved a much higher priority in that country.

The Flixborough Works of Nypro Limited was designed to produce 70,000 tons per year of caprolactam, a basic raw material for the production of nylon. The process uses cyclohexane which has properties similiar to gasoline. Under the process conditions in use at Flixborough (155°C and 7.9 atm), the cyclohexane volatilizes immediately when depressurized to atmospheric conditions.

The process where the accident occurred consisted of six reactors in series. In these reactors, cyclohexane was oxidized to cyclohexanone and then to cyclohexanol using injected air in the presence of a catalyst. The liquid reaction mass was gravity fed through the series of reactors. Each reactor normally contained about 20 tons of cyclohexane.

Several months before the accident occurred, reactor 5 in the series was found to be leaking. Inspection showed a vertical crack in its stainless steel structure. The decision was made to remove the reactor for repairs. An additional decision was made to continue operating by connecting reactor 4 directly to reactor 6 in the series. The loss of the reactor would reduce the yield but enable continued production since unreacted cyclohexane is separated and recycled at a later stage.

The feed pipes connecting the reactors were 28 inches in diameter. Since only 20-inch pipe stock was available at the plant, the connections to reactor 4 and reactor 6 were made using flexible, bellows type piping, as shown in Figure 1-8. It is hypothesized that the bypass pipe section ruptured due to inadequate support and overflexing of the pipe section as a result of internal reactor pressures. Upon rupture of the bypass, an estimated 30 tons of cyclohexane volatilized and formed a large vapor cloud. The cloud was ignited by an unknown source an estimated 45 seconds after the release.

The resulting explosion leveled the entire plant facility, including the administrative offices. A total of 28 people died and 36 others were injured. Eighteen of these fatalities occurred in the main control room when the ceiling collapsed. Loss of life would have been substantially greater had the accident occurred on a weekday when the administrative offices were filled with employees. Damage extended

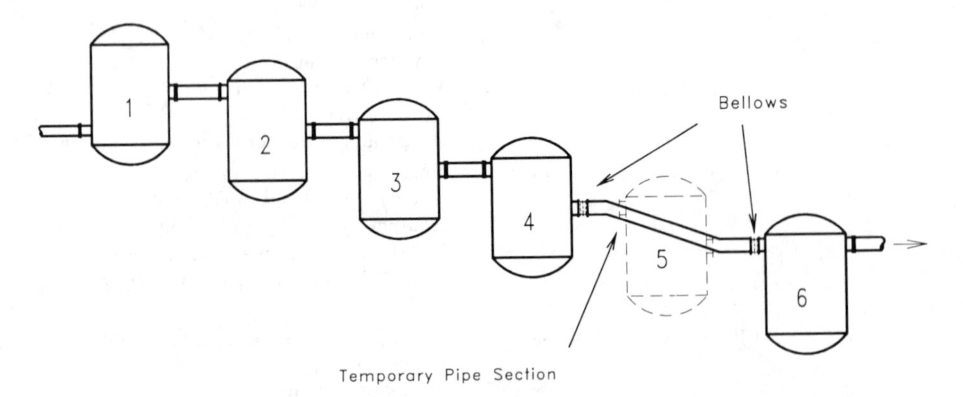

Figure 1-8 A failure of a temporary pipe section replacing reactor 5 caused the Flixborough accident.

to 1821 nearby houses and 167 shops and factories. Fifty-three civilians were reported injured. The resulting fire in the plant burned for over ten days.

This accident could have been prevented by following proper safety procedures. First, the bypass line was installed without a safety review or adequate supervision by experienced engineering personnel. The bypass was sketched on the floor of the machine shop using chalk! Second, the plant site contained excessively large inventories of dangerous compounds. This included 330,000 gallons of cyclohexane, 66,000 gallons of naptha, 11,000 gallons of toluene, 26,400 gallons of benzene, and 450 gallons of gasoline. These inventories contributed to the fires after the initial blast. Finally, the bypass modification was substandard in design. As a rule, any modifications should be of the same quality as the construction of the remainder of the plant.

Bhopal, India

The Bhopal, India accident, on December 3, 1984, has received considerably more attention than the Flixborough accident. This is due to the more than 2000 civilian casualties that resulted.

The Bhopal plant is in the state of Madhya Pradesh in central India. The plant was partially owned by Union Carbide and partially owned locally.

The nearest civilian inhabitants were 1.5 miles away when the plant was constructed. Since the plant represented the dominant source of employment in the area, a shanty town eventually grew around the immediate area.

The plant produced pesticides. An intermediate compound in this process is methyl isocyanate (MIC). MIC is an extremely dangerous compound. It is reactive, toxic, volatile, and flammable. The maximum exposure concentration of MIC for workers over an eight-hour period is 0.02 ppm (parts per million). Individuals exposed to concentrations of MIC vapors above 21 ppm experience severe irritation of the nose and throat. Death at large concentrations of vapor is due to respiratory distress.

MIC demonstrates a number of dangerous physical properties. Its boiling point at atmospheric conditions is 39.1°C and it has a vapor pressure of 348 mm Hg at 20°C. The vapor is about twice as heavy as air, ensuring that the vapors will stay close to the ground once released.

MIC reacts exothermically with water. Although the reaction rate is slow, with inadequate cooling the temperature will increase and the MIC will boil. MIC storage tanks are typically refrigerated to prevent this problem.

The unit using the MIC was not operating due to a local labor dispute. Somehow a storage tank containing a large amount of MIC became contaminated with water or some other substance. A chemical reaction heated the MIC to a temperature past its boiling point. The MIC vapors traveled through a pressure relief system and on into a scrubber and flare system installed to consume the MIC in the event of a release. Unfortunately, the scrubbing and flare systems were not operating, for a variety of reasons. An estimated 25-tons of toxic MIC vapor was released. The toxic cloud spread to the adjacent town, killing over 2000 civilians and injuring an estimated 20,000 more. No plant workers were injured or killed. No plant equipment was damaged.

The exact cause for the contamination of the MIC is not known. If the accident was caused by a problem with the process, a well executed safety review could have identified the problem. The scrubber and flare system should have been fully operational to prevent the release. Inventories of dangerous chemicals, particularly intermediates, should also be minimized.

The reaction scheme used at Bhopal is shown at the top of Figure 1-9 and includes the dangerous intermediate MIC. An alternative reaction scheme is shown on the bottom of the figure that involves a less dangerous chloroformate intermediate. Another solution is to redesign the process to reduce the inventory of hazardous MIC. One such design produces and consumes the MIC in a highly localized area of the process, with an inventory of MIC of less than 20 pounds.

Seveso, Italy

Seveso is a small town of approximately 17,000 inhabitants, 15 miles from Milan, Italy. The plant was owned by the Icmesa Chemical Company. The product was hexachlorophene, a bactericide, with trichlorophenol produced as an intermediate. During normal operation, a very small amount of TCDD (2,3,7,8 tetrachlorodibenzoparadioxin) is produced in the reactor as an undesirable side product.

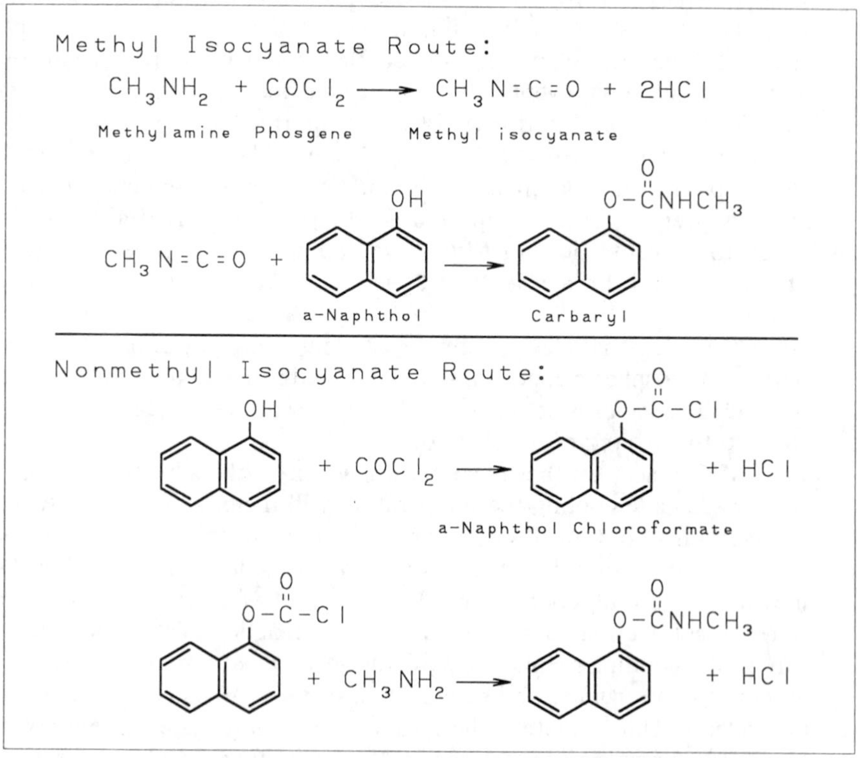

Methyl Isocyanate Route:

$CH_3NH_2 + COCl_2 \longrightarrow CH_3N{=}C{=}O + 2HCl$

Methylamine Phosgene Methyl isocyanate

$CH_3N{=}C{=}O$ + a-Naphthol (OH) ⟶ Carbaryl ($O{-}\overset{O}{\overset{\|}{C}}NHCH_3$)

Nonmethyl Isocyanate Route:

a-Naphthol (OH) + $COCl_2 \longrightarrow$ a-Naphthol Chloroformate ($O{-}\overset{O}{\overset{\|}{C}}{-}Cl$) + HCl

a-Naphthol Chloroformate ($O{-}\overset{O}{\overset{\|}{C}}{-}Cl$) + $CH_3NH_2 \longrightarrow$ $O{-}\overset{O}{\overset{\|}{C}}NHCH_3$ + HCl

Figure 1-9 The upper reaction is the methyl isocyanate route used at Bhopal. The lower reaction suggests an alternate reaction scheme using a less hazardous intermediate. (Adapted from *Chemical and Engineering News,* February 11, 1985, p. 30.)

TCDD is perhaps the most potent toxin known to man. Animal studies have shown TCDD to be fatal in doses as small as 10^{-9} times the body weight. Since it is also insoluable in water, decontamination is very difficult. Nonlethal doses of TCDD result in chloracne, an acne-like disease that can persist for several years.

On July 10, 1976, the trichlorophenol reactor went out of control, resulting in higher than normal operating temperature and increased production of TCDD. An estimated 2 kg of TCDD was released through a relief system in a white cloud over Seveso. A subsequent heavy rain washed the TCDD into the soil. Approximately ten square miles were contaminated.

Due to poor communications with local authorities, civilian evacuation was not started until several days later. By then, over 250 cases of chloracne were reported. Over 600 people were evacuated and an additional 2000 were given blood tests. The most severely contaminated area immediately adjacent to the plant was fenced, the condition it remains in today.

TCDD is so toxic and persistent that for a smaller, but similiar release of TCDD in Duphar, India in 1963, the plant was finally disassembled brick by brick, encased in concrete and dumped into the ocean. Less than 200 grams of TCDD were released, and the contamination was confined to the plant. Of the fifty men assigned to clean up the release, four eventually died from the exposure.

The Seveso and Duphar accidents could have been avoided if proper containment systems had been used to contain the reactor releases. The proper application of fundamental engineering safety principles would have prevented the above accidents. First, by following proper procedures, the initiation steps would not have occurred. Second, by using proper hazard evaluation procedures, the hazards could have been identified and corrected before the accidents occurred.

SUGGESTED READING

General Aspects of Chemical Process Safety

Howard H. Fawcett and William S. Wood, eds., *Safety and Accident Prevention in Chemical Operations,* 2nd Edition (New York: John Wiley and Sons, 1982), Chapter 1.

Frank P. Lees, *Loss Prevention in the Process Industries,* Volume 1 (London: Butterworths, 1980), Chapters 1-5.

Bhopal

Chemical and Engineering News, February 11, 1985, p. 14.

Seveso

Chemical and Engineering News, August 23, 1976, p. 27.

Flixborough

Lees, *Loss Prevention in the Process Industries,* Volume 2, Appendix 1.

General Case Histories

TREVOR A. KLETZ, *What Went Wrong? Case Histories of Process Plant Disasters* (Houston: Gulf Publishing Company, 1985).

LEES, *Loss Prevention in the Process Industries,* Volume 2, Appendix 3.

PROBLEMS

1-1. An employee works in a plant with a FAR of 4. If this employee works a four-hour shift, 200 days per year, what is the expected deaths per person per year?

1-2. Three process units are in a plant. The units have FARs of 0.5, 0.3 and 1.0 respectively.

- **a.** What is the overall FAR for the plant, assuming worker exposure to all three units simultaneously?
- **b.** Assume now that the units are far enough apart that an accident in one would not affect the workers in another. If a worker spends 20% of his time in process area 1, 40% in process area 2, and 40% in process area 3, what is his overall FAR?

1-3. Assuming that a car travels at an average speed of 50 miles per hour, how many miles must be driven before a fatality is expected?

1-4. A worker is told his chances of being killed by a particular process are one in every 500 years. Should the worker be satisfied or alarmed? What is the FAR (assuming normal working hours) and the deaths per person per year? What should his chances be, assuming an average chemical plant?

1-5. A plant employs 1500 full time workers in a process with a FAR of 5. How many industrial related deaths are expected each year?

1-6. Consider Example 1-4. How many hours must be traveled by car for each hour of rock climbing to make the risks of fatality by car equal to the risk of fatality by rock climbing?

1-7. Identify the initiation, propagation and termination steps for the following accident reports.[12] Suggest ways to prevent and contain the accidents.

- **a.** A contractor accidentally cut into a 10-inch propane line operating at 800 psi at a natural gas liquids terminal. The large vapor cloud estimated to cover an area of 44 acres was ignited about 4-5 minutes later by an unknown source. Liquid products from 5 of 26 salt dome caverns fed the fire with an estimated 18,000 to 30,000 gallons of LPGs for almost 6 hours before being blocked in and the fires extinguished. Both engine driven fire pumps failed; one because intense radiated heat damaged its ignition wires and the other because the explosion broke a sight glass fuel gage spilling diesel fuel which ignited, destroying the fire pump engine.
- **b.** An alkylation unit was being started up after shutdown because of an electrical outage. When adequate circulation could not be maintained in a deisobutanizer heater circuit, it was decided to clean the strainer. Workman had depressurized the pipe and removed all but three of the flange bolts when a pressure release blew a black material from the flange, followed by butane vapors. These vapors were carried to a furnace 100 feet away, where they ignited, flashing back to the flange.

[12] *One Hundred Largest Losses*

The ensuing fire exposed a fractionation tower and horizontal receiver drums. These drums exploded, rupturing piplines, which added more fuel. The explosions and heat caused loss of insulation from the 8-foot-by-122-foot fractionator tower, causing it to weaken and fall across two major pipe lanes, breaking piping—which added more fuel to the fire. Extinguishment, achieved basically by isolating the fuel sources, took 2-1/2 hours.

The fault was traced to a 10-inch valve that had been prevented from closing the last 3/4-inch by a fine powder of carbon and iron oxide. When the flange was opened this powder blew out, allowing liquid butane to be released.

1-8. The airline industry claims commercial airline transport has fewer deaths per mile than any other means of transportation. Do the accident statistics support this claim? In 1984 the airline industry posted four deaths per ten-million passenger miles. What additional information is required to compute a FAR? a fatality rate?

Toxicology 2

Due to the quantity and variety of chemicals used by the chemical process industries, chemical engineers must be knowledgeable about

- The way toxicants enter biological organisms,
- The way toxicants are eliminated from biological organisms,
- The effects of toxicants on biological organisms, and,
- Methods to prevent or reduce the entry of toxicants into biological organisms.

The first three areas are related to toxicology. The last area is essentially *industrial hygiene,* a topic to be considered in Chapter 3.

Many years ago, toxicology was defined as "the science of poisons." Unfortunately, the word poison could not be defined adequately. Paracelsus, an early investigator of toxicology during the 1500s, stated the problem: "All substances are poisons; there is none which is not a poison. The right dose differentiates a poison and a remedy." Harmless substances, such as water, can become fatal if delivered to the biological organism in large enough doses. A fundamental principle of toxicology is

> There are no harmless substances, only harmless ways of using substances.

Today, toxicology is more adequately defined as the qualitative and quantitative study of the adverse effects of toxicants on biological organisms. A toxicant can be a chemical or physical agent, including dusts, fibers, noise, and radiation. A good example of a physical agent is asbestos fiber, a known cause of lung damage and cancer.

The *toxicity* of a chemical or physical agent is a property of the agent describing its effect on biological organisms. *Toxic hazard* is the likelihood of damage to biological organisms based on exposure due to transport and other physical factors of usage. The toxic hazard of a substance can be reduced by the application of appropriate industrial hygiene techniques. The toxicity, however, cannot be changed.

2-1 HOW TOXICANTS ENTER BIOLOGICAL ORGANISMS

For higher-order organisms, the path of the chemical agent through the body is well defined. After the toxicant enters the organism it moves into the bloodstream and is eventually eliminated, or it is transported to the target organ. The damage is exerted at the target organ. A common misconception is that damage occurs in the organ where the toxicant is most concentrated. Lead, for instance, is stored in humans mostly in the bone structure, but the damage occurs in many organs. For corrosive chemicals the damage to the organism can occur without absorption or transport through the bloodstream.

Toxicants enter biological organisms by the following routes:

- Ingestion: via mouth into stomach
- Inhalation: via mouth or nose into lungs
- Injection: via cuts into skin
- Dermal Absorption: through skin membrane

All of the above entry routes are controlled by the application of proper industrial hygiene techniques summarized in Table 2-1. These control techniques will be discussed in more detail in Chapter 3 on industrial hygiene. Of the four routes of entry, the inhalation and dermal routes are the most significant to industrial facilities. Inhalation is the easiest to quantify by the direct measurement of airborne concentrations; the usual exposure is by vapor, but small solid and liquid particles can also contribute.

Injection, inhalation and dermal absorption generally result in the toxicant entering the blood stream unaltered. Toxicants entering through ingestion are frequently modified or excreted in the bile.

TABLE 2-1 ENTRY ROUTES FOR TOXICANTS AND METHODS FOR CONTROL

Entry route	Entry organ	Method for control
Ingestion	Mouth or stomach	Enforcement of rules on eating, drinking, and smoking
Inhalation	Mouth or nose	Ventilation, respirators, hoods, and other personal protection equipment
Injection	Cuts in skin	Proper protective clothing
Dermal absorption	Skin	Proper protective clothing

Toxicants that enter by injection and dermal absorption are difficult to measure and quantify. Some toxicants are absorbed very rapidly through the skin.

Figure 2-1 shows the expected blood level concentration as a function of time and route of entry. The blood level concentration is a function of a wide range of parameters; so large variations in this behavior are expected. Injection usually results in the highest blood level concentration, followed by inhalation, ingestion, and absorption. The peak concentration generally occurs earliest with injection, followed by inhalation, ingestion, and absorption.

The gastroinstestinal (GI) tract, the skin, and the respiratory system play significant roles in the various routes of entry.

The Gastrointestinal (GI) Tract

The GI tract plays the most significant role in toxicants entering the body through ingestion. Food or drink is the usual mechanism of exposure. Airborn particles (either solid or liquid) can also lodge in the mucous of the upper respiratory tract and be swallowed.

The rate and selectivity of absorption by the GI tract are highly dependent on many conditions. The type of chemical, its molecular weight, molecule size and shape, acidity, susceptability to attack by intestinal flora, rate of movement through the GI tract, and many other factors affect the rate of absorption.

The Skin

The skin plays important roles in both the dermal absorption and injection routes of entry. Injection includes both entry by absorption through cuts and mechanical

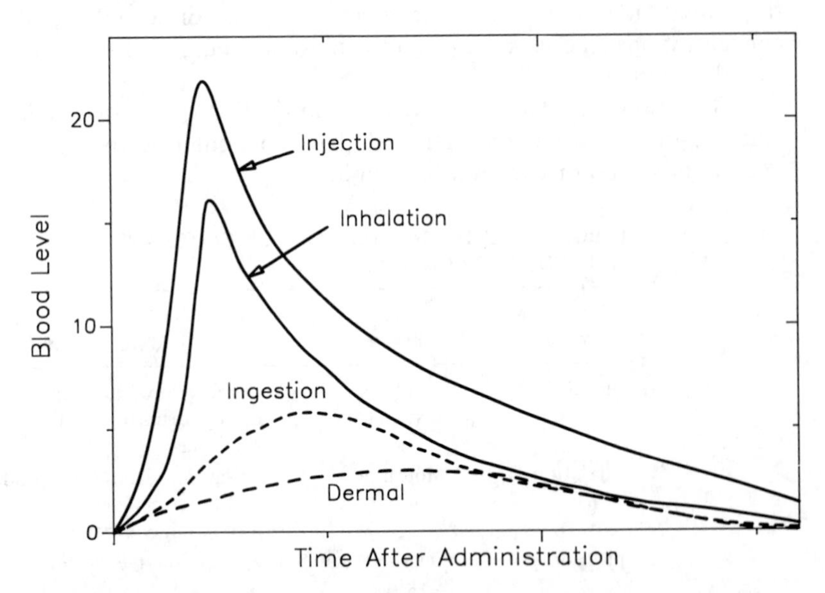

Figure 2-1 Toxic blood level concentration as a function of route of exposure. Wide variations are expected due to rate and extent of absorption, distribution, biotransformation, and excretion.

injection with hypodermic needles. Mechanical injection can occur as a result of improper hypodermic needle storage in a laboratory drawer.

The skin is composed of an outer layer called the stratum corneum. This layer consists of dead, dried cells that are resistant to permeation by toxicants. Absorption also occurs through the hair follicles and sweat glands, but this is normally negligible. The absorption properties of the skin vary as a function of location and the degree of hydration. The presence of water increases the skin hydration and results in increased permeability and absorption.

Most chemicals are not absorbed readily by the skin. A few chemicals, however, do show remarkable skin permeability. Phenol, for example, requires only a small area of skin for the body to absorb an adequate amount to result in death.

The skin on the palm of the hand is thicker than the skin found elsewhere. However, this skin demonstrates increased porosity, resulting in higher toxicant absorption.

Respiratory System

The respiratory system plays a significant role in toxicants entering the body through inhalation.

The main function of the respiratory system is to exchange oxygen and carbon dioxide between the blood and the inhaled air. In one minute a normal person at rest uses an estimated 250 ml of oxygen and expels approximately 200 ml of carbon dioxide. Approximately 8 liters of air are breathed per minute. Only a fraction of the total air within the lung is exchanged with each breath. These demands increase significantly with physical exertion.

The respiratory system is divided into two areas: the upper and lower respiratory system. The upper respiratory system is composed of the nose, sinuses, mouth, pharynx (section between the mouth and esophagus), larynx (the voice box), and the trachea or windpipe. The lower respiratory system is composed of the lungs and its smaller structures, including the bronchi and alveoli. The bronchi tubes carry fresh air from the trachea through a series of branching tubes to the alveoli. The alveoli are small blind air sacs where the gas exchange with the blood occurs. An estimated 300,000,000 alveoli are found in a normal lung. These alveoli contribute a total surface area of approximately 70 square meters. Small capillaries found in the walls of the alveoli transport the blood; an estimated 100 ml is found in the capillaries at any moment.

The upper respiratory tract is responsible for filtering, heating, and humidifying the air. Fresh air brought in through the nose is completely saturated with water and regulated to the proper temperature by the time it reaches the larynx. The mucous lining the upper respiratory tract assists in filtering.

The upper and lower respiratory tracts respond differently to the presence of toxicants. The upper respiratory tract is affected mostly by toxicants that are water soluble. These materials either react or dissolve in the mucous to form acids and bases. Toxicants in the lower respiratory tract affect the alveoli by physically blocking the transfer of gases (as with insoluble dusts) or reacting with the wall of the alveoli to produce corrosive or toxic substances. Phosgene gas, for example, reacts with the water on the alveoli wall to produce HCl and carbon monoxide.

Upper respiratory toxicants include hydrogen halides (hydrogen chloride, hydrogen bromide), oxides (nitrogen oxides, sulfur oxides, sodium oxide), and hydroxides (ammonium hydroxide, dusts of sodium, and potassium hydroxides). Lower respiratory toxicants include monomers (such as acrylonitrile), halides (fluorine, chlorine, bromine), and other miscellaneous substances like hydrogen sulfide, phosgene, methyl cyanide, acrolein, dusts of asbestos, silica, and soot.

Dusts and other insoluble materials present a particular difficulty to the lungs. Particles that enter the alveoli are removed very slowly. For dusts, the following simple rule usually applies: The smaller the dust particles, the further they penetrate into the respiratory system. Particles greater than 5 microns in diameter are usually filtered by the upper respiratory system. Particles with diameters between 2 and 5 microns generally reach the bronchial system. Particles less than 1 micron in diameter may reach the alveoli.

2-2 HOW TOXICANTS ARE ELIMINATED FROM BIOLOGICAL ORGANISMS

Toxicants are eliminated or rendered inactive by the following routes:

- Excretion: through the kidneys, liver, lungs, or others.
- Detoxification: by changing the chemical into something less harmful by biotransformation.
- Storage: in the fatty tissue.

The kidneys are the dominant means of excretion in the human body. They eliminate substances that enter the body by ingestion, inhalation, injection, and dermal absorption. The toxicants are extracted by the kidneys from the bloodstream and are excreted with the urine.

Toxicants that are ingested into the digestive tract are freqently excreted by the liver. In general, chemical compounds with molecular weights greater than about 300 are excreted by the liver into the bile. Compounds with lower molecular weights enter the bloodstream and are excreted by the kidneys. The digestive tract tends to selectively detoxify certain agents while substances that enter through inhalation, injection, or dermal absorption generally arrive in the bloodstream unchanged.

The lungs are also a means for elimination of substances, particularly those that are volatile. Chloroform and alcohol, for example, are excreted partially by this route.

Other routes of excretion are the skin (via sweat), hair, and nails. These routes are usually minor compared to the excretion processes of the kidneys, liver, and lungs.

The liver is the dominant organ in the detoxification process. The detoxification occurs by biotransformation where the chemical agents are transformed by reaction into either harmless or less harmful substances. Biotransformation reactions can also occur in the blood, intestinal tract wall, skin, kidneys, and other organs.

The final mechanism for elimination is storage. This process involves the depositing of the chemical agent mostly in the fatty areas of the organism but also in the bones, blood, liver, and kidney. Storage can create a future problem if the organism's food supply is reduced and the fatty deposits are metabolized; the chemical agents stored will be released into the bloodstream, resulting in possible damage.

For massive exposures to chemical agents, damage can occur to the kidneys, liver, or lungs, significantly reducing the organism's ability to excrete the substance.

2-3 THE EFFECTS OF TOXICANTS ON BIOLOGICAL ORGANISMS

Table 2-2 contains a list of some of the effects or responses from toxic exposure.

The problem is to determine whether exposures have occurred before substantial symptoms are present. This is accomplished by a variety of medical tests. The results from these tests must be compared to a medical baseline study, performed before any exposure. Many chemical companies perform baseline studies on new employees prior to employment.

Respiratory problems are diagnosed using a spirometer. The patient exhales as hard and as fast as possible into the device. The spirometer measures (1) the total volume exhaled called the forced vital capacity (FVC) with units in liters, (2) the forced expired volume measured at 1-second $(FEV)_1$ with units in liters per second, (3) forced expiratory flow in the middle range of the vital capacity (FEV 25-75%) in liters per second, and (4) the ratio of the observed $(FEV)_1$ to FVC $\times$ 100 (FEV_1/FVC%).

Reductions in expiration flow rate are indicative of bronchial disease such as asthma or bronchitis. Reductions in FVC are due to reduction in the lung or chest volume, possibly as a result of fibrosis (an increase in the interstitial fibrous tissue in the lung). The air remaining in the lung after exhalation is called the residual volume (RV). An increase in the RV is indicative of deterioration of the alveoli, possibly due to emphysema. The RV measurement requires a specialized tracer test with helium.

TABLE 2-2 VARIOUS RESPONSES TO TOXICANTS

Effects that are irreversible
Carcinogen causes cancer
Mutagen causes chromosome damage
Reproductive hazard causes damage to reproductive system
Teratogen causes birth defects
Effects that may or may not be reversible
Dermatotoxic affects skin
Hemotoxic affects blood
Hepatotoxic affects liver
Nephrotoxic affects kidneys
Neurotoxic affects nervous system
Pulmonotoxic affects lungs

Nervous system disorders are diagnosed by examination of the patient's mental status, cranial nerve function, motor system reflexes, and sensory systems. An electroencephalogram (EEG) tests higher brain and nervous system functions.

Changes in skin texture, pigmentation, vascularity, and hair and nail appearance are indicative of possible toxic exposures.

Blood counts are also used to determine toxic exposures. Measurements of the red and white cells, hemoglobin content, and platelet count are performed easily and inexpensively. However, blood counts are frequently insensitive to toxic exposure; marked changes are seen only after substantial exposure and damage.

Kidney function is determined by a variety of tests that measure the chemical content and quantity of urine. For early kidney damage, proteins or sugars are found in the urine.

Liver function is diagnosed by a variety of chemical tests on the blood and urine.

2-4 TOXICOLOGICAL STUDIES

A major objective of a toxicological study is to quantify the effects of the suspect toxicant on a target organism. For most toxicological studies animals are used, usually with the hope that the results could be extrapolated to humans. Once the effects of a suspect agent have been quantified, appropriate procedures are established to ensure that the agent is handled properly.

Before undertaking a toxicological study, the following items must be identified.

- the toxicant
- the target or test organism
- the effect or response to be monitored
- the dose range
- the period of the test

The toxicant must be identified with respect to its chemical composition and its physical state. For example, benzene can exist in either liquid or vapor form. Each physical state preferentially enters the body by a different route and requires a different toxicological study.

The test organism can range from a simple single cell up through the higher animals. The selection depends on the effects considered and other factors such as the cost and availability of the test organism. For studies of genetic effects, single cell organisms might be satisfactory. For studies determining the effects on specific organs such as the lungs, kidneys, or liver, higher organisms are a necessity.

The dose units depend on the method of delivery. For substances delivered directly into the organism (by ingestion or injection), the dose is measured in mg of agent per kg of body weight. This enables researchers to apply the results obtained from small animals such as mice (fractions of a kg in body weight) to humans (about 70 kg for males and 60 kg for females). For gaseous airborne substances, the dose is measured in either ppm (parts per million) or mg of agent per cubic meter of air (mg/m^3). For airborne particulates, the dose is measured in mg of agent per cubic meter of air (mg/m^3) or millions of particles per cubic foot (MPPCF).

The period of the test depends on whether long or short term effects are of interest. Acute toxicity is the effect of a single exposure or a series of exposures close together in a short period of time. Chronic toxicity is the effect of multiple exposures occurring over a long period of time. Chronic toxicity studies are difficult to perform due to the time involved; most toxicological studies are based on acute exposures. The toxicological study can be complicated by latency, an exposure that results in a delayed response.

2-5 DOSE VERSUS RESPONSE

Biological organisms respond differently to the same dose of a toxicant. These differences are a result of age, sex, weight, diet, general health, and other factors. For example, consider the effects of an irritant vapor on human eyes. Given the same dose of vapors, some individuals will barely notice any irritation (weak or low response) while other individuals will be severely irritated (high response).

Consider a toxicological test run on a large number of individuals. Each individual is exposed to the same dose and the response is recorded. A plot of the type shown in Figure 2-2 is prepared with the data. The fraction or percentage of individuals experiencing a specific response is plotted. Curves of the form shown in

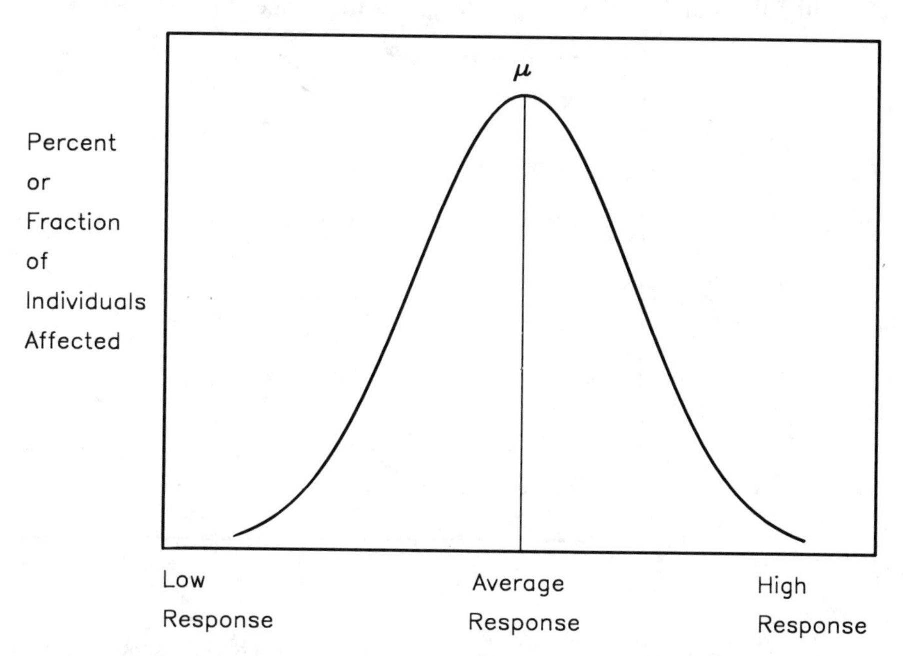

Figure 2-2 A Gaussian or normal distribution representing the biological response to exposure to a toxicant.

Figure 2-2 are frequently represented by a normal or Gaussian distribution given by the equation

$$f(x) = \frac{1}{\sigma\sqrt{2\pi}} e^{-\frac{1}{2}\left(\frac{x-\mu}{\sigma}\right)^2} \tag{2-1}$$

where $f(x)$ is the probability (or fraction) of individuals experiencing a specific response, x is the response, σ is the standard deviation, and μ is the mean.

The standard deviation and mean characterize the shape and location of the normal distribution curve, respectively. They are computed from the original data $f(x_i)$ using the equations:

$$\mu = \frac{\sum_{i=1}^{n} x_i f(x_i)}{\sum_{i=1}^{n} f(x_i)} \tag{2-2}$$

$$\sigma^2 = \frac{\sum_{i=1}^{n} (x_i - \mu)^2 f(x_i)}{\sum_{i=1}^{n} f(x_i)} \tag{2-3}$$

where n is the number of data points. The quantity σ^2 is called the variance.

The mean determines the location of the curve with respect to the x-axis while the standard deviation determines the shape. Figure 2-3 shows the effect of

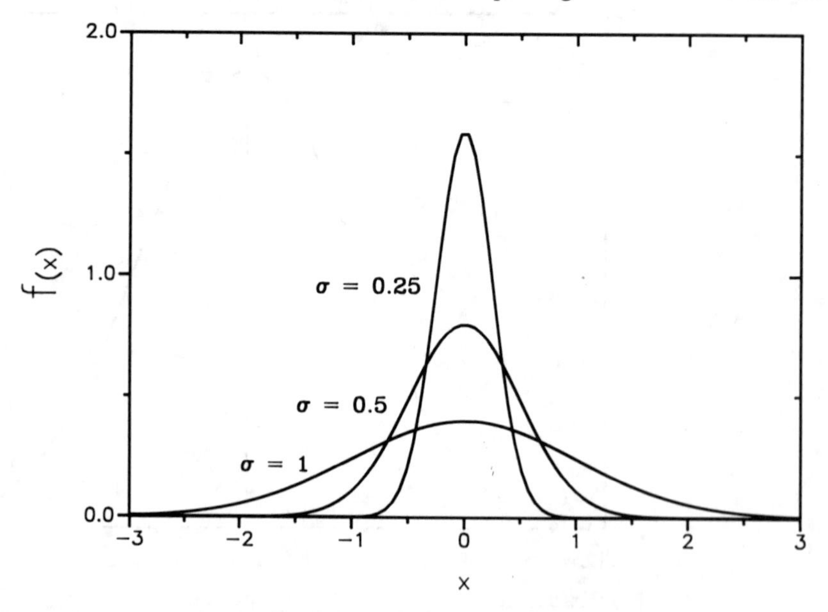

Figure 2-3 Effect of the standard deviation on a normal distribution with a mean of 0. The distribution becomes more pronounced around the mean as the standard deviation decreases.

the standard deviation on the shape. As the standard deviation decreases, the distribution curve becomes more pronounced around the mean value.

The area under the curve of Figure 2-2 represents the percentage of individuals affected for a specified response interval. In particular, the response interval within one standard deviation of the mean represents 68% of the individuals, as shown in Figure 2-4. A response interval of two standard deviations represents 95.5% of the total individuals. The area under the entire curve represents 100% of the individuals

Example 2-1

Seventy-five people are tested for skin irritation due to a specific dose of a substance. The responses are recorded on a scale from 0 to 10, with 0 indicating no response and 10 indicating a high response. The number of individuals exhibiting a specific response is given below.

Response	Number of individuals Affected
0	0
1	5
2	10
3	13
4	13
5	11
6	9
7	6
8	3
9	3
10	2
	75

1. Plot a histogram of the number of individuals affected versus the response.
2. Determine the mean and the standard deviation.
3. Plot the normal distribution on the histogram of the original data.

Solution

1. The histogram is shown in Figure 2-5. The number of individuals affected is plotted versus the response. An alternate method is to plot the percentage of individuals versus the response.
2. The mean is computed using Equation 2-2.

$$\mu = \frac{0\times0+1\times5+2\times10+3\times13+4\times13+5\times11+6\times9+7\times6+8\times3+9\times3+10\times2}{75}$$

$$= \frac{338}{75} = 4.51$$

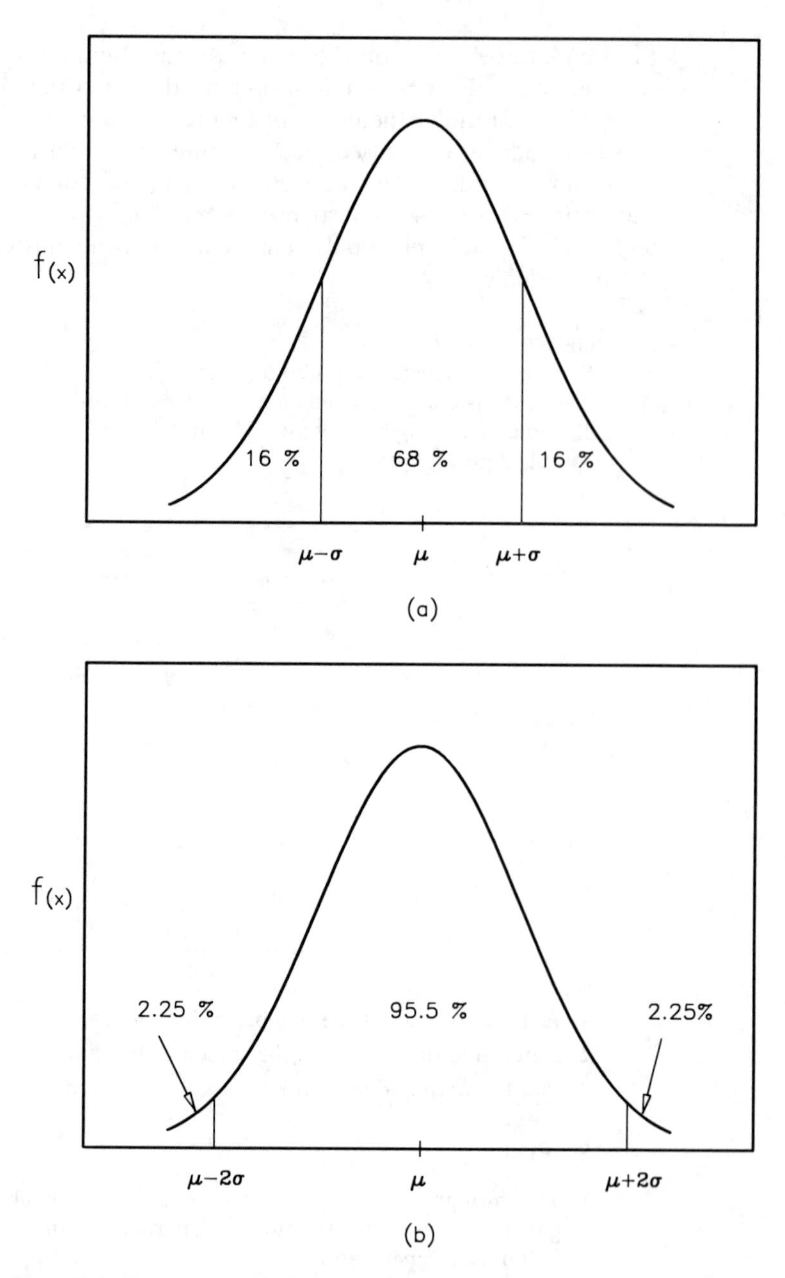

Figure 2-4 Percentage of individuals affected based on a response between one and two standard deviations of the mean.

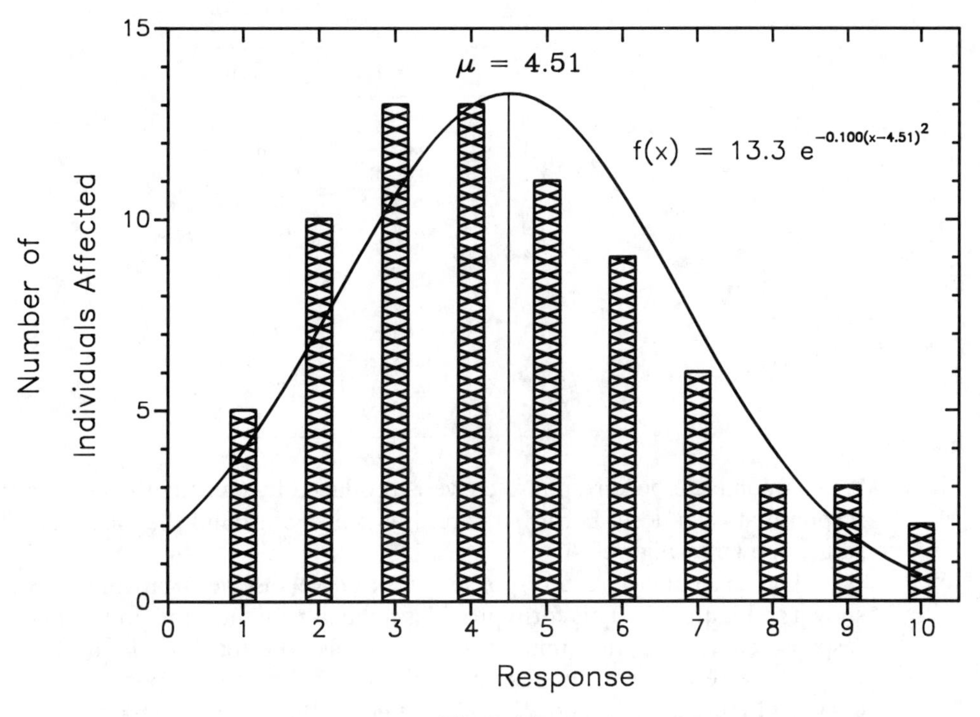

Figure 2-5

The standard deviation is computed using Equation 2-3.

$$\sigma^2 = [(1 - 4.51)^2(5) + (2 - 4.51)^2(10) + (3 - 4.51)^2(13) + (4 - 4.51)^2(13) + (5 - 4.51)^2(11) + (6 - 4.51)^2(9) + (7 - 4.51)^2(6) + (8 - 4.51)^2(3) + (9 - 4.51)^2(3) + (10 - 4.51)^2(2)] / 75 = 374.7 / 75 = 5.00$$

$$\sigma = \sqrt{\sigma^2} = \sqrt{5.00} = 2.24$$

c. The normal distribution is computed using Equation 2-1. Substituting the mean and standard deviations:

$$f(x) = \frac{1}{(2.24)\sqrt{6.28}} e^{-\frac{1}{2}\left(\frac{x - 4.51}{2.24}\right)^2}$$

$$f(x) = 0.178\ e^{-0.100\ (x-4.51)^2}$$

The distribution is converted to a function representing the number of individuals affected by multiplying by the total number of individuals, in this case 75. The corresponding values are shown in Table 2-3 and on Figure 2-5.

The toxicological experiment is repeated for a number of different doses and normal curves similiar to Figure 2-3 are drawn. The standard deviation and mean response are determined from the data for each dose.

TABLE 2-3

x	$f(x)$	$75\ f(x)$
0	0.0232	1.74
1	0.0519	3.89
2	0.0948	7.11
3	0.1417	10.6
4	0.173	13.0
4.51	0.178	13.3
5	0.174	13.0
6	0.143	10.7
7	0.096	7.18
8	0.0527	3.95
9	0.0237	1.78
10	0.00874	0.655

A complete dose-response curve is produced by plotting the cumulative mean response at each dose. Error bars are drawn at $\pm\sigma$ around the mean. A typical result is shown in Figure 2-6.

For convenience, the response is plotted versus the logarithm of the dose as shown in Figure 2-7. This form provides a much straighter line in the middle of the response curve than the simple response versus dose form of Figure 2-6.

If the response of interest is death or lethality, the response versus log dose curve of Figure 2-7 is called a lethal dose curve. For comparison purposes, the dose that results in 50% lethality of the subjects is frequently reported. This is called the LD_{50} dose (lethal dose for 50% of the subjects). Other values such as LD_{10} or LD_{90} are sometimes also reported. For gases, LC (for lethal concentration) data are used.

If the response to the chemical or agent is minor and reversible (such as minor eye irritation), the response-log dose curve is called the effective dose, or ED curve. Values for ED_{50}, ED_{10} and so forth are also used.

Finally, if the response to the agent is toxic (an undesirable response that is not lethal but is irreversible, such as liver or lung damage), the response-log dose curve is called the toxic dose, or TD curve.

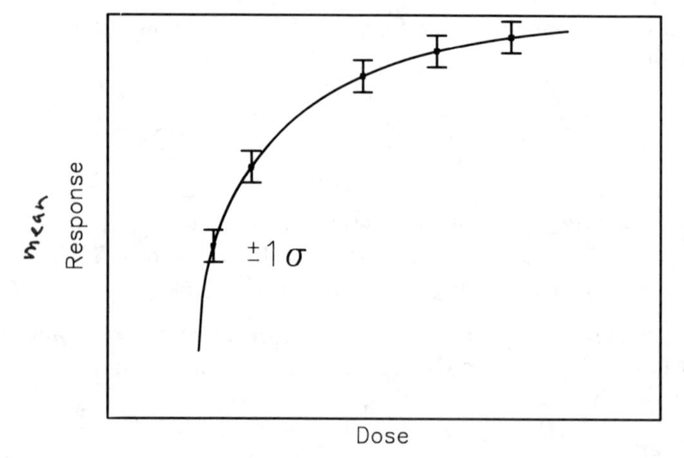

Figure 2-6 Dose-Response Curve. The bars around the data points represent the standard deviation in response to a specific dose.

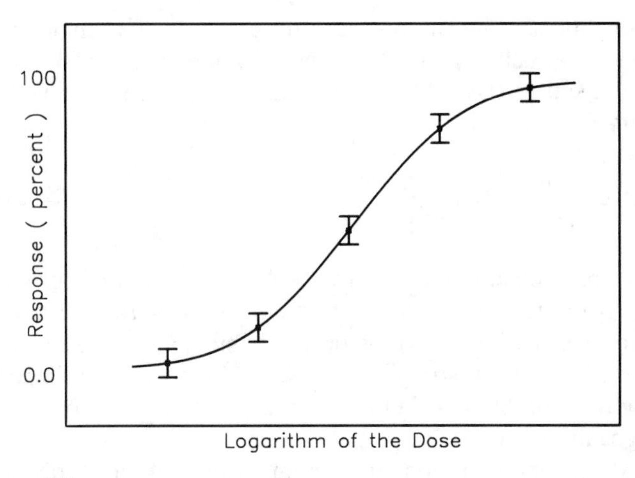

Figure 2-7 Response versus log dose curve. This form presents a much straighter function than Figure 2-6.

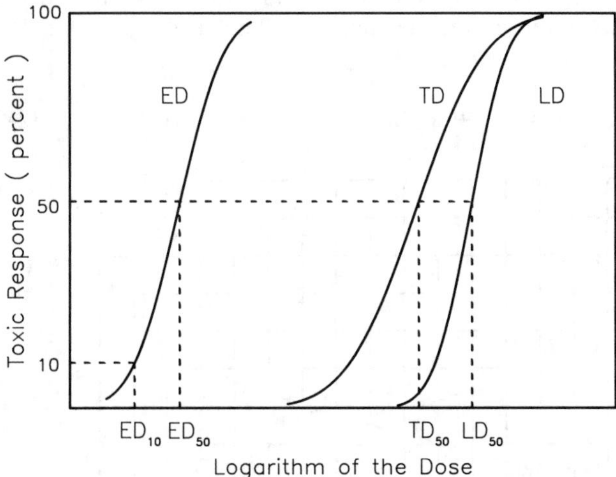

Figure 2-8 The various types of response vs. log dose curves. ED = Effective Dose, TD = Toxic Dose, LD = Lethal Dose. For gases, LC, for lethal concentration is used.

The relationship between the various types of response-log dose curves is shown in Figure 2-8.

Most often response-dose curves are developed using acute toxicity data. Chronic toxicity data is usually considerably different. Furthermore, the data is complicated by differences in group age, sex, and method of delivery. If several chemicals are involved, the toxicants might interact additively (the combined effect is the sum of the individual effects), synergistically (the combined effect is more than the individual effects), potentiately (presence of one increases the effect of the other), or antagonistically (both counteract each other).

2-6 MODELS OF DOSE: RESPONSE CURVES

Response versus dose curves can be drawn for a wide variety of exposures, including exposure to heat, pressure, radiation, impact, and sound, to name a few. For computational purposes, the response versus dose curve is not very convenient; an analytical equation is preferred.

Many methods exist for representing the response-dose curve[1]. For single exposures, the probit (probit = probability unit) method is particularly suited, providing a straight-line equivalent to the response-dose curve. The probit variable Y is related to the probability P by:[2]

$$P = \frac{1}{(2\pi)^{1/2}} \int_{-\infty}^{Y-5} \exp\left(-\frac{u^2}{2}\right) du \tag{2-4}$$

Equation 2-4 provides a relationship between the probability P and the probit variable Y. This relationship is plotted in Figure 2-9 and tabulated in Table 2-4.

The probit relationship of Equation 2-4 transforms the sigmoid shape of the normal response versus dose curve into a straight line when plotted using a linear probit scale, as shown in Figure 2-10. Standard curve fitting techniques are used to determine the best fit straight line.

Table 2-5 lists a variety of probit equations for a number of different types of exposures. The causative factor represents the dose, V. The probit variable Y is computed from:

$$Y = k_1 + k_2 \ln V \tag{2-5}$$

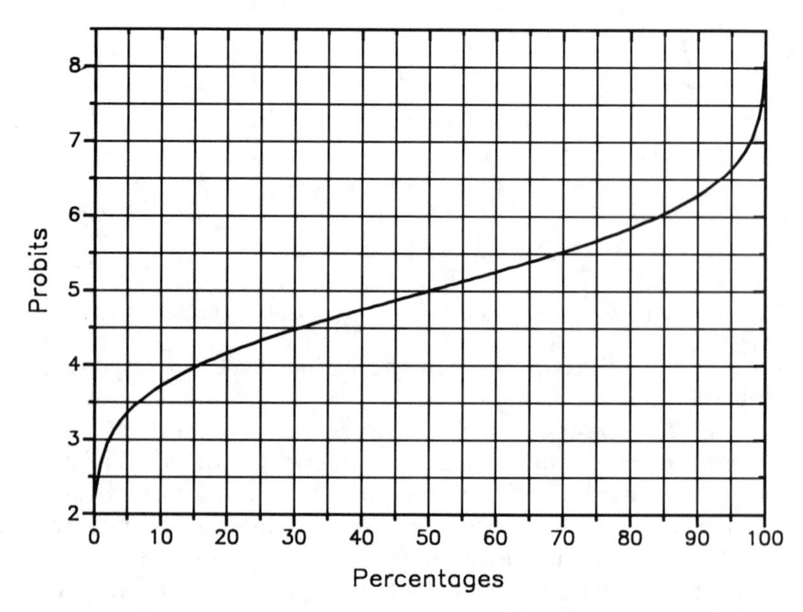

Figure 2-9 The relationship between percentages and probits. (D. J. Finney, *Probit Analysis,* 3rd Ed., 1971, p. 23. Reprinted by permission of Cambridge University Press, London.)

[1]Phillip L Williams and James L. Burson, Eds., *Industrial Toxicology, Safety and Health Applications in the Workplace* (New York: Van Nostrand Reinhold Company, 1985), p. 379.

[2]D. J. Finney, *Probit Analysis* (Cambridge: Cambridge University Press, 1971), p. 23.

TABLE 2-4 THE TRANSFORMATION FROM PERCENTAGES TO PROBITS.*

%	0	1	2	3	4	5	6	7	8	9
0	—	2.67	2.95	3.12	3.25	3.36	3.45	3.52	3.59	3.66
10	3.72	3.77	3.82	3.87	3.92	3.96	4.01	4.05	4.08	4.12
20	4.16	4.19	4.23	4.26	4.29	4.33	4.36	4.39	4.42	4.45
30	4.48	4.50	4.53	4.56	4.59	4.61	4.64	4.67	4.69	4.72
40	4.75	4.77	4.80	4.82	4.85	4.87	4.90	4.92	4.95	4.97
50	5.00	5.03	5.05	5.08	5.10	5.13	5.15	5.18	5.20	5.23
60	5.25	5.28	5.31	5.33	5.36	5.39	5.41	5.44	5.47	5.50
70	5.52	5.55	5.58	5.61	5.64	5.67	5.71	5.74	5.77	5.81
80	5.84	5.88	5.92	5.95	5.99	6.04	6.08	6.13	6.18	6.23
90	6.28	6.34	6.41	6.48	6.55	6.64	6.75	6.88	7.05	7.33
%	0.0	0.1	0.2	0.3	0.4	0.5	0.6	0.7	0.8	0.9
99	7.33	7.37	7.41	7.46	7.51	7.58	7.65	7.75	7.88	8.09

*D. J. Finney, *Probit Analysis,* 1971, p. 25. Reprinted by permission of Cambridge University Press.

TABLE 2-5 PROBIT CORRELATIONS FOR A VARIETY OF EXPOSURES. THE CAUSATIVE VARIABLE IS REPRESENTATIVE OF THE MAGNITUDE OF THE EXPOSURE.*

Type of injury or damage	Causative variable	Probit parameters k_1	k_2
Fire:			
Burn deaths from flash fire	$t_e I_e^{4/3}/10^4$	−14.9	2.56
Burn deaths from pool burning	$tI^{4/3}/10^4$	−14.9	2.56
Explosion:			
Deaths from lung hemorrhage	p^o	−77.1	6.91
Eardrum ruptures	p^o	−15.6	1.93
Deaths from impact	J	−46.1	4.82
Injuries from impact	J	−39.1	4.45
Injuries from flying fragments	J	−27.1	4.26
Structural damage	p^o	−23.8	2.92
Glass breakage	p^o	−18.1	2.79
Toxic release:			
Chlorine deaths	$\Sigma C^{2.75}T$	−17.1	1.69
Chlorine injuries	C	−2.40	2.90
Ammonia deaths	$\Sigma C^{2.75}T$	−30.57	1.385

t_e = effective time duration (s)
I_e = effective radiation intensity (W/m^2)
t = time duration of pool burning (sec)
I = radiation intensity from pool burning (W/m^2)
p^o = peak overpressure (N/m^2)
J = impulse (N s/m^2)
C = concentration (ppm)
T = time interval (min)

*Selected from Frank P. Lees, *Loss Prevention in the Process Industries,* Butterworths, London, 1986, p. 208.

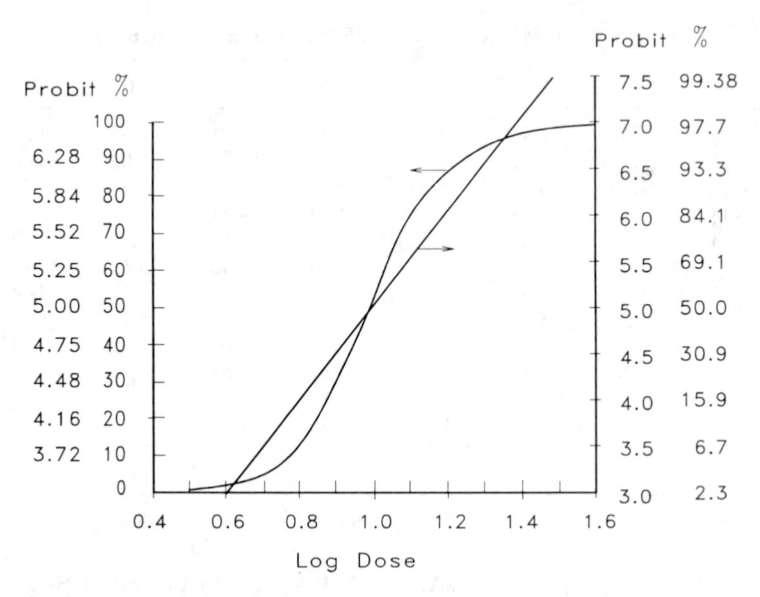

Figure 2-10 The probit transformation converts the sigmoidal response vs. log dose curve above into a straight line when plotted on a linear probit scale. (D. J. Finney, *Probit Analysis,* 3rd ed., 1971, p. 24. Reprinted by permission of Cambridge University Press, London.)

Example 2-2

Eisenberg[3] reports the following data on the effect of explosion peak overpressures on eardrum rupture in humans.

Percentage affected	Peak overpressure (N/m^2)
1	16,500
10	19,300
50	43,500
90	84,300

Confirm the probit correlation for this type of exposure as shown in Table 2-5.

Solution The percentage is converted to a probit variable using Table 2-4. The results are:

Percentage	Probit
1	2.67
10	3.72
50	5.00
90	6.28

[3]N. A. Eisenberg, *Vulnerability Model: A Simulation System for Assessing Damage Resulting from Marine Spills,* NTIS Report AD-A015-245 (Springfield, VA: National Technical Information Service, 1975).

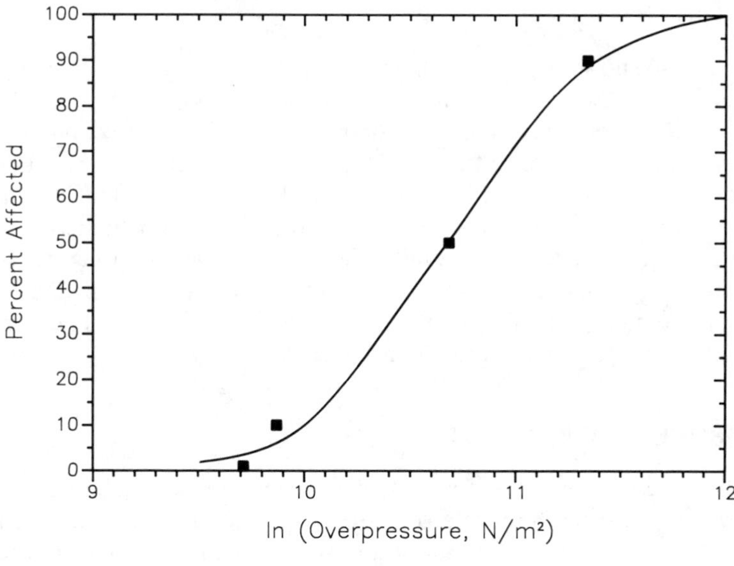

Figure 2-11

Figure 2-11 is a plot of the percentage affected versus the natural log of the peak overpressure. This demonstrates the classical sigmoid shape of the response versus log dose curve. Figure 2-12 is a plot of the probit variable (with a linear probit scale) versus the log of the peak overpressure. The straight line verifies the values reported in Table 2-5. The sigmoid curve of Figure 2-11 is drawn after converting the probit correlation back to percentages.

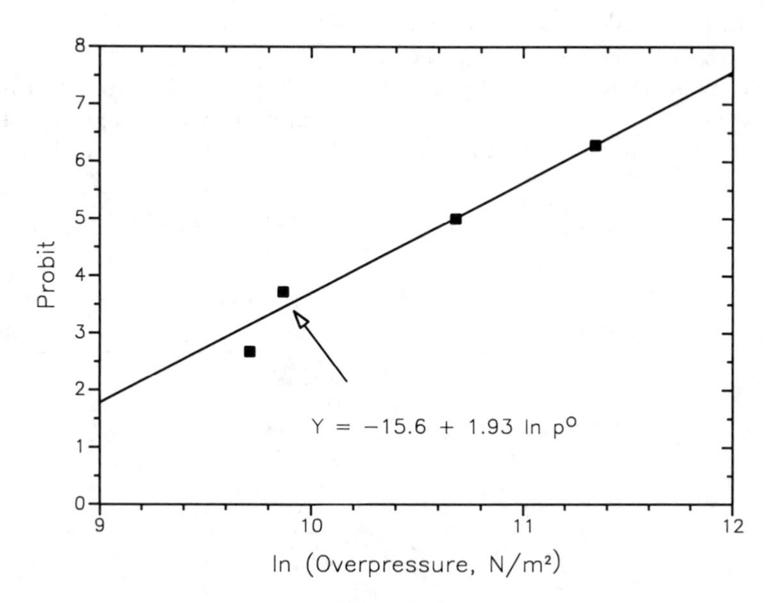

Figure 2-12

2-7 RELATIVE TOXICITY

Table 2-6 shows the Hodge-Sterner table for the degree of toxicity. This table covers a range of doses from 1.0 mg/kg to 15,000 mg/kg.

Toxicants are compared for relative toxicity based on the LD, ED, or TD curves. If the response - dose curve for chemical A is to the right of the response - dose curve for chemical B, then chemical A is more toxic. Care must be taken when comparing two response - dose curves when partial data are available. If the slopes of the curves differ substantially, the situation shown in Figure 2-13 might occur. If only a single data point is available in the upper part of the curves, it might appear that chemical A is always more toxic than chemical B. The complete data shows that chemical B is more toxic at lower doses.

2-8 THRESHOLD LIMIT VALUES

The lowest value on the response versus dose curve is called the threshold dose. Below this dose the body is able to detoxify and eliminate the agent without any detectable effects. In reality the response is only identically zero when the dose is zero but for small doses the response is not detectable.

The American Conference of Governmental Industrial Hygienists (ACGIH) has established threshold doses called threshold limit values (TLVs) for a large number of chemical agents. The TLV refers to airborne concentrations that correspond to conditions where no adverse effects are normally expected during a worker's lifetime. The exposure occurs only during normal working hours, eight hours per day and five days per week. The TLV was formerly called the maxiumum allowable concentration (MAC).

There are three different types of TLVs (TLV-TWA, TLV-STEL and TLV-C) with precise definitions provided in Table 2-7. More TLV-TWA data is available than TWA-STEL or TLV-C data.

The United States Occupational Safety and Health Administration (OSHA) has defined their own threshold dose called a permissible exposure level (PEL). The PEL values follow the TLV-TWA of the ACGIH very closely. However, the

TABLE 2-6 HODGE - STERNER TABLE FOR DEGREE OF TOXICITY.*

Experimental LD_{50} dose per kg of body weight	Degree of toxicity	Probable lethal dose for a 70 kg person
< 1.0 mg	Dangerously toxic	A taste
1.0 - 50 mg	Seriously toxic	A teaspoonful
50 - 500 mg	Highly toxic	An ounce
0.5 - 5 gm	Moderately toxic	A pint
5 - 15 gm	Slightly toxic	A quart
> 15 gm	Extremely low toxicity	More than a quart

*N. Irving Sax, *Dangerous Properties of Industrial Materials,* Van Nostrand Reinhold Company, New York, 1984, p. 1.

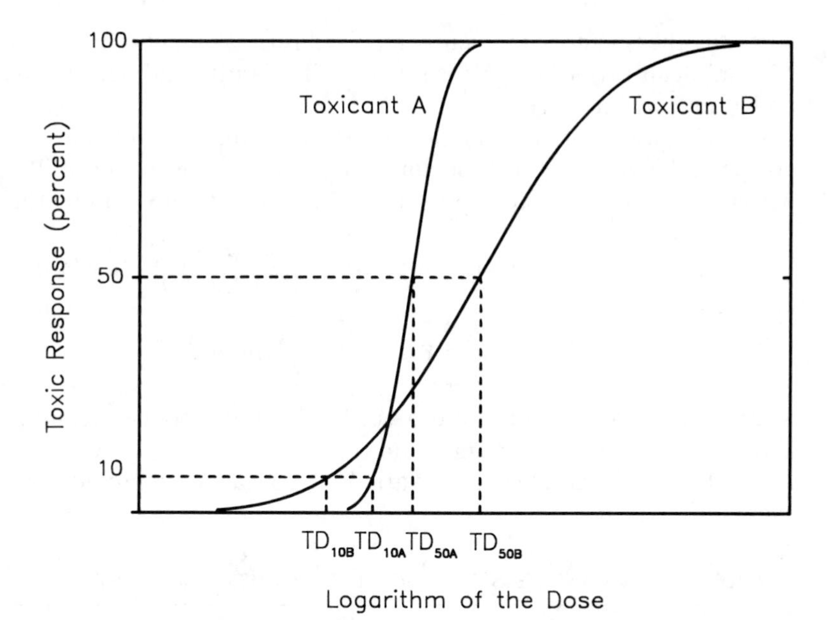

Figure 2-13 Two toxicants with differing relative toxicities at different doses. Toxicant A is more toxic at high doses while toxicant B is more toxic at low doses. (Robert C. James, "General Principles of Toxicology," *Industrial Hygiene, Safety and Health Applications in the Workplace,* ed. Phillip L. Williams and James L. Burson, 1985, p. 14. Reprinted by permission of Van Nostrand Reinhold Company, New York.)

TABLE 2-7 DEFINITIONS FOR THRESHOLD LIMIT VALUES (TLVS)

TLV-TWA:	Time weighted average for a normal 8-hour workday or 40-hour workweek, to which nearly all workers can be exposed, day after day, without adverse effects. Excursions above the limit are allowed if compensated by excursions below the limit.
TLV-STEL:	Short-term exposure limit. The maximum concentration to which workers can be exposed for a period of up to 15-minutes continuously without suffering (1) intolerable irritation, (2) chronic or irreversible tissue change, (3) narcosis of sufficient degree to increase accident proneness, impair self-rescue, or materially reduce worker efficiency, provided that no more than four excursions per day are permitted, with at least 60-minutes between exposure periods, and provided that the daily TLV-TWA is not exceeded.
TLV-C:	Ceiling limit. The concentration which should not be exceeded, even instantaneously.

NOTE: TLVs should not be used for (1) a relative index of toxicity, (2) air pollution work, or (3) assessment of toxic hazard from continuous, uninterrupted exposure.

PEL values are not as numerous and are not updated as frequently. The TLV values are often somewhat more conservative.

For some toxicants (particularly carcinogens) exposures at any level are not permitted. These toxicants have zero thresholds.

Another quantity frequently reported is the amount immediately dangerous to life and health (IDLH). Exposures to this quantity and above should be avoided under any circumstances.

TLVs are reported using ppm (parts per million by volume), mg/m^3 (mg of vapor per cubic meter of air), or, for dusts, in mg/m^3 or mppcf (millions of particles per cubic foot of air). For vapors, mg/m^3 is converted to ppm by the equation:

$$C_{\text{ppm}} = \text{Conc. in ppm} = \frac{22.4}{M}\left(\frac{T}{273}\right)\left(\frac{1}{P}\right)(\text{mg/m}^3)$$
$$= 0.08205\left(\frac{T}{PM}\right)(\text{mg/m}^3) \tag{2-6}$$

where T is the temperature in degrees K, P is the absolute pressure in atm, and M is the molecular weight in gm/gm-mole.

TLV and PEL values for a variety of toxicants are provided in Table 2-8.

TABLE 2-8 TLVS AND PELS FOR A VARIETY OF CHEMICAL SUBSTANCES

	Threshold Limit Value[1] Time Weighted Average		OSHA Permissible Exposure Level (PEL)	
Substance	ppm	mg/m^3, 25°C	ppm	mg/m^3, 25°C
Acetaldehyde	100	180	100	180
Acetic acid	10	25	10	25
Acetone	750	1780	750	1780
Acrolein	0.1	0.25	0.1	0.25
Acrylic acid (*skin*)	2	6		
Acrylonitrile* (*skin*)	2	4.5	2	4.5
Ammonia	25	18	25	18
Aniline (*skin*)	2	8	2	8
Arsine	0.05	0.2	0.05	0.2
Benzene*	10	30	10	30
Biphenyl	0.2	1.5	0.2	1.5
Bromine	0.1	0.7	0.1	0.7
Butane	800	1900		
Caprolactum (*vapor*)	0.22	1		
Carbon dioxide	5000	9000	5000	9000
Carbon monoxide	50	55	35	38
Carbon tetrachloride* (*skin*)	5	30	2	12
Chlorine	0.5	1.5	0.5	1.5
Chloroform*	10	50	2	10
Cyclohexane	300	1015	300	1015
Cyclohexanol (*skin*)	50	200	50	200
Cyclohexanone (*skin*)	25	100	25	100
Cyclohexene	300	1010	300	1010
Cyclopentane	600	1720		
Diborane	0.1	0.1	0.1	0.1

TABLE 2-8 TLVS AND PELS FOR A VARIETY OF CHEMICAL SUBSTANCES (continued)

Substance		Threshold Limit Value[1] Time Weighted Average		OSHA Permissible Exposure Level (PEL)	
		ppm	mg/m^3, 25°C	ppm	mg/m^3, 25°C
1,1 Dichloroethane		200	810	100	400
1,2 Dichloroethylene		200	790	200	790
Diethylamine		10	30	10	30
Diethyl ketone		200	705		
Dimethylamine		10	18	10	18
Dioxane (*skin*)		25	90	25	90
Ethyl acetate		400	1400	400	1400
Ethylamine		10	18	10	18
Ethyl benzene		100	435	100	435
Ethyl bromide		200	890	200	890
Ethyl chloride		1000	2600	1000	2600
Ethylene dichloride		10	40	1	4
Ethylene oxide*		1	2	1	2
Ethyl ether		400	1200	400	1200
Ethyl mercaptan		0.5	1	0.5	1
Fluorine		1	2	0.1	0.2
Formaldehyde*		1	1.5	1	1.5
Formic acid		5	9	5	9
Furfural (*skin*)		2	8	2	8
Gasoline		300	900		
Heptane		400	1600	400	1600
Hexachloroethane		1	10	1	10
Hexane		50	180	50	180
Hydrogen chloride	TLV-C:	5	7	5	7
Hydrogen cyanide (*skin*)	TLV-C:	10	10	10	10
Hydrogen fluoride	TLV-C:	3	2.5	3	2.5
Hydrogen peroxide		1	1.5	1 (90%)	1.4 (90%)
Hydrogen sulfide		10	14	10	14
Iodine	TLV-C:	0.1	1	0.1	1
Isobutyl alcohol		50	150	50	150
Isopropyl alcohol		400	980	400	980
Isopropyl ether		250	1050	250	1050
Ketene		0.5	0.9	0.5	0.9
Maleic anhydride		0.25	1	0.25	1
Methyl acetate		200	610	200	610
Methyl acetylene		1000	1650	1000	1650
Methyl alcohol		200	260	200	260
Methylamine		10	12	10	12
Methyl bromide (*skin*)		5	20	5	20
Methyl chloride		50	105	50	105
Methylene chloride*		50	175	50	175
Methyl ethyl ketone		200	590	200	590
Methyl formate		100	250	100	250

TABLE 2-8 TLVS AND PELS FOR A VARIETY OF CHEMICAL SUBSTANCES (continued)

Substance		Threshold Limit Value[1] Time Weighted Average		OSHA Permissible Exposure Level (PEL)	
		ppm	mg/m^3, 25°C	ppm	mg/m^3, 25°C
Methyl isocyanate (*skin*)		0.02	0.05	0.02	0.05
Methyl mercaptan		0.5	1	0.5	1
Napthalene		10	50	10	50
Nitric acid		2	5	2	5
Nitric oxide		25	30	25	30
Nitrobenzene (*skin*)		1	5	1	5
Nitrogen dioxide		3	6	3	6
Nitromethane		100	250	100	250
Nonane		200	1050		
Octane		300	1450	300	1450
Oxalic acid			1		1
Ozone	TLV-C:	0.1	0.2	0.1	0.2
Pentane		600	1800	600	1800
Phenol (*skin*)		5	19	5	19
Phosgene		0.1	0.4	0.1	0.4
Phosphine		0.3	0.4	0.3	0.4
Phosphoric acid			1		1
Phthalic anhydride		1	6	1	6
Pyridine		5	15	5	15
Styrene, monomer (*skin*)		50	215	50	215
Sulfur dioxide		2	5	2	5
Toluene (*skin*)		100	375	100	375
Trichloroethylene		50	270	50	270
Triethylamine		10	40	10	40
Turpentine		100	560	100	560
Vinyl acetate		10	30		
Vinyl chloride**		5	10	5	10
Xylene		100	435	100	435

*possible carcinogen

**human carcinogen

[1]*Documentation of the Threshold Limit Values and Biological Exposure Indices,* 5th ed. (Cincinnati: American Conference of Governmental Industrial Hygienists, 1986).

The ACGIH clearly points out that the TLV values should not be used as a relative index of toxicity (see Figure 2-8), should not be used for air pollution work, and cannot be used to assess the impact of continuous exposures to toxicants. The TLV assumes that workers are only exposed during a normal eight-hour workday.

Every effort must be made to reduce worker exposures to toxicants to below the PEL and lower if possible.

SUGGESTED READING

Toxicology

HOWARD H. FAWCETT and WILLIAM S. WOOD, eds., *Safety and Accident Prevention in Chemical Operations,* 2nd ed. (New York: John Wiley and Sons, 1982), Chapters 14, 15, 25.

N. IRVING SAX, *Dangerous Properties of Industrial Materials,* 6th Edition (New York: Van Nostrand Reinhold Company, 1984), Section 1.

PHILLIP L. WILLIAMS and JAMES L. BURSON, eds., *Industrial Toxicology, Safety and Health Applications in the Workplace* (New York: Van Nostrand Reinhold Company, 1985).

Probit Analysis

D. J. FINNEY, *Probit Analysis* (Cambridge: Cambridge University Press, 1971).

FRANK P. LEES, *Loss Prevention in the Process Industries* (London: Butterworths, 1986), p. 207

Threshold Limit Values

Documentation of the Threshold Limit Values and Biological Exposure Indices, 5th ed. (Cincinnati: American Conference of Governmental Industrial Hygienists, 1986).

PROBLEMS

2-1. Derive Equation 2-6.

2-2. Finney[4] reports the data of Martin[5] involving the toxicity of rotenone to the insect species *Macrosiphoniella sanborni.* The rotenone was applied in a medium of 0.5% saponin, containing 5% alcohol. The insects were examined and classified one day after spraying.

Dose of rotenone (mg/l)	No. of insects	Number affected
10.2	50	44
7.7	49	42
5.1	46	24
3.8	48	16
2.6	50	6
0	49	0

[4]D. J. Finney, *Probit Analysis* (Cambridge: Cambridge University Press, 1971), p. 20.

[5]J. T. Martin, "The Problem of the Evaluation of Rotenone-containing Plants. VI. The Toxicity of l-elliptone and of Poisons Applied Jointly, with Further Observations on the Rotenone Equivalent Method of Assessing the Toxicity of Derris Root," *Ann. Appl. Biol.,* 29(1942), pp 69-81.

a. From the data given plot the percent of insects affected versus the log of the dose.

b. Convert the data to a probit variable and plot the probit versus the log of the dose. If the result is linear, determine a straight line that fits the data. Compare the probit and number of insects affected predicted by the straight-line fit to the actual data.

2-3. A blast produces a peak overpressure of 47,000 N/m^2. What fraction of structures will be damaged by exposure to this overpressure? What fraction of people exposed will die due to lung hemorrhage? What fraction will have eardrums ruptured? What conclusions about the effects of this blast can be drawn?

2-4. The peak overpressure expected due to the explosion of a tank in a plant facility is approximated by the equation

$$\log P = 4.2 - 1.8 \log r$$

where P is the overpressure in psi and r is the distance from the blast in feet. The plant employs 500 people who work in an area from 10 to 500 feet from the potential blast site. Estimate the number of fatalities expected due to lung hemorrhage and the number of eardrums ruptured as a result of this blast. Be sure to state any additional assumptions.

2-5. A certain volatile substance evaporates from an open container into a room of volume 1000 ft^3. The evaporation rate is determined to be 100 mg/min. If the air in the room is assumed to be well mixed, how many ft^3/min of fresh air must be supplied to insure that the concentration of the volatile is maintained below its TLV of 100 ppm? The temperature is 77°F and the pressure is 1 atm. Assume a volatile species molecular weight of 100. Under most circumstances the air in a room cannot be assumed to be well mixed. How would poor mixing affect the quantity of air required?

2-6. In Example 2-1, Part c, the data were represented by the normal distribution function

$$f(x) = 0.178\, e^{-0.100(x-4.51)^2}$$

Use this distribution function to determine the fraction of individuals demonstrating a response between the range of 2.5 to 7.5.

Industrial Hygiene 3

Industrial hygiene is a science devoted to the identification, evaluation, and control of occupational conditions which cause sickness and injury. Industrial hygienists are also responsible for selecting and using instrumentation to monitor the workplace during the identification and control phases of industrial hygiene projects.

Typical projects involving industrial hygiene are:

- Monitoring of toxic airborne vapor concentrations,
- Reduction of toxic airborne vapors through the use of ventilation,
- Selection of proper personal protective equipment to prevent worker exposure,
- Development of procedures for the handling of hazardous materials, and
- Monitoring and reduction of noise, heat, radiation, and other physical factors to insure that workers are not exposed to harmful levels.

The three phases in any industrial hygiene project are *identification*, *evaluation,* and *control*.

- Identification: determination of the presence or possibility of workplace exposures.
- Evaluation: determination of the magnitude of the exposure.
- Control: application of appropriate technology to reduce workplace exposures to acceptable levels.

In chemical plants and laboratories, the industrial hygienist works closely with safety professionals as an integral part of a safety and loss prevention program. After identifying and evaluating the hazards, the industrial hygienist makes recommendations relevant to control techniques. The industrial hygienist, safety professionals, and plant operations personnel work together to insure that the control measures are applied and maintained. It has been clearly demonstrated that very toxic chemicals can be handled safely when principles of industrial hygiene are appropriately applied.

3-1 GOVERNMENT REGULATIONS

In 1970, the United States Congress enacted a health and safety law which continues to have a significant impact on the practices of industrial hygiene in the chemical industry: The Occupational Safety and Health Act of 1970 (OSHAct). To appreciate the significance of the OSHAct it is helpful to review regulations and practices[1] before 1970.

Prior to 1936, regulations concerning occupational health were poorly administered by state and local governmental agencies. During this era, staffs and funds were too small to carry out effective programs. In 1936, the federal government enacted the Walsh-Healy Act to establish federal safety and health standards for activities relating to federal contracts. This 1936 act also initiated significant research related to the cause, recognition, and control of occupational disease. The concepts promulgated by the Walsh-Healy Act, although not adequate by today's standards, were the forerunners in the development of our current occupational health and safety regulations.

During the period between 1936 and 1970, a number of states enacted their own safety and health regulations. Although some progress was made, these regulations were never sufficiently supported to carry out a satisfactory program. This produced relatively inconsistent and ineffective results.

The OSHAct of 1970 was developed to solve these problems and to give a nationally consistent program with the funding necessary to manage it effectively. This act defined clear procedures for establishing regulations, conducting investigations for compliance, and developing and maintaining safety and health records.

As a result of the OSHAct, sufficient funding was committed to create and support the

- Occupational Safety and Health Administration (OSHA) to manage and administer the government's responsibilities specified in the OSHAct, and the
- National Institute for Occupational Safety and Health (NIOSH) to conduct research and technical assistance programs for improving the protection and maintenance of workers' health.

Examples of NIOSH responsibilities include (1) measuring health effects of exposure in the work environment, (2) developing criteria for handling toxic materials, (3) establishing safe levels of exposure, and (4) training professionals for administering the programs of the act.

[1]J. B. Olishifski, ed., *Fundamentals of Industrial Hygiene,* 2nd Edition (Chicago: National Safety Council, 1979), pp. 758-777.

NIOSH develops data and information regarding hazards, and OSHA uses this data to promulgate standards. Some standards, particularly relevant to the chemical industry, are shown in Table 3-1.

The OSHAct gives the employers the responsibility to provide safe and healthy working conditions for its employees. OSHA is authorized, however, to conduct inspections, and when violations of the safety and health standards are found, they may issue citations and financial penalties. Highlights of OSHA enforcement rights are illustrated in Table 3-2.

The implications, interpretations, and applications of the OSHAct will continue to develop as the standards are promulgated. Especially within the chemical industry, these standards will continue to create an environment for improving process designs and process conditions relevant to the safety and health of workers and the surrounding communities.

Government regulation will continue to be a significant part of the practice of chemical process safety. Since the OSHAct, substantial new legislation controlling the workplace and community environment has been enacted. Table 3-3 provides a summary of relevent safety legislation and Figure 3-1 shows how the amount of legislation has increased. A description of this legislation is well beyond the scope and goals of this textbook. However, it is important that chemical engineers be aware of the law to insure that their facilities comply.

3-2 IDENTIFICATION

One of the major responsibilities of the industrial hygienist is to identify and solve potential health problems within plants. Chemical process technology, however, is

TABLE 3-1 A FEW OCCUPATIONAL SAFETY AND HEALTH STANDARDS ESTABLISHED BY OSHA

Number	Title
1910.94	Ventilation
1910.95	Occupational noise exposures
1910.1000	Air contaminants
1910.1001	Asbestos
1910.1003	4-Nitrobiphenyl
1910.1008	bis-Chloromethyl ether
1910.1017	Vinyl chloride
1910.1028	Benzene

TABLE 3-2 HIGHLIGHTS OF OSHA'S RIGHT OF ENFORCEMENT

Employers must admit OSHA compliance officers into their plant sites for safety inspections with no advance notice. A search warrant may be required to show probable cause.
OSHA's right of inspection includes safety and health records.
Criminal penalties can be invoked.
OSHA officers finding conditions of imminent danger may request plant shutdowns.

TABLE 3-3 FEDERAL LEGISLATION RELEVANT TO CHEMICAL PROCESS SAFETY.**

Date	Abbreviation	Act
1899	RHA	River and Harbor Act
1906*	FDCA	Federal Food, Drug and Cosmetic Act
1947*	FIFRA	Federal Insecticide, Fungicide, and Rodenticide Act
1952	DCA	Dangerous Cargo Act
1952*	FWPCA	Federal Water Pollution Control Act
1953*	FFA	Flammable Fabrics Act
1954	AEA	Atomic Energy Act
1956*	FWA	Fish and Wildlife Act of 1956
1960*	FHSA	Federal Hazardous Substances Labeling Act
1965*	SWDA	Solid Waste Disposal Act
1966	MNMSA	Metal and Non-Metallic Mine Safety Act
1969	NEPA	National Environmental Policy Act
1969	CMHSA	Federal Coal Mine Health and Safety Act
1970*	CAA	Clean Air Act
1970	PPPA	Poison Prevention Packaging Act of 1970
1970	WQI	Water Quality Improvement Act of 1970
1970	RSA	Federal Railroad Safety Act of 1970
1970	RRA	Resource Recovery Act of 1970
1970	OSHA	Occupational Safety and Health Act
1972	NCA	Noise Control Act of 1972
1972	FEPCA	Federal Environmental Pollution Control Act
1972	HMTA	Hazardous Materials Transportation Act
1972	CPSA	Consumer Product Safety Act
1972	MPRSA	Marine Protection, Research and Sanctuary Act of 1972
1972*	CWA	Clean Water Act
1972*	CZMA	Coastal Zone Management Act
1973	ESA	Endangered Species Act of 1973
1974*	SDWA	Safe Drinking Water Act
1974	TSA	Transportation Safety Act of 1974
1974*	ESECA	Energy Supply and Environmental Coordination Act
1976	TSCA	Toxic Substances Control Act
1976*	RCRA	Resource Conservation and Recovery Act
1977	FMSHA	Federal Mine Safety and Health Act
1977	SMCRA	Surface Mine Control and Reclamation Act
1978	UMTCA	Uranium Mill Tailings Control Act
1978	PTSA	Port and Tanker Safety Act
1980	CERCLA	Comprehensive Environmental Response, Compensation and Liability Act of 1980 (Superfund)

* Revised or amended

**Adapted from Lewis J. Cralley and Lester V. Cralley, eds., *Industrial Hygiene Aspects of Plant Operations,* Vol. 2, 1984, p. 13.

so complex that this task requires the concerted efforts of industrial hygienists, process designers, operators, laboratory personnel, and management. The industrial hygienist helps the effectiveness of the overall program by working with these plant personnel. For these reasons, industrial hygiene (particularly identification) must be a part of the education process of chemists, engineers, and managers.

Many hazardous chemicals are handled safely on a daily basis within chemical plants. To achieve this operating success, *all* of the potential hazards must be iden-

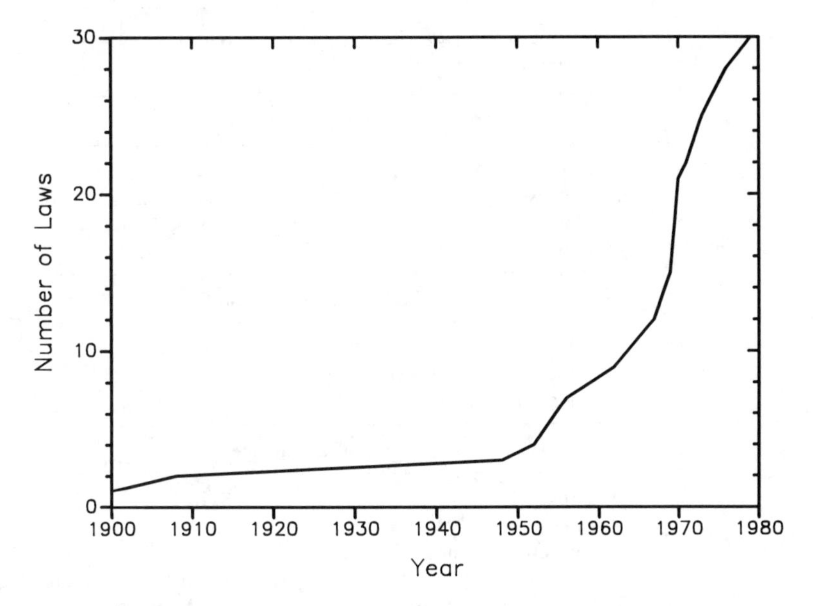

Figure 3-1 Number of federal legislative bills passed relevant to chemical process safety. (Adapted from Lewis J. Cralley and Lester V. Cralley, eds., *Industrial Hygiene Aspects of Plant Operations,* Vol. 2, 1984, p. 13.)

tified and controlled. When handling toxic and/or flammable chemicals, the potentially hazardous conditions may be numerous—in large plants there may be thousands. To be safe under these conditions requires discipline, skill, concern, and attention to detail.

The identification step requires a thorough study of the chemical process, operating conditions, and operating procedures. The sources of information include process design descriptions, operating instructions, safety reviews, equipment vendor descriptions, information from chemical suppliers, and information from operating personnel. The quality of this identification step is often a function of the number of resources used and the quality of the questions asked. The different resources may have different operating and technical emphases unique to pieces of equipment or specific chemicals. In this identification step, it is often necessary to collate and integrate the available information to identify new potential problems due to the combined effects of multiple exposures.

During the identification step, the potential hazards and methods of contact are identified and recorded. As illustrated in Table 3-4, the potential hazards are numerous, especially since those listed can also act in combination. This list of potential hazards, together with the required data for hazard identification (see Table 3-5) are commonly used during the identification step of industrial hygiene projects.

A determination of the potential for hazards to result in an accident (risk assessment) is frequently a part of the identification step (see Chapter 11). This list of potential hazards and their risk is used during the evaluation and control phase of the project. Resources for evaluating the hazards and developing control methods

TABLE 3-4 IDENTIFICATION OF POTENTIAL HAZARDS[1]

Potential hazards	
Liquids	Noise
Vapors	Radiation
Dusts	Temperature
Fumes	Mechanical
Entry mode of toxicants	
Inhalation	Ingestion
Body Absorption (skin or eyes)	
Injection	
Potential damage	
Lungs	Skin
Ears	Eyes
Nervous system	Liver
Kidneys	Reproductive organs
Circulatory system	Other organs

[1]Olishifski, *Fundamentals of Industrial Hygiene,* pp. 24-26.

TABLE 3-5 DATA USEFUL FOR HEALTH IDENTIFICATION

Threshold limit values (TLV).
Odor threshold for vapors.
Physical state.
Vapor pressure of liquids.
Sensitivity of chemical to temperature or impact.
Rates and heats of reaction.
Hazardous by-products.
Reactivity with other chemicals.
Explosive concentrations of chemicals, dusts, and vapors.
Noise levels of equipment.
Types and degree of radiation.

are allocated on a priority basis, giving the appropriate time and attention to the most significant hazards.

Material Safety Data Sheets (MSDSs)

One of the most important references used during an industrial hygiene study involving toxic chemicals is the material safety data sheet, or MSDS. A sample MSDS is shown in Figure 3-2. The MSDS lists the physical properties of a substance that may be required to determine the potential hazards of the substance.

MSDSs are available from (1) the chemical manufacturer, (2) a commercial source, or (3) a private library developed by the chemical plant.

MATERIAL SAFETY DATA SHEET

S-210
31:9203

Material Safety Data Sheet
May be used to comply with
OSHA's Hazard Communication Standard,
29 CFR 1910.1200. Standard must be
consulted for specific requirements.

U.S. Department of Labor
Occupational Safety and Health Administration
(Non-Mandatory Form)
Form Approved
OMB No. 1218-0072

IDENTITY *(As Used on Label and List)*

Note: Blank spaces are not permitted. If any item is not applicable, or no information is available, the space must be marked to indicate that.

Section I

Manufacturer's Name	Emergency Telephone Number
Address *(Number, Street, City, State, and ZIP Code)*	Telephone Number for Information
	Date Prepared
	Signature of Preparer *(optional)*

Section II — Hazardous Ingredients/Identity Information

Hazardous Components (Specific Chemical Identity; Common Name(s))	OSHA PEL	ACGIH TLV	Other Limits Recommended	% *(optional)*

Section III — Physical/Chemical Characteristics

Boiling Point		Specific Gravity (H_2O = 1)	
Vapor Pressure (mm Hg.)		Melting Point	
Vapor Density (AIR = 1)		Evaporation Rate (Butyl Acetate = 1)	
Solubility in Water			
Appearance and Odor			

Section IV — Fire and Explosion Hazard Data

Flash Point (Method Used)	Flammable Limits	LEL	UEL
Extinguishing Media			
Special Fire Fighting Procedures			
Unusual Fire and Explosion Hazards			

(Reproduce locally)

OSHA 174, Sept. 1985

Figure 3-2 Material safety data sheet. Most companies use their own MSDS format.

Section V — Reactivity Data

Stability	Unstable		Conditions to Avoid
	Stable		

Incompatibility (*Materials to Avoid*)

Hazardous Decomposition or Byproducts

Hazardous Polymerization	May Occur		Conditions to Avoid
	Will Not Occur		

Section VI — Health Hazard Data

Route(s) of Entry: Inhalation? Skin? Ingestion?

Health Hazards (*Acute and Chronic*)

Carcinogenicity: NTP? IARC Monographs? OSHA Regulated?

Signs and Symptoms of Exposure

Medical Conditions
Generally Aggravated by Exposure

Emergency and First Aid Procedures

Section VII — Precautions for Safe Handling and Use

Steps to Be Taken in Case Material Is Released or Spilled

Waste Disposal Method

Precautions to Be Taken in Handling and Storing

Other Precautions

Section VIII — Control Measures

Respiratory Protection (*Specify Type*)

Ventilation	Local Exhaust	Special
	Mechanical (*General*)	Other

Protective Gloves	Eye Protection

Other Protective Clothing or Equipment

Work/Hygienic Practices

Page 2

* USGPO 1986-491-529/45775

Figure 3-2 (continued)

The industrial hygienist or safety professional must interpret the physical and toxicological properties to determine the hazards associated with a chemical. These properties are also used to develop a strategy for the proper control and handling of these chemicals.

Example 3-1

A survey of a laboratory is made and the following chemical species are identified:

Sodium chloride	Toluene
Hydrochloric acid	Phenol
Sodium hydroxide	Benzene
Ether	

Identify the potential hazards in this laboratory.

Solution Sax[2] provides the technical information required to solve this problem. The following summarizes the results.

Sodium chloride	Common table salt, no hazard.
Toluene	Clear, colorless liquid with a slight fire hazard and moderate explosion hazard. Entry into the body is mostly by vapor inhalation. Acute and chronic exposures occur with concentrations greater than 200 ppm. Irritant to skin and eyes.
Hydrochloric acid	Clear, colorless liquid with no fire or explosion hazard. It is a moderate irritant to the skin, eyes, and mucous membranes and by ingestion and inhalation. Throat irritation occurs with concentrations of 35 ppm. Highly reactive with a wide variety of substances.
Phenol	A white, crystalline mass that is most frequently found in solution form. It is a moderate fire hazard. Emits toxic fumes when heated. Absorbed readily through the skin. Exposures to skin areas of as small as 64 in^2 have resulted in death in less than one hour.
Sodium hydroxide	A skin and eye irritant. Corrosive action upon all body tissues. Reacts violently with a number of substances.
Benzene	Clear, colorless liquid with a dangerous fire hazard and a moderate explosion hazard. It is a possible carcinogen. Entry into the body is mostly by inhalation but it is also absorbed through the skin. High concentrations produce a narcotic-type effect.
Ether	A wide variety of organic compounds that are mostly narcotic in effect. Large doses can cause death. Most ethers are dangerously flammable and explosive.

3-3 EVALUATION

The evaluation phase determines the extent and degree of employee exposure to toxicants and physical hazards in the workplace environment.

[2] N. Irving Sax, *Dangerous Properties of Industrial Materials,* 6th Ed. (New York: Van Nostrand Reinhold Co, 1984).

During the evaluation phase the various types of existing control measures and their effectiveness are also studied. Control techniques are presented in more detail in Section 3-4.

During the evaluation study, the likelihood of large and small leaks must be considered. Sudden exposures to high concentrations, via large leaks, may lead to immediate acute effects such as unconsciousness, burning eyes, fits of coughing, etc. There is rarely lasting damage to the individual if removed promptly from the contaminated area. In this case, a ready access to a clean environment is important.

Chronic effects, however, arise from repeated exposures to low concentrations, mostly by small leaks. Many toxic chemical vapors are colorless and odorless (or the toxic concentration might be below the odor threshold). Small leaks of these substances might not become obvious for months or even years. There may be permanent and serious impairments from such exposures. Special attention must be directed toward preventing and controlling low concentrations of toxic gases. In these circumstances some provision for continuous evaluation is necessary; that is, continuous or frequent and periodic sampling and analysis is important.

To establish the effectiveness of existing controls, samples are taken to determine the workers' exposure to conditions that may be harmful. If problems are evident, controls must be implemented immediately; temporary controls such as personal protective equipment can be used. Longer term and permanent controls are subsequently developed.

After obtaining this exposure data, it is necessary to compare actual exposure levels to acceptable occupational health standards such as TLVs, PELs, IDLHs, and the like. These standards together with the actual concentrations are used to identify the potential hazards requiring better or more control measures.

Evaluating Exposures to Volatile Toxicants by Monitoring

A direct method for determining worker exposures is by continuously monitoring the air concentations of toxicants on-line in a work environment. For continuous concentration data $C(t)$ the TWA (time-weighted average concentration) is computed using the equation

$$TWA = \frac{1}{8} \int_0^{t_w} C(t)\, dt \tag{3-1}$$

where $C(t)$ is the concentation, in ppm or mg/m^3, of the chemical in the air and t_w is the worker shift time in hours. The integral is always divided by eight hours, independent of the length of time actually worked in the shift. Thus, if a worker is exposed for twelve hours to a concentration of chemical equal to the TLV-TWA, then the TLV-TWA has been exceeded, since the computation is normalized to eight hours.

Continuous monitoring is not the usual situation since most facilities do not have the necessary equipment available.

The more usual case is for intermittent samples to be obtained representing worker exposures at fixed points in time. If we assume that the concentration C_i is fixed (or averaged) over the period of time T_i, the TWA is computed by

$$TWA = \frac{C_1T_1 + C_2T_2 + \ldots + C_nT_n}{8 \text{ hours}} \tag{3-2}$$

All monitoring systems have drawbacks because (1) the workers move in and out of the exposed workplace, and (2) the concentration of toxicant may vary at different locations in the work area. Industrial hygienists play an important role in the selection and placement of workplace monitoring equipment and the interpretation of the data.

If more than one chemical is present in the workplace one procedure is to assume that the effects of the toxicants are additive (unless other information to the contrary is available). The combined exposures from multiple toxicants with different TLV-TWAs is determined from the equation

$$\sum_{i=1}^{n} \frac{C_i}{(TLV\text{-}TWA)_i} \tag{3-3}$$

where n is the total number of toxicants, C_i is the concentration of chemical i with respect to the other toxicants, and $(TLV\text{-}TWA)_i$ is the TLV-TWA for chemical species i. If this sum exceeds unity, then the workers are over-exposed.

The mixture TLV-TWA can be computed from

$$(TLV\text{-}TWA)_{\text{mix}} = \frac{\sum_{i=1}^{n} C_i}{\sum_{i=1}^{n} \frac{C_i}{(TLV\text{-}TWA)_i}} \tag{3-4}$$

If the sum of the concentrations of the toxicants in the mixture exceeds this amount, then the workers are over-exposed.

For mixtures of toxicants with different effects (such as an acid vapor mixed with lead fume) the TLVs cannot be assumed to be additive.

Example 3-2

Air contains 5 ppm of diethylamine (TLV-TWA of 10 ppm), 20 ppm of cyclohexanol (TLV-TWA of 50 ppm), and 10 ppm of propylene oxide (TLV-TWA of 20 ppm). What is the mixture TLV-TWA and has this been exceeded?

Solution From Equation 3-4,

$$(TLV\text{-}TWA)_{\text{mix}} = \frac{5 + 20 + 10}{\frac{5}{10} + \frac{20}{50} + \frac{10}{20}}$$

$$= 25 \text{ ppm}$$

The total mixture concentration = 5 + 20 + 10 = 35 ppm.
The workers are overexposed under these circumstances.
An alternate approach is to use Equation 3-3.

$$\sum_{i=1}^{3} \frac{C_i}{(TLV\text{-}TWA)_i} = \frac{5}{10} + \frac{20}{50} + \frac{10}{20} = 1.40$$

Since this quantity is greater than 1 the TLV-TWA has been exceeded.

Example 3-3

Determine the eight-hour TWA worker exposure if the worker is exposed to toluene vapors as follows.

Duration of exposure (hours)	Measured concentration (ppm)
2	110
2	330
4	90

Solution Using Equation 3-2

$$TWA = (C_1T_1 + C_2T_2 + C_3T_3) / 8$$

$$TWA = [110(2) + 330(2) + 90(4)] / 8 = 155 \text{ ppm}$$

Since the TLV for toluene is 100 ppm, the worker is overexposed. Additional control measures need to be developed. On a temporary and immediate basis, all employees working in this environment need to wear the appropriate respirators.

Example 3-4

Determine the mixture TLV, at 25°C. and 1 atm pressure, of a mixture derived from the following liquid.

Component	Mole percent	Species TLV (ppm)
Heptane	50	400
Toluene	50	100

Solution The solution requires the concentration of the heptane and toluene in the vapor phase. Assuming that the composition of the liquid does not change as it evaporates (the quantity is large), the vapor composition is computed using standard vapor-liquid equilibrium calculations. Assuming Raoult's and Dalton's laws apply to this system under these conditions, the vapor composition is determined directly from the saturation vapor pressures of the pure components. Himmelblau[3] provides the following data at the specified temperature,

$$P^{\text{sat}}_{\text{heptane}} = 46.4 \text{ mm Hg}$$

$$P^{\text{sat}}_{\text{toluene}} = 28.2 \text{ mm Hg}$$

Using Raoult's law the partial pressures in the vapor is determined,

$$p_i = x_i\, P_i^{\text{sat}}$$

$$p_{\text{heptane}} = (0.5)(46.4 \text{ mm Hg}) = 23.2 \text{ mm Hg}$$

$$p_{\text{toluene}} = (0.5)(28.2 \text{ mm Hg}) = 14.1 \text{ mm Hg}$$

[3]David M. Himmelblau, *Basic Principles and Calculations in Chemical Engineering,* 4th ed. (Englewood Cliffs: Prentice-Hall, 1982), p. 591.

The total pressure of the toxicants is (23.2 + 14.1) = 37.3 mm Hg. From Dalton's law the mole fractions on a toxicant basis are

$$y_{\text{heptane}} = \frac{23.2 \text{ mm Hg}}{37.3 \text{ mm Hg}} = 0.622$$

$$y_{\text{toluene}} = 1 - 0.622 = 0.378$$

The mixture TLV is computed using Equation 3-4.

$$TLV_{\text{mix}} = \frac{1}{\frac{0.622}{400} + \frac{0.378}{100}} = 187 \text{ ppm}$$

Since the vapor will always be the same concentration, the TLVs for the individual species **in the mixture** are

$$TLV_{\text{heptane}} = (0.622)(187 \text{ ppm}) = 116 \text{ ppm}$$

$$TLV_{\text{toluene}} = (0.378)(187 \text{ ppm}) = 71 \text{ ppm}$$

If the actual concentration exceeds these levels, more control measures will be needed. For mixtures of vapors, the individual species TLVs in the mixture are significantly reduced from the TLVs of the pure substance.

Evaluation of Worker Exposures to Dusts

Industrial hygiene studies include any contaminant which may cause health injuries; dusts, of course, fit this category. Toxicological theory teaches that dust particles which present the greatest hazard to the lungs are normally in the respirable particle size range of 0.2 to 0.5 microns (see Chapter 2). Particles larger than 0.5 micron are usually unable to penetrate the lungs, while those smaller than 0.2 micron settle out too slowly and are mostly exhaled with the air.

The main reason for sampling for atmospheric particulates is to estimate the concentrations which are inhaled and deposited in the lungs. Sampling methods and the interpretation of data relevant to health hazards is relatively complex; industrial hygienists, who are specialists in this technology, should be consulted when confronted with this type of problem.

Dust evaluation calculations are performed in an identical manner to volatile vapors. Instead of using ppm as a concentration unit, mg/m^3 or mppcf (millions of particles per cubic foot) is more convenient.

Example 3-5

Determine the TLV for a uniform mixture of dusts containing

Type of dust	Concentration (wt %)	TLV (mppcf)
Nonasbestiform talc	70	20
Quartz	30	2.7

Solution From Equation 3-4

$$\text{TLV of mixture} = \frac{1}{\dfrac{C_1}{TLV_1} + \dfrac{C_2}{TLV_2}}$$

$$= 1/(.70/20 + .30/2.7)$$

$$= 6.8 \text{ mppcf}$$

Special control measures will be required when the actual particle count (of the size range specified in the standards or by an industrial hygienist) exceeds 6.8 mppcf.

Evaluating Worker Exposures to Noise

Noise problems are common in chemical plants; this type problem is also evaluated by industrial hygienists. If a noise problem is suspected, the industrial hygienist should immediately make the appropriate noise measurements and develop recommendations.

Noise levels are measured in decibels. A decibel is a relative logarithmic scale used to compare the intensities of two sounds. If one sound is at intensity I and another sound at intensity I_o, then the difference in intensity levels in decibels is given by

$$\text{Noise Intensity (dB)} = -10 \log_{10}\left(\frac{I}{I_o}\right) \tag{3-5}$$

Thus, a sound ten times as intense as another has an intensity level ten dB greater.

An absolute sound scale (in dBA for absolute decibels) is defined by establishing an intensity reference. For convenience, the hearing threshold is set at 0 dBA. Table 3-6 contains dBA levels for a variety of common activities.

Some permissible noise exposure levels for single sources are provided in Table 3-7.

Noise evaluation calculations are performed identically to calculations for vapors except that dBA is used instead of ppm and hours of exposure is used instead of concentration.

Example 3-6

Determine if the following noise level is permissible with no additional control features.

Noise level (dBA)	Duration (hours)	Maximum allowed (hours)
85	3.6	no limit
95	3.0	4
110	0.5	0.5

Solution From Equation 3-3

$$\sum_{i=1}^{3} \frac{C_i}{(TLV\text{-}TWA)_i} = \frac{3.6}{\text{no limit}} + \frac{3}{4} + \frac{0.5}{0.5} = 1.75$$

Since the sum exceeds 1.0, employees in this environment are immediately required to wear ear protection. On a longer term basis, noise reduction control methods should be developed for the specific pieces of equipment with excessive noise levels.

Estimating Worker Exposures to Toxic Vapors

The best procedure to determine exposures to toxic vapors is to measure the vapor concentrations directly. For design purposes, estimates of vapor concentrations are

TABLE 3-6 SOUND INTENSITY LEVELS FOR A VARIETY OF COMMON ACTIVITIES[1]

Sound	Intensity level (dB)
Painful	120
Riveting	95
Factory	90
Noisy office	80
Busy street traffic	70
Conversational speech	65
Private office	50
Average residence	40
Recording studio	30
Whisper	20
Rustle of leaves	10
Hearing threshold	0

[1]George Shortley and Dudley Williams, *Elements of Physics,* 4th ed. (Englewood Cliffs: Prentice Hall, 1965), p. 455, and Richard A. Wadden and Peter A. Scheff, *Engineering Design for the Control of Workplace Hazards* (New York: McGraw-Hill Book Company, 1987), p. 665.

TABLE 3-7 PERMISSIBLE NOISE EXPOSURES[1]

Sound level (dBA)	Maximum exposure hours
90	8
92	6
95	4
97	3
100	2
102	1.5
105	1
110	0.5
115	0.25

[1]N. Irving Sax, *Dangerous Properties of Industrial Materials,* 4th ed. (New York: Van Nostrand Reinhold Company, 1975), p. 118.

frequently required in enclosed spaces, above open containers, where drums are filled, and in the area of spills.

Consider the enclosed volume shown in Figure 3-3. This enclosure is ventilated by a constant volume airflow. Volatile vapors are evolved within the enclosure. An estimate of the concentration of volatile in the air is required.

Let

C = concentration of volatile vapor in the enclosure (mass/volume)

V = volume of the enclosure (volume)

Q_v = ventilation rate (volume/time)

k = nonideal mixing factor (unitless)

Q_m = evolution rate of volatile material (mass/time)

The nonideal mixing factor, k, accounts for conditions in the enclosure less than well-mixed. It follows that,

$$\text{Total mass of volatile in volume} = VC$$

$$\text{Accumulation of mass of volatile} = \frac{d(VC)}{dt} = V\frac{dC}{dt}$$

$$\text{Mass rate of volatile due to evolution} = Q_m$$

$$\text{Mass rate of volatile out} = kQ_vC$$

Since accumulation = mass in − mass out, the dynamic mass balance on the volatile species is

$$V\frac{dC}{dt} = Q_m - kQ_vC \tag{3-6}$$

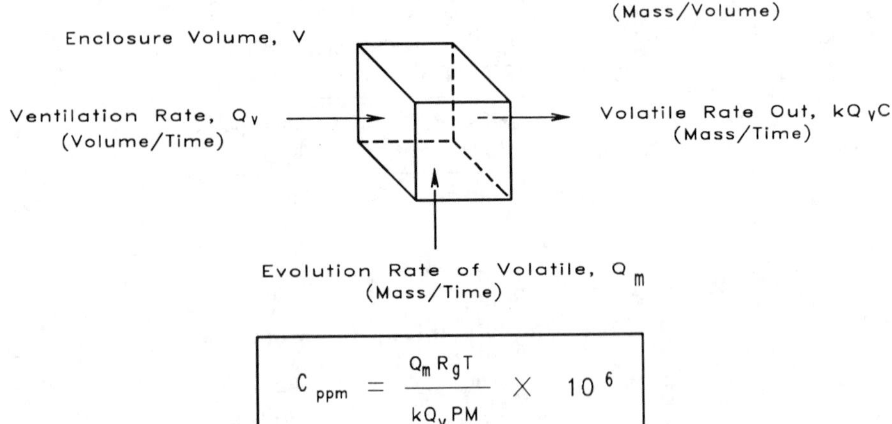

Figure 3-3 Mass balance on volatile vapor in an enclosure.

At steady state the accumulation term is zero and Equation 3-6 is solved for C,

$$C = \frac{Q_m}{kQ_v} \tag{3-7}$$

Equation 3-7 is converted to the more convenient concentration units of ppm by direct application of the ideal gas law. Let m represent mass, ρ represent density, and the subscripts v and b denote the volatile and bulk gas species, respectively. Then,

$$C_{\text{ppm}} = \frac{V_v}{V_b} \times 10^6 = \left(\frac{m_v/\rho_v}{V_b}\right) \times 10^6 = \left(\frac{m_v}{V_b}\right)\left(\frac{R_g T}{PM}\right) \times 10^6 \tag{3-8}$$

where R_g is the ideal gas constant, T the absolute ambient temperature, P the absolute pressure, and M the molecular weight of the volatile species. The term $\left(\frac{m_v}{V_b}\right)$ is identical to the concentration of volatile computed using Equation 3-7. Substituting Equation 3-7 into Equation 3-8,

$$\boxed{C_{\text{ppm}} = \frac{Q_m R_g T}{kQ_v PM} \times 10^6} \tag{3-9}$$

Equation 3-9 is used to determine the average concentration (in ppm) of any volatile species in an enclosure given a source term Q_m and a ventilation rate Q_v. It can be applied to the following types of exposures:

- worker standing near a pool of volatile liquid
- worker standing near an opening to a storage tank
- worker standing near an open container of volatile liquid.

Equation 3-9 includes the following important assumptions.

- The concentration calculated is an average concentration in the enclosure. Localized conditions could result in significantly higher concentrations; workers directly above an open container might be exposed to higher concentrations.
- A steady state condition is assumed; that is, the accumulation term in the mass balance is zero.

The nonideal mixing factor varies from 0.1 to 0.5 for most practical situations.[4] For perfect mixing $k = 1$.

Example 3-7

An open toluene container in an enclosure is weighed as a function of time and it is determined that the average evaporation rate is 0.1 gm/min. The ventilation rate is 100 ft^3/min. The temperature is 80°F and the pressure is 1 atm. Estimate the concentration of toluene vapor in the enclosure and compare to the TLV for toluene of 100 ppm.

[4] R. Craig Matthiessen, "Estimating Chemical Exposure Levels in the Workplace," *Chemical Engineering Progress,* April, 1986, p. 30.

Solution Since the value of k is not known directly, it must be used as a parameter. From Equation 3-9,

$$kC_{\text{ppm}} = \frac{Q_m R_g T}{Q_v PM} \times 10^6$$

From the data provided,

$$Q_m = 0.1 \text{ gm/min} = 2.20 \times 10^{-4} \text{ lb}_m\text{/min}$$

$$R_g = 0.7302 \text{ ft}^3 \text{ atm/lb-mole °R}$$

$$T = 80° \text{ F} = 540° \text{ R}$$

$$Q_v = 100 \text{ ft}^3\text{/min}$$

$$M = 92 \text{ lb}_m\text{/lb-mole}$$

$$P = 1 \text{ atm}$$

Substituting into the equation above,

$$kC_{\text{ppm}} = \frac{(2.20 \times 10^{-4} \text{ lb}_m\text{/min})(0.7302 \text{ ft}^3\text{atm/lb-mole°R})(540\text{°R})}{(100 \text{ ft}^3\text{/min})(1 \text{ atm})(92 \text{ lb}_m\text{/lb-mole})} \times 10^6$$

$$kC_{\text{ppm}} = 9.43 \text{ ppm}$$

Since k varies from 0.1 to 0.5, the concentration is expected to vary from 18.9 to 94.3 ppm. Actual vapor sampling is recommended to insure that the TLV is not exceeded.

Estimating the vaporization rate of a liquid. Liquids with high saturation vapor pressures evaporate faster. As a result, the evaporation rate (mass/time) is expected to be a function of the saturation vapor pressure. In reality, for vaporization into stagnant air, the vaporization rate is proportional to the difference between the saturation vapor pressure and the partial pressure of the vapor in the stagnant air; that is,

$$Q_m \; \alpha \; (P^{\text{sat}} - p) \tag{3-10}$$

where P^{sat} is the saturation vapor pressure of the pure liquid at the temperature of the liquid and p is the partial pressure of the vapor in the bulk stagnant gas above the liquid.

A more generalized expression for the vaporization rate is available.[5]

$$Q_m = \frac{MKA(P^{\text{sat}} - p)}{R_g T_L} \tag{3-11}$$

where Q_m is the evaporation rate (mass/time)

M is the molecular weight of the volatile substance,

K is a mass transfer coefficient (length/time) for an area A

R_g is the ideal gas constant

T_L is the absolute temperature of the liquid.

[5]Steven R. Hanna and Peter J. Drivas, *Guidelines for the Use of Vapor Cloud Dispersion Models* (New York: American Institute of Chemical Engineers, 1987).

For many situations, $P^{sat} >> p$ and Equation 3-11 is simplified to,

$$Q_m = \frac{MKAP^{sat}}{R_g T_L} \quad (3\text{-}12)$$

Equation 3-12 is used to estimate the vaporization rate of volatile from an open vessel or from a spill of liquid.

The vaporization rate or source term, determined by Equation 3-12, is used in Equation 3-9 to estimate the concentration in ppm of volatile in an enclosure due to evaporation of a liquid,

$$C_{ppm} = \frac{KATP^{sat}}{kQ_v PT_L} \times 10^6 \quad (3\text{-}13)$$

For most situations, $T = T_L$ and Equation 3-13 is simplified to,

$$C_{ppm} = \frac{KAP^{sat}}{kQ_v P} \times 10^6 \quad (3\text{-}14)$$

The gas mass-transfer coefficient is estimated using the relationship[6]

$$K = aD^{2/3} \quad (3\text{-}15)$$

where a is a constant and D is the gas-phase diffusion coefficient. Equation 3-15 is used to determine the ratio of the mass transfer coefficients between the species of interest, K, and a reference species, K_o,

$$\frac{K}{K_0} = \left(\frac{D}{D_0}\right)^{2/3} \quad (3\text{-}16)$$

The gas-phase diffusion coefficients are estimated from the molecular weights, M, of the species,[7]

$$\frac{D}{D_0} = \sqrt{\frac{M_0}{M}} \quad (3\text{-}17)$$

Equation 3-17 is combined with Equation 3-16,

$$K = K_o \left(\frac{M_o}{M}\right)^{1/3} \quad (3\text{-}18)$$

Water is most frequently used as a reference substance with a mass transfer coefficient[8] of 0.83 cm/sec.

Example 3-8

A large open tank with a five-foot diameter contains toluene. Estimate the evaporation rate from this tank assuming a temperature of 77°F and a pressure of 1 atm. If the ventilation rate is 3000 ft^3/min, estimate the concentration of toluene in this workplace enclosure.

[6]Louis. J. Thibodeaux, *Chemodynamics* (New York: John Wiley and Sons, 1979), p. 85.

[7]Gordon M. Barrow, *Physical Chemistry,* 2nd ed. (New York: McGraw-Hill Book Company, 1966), p. 19.

[8]Matthiessen, "Estimating Chemical Exposure," p. 33.

Solution The molecular weight of toluene is 92. The mass transfer coefficent is estimated from Equation 3-18 using water as a reference.

$$K = (0.83 \text{ cm/sec})\left(\frac{18}{92}\right)^{1/3} = 0.482 \text{ cm/sec} = 0.949 \text{ ft/min}$$

The saturation vapor pressure is given in Example 3-4.

$$P^{\text{sat}}_{\text{toluene}} = 28.2 \text{ mm Hg} = 0.0371 \text{ atm}$$

The pool area is

$$A = \frac{\pi d^2}{4} = \frac{(3.14)(5 \text{ ft})^2}{4} = 19.6 \text{ ft}^2$$

The evaporation rate is computed using Equation 3-12,

$$Q_{\text{m}} = \frac{MKAP^{\text{sat}}}{R_g T_{\text{L}}}$$

$$= \frac{(92 \text{ lb}_{\text{m}}/\text{lb-mole})(0.949 \text{ ft/min})(19.6 \text{ ft}^2)(0.0371 \text{ atm})}{(0.7302 \text{ ft}^3\text{atm/lb-mole}^\circ\text{R})(537^\circ\text{R})}$$

$$Q_{\text{m}} = 0.162 \text{ lb}_{\text{m}}/\text{min}$$

The concentration is estimated using Equation 3-14 with k as a parameter.

$$kC_{\text{ppm}} = \frac{KAP^{\text{sat}}}{Q_{\text{v}}P} \times 10^6$$

$$= \frac{(0.949 \text{ ft/min})(19.6 \text{ ft}^2)(0.0371 \text{ atm})}{(3000 \text{ ft}^3/\text{min})(1 \text{ atm})} \times 10^6$$

$$kC_{\text{ppm}} = 230 \text{ ppm}$$

The concentration will range from 460 to 2300 ppm, depending on the value of k. Since the TLV for toluene is 100 ppm, additional ventilation is recommended, or the amount of exposed surface area should be reduced. The amount of ventilation required to reduce the worst case concentration (2300 ppm) to 100 ppm is

$$Q_{\text{v}} = (3000 \text{ ft}^3/\text{min})\left(\frac{2300 \text{ ppm}}{100 \text{ ppm}}\right) = 69{,}000 \text{ ft}^3/\text{min}$$

This represents an impractical level of general ventilation. Potential solutions to this problem include containing the toluene in a closed vessel or using local ventilation at the vessel opening.

Estimating worker exposures during vessel filling operations. For vessels being filled with liquid, volatile emissions are generated from two sources, as shown in Figure 3-4. These sources are

- evaporation of the liquid, represented by Equation 3-14 and
- displacement of the vapor in the vapor space by the liquid filling the vessel.

The net generation of volatile is the sum of the two sources.

$$Q_{\text{m}} = (Q_{\text{m}})_1 + (Q_{\text{m}})_2 \tag{3-19}$$

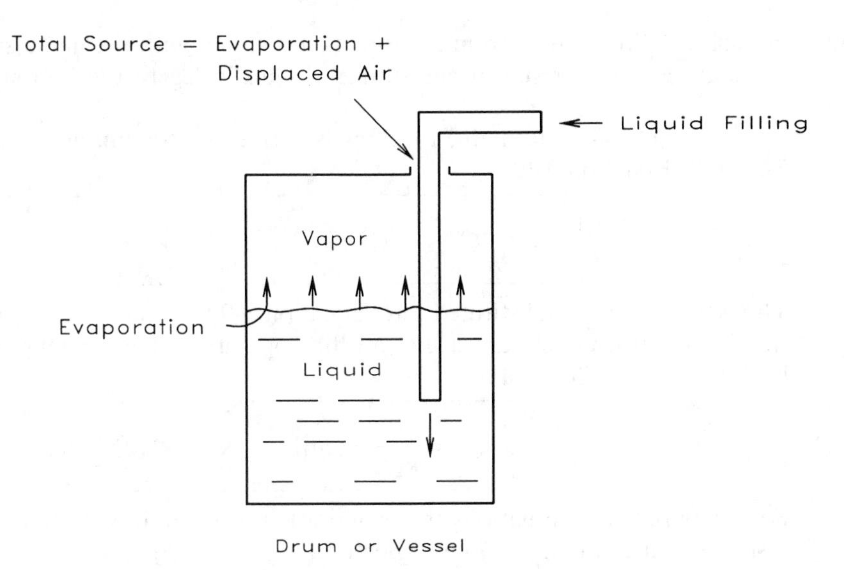

Figure 3-4 Evaporation and displacement from a filling vessel.

where $(Q_m)_1$ represents the source due to evaporation and $(Q_m)_2$ represents the source due to displacement.

The source term $(Q_m)_1$ is computed using Equation 3-12. $(Q_m)_2$ is determined by assuming that the vapor is completely saturated with volatile. An adjustment is introduced later for less than saturated conditions. Let

V_c = volume of the container (volume)

r_f = constant filling rate of the vessel (time^{-1})

P^{sat} = saturation vapor pressure of the volatile liquid

T_L = absolute temperature of the container and liquid

It follows that $r_f V_c$ is the volumetric rate of bulk vapor being displaced from the drum (volume/time). Also, if ρ_v is the density of the volatile vapor, $r_f V_c \rho_v$ is the mass rate of volatile displaced from the container (mass/time). Using the ideal-gas law,

$$\rho_v = \frac{MP^{sat}}{R_g T_L} \tag{3-20}$$

and it follows that

$$(Q_m)_2 = \frac{MP^{sat}}{R_g T_L} r_f V_c \tag{3-21}$$

Equation 3-21 can be modified for container vapors that are not saturated with volatile. Let ϕ represent this adjustment factor, then,

$$(Q_m)_2 = \frac{MP^{sat}}{R_g T_L} \phi r_f V_c \tag{3-22}$$

For splash filling (filling from the top of a container with the liquid splashing to the bottom), $\phi = 1$. For subsurface filling[9] (by a dip leg to the bottom of the tank), $\phi = 0.5$.

The net source term due to filling is derived by combining Equations 3-12 and 3-22 with Equation 3-19.

$$Q_m = (Q_m)_1 + (Q_m)_2 = \frac{MP^{sat}}{R_g T_L}(\phi r_f V_c + KA) \tag{3-23}$$

This source term is substituted into Equation 3-9 to compute the vapor concentration in ppm in an enclosure due to a filling operation. The assumption that $T = T_L$ is also invoked. The result is

$$C_{ppm} = \frac{P^{sat}}{kQ_v P}(\phi r_f V_c + KA) \times 10^6 \tag{3-24}$$

For many practical situations the evaporation term, KA, is much smaller than the displacement term and can be neglected.

Example 3-9

Railroad cars are being splash-filled with toluene. The 10,000-gallon cars are being filled at the rate of one every eight hours. The filling hole in the tank car is four inches in diameter. Estimate the concentration of toluene vapor as a result of this filling operation. The ventilation rate is estimated at 3000 ft³/min. The temperature is 77°F and the pressure is 1 atm.

Solution The concentration is estimated using Equation 3-24. From Example 3-8, $K = 0.949$ ft/min and $P^{sat} = 0.0371$ atm. The area of the filling hole is

$$A = \frac{\pi d^2}{4} = \frac{(3.14)(4\ \text{in})^2}{(4)(144\ \text{in}^2/\text{ft}^2)} = 0.0872\ \text{ft}^2$$

Thus,

$$KA = (0.949\ \text{ft/min})(0.0872\ \text{ft}^2) = 0.0827\ \text{ft}^3/\text{min}$$

The filling rate, r_f, is

$$r_f = \left(\frac{1}{8\ \text{hours}}\right)\left(\frac{1\ \text{hour}}{60\ \text{min}}\right) = 0.00208\ \text{min}^{-1}$$

For splash filling, the nonideal filling factor, ϕ, is 1.0 The displacement term in Equation 3-24 is

$$\phi r_f V_c = (1.0)(0.00208\ \text{min}^{-1})(10{,}000\ \text{gal})\left(\frac{7.48\ \text{ft}^3}{1\ \text{gal}}\right) = 2.78\ \text{ft}^3/\text{min}$$

As expected, the evaporation term is small compared to the displacement term. The concentration is computed from Equation 3-24 using k as a parameter.

$$kC_{ppm} = \frac{P^{sat}\phi r_f V_c}{Q_v P} = \frac{(0.0371\ \text{atm})(2.78\ \text{ft}^3/\text{min})}{(3{,}000\ \text{ft}^3/\text{min})(1\ \text{atm})} \times 10^6$$

$$= 34.4\ \text{ppm}$$

[9]Matthiessen, "Estimating Chemical Exposure," p. 33.

The actual concentration could range from 69 to 344 ppm depending on the value of k. Sampling to insure that the concentration is below 100 ppm is recommended. For subsurface filling, $\phi = 0.5$ and the concentration range is reduced to 35 to 172 ppm.

3-4 CONTROL

After potential health hazards are identified and evaluated, the appropriate control techniques must be developed and installed. This requires the application of appropriate technology for reducing workplace exposures.

The types of control techniques used in the chemical industry are illustrated in Table 3-8.

Designing control methods is a very important and creative task. During the design process, the designer must pay particular attention to insure the newly designed control technique provides the desired control, and the new control technique itself does not create another hazard, sometimes even more hazardous than the original problem.

TABLE 3-8 CHEMICAL PLANT CONTROL TECHNIQUES

Type and explanation	Typical techniques
Substitution Use chemicals and equipment which are less hazardous.	* Use mechanical pump seals vs. packing. * Use welded pipe vs. flanged sections. * Use solvents that are less toxic. * Use mechanical gauges vs. mercury. * Use chemicals with higher flash points, boiling points, and other less hazardous properties. * Use water as a heat transfer fluid instead of hot oil.
Attenuation Use chemicals under conditions which make them less hazardous	* Use vacuum to reduce boiling point. * Reduce process temperature and pressures. * Refrigerate storage vessels. * Dissolve hazardous material in safe solvent. * Operate at conditions where reactor runaway is not possible.
Isolation Isolate equipment and/or sources of hazard.	* Place control rooms away from operations. * Separate pump rooms from other rooms. * Accoustically insulate noisy lines and equipment. * Barricade control rooms and tanks.
Intensification Reduce quantity of chemical.	* Change from large batch reactor to smaller continuous reactor. * Reduce storage inventory of raw materials. * Improve control to reduce inventory of hazardous intermediate chemicals. * Reduce process hold-up.

TABLE 3-8 CHEMICAL PLANT CONTROL TECHNIQUES (continued)

Type and explanation	Typical techniques
Enclosures Enclose room or equipment and place under negative pressure.	* Enclose hazardous operations like sample points. * Seal rooms, sewers, ventilation, and the like. * Use analyzers and instruments to observe inside equipment. * Shield high temperature surfaces. * Pneumatically convey dusty material.
Local Ventilation Contain and exhaust hazardous substances.	* Use properly designed hoods. * Use hoods for charging and discharging. * Use ventilation at drumming station. * Use local exhaust at sample points. * Keep exhaust systems under negative pressure
Dilution Ventilation Design ventilation systems to control low level toxics	* Design locker rooms with good ventilation and special areas or enclosures for contaminated clothing. * Design ventilation to isolate operations from rooms and offices. * Design filter press rooms with directional ventilation.
Wet Methods Use wet methods to minimize contamination with dusts.	* Clean vessels chemically vs. sand blasting. * Use water sprays for cleaning. * Clean areas frequently. * Use water sprays to shield trenches or pump seals.
Good Housekeeping Keep toxics and dusts contained.	* Use dikes around tanks and pumps. * Provide water and steam connections for area washing. * Provide lines for flushing and cleaning. * Provide well-designed sewer system with emergency containment.
Personal Protection As last line of defense.	* Use safety glasses and face shields. * Use aprons, arm shields, and space suits. * Wear appropriate respirators; airline respirators are required when oxygen concentration is less than 19.5%.

The two major control techniques are environmental controls and personal protection. Environmental control reduces exposure by reducing the concentration of toxic in the workplace environment. This includes substitution, isolation, enclosure, local ventilation, dilution ventilation, wet methods, and good housekeeping, as discussed previously. Personal protection prevents or reduces exposure by providing a barrier between the worker and the workplace environment. This barrier is usu-

TABLE 3-9 PERSONAL PROTECTIVE EQUIPMENT, NOT INCLUDING RESPIRATORS.[1]

Type	Description
Hard hat	Protects head from falling equipment and bumps.
Safety glasses	Impact resistent lenses.
Chemical splash goggles gas tight	Suitable for liquids and fumes.
Steel-toed safety shoes	Protects against dropped equipment.
Wrap around face shield	Fiberglass, resistant to most chemicals.
Vinyl apron	Resists most chemicals.
Splash suit	Viton or butyl rubber for nonflammable exposures.
Umbilical cord suit	Used with external air supply.
Rubber oversleeves	Protects forearms.
PVC coated gloves	Resists acids and bases.
PVC and nitrile knee boots	Resists acids, oils, and greases.
Ear plugs	Protects against high noise levels.

[1]*Lab Safety Supply Catalog* (Janesvelle, Wisconsin 53547-1368) and *Best Safety Directory* (A.M. Best Company, Oldwick, New Jersey 08858). Manufacturers' technical specifications must always be consulted.

ally worn by the worker, hence the designation "personal." Typical types of personal protective equipment are listed in Table 3-9.

Respirators

Respirators are routinely found in chemical laboratories and plants. Respirators should only be used

- on a temporary basis, until regular control methods can be implemented.
- as emergency equipment, to insure worker safety in the event of an accident.
- as a "last resort," in the event that environmental control techniques are unable to provide satisfactory protection.

Respirators always compromise worker ability. A worker with a respirator is unable to perform or respond as well as a worker without one. Various types of respirators are listed in Table 3-10.

Respirators can be used improperly and/or can be damaged to the extent that they do not provide the needed protection. OSHA and NIOSH have developed standards for using respirators,[10] including fit testing (to insure that the device does not leak excessively), periodic inspections (to insure that the equipment works properly), specified use applications (to insure that the equipment is used for the correct job), training (to insure that it is used properly), and record keeping (to insure that the program is operating efficiently). All industrial users of respirators are legally bound to understand and fulfill these OSHA requirements.

[10]*NIOSH Respirator Decision Logic* (Washington, DC: U.S. Department of Health and Human Services, DHHS-NIOSH Publication No. 87-1-8, May 1987).

TABLE 3-10 RESPIRATORS USEFUL TO CHEMICAL INDUSTRY

Type	Examples of commerical brand	Limitations
Mouth and nose dust mask	3M 8710 Dust[1]	$O_2 > 19.5\%$ Single use; PEL > 0.05 mg/m^3
Mouth and nose with chemical cartridge	MSA Comfo II Cartridge[2]	$O_2 > 19.5\%$; GMA Cartridge (black) for 1000 ppm of organic vapors; GMC Cartridge (yellow) for 10 ppm Cl_2; and 50 ppm of HCl or SO_2
Full face mask with chemical canister	MSA Industrial Gas Mask[3]	$O_2 > 19.5\%$; GMA Canister (black) for 2% organics; GMP Canister (olive) for 0.05% pesticides.
Self-contained breathing apparatus (SCBA)	Scott Air PAK II[4]	Good for toxic or noxious gases; air tank lasts approx. 20 minutes.

[1]*3M Company Catalog* (220-7W-02, St. Paul, MN 55144).
[2]*MSA Catalog* (600 Penn Center Blv., Pittsburgh, PA, 15235).
[3]*MSA Catalog.*
[4]*Scott Aviation Catalog* (225 Erie St., Lancaster, NY, 14086).

Ventilation

For environmental control of airborne toxic material the most common method of choice is ventilation, due to the following reasons.

- Ventilation can quickly remove dangerous concentrations of flammable and toxic materials.
- Ventilation can be highly localized, reducing the quantity of air moved and the equipment size.
- Ventilation equipment is readily available and can be easily installed.
- Ventilation equipment can be added to an existing facility.

The major disadvantage of ventilation is the operating costs. Substantial electrical energy may be needed to drive the potentially large fans and the cost to heat or cool the large quantities of fresh air can be large. These operating costs need to be considered when evaluating alternatives.

Ventilation is based on two principles.

- Dilute the contaminant below the target concentration.
- Remove the contaminant before workers are exposed.

Ventilation systems are composed of fans and ducts. The fans produce a small pressure drop (less than 0.1 psi) that moves the air. The best system is a negative pressure system, with the fans located at the exhaust end of the system, pulling air out. This insures that leaks in the system will draw air in from the workplace rather than expel contaminated air from the ducts into the workplace. This is shown in Figure 3-5.

There are two types of ventilation techniques: local and dilution ventilation.

Local ventilation. The most common example of local ventilation is the hood. A hood is a device that either completely encloses the source of contaminant and/or moves the air in such a fashion to carry the contaminant to an exhaust device. There are several types of hoods.

- The *enclosed hood* completely contains the source of contaminant.
- The *exterior hood* continuously draws contaminants into an exhaust from some distance away.

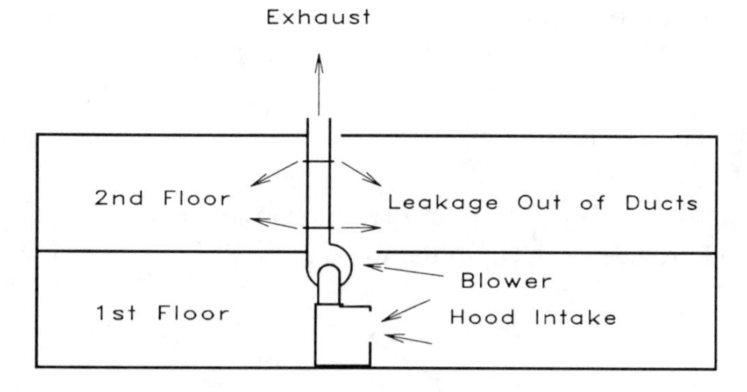

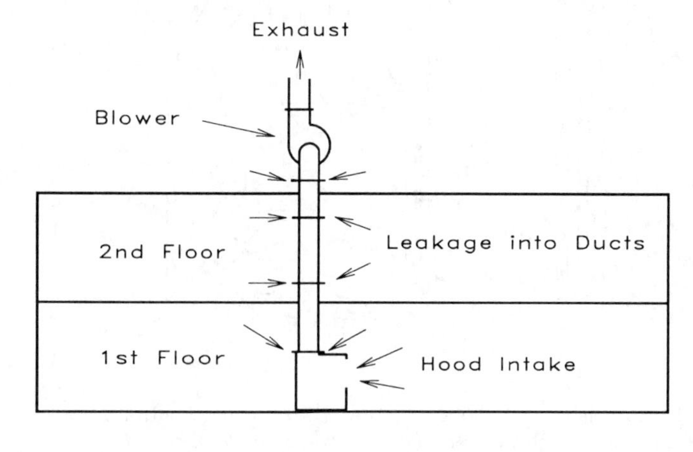

Figure 3-5 The difference between a positive and negative pressure ventilation system. The negative pressure system assures that contaminants do not leak into workplace environments.

- The *receiving hood* is an exterior hood that uses the discharge motion of the contaminant for collection.
- The *push-pull hood* uses a stream of air from a supply to push contaminants toward an exhaust system.

The most common example of an enclosed hood is the laboratory hood. A standard laboratory utility hood is shown in Figure 3-6. Fresh air is drawn through the window area of the hood and is removed out the top through a duct. The airflow profiles within the hood are highly dependent on the location of the window sash. It is important to keep the sash open a few inches, minimally, to insure adequate fresh air. Likewise, the sash should never be fully opened since contaminants might escape. The baffle at the rear of the hood insures that contaminants are removed from the working surface and the rear lower corner.

Another type of laboratory hood is the bypass hood shown in Figure 3-7. For this design, bypass air is supplied through a grill at the top of the hood. This insures the availability of fresh air to sweep out contaminants in the hood. The bypass air supply is reduced as the hood sash is opened.

The advantages of enclosed hoods are that they

- completely eliminate exposure to workers,
- require minimal airflow,
- provide a containment device in the event of fire or explosion, and
- a sliding door on hood provides a shield to the worker.

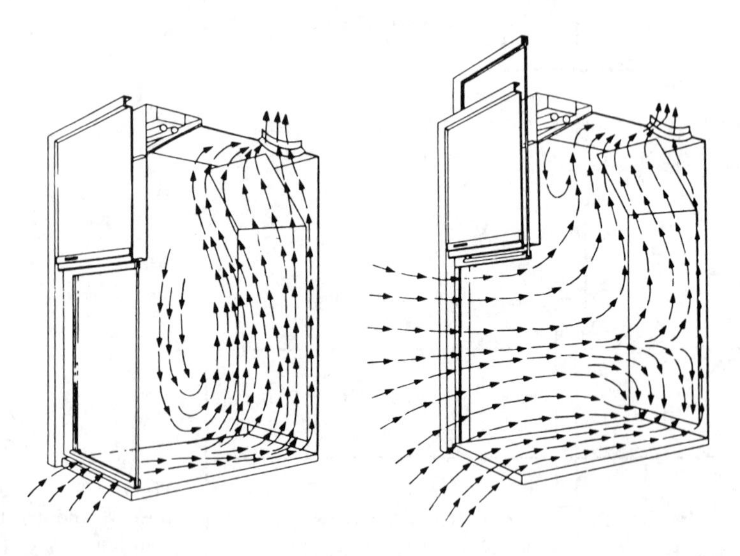

Figure 3-6 Standard utility laboratory hood. Airflow patterns and control velocity are dependent on sash height. (N. Irving Sax, *Dangerous Properties of Industrial Materials,* 4th ed., 1975, p. 74. Reprinted by permission of Van Nostrand Reinhold Company, New York.)

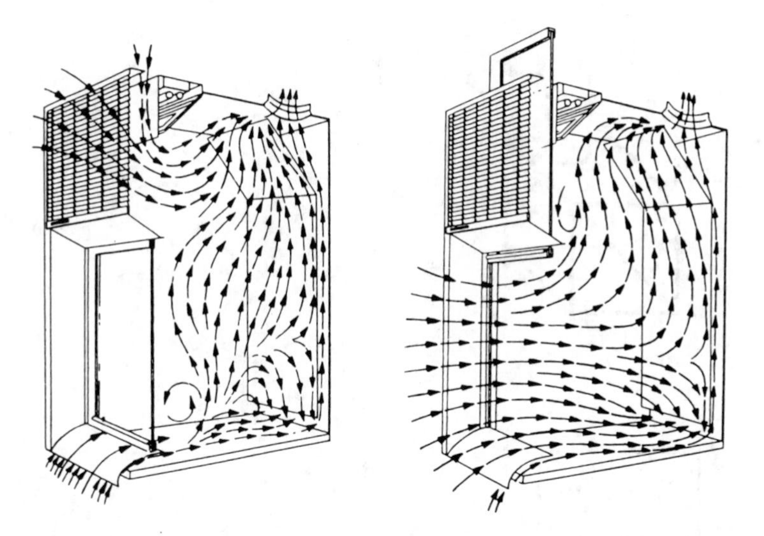

Figure 3-7 Standard bypass laboratory hood. The bypass air is controlled by the height of the sash. (N. Irving Sax, *Dangerous Properties of Industrial Materials,* 4th ed., 1975, p. 75. Reprinted by permission of Van Nostrand Reinhold Company, New York.)

The disadvantages to hoods are that they

- limit workspace and
- can only be used for small, benchscale or pilot plant equipment.

Most hood calculations assume plug flow. For a duct of cross-sectional area A and average air velocity $\bar{u}$ (distance/time), the volume of air moved per unit time, Q_v, is computed from

$$Q_v = A\bar{u} \tag{3-25}$$

For a rectangular duct of width W and length L, Q_v is determined using the following equation.

$$Q_v = LW\bar{u} \tag{3-26}$$

Consider the simple box-type enclosed hood shown in Figure 3-8. The design strategy is to provide a fixed velocity of air at the opening of the hood. This face or control velocity (referring to the face of the hood) will insure that contaminants do not exit from the hood.

The control velocity required depends on the toxicity of the material, the depth of the hood, and evolution rate of contaminant. Shallower hoods need higher control velocities to prevent contaminants from exiting the front. However, experience has shown that higher velocities can lead to the formation of a turbulent eddy from the bottom of the sash; backflow of contaminated air is possible. For general operation a control velocity between 100 to 125 feet per minute (fpm) is suggested.

Instruments are available for measuring the airflow velocity at specific points of the hood window opening. Testing is an OSHA requirement.

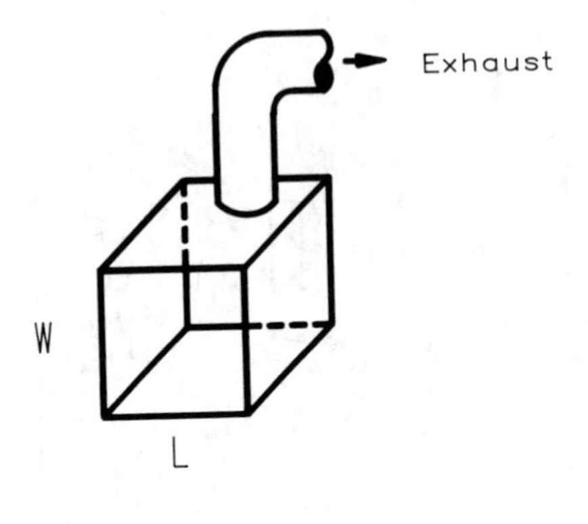

$$Q_v = LW\bar{u}$$

Q_v = Volumetric Flow Rate, Volume/Time

L = Length

W = Width

$\bar{u}$ = Required Control Velocity

Figure 3-8 Determining the total volumetric air flow rate for a box-type hood. For general operation a control velocity of between 100 to 125 feet per minute (fpm) is desired.

The airflow velocity is a function of the sash height and the blower speed. Arrows are frequently used to indicate the proper sash height to insure a specified face velocity.

Design equations are available for a wide variety of hood and duct shapes.[11]

Other types of local ventilation methods include "elephant trunks" and free-hanging canopies and plenums. The elephant trunk is simply a flexible vent duct that is positioned near a source of contaminant. It is most frequently used for loading and unloading toxic materials from drums and vessels. Free-hanging canopies and plenums can be either fixed in position or attached to a flexible duct to enable movement. These methods will most likely expose workers to toxicants, but in diluted amounts.

Dilution ventilation. If the contaminant cannot be placed in a hood and must be used in an open area or room, dilution ventilation is necessary. Unlike hood ventilation where the airflow prevents worker exposure, dilution ventilation always exposes the worker, but in amounts diluted by fresh air. Dilution ventilation always requires more airflow than local ventilation; operating expenses can be substantial.

Equations 3-9, 3-12, and 3-14 are used to compute the ventilation rates required. Table 3-11 lists values for k, the non-ideal mixing factor used with these equations.

[11] *Industrial Ventilation, A Manual of Recommended Practice,* 19th ed. (Cincinnati: American Conference of Governmental Industrial Hygienists, 1986)

TABLE 3-11 NONIDEAL MIXING FACTOR, k, FOR VARIOUS DILUTION VENTILATION CONDITIONS.[1]

Vapor concentration (ppm)	Dust concentration (mppcf)	Mixing factor			
		Ventilation condition			
		Poor	Average	Good	Excellent
over 500	50	1/7	1/4	1/3	1/2
101 - 500	20	1/8	1/5	1/4	1/3
0 - 100	5	1/11	1/8	1/7	1/6

[1]N. Irving Sax, *Dangerous Properties,* 6th Ed., p. 29. The values reported here are the *reciprocal* of Sax's values.

For exposures to multiple sources, the dilution air requirement is computed for each individual source. The total dilution requirement is the sum of the individual dilution requirements.

The following restrictions should be considered before implementing dilution ventilation:

- the contaminant must not be highly toxic,
- the contaminant must be evolved at a uniform rate,
- workers must remain a suitable distance from the source to insure proper dilution of the contaminant, and
- scrubbing systems must not be required to treat the air prior to exhaust into the environment.

Example 3-10

Xylene is used as a solvent in paint. A certain painting operation evaporates an estimated three gallons of xylene in an eight-hour shift. The ventilation quality is rated as average. Determine the quantity of dilution ventilation air required to maintain the xylene concentration below 100 ppm, the TLV-TWA. Also compute the air required if the operation is carried out in an enclosed hood with an opening of 50 ft^2 and a face velocity of 100 ft/min. The temperature is 77°F and the pressure is 1 atm. The specific gravity of the xylene is 0.864 and its molecular weight is 106.

Solution The evaporation rate of xylene is

$$Q_m = \left(\frac{3 \text{ gallons}}{8 \text{ hours}}\right)\left(\frac{1 \text{ hour}}{60 \text{ min}}\right)\left(\frac{0.1337 \text{ ft}^3}{1 \text{ gallon}}\right)\left(\frac{62.4 \text{ lb}_m}{\text{ft}^3}\right)(0.864)$$

$$Q_m = 0.0450 \text{ lb}_m/\text{min}$$

From Table 3-11, for average ventilation and a vapor concentration of 100 ppm, $k = 1/8 = 0.125$. With Equation 3-9, we solve for Q_v.

$$Q_v = \frac{Q_m R_g T}{k C_{ppm} PM} \times 10^6$$

$$= \frac{(0.0450 \text{ lb}_m/\text{min})(0.7302 \text{ ft}^3\text{atm/lb-mole °R})(537 \text{ °R})}{(0.125)(100 \text{ ppm})(1 \text{ atm})(106 \text{ lb}_m/\text{lb-mole})} \times 10^6$$

$$= 13{,}300 \text{ ft}^3/\text{min required dilution air.}$$

For a hood with an open area of 50 ft^2, using Equation 3-25, and assuming a required control velocity of 100 fpm,

$$Q_v = A\bar{u} = (50 \text{ ft}^2)(100 \text{ ft/min}) = 5000 \text{ ft}^3\text{/min}$$

The hood requires significantly less airflow than dilution ventilation and prevents worker exposure completely.

SUGGESTED READING

Industrial Hygiene

LEWIS J. CRALLEY and LESTER V. CRALLEY, eds., *Industrial Hygiene Aspects of Plant Operations,* Vol. 1-3 (New York: MacMillan Publishing Company, 1984).

J. B. OLISHIFSKI, ed., *Fundamentals of Industrial Hygiene,* 2nd ed. (Chicago: National Safety Council, 1979).

N. IRVING SAX, *Dangerous Properties of Industrial Materials,* 6th ed. (New York: Van Nostrand Reinhold Company, 1984), Sections 2 and 3.

RICHARD A. WADDEN and PETER A. SCHEFF, eds., *Engineering Design for the Control of Workplace Hazards* (New York: McGraw-Hill Book Company, 1987).

Ventilation

Industrial Ventilation, A Manual of Recommended Practice, 19th ed. (Cincinnati: American Conference of Governmental Industrial Hygienists, 1986).

WADDEN and SCHEFF, *Engineering Design,* Chapter 5.

PROBLEMS

3-1. A process plant inventories the following chemicals:

Vinyl chloride	Methyl ethyl ketone
Ethylene oxide	Styrene
Cyclohexane	

Determine the hazards associated with these chemicals. What additional information might you request to perform an appropriate assessment of the risk associated with these chemicals?

3-2. The TLV-TWA for a substance is 150 ppm. A worker begins a work shift at 8 A.M. and completes the shift at 5 P.M. A one hour lunch break is included between 12 noon and 1 P.M. where it can be assumed that no exposure to the chemical occurs.

The data were taken in the work area at the times indicated. Has the worker exceeded the TLV specification?

Time	Concentration ppm
8:10 A.M.	110
9:05 A.M.	130
10:07 A.M.	143
11:20 A.M.	162
12:12 P.M.	142
1:17 P.M.	157
2:03 P.M.	159
3:13 P.M.	165
4:01 P.M.	153
5:00 P.M.	130

3-3. Air contains 4 ppm of carbon tetrachloride and 25 ppm of 1,1-dichloroethane. Compute the mixture TLV and determine if this value has been exceeded.

3-4. A substance has a TLV-TWA of 200 ppm, a TLV-STEL of 250 ppm, and a TLV-C of 300 ppm. The data below were taken in a work area. A worker on an eight hour shift is exposed to this toxic vapor. Is the exposure within compliance? If not, what are the violations? Assume that the worker is at lunch between the hours of 12 noon to 1 P.M. and is not exposed to the chemical during that time.

Time	Concentration ppm
8:01 A.M.	185
9:17 A.M.	240
10:05 A.M.	270
11:22 A.M.	230
12:08 P.M.	190
1:06 P.M.	150
2:05 P.M.	170
3:09 P.M.	165
4:00 P.M.	160
5:05 P.M.	130

3-5. Sax[12] provides the following working equation for determining the dilution air requirements due to evaporation of a solvent,

$$CFM = \frac{(3.87 \times 10^8)\,(\text{lb}_\text{m} \text{ of liquid evaporated / min})}{(\text{Molecular weight})\,(TLV)\,(k)}$$

where CFM is the ft^3/min of dilution air required. Show that this equation is the same as Equation 3-9. What assumptions are inherent in the above equation?

[12]Sax, *Dangerous Properties,* 6th Ed., p. 28

Problems 3-6 through 3-11 apply to toluene and benzene. The following data are available for these materials:

	Benzene	Toluene
Formula:	C_6H_6	C_7H_8
Molecular weight:	78.11	92.13
Specific gravity:	0.8794	0.866
TLV (ppm):	10	100

Saturation Vapor Pressures

$$\ln(P^{sat}) = A - \frac{B}{C + T}$$

where P^{sat} is the saturation vapor pressure in mm Hg, T is the temperature in K, and A, B, and C are the constants, given below.

	A	B	C
Benzene	15.9008	2788.51	−52.36
Toluene	16.0137	3096.52	−53.67

3-6. Compute the concentration (in ppm) of the saturated vapor with air above a solution of pure toluene. Compute the concentration (in ppm) of the equilibrium vapor with air above a solution of 50 mole percent toluene and benzene. The temperature is 80°F and the total pressure is 1 atm.

3-7. Compute the density of pure air and the density of air contaminated with 100 ppm benzene. Do the densities of these two gases differ enough to insure a higher concentration on floors and other low spots? The temperature is 70°F and the pressure is 1 atm.

3-8. Equations 3-12 and 3-14 represent the evaporation of a pure liquid. Modify these equations to represent the evaporation of a mixture of ideal, miscible liquids.

3-9. Benzene and toluene form an ideal liquid mixture. A mixture composed of 50 mole percent benzene is used in a chemical plant. Temperature: 80°F, pressure: 1 atm.

a. Determine the mixture TLV.

b. Determine the evaporation rate per unit area for this mixture.

c. A drum with a two-inch diameter bung is used to contain the mixture. Determine the ventilation rate required to maintain the vapor concentration below the TLV. The ventilation quality within the vicinity of this operation is average.

3-10. A drum contains 42 gallons of toluene. If the lid of the drum is left opened (lid diameter = 3 ft), determine the time required to evaporate all of the toluene in the drum. The temperature is 85°F. Estimate the concentration of toluene (in ppm) near the drum if the local ventilation rate is 1000 ft^3/min. Pressure: 1 atm.

3-11. A certain plant operation evaporates 2 pint/hour of toluene and 1 pint/8-hour shift of benzene. Determine the ventilation rate required to maintain the vapor concentration below the TLV. Temperature: 80°F, pressure: 1 atm.

3-12. Equations 3-12 and 3-14 can be applied to nonenclosed exposures by using an effective ventilation rate. The effective ventilation rate for outside exposures has been estimated at 3000 ft^3/min.[14]

A worker is standing near an open manway of a tank containing 2-butoxyethanol (molecular weight = 118). The manway area is 7 ft^2. Estimate the concentration (in ppm) of the vapor near the manway opening. The vapor pressure of the 2-butoxyethanol is 0.6 mm Hg.

3-13. Fifty-five gallon drums are being filled with 2-butoxyethanol. The drums are being splash filled at the rate of 30 drums per hour. The bung opening through which the drums are being filled has an area of 8 cm^2. Estimate the ambient vapor concentration if the ventilation rate is 3000 ft^3/min. The vapor pressure of 2-butoxyethanol is 0.6 mm Hg at these conditions.

3-14. A gasoline tank in a standard automobile contains about 14-gallons of gasoline and can be filled in about three minutes. The molecular weight of gasoline is approximately 94 and its vapor pressure at 77°F is 4.6 psi. Estimate the concentration (in ppm) of gasoline vapor as a result of this filling operation. Assume a ventilation rate of 3000 ft^3/min. The TLV for gasoline is 300 ppm.

3-15. A six-inch diameter elephant trunk is used to remove contaminants near the open bung of a drum during a filling operation. The air velocity required at the end of the elephant trunk is 100 ft/min. Compute the volumetric flow rate of air required.

3-16. To reduce air pollution, gasoline filling stations are installing scavenger systems to remove the gasoline vapors ejected from the automobile tank during the filling operation. This is accomplished by an elephant trunk ventilation system installed as part of the filler hose.

Assume an average automobile tank size of 14 gallons. If the vapor in the tank is saturated with gasoline vapor at a vapor pressure of 4.6 psi at these conditions, how many gallons of gasoline are recovered free for the station owner with each fill-up? For 10,000 gallons of delivered gasoline, how many gallons are recovered? The molecular weight of gasoline is about 94 and its liquid specific gravity is 0.7.

3-17. Normal air contains about 21% oxygen by volume. The human body is very sensitive to reductions in oxygen concentration; concentrations below 19.5% are dangerous and concentrations below 16% can cause distress. Respiratory equipment without self-contained air supplies must never be used in atmospheres below 19.5% oxygen.

A storage tank of 1000 ft^3 capacity must be cleaned prior to reuse. Proper procedures must be used to insure that the oxygen concentration of the air within the tank is adequate.

Compute the cubic feet of additional nitrogen at 77°F and 1 atm that will reduce the oxygen concentration within the tank to 19.5%? 16%? Oxygen concentrations within tanks and enclosures can be reduced significantly by small amounts of inert!

3-18. A laboratory hood has an opening four feet in length by three feet in height. The hood depth is 18 inches. This hood will be used for an operation involving trichloroethylene (TCE) (*TLV-TWA*: 50 ppm). The TCE will be used in liquid form at room temperature. Determine an appropriate control velocity for this hood and calculate the total air flow rate.

3-19. It is desired to operate the hood of Problem 3-18 so that the vapor concentration in the hood plenum is below the lower explosion limit of 12.5% by volume. Estimate the minimum control velocity required to achieve this objective. The amount of TCE evaporated within the hood is 5.3 lb per hour. The molecular weight of TCE is 131.4. The temperature is 70°F and the pressure is 1 atm.

[14]Matthieson, "Estimating Chemical Exposure," p 33.

Source Models 4

Most accidents in chemical plants result in spills of toxic, flammable, and explosive materials. For example, material is released from holes and cracks in tanks and pipes, from leaks in flanges, pumps, and valves, and a large variety of other sources.

Source models represent the material release process. They provide useful information for determining the consequences of an accident, including the rate of material release, the total quantity released, and the physical state of the material. This information is valuable for evaluating new process designs, process improvements and the safety of existing processes. Alternatives must be considered if the source models predict unacceptable release characteristics.

Source models are constructed from fundamental or empirical equations representing the physico-chemical processes occurring during the release of materials. For a reasonably complex plant, many source models are needed to describe the release. Some development and modification of the original models is normally required to fit the specific situation. Frequently the results are only estimates since the physical properties of the materials are not adequately characterized, or the physical processes themselves are not completely understood. If uncertainty exists, the parameters should be selected to maximize the release rate and quantity. This insures that a design is "on the safe side."

Release mechanisms are classified into wide and limited aperture releases. In the wide aperture case, a large hole develops in the process unit, releasing a substantial amount of material in a very short time. An excellent example is the overpressuring and explosion of a storage tank. For the limited aperture case, material is released at a slow enough rate that upstream conditions are not immediately affected; the assumption of constant upstream pressure is frequently valid.

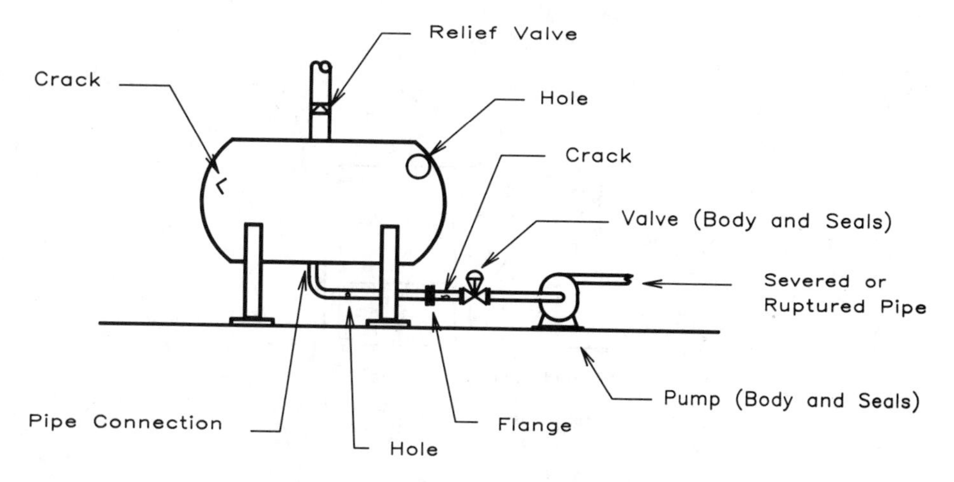

Figure 4-1 Various types of limited aperture releases.

Limited aperture releases are conceptualized in Figure 4-1. For these releases, material is ejected from holes and cracks in tanks and pipes, leaks in flanges, valves, and pumps, and severed or ruptured pipes. Relief systems, designed to prevent the overpressuring of tanks and process vessels, are also sources of released material.

Figure 4-2 shows how the physical state of the material affects the release mechanism. For gases or vapors stored in a tank, a leak results in a jet of gas or vapor. For liquids, a leak below the liquid level in the tank results in a stream of escaping liquid. If the liquid is stored under pressure above its atmospheric boiling point, a leak below the liquid level will result in a stream of liquid flashing partially into vapor. Small liquid droplets or aerosols might also form from the flashing stream, with the possibility of transport away from the leak by wind currents. A leak in the vapor space above the liquid can result in either a vapor stream or a two-phase stream composed of vapor and liquid depending on the physical properties of the material.

There are several basic source models that are used repeatedly and will be developed in detail here. These source models are

- Flow of liquids through a hole
- Flow of liquids through a hole in a tank
- Flow of liquids through pipes
- Flow of vapor through holes
- Flow of vapor through pipes
- Flashing liquids
- Liquid pool evaporation or boiling

Other source models, specific to certain material, will be introduced in subsequent chapters.

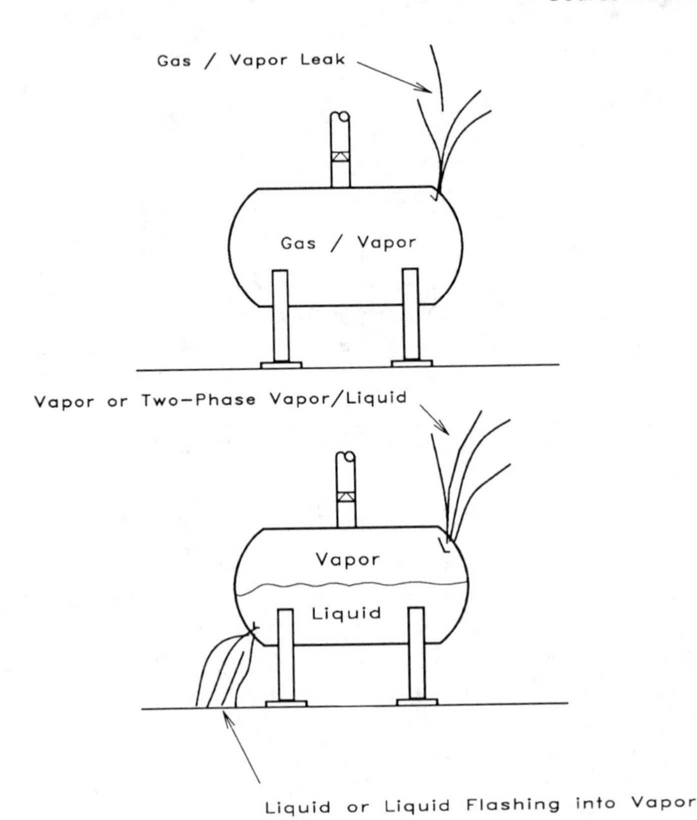

Figure 4-2 Vapor and liquid are ejected from process units in either single or two-phase states.

4-1 FLOW OF LIQUID THROUGH A HOLE

A mechanical energy balance describes the various energy forms associated with flowing fluids,

$$\int \frac{dP}{\rho} + \Delta\left(\frac{\bar{u}^2}{2\alpha g_c}\right) + \frac{g}{g_c}\Delta z + F = -\frac{W_s}{\dot{m}} \tag{4-1}$$

where

P is the pressure (force/area)
ρ is the fluid density (mass/volume)
$\bar{u}$ is the average instantaneous velocity of the fluid (length/time)
g_c is the gravitational constant (length mass/force time^2)
α is the unitless velocity profile correction factor with the following values,
 = 0.5 for laminar flow
 = 1.0 for plug flow
 → 1.0 for turbulent flow

g is the acceleration due to gravity (length/time^2)
z is the height above datum (length)
F is the net frictional loss term (length force/mass)
W_s is the shaft work (force length)
$\dot{m}$ is the mass flow rate (mass/time)

The Δ function represents the final minus the initial state.

For incompressible liquids the density is constant and,

$$\int \frac{dP}{\rho} = \frac{\Delta P}{\rho} \tag{4-2}$$

Consider a process unit that develops a small hole, as shown in Figure 4-3. The pressure of the liquid contained within the process unit is converted to kinetic energy as the fluid escapes through the leak. Frictional forces between the moving liquid and the wall of the leak converts some of the kinetic energy of the liquid into thermal energy, resulting in a reduced velocity.

For this limited aperture release, assume a constant gauge pressure, P_g, within the process unit. The external pressure is atmospheric; so $\Delta P = P_g$. The shaft work is zero and the velocity of the fluid within the process unit is assumed negligible.

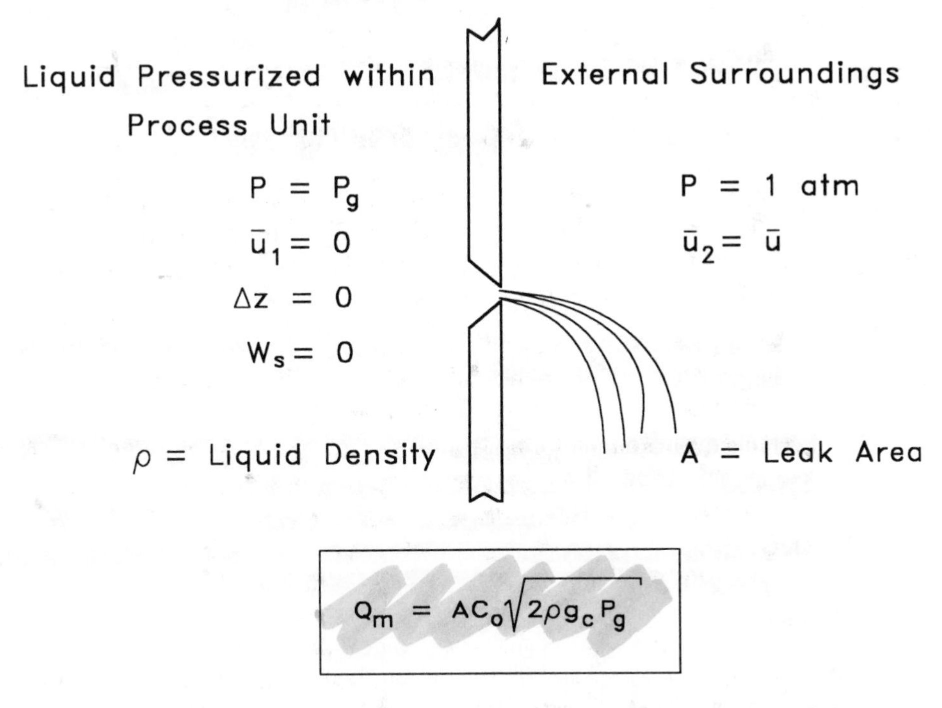

Figure 4-3 Liquid escaping through a hole in a process unit. The energy of the liquid due to its pressure in the vessel is converted to kinetic energy, with some frictional flow losses in the hole.

The change in elevation of the fluid during the discharge through the hole is also negligible; so $\Delta z = 0$. The frictional losses in the leak are approximated by a constant discharge coefficient, C_1, defined as

$$-\frac{\Delta P}{\rho} - F = C_1^2\left(-\frac{\Delta P}{\rho}\right) \tag{4-3}$$

The above modifications are substituted into the mechanical energy balance, Equation 4-1, to determine $\bar{u}$, the average discharge velocity from the leak,

$$\bar{u} = C_1\sqrt{\alpha}\sqrt{\frac{2g_c P_g}{\rho}} \tag{4-4}$$

A new discharge coefficient, C_o, is defined as,

$$C_o = C_1\sqrt{\alpha} \tag{4-5}$$

The resulting equation for the velocity of fluid exiting the leak is

$$\boxed{\bar{u} = C_o\sqrt{\frac{2g_c P_g}{\rho}}} \tag{4-6}$$

The mass flow rate, Q_m, due to a hole of area A is given by

$$\boxed{Q_m = \rho\bar{u}A = AC_o\sqrt{2\rho g_c P_g}} \tag{4-7}$$

The total mass of liquid spilled is dependent on the total time the leak is active.

The discharge coefficient, C_o, is a complicated function of the Reynolds number of the fluid escaping through the leak and the diameter of the hole. The following guidelines are suggested.[1]

1. For sharp-edged orifices and for Reynolds numbers greater than 30,000, C_o approaches the value 0.61. For these conditions, the exit velocity of the fluid is independent of the size of the hole.
2. For a well-rounded nozzle the discharge coefficient approaches unity.
3. For short sections of pipe attached to a vessel (with a length-diameter ratio not less than 3), the discharge coefficient is approximately 0.81.
4. For cases where the discharge coefficient is unknown or uncertain, use a value of 1.0 to maximize the computed flows.

Example 4-1

At 1 P.M. the plant operator notices a drop in pressure in a pipeline transporting benzene. The pressure is immediately restored to 100 psig. At 2:30 P.M. a 1/4-inch diameter leak is found in the pipeline and immediately repaired. Estimate the total amount of benzene spilled. The specific gravity of benzene is 0.8794.

[1]Frank P. Lees, *Loss Prevention in the Process Industries* (London: Butterworths. 1986), p. 417.

Solution The drop in pressure observed at 1 P.M. is indicative of a leak in the pipeline. The leak is assumed to be active between 1 P.M. and 2:30 P.M., a total of 90 minutes. The area of the hole is

$$A = \frac{\pi d^2}{4} = \frac{(3.14)(0.25\text{ in})^2(1\text{ ft}^2/144\text{ in}^2)}{4}$$

$$= 3.41 \times 10^{-4}\text{ ft}^2$$

The density of the benzene is,

$$\rho = (0.8794)(62.4\text{ lb}_\text{m}/\text{ft}^3) = 54.9\text{ lb}_\text{m}/\text{ft}^3$$

The leak mass flow rate is given by Equation 4-7. A discharge coefficient of 0.61 is assumed for this orifice-type leak.

$$Q_\text{m} = AC_\text{o}\sqrt{2\rho g_\text{c} P_\text{g}}$$

$$= (3.41 \times 10^{-4}\text{ ft}^2)(0.61) \times \sqrt{(2)\left(54.9\frac{\text{lb}_\text{m}}{\text{ft}^3}\right)\left(32.17\frac{\text{ft lb}_\text{m}}{\text{lb}_\text{f}\text{ s}^2}\right)\left(100\frac{\text{lb}_\text{f}}{\text{in}^2}\right)\left(144\frac{\text{in}^2}{\text{ft}^2}\right)}$$

$$= 1.48\text{ lb}_\text{m}/\text{s}$$

The total quantity of benzene spilled is

$$= (1.48\text{ lb}_\text{m}/\text{s})(90\text{ min})(60\text{ s/min}) = 7990\text{ lb}_\text{m}$$

$$\boxed{= 1090\text{ gallons}}$$

4-2 FLOW OF LIQUID THROUGH A HOLE IN A TANK

A storage tank is shown in Figure 4-4. A hole develops at a height h_L below the fluid level. The flow of liquid through this hole is represented by the mechanical energy balance, Equation 4-1, and the incompressible assumption as shown in Equation 4-2.

The gauge pressure on the tank is P_g and the external gauge pressure is atmospheric, or 0. The shaft work, W_s, is zero and the velocity of the fluid in the tank is zero.

A dimensionless discharge coefficient, C_1, is defined as follows.

$$-\frac{\Delta P}{\rho} - \frac{g}{g_\text{c}}\Delta z - F = C_1^2\left(-\frac{\Delta P}{\rho} - \frac{g}{g_\text{c}}\Delta z\right) \tag{4-8}$$

The mechanical energy balance, Equation 4-1, is solved for $\bar{u}$, the average instantaneous discharge velocity from the leak

$$\bar{u} = C_1\sqrt{\alpha}\sqrt{2\left(\frac{g_\text{c}P_\text{g}}{\rho} + gh_\text{L}\right)} \tag{4-9}$$

where h_L is the liquid height above the leak. A new discharge coefficient, C_o, is defined as

$$C_\text{o} = C_1\sqrt{\alpha} \tag{4-10}$$

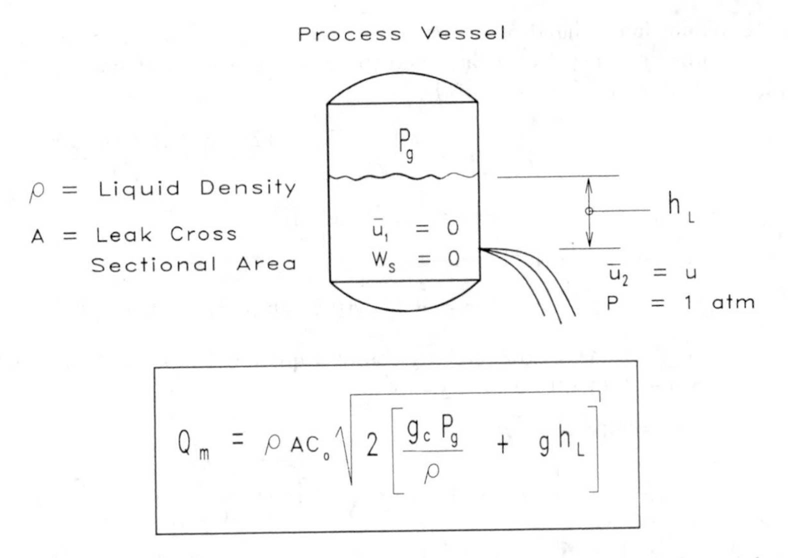

Figure 4-4 An orifice-type leak in a process vessel. The energy due to the pressure of the fluid height above the leak is converted to kinetic energy as the fluid exits through the hole. Some energy is lost due to frictional fluid flow.

The resulting equation for the instantaneous velocity of fluid exiting the leak is

$$\bar{u} = C_o \sqrt{2\left(\frac{g_c P_g}{\rho} + g h_L\right)} \tag{4-11}$$

The instantaneous mass flow rate, Q_m, due to a hole of area A is given by

$$\boxed{Q_m = \rho \bar{u} A = \rho A C_o \sqrt{2\left(\frac{g_c P_g}{\rho} + g h_L\right)}} \tag{4-12}$$

As the tank empties, the liquid height decreases and the velocity and mass flow rate will decrease.

Assume that the gauge pressure, P_g, on the surface of the liquid is constant. This would occur if the vessel were padded with an inert gas to prevent explosion or were vented to the atmosphere. For a tank of constant cross-sectional area A_t, the total mass of liquid in the tank above the leak is,

$$m = \rho A_t h_L \tag{4-13}$$

The rate of change of mass within the tank is

$$\frac{dm}{dt} = -Q_m \tag{4-14}$$

where Q_m is given by Equation 4-12. By substituting Equations 4-12 and 4-13 into Equation 4-14 and assuming constant tank cross section and liquid density, a differential equation representing the change in the fluid height is obtained.

$$\frac{dh_L}{dt} = -\frac{C_o A}{A_t}\sqrt{2\left(\frac{g_c P_g}{\rho} + g h_L\right)} \tag{4-15}$$

Equation 4-15 is rearranged and integrated from an initial height h_L^o to any height h_L.

$$\int_{h_L^o}^{h_L} \frac{dh_L}{\sqrt{\frac{2g_c P_g}{\rho} + 2gh_L}} = -\frac{C_o A}{A_t}\int_0^t dt \tag{4-16}$$

This equation is integrated to

$$\frac{1}{g}\sqrt{\frac{2g_c P_g}{\rho} + 2gh_L} - \frac{1}{g}\sqrt{\frac{2g_c P_g}{\rho} + 2gh_L^o} = -\frac{C_o A}{A_t}t \tag{4-17}$$

Solving for h_L, the liquid level height in the tank, yields

$$h_L = h_L^o - \frac{C_o A}{A_t}\sqrt{\frac{2g_c P_g}{\rho} + 2gh_L^o}\; t + \frac{g}{2}\left(\frac{C_o A}{A_t}t\right)^2 \tag{4-18}$$

wrong

Equation 4-18 is substituted into Equation 4-12 to obtain the mass discharge rate at any time t.

$$Q_m = \rho C_o A\sqrt{2\left(\frac{g_c P_g}{\rho} + gh_L^o\right)} - \frac{\rho g C_o^2 A^2}{A_t}t \tag{4-19}$$

The first term on the RHS of Equation 4-19 is the initial mass discharge rate at $h_L = h_L^o$.

The time for the vessel to empty to the level of the leak, t_e, is found by solving Equation 4-18 for t after setting $h_L = 0$.

$$t_e = \frac{1}{C_o g}\left(\frac{A_t}{A}\right)\left[\sqrt{2\left(\frac{g_c P_g}{\rho} + gh_L^o\right)} - \sqrt{\frac{2g_c P_g}{\rho}}\right] \tag{4-20}$$

If the vessel is at atmospheric pressure, $P_g = 0$ and Equation 4-20 reduces to

$$t_e = \frac{1}{C_o g}\left(\frac{A_t}{A}\right)\sqrt{2gh_L^o} \tag{4-21}$$

Example 4-2

A cylindrical tank 20-feet high and 8-feet in diameter is used to store benzene. The tank is padded with nitrogen to a constant, regulated pressure of one atm gauge to prevent explosion. The liquid level within the tank is presently at 17 feet. A one-inch puncture occurs in the tank five feet off the ground due to the careless driving of a fork lift truck. Estimate (a) the gallons of benzene spilled, (b) the time required for the benzene to leak out, and (c) the maximum mass flow rate of benzene through the leak. The specific gravity of benzene at these conditions is 0.8794.

Solution The density of the benzene is

$$\rho = (0.8794)(62.4\ \text{lb}_\text{m}/\text{ft}^3)$$
$$= 54.9\ \text{lb}_\text{m}/\text{ft}^3$$

The area of the tank is

$$A_t = \frac{\pi d^2}{4} = \frac{(3.14)(8\ \text{ft})^2}{4} = 50.2\ \text{ft}^2$$

The area of the leak is

$$A = \frac{(3.14)(1\ \text{in})^2(1\ \text{ft}^2/144\ \text{in}^2)}{4} = 5.45 \times 10^{-3}\ \text{ft}^2$$

The gauge pressure is

$$P_\text{g} = (1\ \text{atm})(14.7\ \text{lb}_\text{f}/\text{in}^2)(144\ \text{in}^2/\text{ft}^2) = 2.12 \times 10^3\ \text{lb}_\text{f}/\text{ft}^2$$

a. The volume of benzene above the leak is

$$V = A_t h_\text{L}^\text{o} = (50.2\ \text{ft}^2)(17\ \text{ft} - 5\ \text{ft})(7.48\ \text{gal}/\text{ft}^3) = 4{,}506\ \text{gallons}$$

This is the total benzene that will leak out.

b. The length of time for the benzene to leak out is given by Equation 4-20

$$t_\text{e} = \frac{1}{C_\text{o} g}\left(\frac{A_t}{A}\right)\left[\sqrt{2\left(\frac{g_\text{c} P_\text{g}}{\rho} + g h_\text{L}^\text{o}\right)} - \sqrt{\frac{2 g_\text{c} P_\text{g}}{\rho}}\right]$$

$$= \frac{1}{(0.61)(32.17\ \text{ft/s}^2)}\left(\frac{50.2\ \text{ft}^2}{5.45 \times 10^{-3}\ \text{ft}^2}\right)$$

$$\times \left[\left[\frac{(2)(32.17\ \text{ft-lb}_\text{m}/\text{lb}_\text{f}\text{-s}^2)(2.12 \times 10^3\ \text{lb}_\text{f}/\text{ft}^2)}{54.9\ \text{lb}_\text{m}/\text{ft}^3}\right.\right.$$

$$\left.\left. + (2)(32.17\ \text{ft/s}^2)(12\ \text{ft})\right]^{1/2} - \sqrt{2484\ \text{ft}^2/\text{s}^2}\right]$$

$$= 469\ \text{s}^2/\text{ft}(7.22\ \text{ft}^2/\text{s}^2) = 3386\ \text{s} = 56.4\ \text{minutes}$$

This appears to be more than adequate time to stop the leak or to invoke an emergency procedure to reduce the impact of the leak. However, the maximum discharge occurs when the hole is first opened.

c. The maximum discharge occurs at $t = 0$ at a liquid level of 17.0 feet. Equation 4-19 is used to compute the mass flow rate.

$$Q_\text{m} = \rho A C_\text{o} \sqrt{2\left(\frac{g_\text{c} P_\text{g}}{\rho} + g h_\text{L}^\text{o}\right)}$$

$$= (54.9\ \text{lb}_\text{m}/\text{ft}^3)(5.45 \times 10^{-3}\ \text{ft}^2)(0.61)\sqrt{3.26 \times 10^3\ \text{ft}^2/\text{s}^2}$$

$$\boxed{Q_\text{m} = 10.4\ \text{lb}_\text{m}/\text{s}}$$

4-3 FLOW OF LIQUIDS THROUGH PIPES

A pipe transporting liquid is shown in Figure 4-5. A pressure gradient across the pipe is the driving force for the movement of liquid. Frictional forces between the

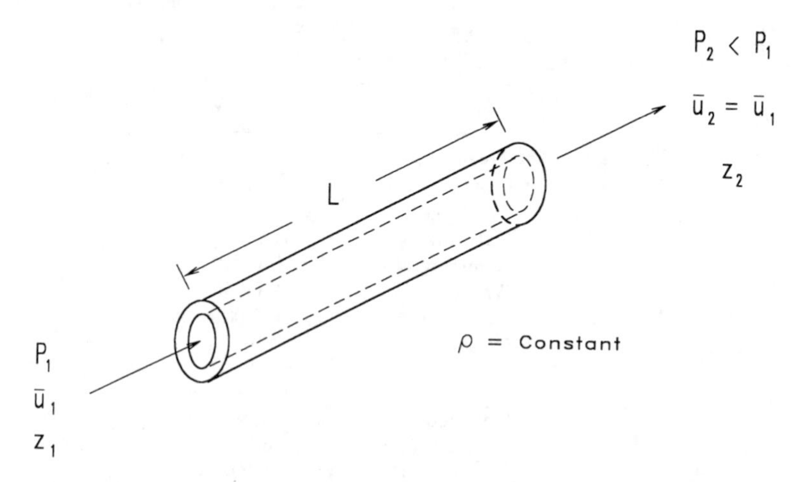

Figure 4-5 Liquid flowing through a pipe. The frictional flow losses between the fluid and the pipe wall result in a pressure drop across the pipe length. Kinetic energy changes are frequently negligible.

liquid and the wall of the pipe converts kinetic energy into thermal energy. This results in a decrease in the liquid velocity and a decrease in the liquid pressure.

Flow of incompressible liquids through pipes is described by the mechanical energy balance, Equation 4-1, combined with the incompressible fluid assumption, Equation 4-2. The net result is

$$\frac{\Delta P}{\rho} + \frac{\Delta \bar{u}^2}{2\alpha g_c} + \frac{g}{g_c}\Delta z + F = -\frac{W_s}{\dot{m}} \tag{4-22}$$

The friction term, F, is the sum of all of the frictional elements in the piping system. For a straight pipe, without valves or fittings, F is given by

$$F = \frac{2fL\bar{u}^2}{g_c d} \tag{4-23}$$

where

f is the Fanning friction factor (no units)
L is the length of the pipe
d is the diameter of the pipe (length).

The Fanning friction factor, f, is a function of the Reynolds number, Re, and the roughness of the pipe, ε. Table 4-1 provides values of ε for various types of clean pipe. Figure 4-6 is a plot of the Fanning friction factor versus Reynolds number with the pipe roughness, ε/d, as a parameter. Figure 4-7 presents the data of Figure 4-6 in a form useful for certain types of calculations (see Example 4-3).

For laminar flow, the Fanning friction factor is given by

$$f = \frac{16}{Re} \tag{4-24}$$

TABLE 4-1 ROUGHNESS FACTOR, ε, FOR CLEAN PIPES[1]

Pipe material	ε, mm
Riveted steel	1-10
Concrete	0.3-3
Cast iron	0.26
Galvanized iron	0.15
Commercial steel	0.046
Wrought iron	0.046
Drawn tubing	0.0015
Glass	0
Plastic	0

[1]Selected from Octave Levenspiel, *Engineering Flow and Heat Exchange* (New York: Plenum Press, 1984), p. 22.

For turbulent flow, the data shown on Figure 4-6 are represented by the Colebrook equation,

$$\frac{1}{\sqrt{f}} = -4 \log\left(\frac{1}{3.7}\frac{\varepsilon}{d} + \frac{1.255}{Re\sqrt{f}}\right) \tag{4-25}$$

An alternate form of Equation 4-25, useful for determining the Reynolds number the friction factor, f, is

$$\frac{1}{Re} = \frac{\sqrt{f}}{1.255}\left[10^{-0.25/\sqrt{f}} - \frac{1}{3.7}\frac{\varepsilon}{d}\right] \tag{4-26}$$

For fully developed turbulent flow in rough pipes, f is independent of the Reynolds number as shown by the nearly constant friction factors at high Reynolds number on Figure 4-6. For this case Equation 4-26 is simplified to

$$\frac{1}{\sqrt{f}} = 4 \log\left(3.7\frac{d}{\varepsilon}\right) \tag{4-27}$$

For smooth pipes, $\varepsilon = 0$ and Equation 4-25 reduces to

$$\frac{1}{\sqrt{f}} = 4 \log \frac{Re\sqrt{f}}{1.255} \tag{4-28}$$

Finally, for smooth pipe with the Reynolds number less than 100,000, the following Blasius approximation to Equation 4-28 is useful.

$$f = 0.079\, Re^{-1/4} \tag{4-29}$$

For piping systems composed of fittings, elbows, valves, and other assorted hardware, the pipe length is adjusted to compensate for the additional friction losses due to these fixtures. The equivalent pipe length is defined as

$$L_{\substack{\text{equiv}\\ \text{total}}} = L_{\substack{\text{straight}\\ \text{pipe}}} + \sum L_{\text{equiv}} \tag{4-30}$$

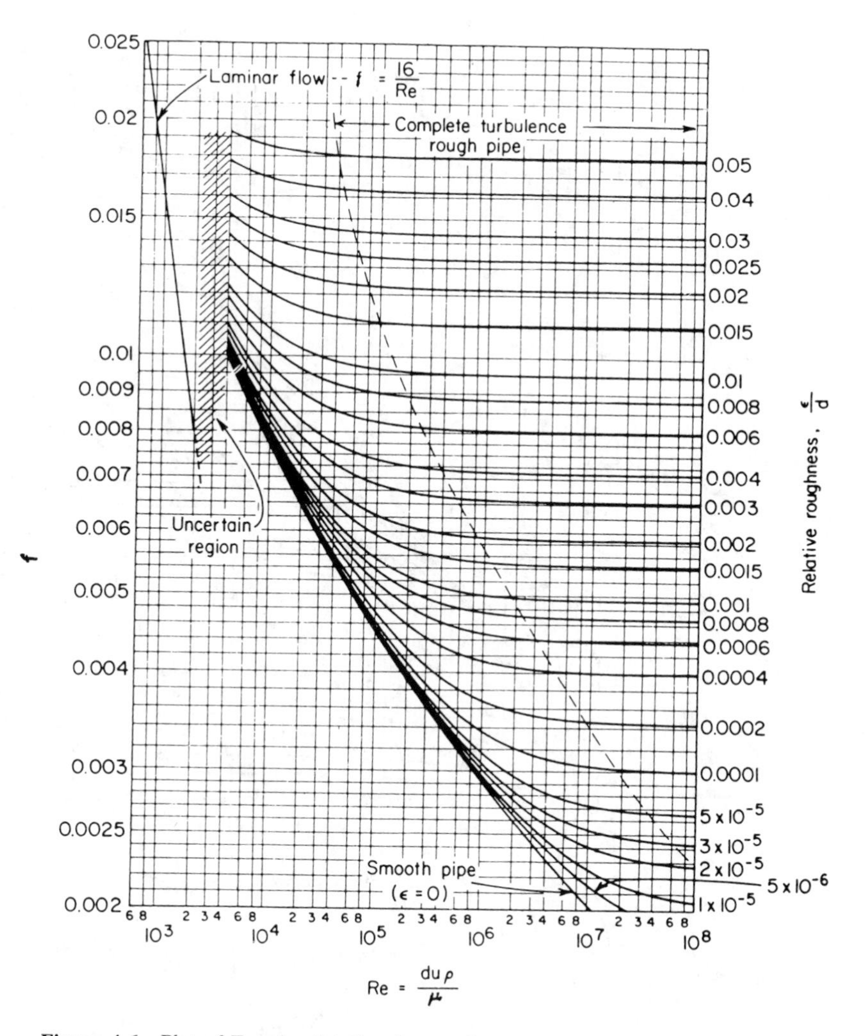

Figure 4-6 Plot of Fanning friction factor, f, versus Reynolds number. (Octave Levenspiel, *Engineering Flow and Heat Exchange,* 1984, p. 20. Reprinted by permission of Plenum Press.)

where the summation is over all of the valves, unions, elbows, and so on within the piping system. Table 4-2 provides selected values for the equivalent lengths. Note that Table 4-2 includes corrections for contractions and expansions in the piping system.

For many problems associated with pipe flow the contribution due to the kinetic energy term in the mechanical energy balance is negligible and can be ignored. A typical procedure is to assume it is negligible and then check the validity of the assumption at the completion of the calculation.

For problems involving laminar flow, the solution is always direct. Turbulent flow problems with an unknown pipe diameter, d, always require a trial-and-error solution. Other types of turbulent flow problems might be direct or trial-and-error depending on the work and kinetic energy terms.

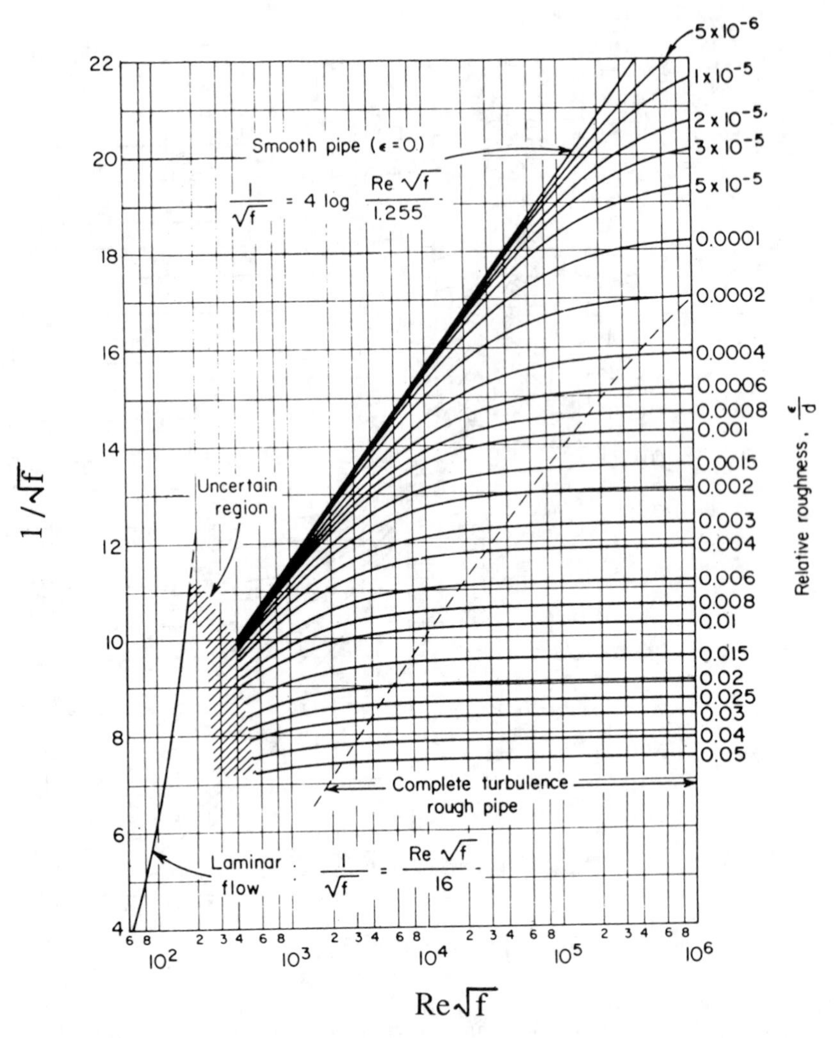

Figure 4-7 Plot of $1/\sqrt{f}$, verses $Re\sqrt{f}$. This form is convenient for certain types of problems. (see Example 4-2.) (Octave Levenspiel, *Engineering Flow and Heat Exchange,* 1984, p. 21. Reprinted by permission of Plenum Press, NY.)

Example 4-3

Water contaminated with small amounts of hazardous waste is gravity drained out of a large storage tank through a straight, commercial steel pipe 100 mm in ID. The pipe is 100 m long with a gate valve near the tank. The entire pipe assembly is mostly horizontal. If the liquid level in the tank is 5.8 m above the pipe outlet, and the pipe is accidently severed 33 m from the tank, compute the flow rate of material escaping from the pipe.

Solution The draining operation is shown in Figure 4-8. Assuming negligible KE changes, no pressure changes, and no shaft work, the mechanical energy balance, Equation 4-22, applied between points 1 and 2, reduces to

$$\frac{g}{g_c}\Delta z + F = 0$$

TABLE 4-2 EQUIVALENT PIPE LENGTHS FOR VARIOUS PIPE FITTINGS (TURBULENT FLOW ONLY).

Pipe Fitting	L_{equiv}/d
Globe valve, wide open	~300
Gate valve, wide open	~7
3/4 open	~40
1/2 open	~200
1/4 open	~900
90° elbow, standard	30
45° elbow, standard	15
Tees	
Used as elbow, entering the stem	90
Used as elbow, entering one of two sides	60
Straight through	20
Pipe connections to vessels	
Ordinary, pipe flush with wall	16
Borda, pipe protruding into vessel	30
Rounded entrance, union, coupling	~0
Sudden enlargement from d to D	
Laminar flow in d:	$\frac{Re}{32}\left[1 - \left(\frac{d^2}{D^2}\right)\right]^2$
Turbulent flow in d:	$\frac{f_{(\text{in } d)}}{4}\left[1 - \left(\frac{d^2}{D^2}\right)\right]^2$
Sudden contraction from D to d (except choked gas flow)	
Laminar flow in d:	$\frac{Re}{160}\left[1.25 - \left(\frac{d^2}{D^2}\right)\right]$
Turbulent flow in d:	$\frac{f_{(\text{in } d)}}{10}\left[1.25 - \left(\frac{d^2}{D^2}\right)\right]$

Selected from Octave Levenspiel, *Engineering Flow and Heat Exchange* (New York: Plenum Press, 1984), p. 25.

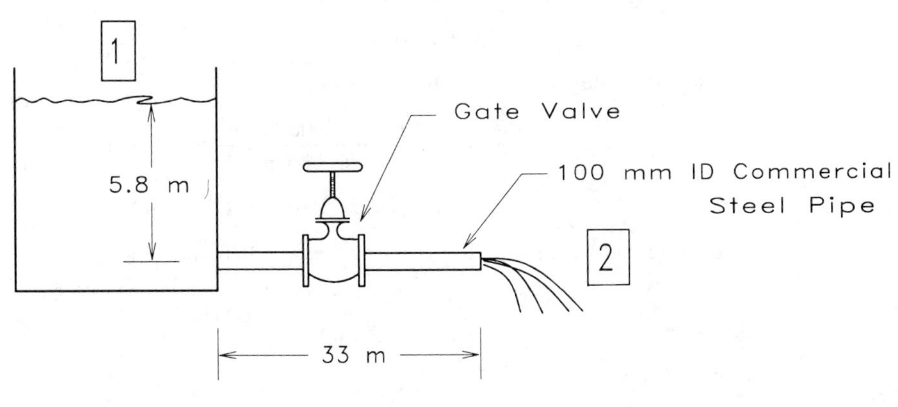

Figure 4-8

For water,

$$\mu = 1.0 \times 10^{-3} \text{ kg/m s}$$

$$\rho = 1000 \text{ kg/m}^3$$

The frictional loss term, F, is given by Equation 4-23,

$$F = \frac{2fL\bar{u}^2}{g_c d}$$

The equivalent pipe length is computed from Equation 4-30. The equivalent lengths for the pipe connected to the vessel and the gate valve are required. These are available in Table 4-2, assuming turbulent flow in the pipe. For the connection to the vessel, from Table 4-2,

$$\frac{L}{d} = 16 \Rightarrow L = (16)(100 \text{ mm}) = 1600 \text{ mm} = 1.6 \text{ m}$$

For a fully open gate valve, from Table 4-2

$$\frac{L}{d} = 7 \Rightarrow L = (7)(100 \text{ mm}) = 700 \text{ mm} = 0.7 \text{ m}$$

Then

$$L_{\substack{\text{equiv}\\ \text{total}}} = L_{\substack{\text{horizontal}\\ \text{pipe}}} + \sum L_{\text{equiv}}$$

$$= 33 \text{ m} + 1.6 \text{ m} + 0.7 \text{ m} = 35.3 \text{ m}$$

and

$$F = \frac{2f(35.3 \text{ m})\bar{u}^2}{(0.1 \text{ m})} = 706\, f\bar{u}^2$$

Substituting into the mechanical energy balance above and solving for $\bar{u}$,

$$g\Delta z + 706\, f\bar{u}^2 = 0$$

$$706\, f\bar{u}^2 = -g\Delta z = -(9.8 \text{ m/s}^2)(-5.8 \text{ m}) = 56.8 \text{ m}^2/\text{s}^2$$

$$\bar{u}^2 = 0.08051/f$$

$$\bar{u} = 0.284/\sqrt{f}$$

The Reynolds number is given by

$$Re = \frac{d\bar{u}\rho}{\mu} = \frac{(0.1 \text{ m})(\bar{u})(1000 \text{ kg/m}^3)}{1.0 \times 10^{-3} \text{ kg/m s}} = 1.0 \times 10^5\, \bar{u}$$

$$= 2.84 \times 10^4/\sqrt{f} \Rightarrow Re\sqrt{f} = 2.84 \times 10^4$$

For commercial steel pipe, from Table 4-1, $\varepsilon = 0.046$ and

$$\frac{\varepsilon}{d} = \frac{0.046 \text{ mm}}{100 \text{ mm}} = 0.00046$$

From Figure 4-7,

$$\frac{1}{\sqrt{f}} = 15.1 \Rightarrow \sqrt{f} = 0.0662 \Rightarrow f = 0.00438$$

The velocity of the fluid is

$$\bar{u} = 0.284/0.0662 = 4.29 \text{ m/s}$$

The cross-sectional area of the pipe is

$$A = \frac{\pi d^2}{4} = \frac{(3.14)(0.1\text{ m})^2}{4} = 0.00785\text{ m}^2$$

The mass flowrate from the release is found by

$$Q_m = \rho \bar{u} A = (1000\text{ kg/m}^3)(4.29\text{ m/s})(0.00785\text{ m}^2) = 33.7\text{ kg/s}$$

The kinetic energy term must be checked. Its value is

$$KE = \frac{\bar{u}^2}{2g_c} = \frac{(4.29\text{ m/s})^2}{2} = 9.20$$

This is compared to the frictional loss term, F.

$$F = 706 f \bar{u}^2 = 706(0.00438)(4.29\text{ m/s})^2 = 57.0$$

Thus, the assumption of negligible KE is not valid for this case. The liquid has zero initial velocity and is accelerated to a high velocity, resulting in a significant KE term. For problems involving liquids flowing in pipes with an initial and final velocity, the KE changes are usually negligible.

Start over from the mechanical energy balance, but this time include the KE term,

$$\frac{g}{g_c}\Delta z + \frac{\bar{u}^2}{2g_c} + 706 f \bar{u}^2 = 0$$

Solving for $\bar{u}$,

$$\bar{u}^2 = \frac{g\Delta z}{1/2 + 706f} = \frac{(9.8\text{ m/s}^2)(5.8\text{ m})}{(0.5 + 706f)} = \frac{56.84}{0.5 + 706f}$$

The solution to this equation requires a trial and error procedure since f is a function of $\bar{u}$. The procedure is,

a. Guess a value for the friction factor, f,

b. Determine average velocity, $\bar{u}$, from above equation,

c. Determine Reynolds number, Re,

d. Compute f from Colebrook equation, Equation 4-25, and

e. Iterate until value of f converges.

This procedure produces the following results.

f	$\bar{u}$	Re	Computed f
0.004	4.13	4.13×10^5	0.00439
0.00439	3.97	3.97×10^5	0.00438

The last result is close enough. The mass flow rate is

$$Q_m = \rho \bar{u} A = (1000\text{ kg/m}^3)(3.97\text{ m/s})(0.00785\text{ m}^2)$$

$$\boxed{= 31.1\text{ kg/s}}$$

This represents a significant flow rate. Assuming a 15-minute emergency response period to stop the release, a total of 28,000 kg of hazardous waste will be spilled. In addition to the material released by the flow, the liquid contained within the pipe between the valve and the rupture will also spill. An alternate system must be designed to limit the release. This could include a reduction in the emergency response period, replacement of the pipe by one with a smaller diameter or modification of the piping system to include additional control valves to stop the flow.

4-4 FLOW OF VAPOR THROUGH HOLES

For flowing liquids the kinetic energy changes are frequently negligible and the physical properties (particularly the density) are constant. For flowing gases and vapors these assumptions are only valid for small pressure changes ($P_1/P_2 < 2$) and low velocities ($< 0.3 \times$ speed of sound in gas). Energy contained within the gas or vapor as a result of its pressure is converted into kinetic energy as the gas or vapor escapes and expands through the hole. The density, pressure, and temperature change as the gas or vapor exits through the leak.

Gas and vapor discharges are classified into throttling and free expansion releases. For throttling releases, the gas issues through a small crack with large frictional losses; very little of the energy inherent with the gas pressure is converted to kinetic energy. For free expansion releases, most of the pressure energy is converted to kinetic energy; the assumption of isentropic behavior is usually valid.

Source models for throttling releases require detailed information on the physical structure of the leak; they will not be considered here. Free expansion release source models require only the diameter of the leak.

A free expansion leak is shown in Figure 4-9. The mechanical energy balance, Equation 4-1, describes the flow of compressible gases and vapors. Assuming negligible potential energy changes and no shaft work results in a reduced form of the mechanical energy balance describing compressible flow through holes.

$$\int \frac{dP}{\rho} + \Delta\left(\frac{\bar{u}^2}{2\alpha g_c}\right) + F = 0 \tag{4-31}$$

A discharge coefficient, C_1, is defined in a similiar fashion to the coefficient defined in Section 4-1.

$$-\int \frac{dP}{\rho} - F = C_1^2\left(-\int \frac{dP}{\rho}\right) \tag{4-32}$$

Equation 4-32 is combined with Equation 4-31 and integrated between any two convenient points. An initial point (denoted by subscript o) is selected where the velocity is zero and the pressure is P_o. The integration is carried to any arbitrary final point (denoted without a subscript). The result is

$$C_1^2 \int_{P_o}^{P} \frac{dP}{\rho} + \frac{\bar{u}^2}{2\alpha g_c} = 0 \tag{4-33}$$

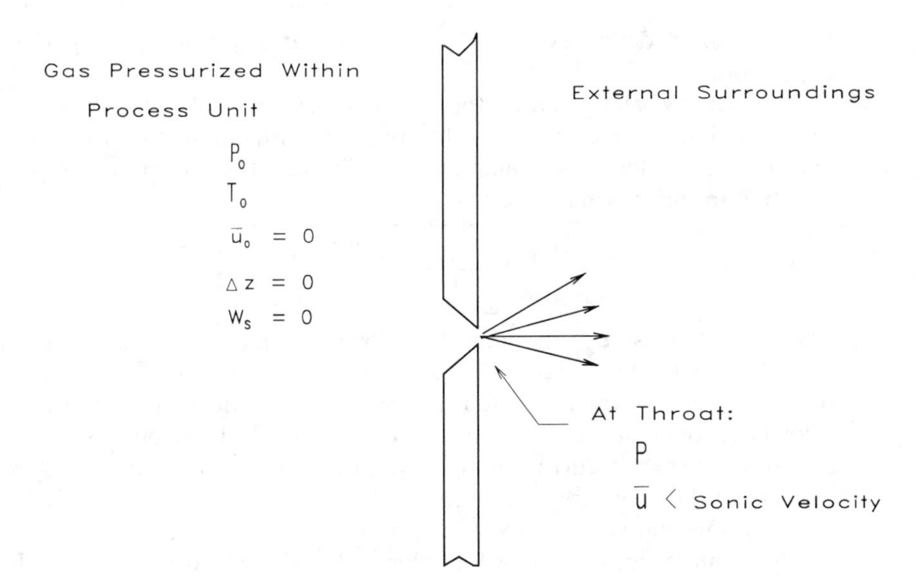

Figure 4-9 A free expansion gas leak. The gas expands isentropically through the hole. The gas properties (P, T) and velocity change during the expansion.

For any ideal gas undergoing an isentropic expansion,

$$Pv^{\gamma} = \frac{P}{\rho^{\gamma}} = \text{constant} \tag{4-34}$$

where γ is the ratio of the heat capacities, $\gamma = C_p/C_v$. Substitution of Equation 4-34 into Equation 4-33, defining a new discharge coefficient C_o identical to Equation 4-5 and integrating results in an equation representing the velocity of the fluid at any point during the isentropic expansion

$$\bar{u}^2 = 2g_c C_o^2 \frac{\gamma}{\gamma - 1} \frac{P_o}{\rho_o} \left[1 - \left(\frac{P}{P_o} \right)^{(\gamma-1)/\gamma} \right] = \frac{2g_c C_o^2 R_g T_o}{M} \frac{\gamma}{\gamma - 1} \left[1 - \left(\frac{P}{P_o} \right)^{(\gamma-1)/\gamma} \right] \tag{4-35}$$

The second form incorporates the ideal gas law for the initial density, ρ_o. R_g is the ideal gas constant and T_o is the temperature of the source. Using the continuity equation

$$Q_m = \rho \bar{u} A \tag{4-36}$$

and the ideal gas law for isentropic expansions in the form

$$\rho = \rho_o \left(\frac{P}{P_o} \right)^{1/\gamma} \tag{4-37}$$

results in an expression for the mass flowrate.

$$Q_m = C_o A P_o \sqrt{\frac{2g_c M}{R_g T_o} \frac{\gamma}{\gamma - 1} \left[\left(\frac{P}{P_o} \right)^{2/\gamma} - \left(\frac{P}{P_o} \right)^{(\gamma+1)/\gamma} \right]} \tag{4-38}$$

Equation 4-38 describes the mass flowrate at any point during the isentropic expansion.

For many safety studies, the maximum flowrate of vapor through the hole is required. This is determined by differentiating Equation 4-38 with respect to P/P_o and setting the derivative equal to zero. The result is solved for the pressure ratio resulting in the maximum flow.

$$\frac{P_{choked}}{P_o} = \left(\frac{2}{\gamma + 1}\right)^{\gamma/(\gamma-1)} \tag{4-39}$$

The choked pressure is the maximum downstream pressure resulting in maximum flow through the hole or pipe. For downstream pressures *less* than P_{choked} the following statements are valid: (1) the velocity of the fluid at the throat of the leak is the velocity of sound at the prevailing conditions and (2) the velocity and mass flowrate cannot be increased further by reducing the downstream pressure; they are independent of the downstream conditions. This type of flow is called *choked, critical,* or *sonic flow* and is illustrated in Figure 4-10.

An interesting feature of Equation 4-39 is that for ideal gases the choked pressure is a function only of the heat capacity ratio γ. Thus,

Gas	γ	P_{choked}
Monotonic	$\cong 1.67$	$0.487\ P_o$
Diatomic and air	$\cong 1.40$	$0.528\ P_o$
Triatomic	$\cong 1.32$	$0.542\ P_o$

For an air leak to atmospheric conditions ($P_{choked} = 14.7$ psia), if the upstream pressure is greater than $14.7/0.528 = 27.8$ psia, or 13.1 psig, the flow will be choked and maximized through the leak. Conditions leading to choked flow are very common in the process industries.

The maximum flow is determined by substituting Equation 4-39 into Equation 4-38,

$$(Q_m)_{choked} = C_o A P_o \sqrt{\frac{\gamma g_c M}{R_g T_o}\left(\frac{2}{\gamma + 1}\right)^{(\gamma+1)/(\gamma-1)}} \tag{4-40}$$

where M is the molecular weight of the escaping vapor or gas, T_o the temperature of the source, and R_g the ideal gas constant.

For sharp-edged orifices with Reynolds numbers greater than 30,000 (and not choked), a constant discharge coefficient, C_o, of 0.61 is indicated. However, for choked flows, the discharge coefficient increases as the downstream pressure decreases.[2] For these flows and for situations where C_o is uncertain, a conservative value of 1.0 is recommended.

Example 4-4

A 0.1 inch hole forms in a tank containing nitrogen at 200 psig and 80°F. Determine the mass flowrate through this leak.

[2]Robert H. Perry and Cecil H. Chilton, *Chemical Engineers Handbook,* 5th ed. (New York: McGraw Hill Book Company, 1973), p. 5-14.

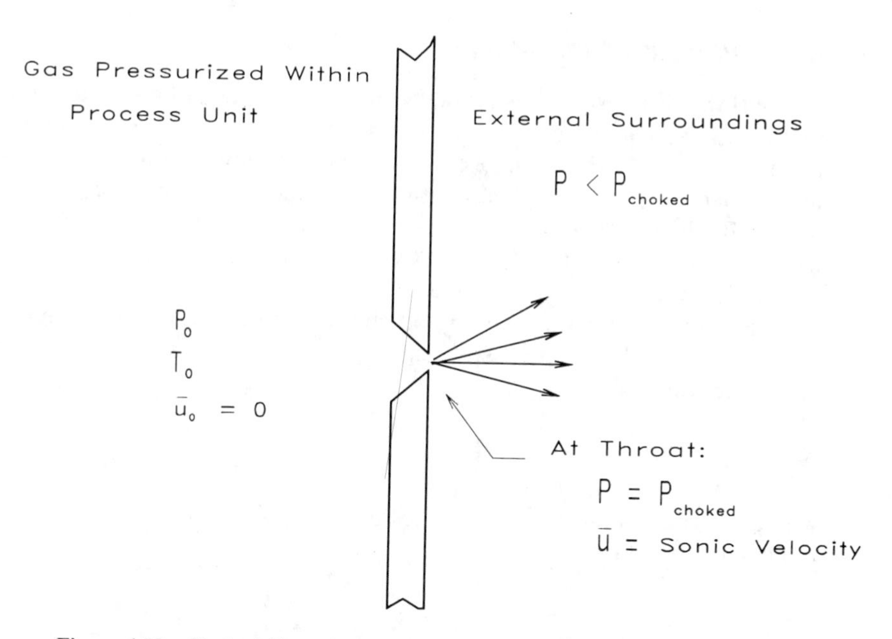

Figure 4-10 Choked flow of gas through a hole. The gas velocity is sonic at the throat. The mass flow rate is independent of the downstream pressure.

Solution For the diatomic gas nitrogen, $\gamma = 1.4$. Thus,

$$P_{\text{choked}} = 0.528\ (200 + 14.7)\ \text{psia} = 113.4\ \text{psia}$$

An external pressure less than 113.4 psia will result in choked flow through the leak. Since the external pressure is atmospheric in this case, choked flow is expected and Equation 4-40 applies. The area of the hole is

$$A = \frac{\pi d^2}{4} = \frac{(3.14)\,(0.1\ \text{in})^2(1\ \text{ft}^2/144\ \text{in}^2)}{4} = 5.45 \times 10^{-5}\ \text{ft}^2$$

The discharge coefficient, C_o, is assumed to be 1.0. Also,

$$P_o = 200 + 14.7 = 214.7\ \text{psia}$$

$$T_o = 80 + 460 = 540°\text{R}$$

$$\left(\frac{2}{\gamma+1}\right)^{(\gamma+1)/(\gamma-1)} = \left(\frac{2}{2.4}\right)^{2.4/0.4} = (0.833)^{6.00} = 0.335$$

Then, using Equation 4-40,

$$(Q_m)_{\text{choked}} = C_o A P_o \sqrt{\frac{\gamma g_c M}{R_g T_o}\left(\frac{2}{\gamma + 1}\right)^{(\gamma+1)/(\gamma-1)}}$$

$$= (1.0)\,(5.45 \times 10^{-5}\ \text{ft}^2)\,(214.7\ \text{lb}_f/\text{in}^2)\,(144\ \text{in}^2/\text{ft}^2)$$

$$\times \sqrt{\frac{(1.4)\,(32.17\ \text{ft lb}_m/\text{lb}_f\text{s}^2)\,(28\ \text{lb}_m/\text{lb-mole})}{(1545\ \text{ft lb}_f/\text{lb-mole°R})\,(540°\text{R})}(0.335)}$$

$$= 1.685\ \text{lb}_f\sqrt{5.064 \times 10^{-4}\ \text{lb}_m^2/\text{lb}_f^2\ \text{s}^2}$$

$$\boxed{(Q_m)_{\text{choked}} = 3.79 \times 10^{-2}\ \text{lb}_m/\text{s}}$$

4-5 FLOW OF VAPOR THROUGH PIPES

Vapor flow through pipes is modelled using two special cases: adiabatic or isothermal behavior. The adiabatic case corresponds to rapid vapor flow through an insulated pipe. The isothermal case corresponds to flow through an uninsulated pipe maintained at a constant temperature; an underwater pipeline is an excellent example. Real vapor flows behave somewhere between the adiabatic and isothermal cases. Unfortunately, the "real" case is very difficult to model and no generalized and useful equations are available.

For both the isothermal and adiabatic cases it is convenient to define a Mach number as the ratio of the gas velocity to the velocity of sound in the gas at the prevailing conditions.

$$Ma = \frac{\bar{u}}{a} \tag{4-41}$$

where a is the velocity of sound. The velocity of sound is determined using the thermodynamic relationship,

$$a = \sqrt{g_c \left(\frac{\partial P}{\partial \rho}\right)_s} \tag{4-42}$$

which, for an ideal gas, is equivalent to,

$$a = \sqrt{\gamma g_c R_g T/M} \tag{4-43}$$

which demonstrates that, for ideal gases the sonic velocity is a function of temperature only. For air at 20°C the velocity of sound is 344 m/s (1129 ft/s).

Adiabatic Flows

An adiabatic pipe containing a flowing vapor is shown in Figure 4-11. For this particular case the outlet velocity is less than the sonic velocity. The flow is driven by a pressure gradient across the pipe. As the gas flows through the pipe it expands due to a decrease in pressure. This expansion leads to an increase in velocity and an increase in the kinetic energy of the gas. The kinetic energy is extracted from the thermal energy of the gas; a decrease in temperature occurs. However, frictional forces are present between the gas and the pipe wall. These frictional forces increase the temperature of the gas. Depending on the magnitude of the kinetic and frictional energy terms either an increase or decrease in the gas temperature is possible.

The mechanical energy balance, Equation 4-1, also applies to adiabatic flows. For this case it is more conveniently written in the form

$$\frac{dP}{\rho} + \frac{\bar{u}\,d\bar{u}}{\alpha g_c} + \frac{g}{g_c}dz + dF = -\frac{\delta W_s}{m} \tag{4-44}$$

The following assumptions are valid for this case:

$$\frac{g}{g_c}dz \approx 0$$

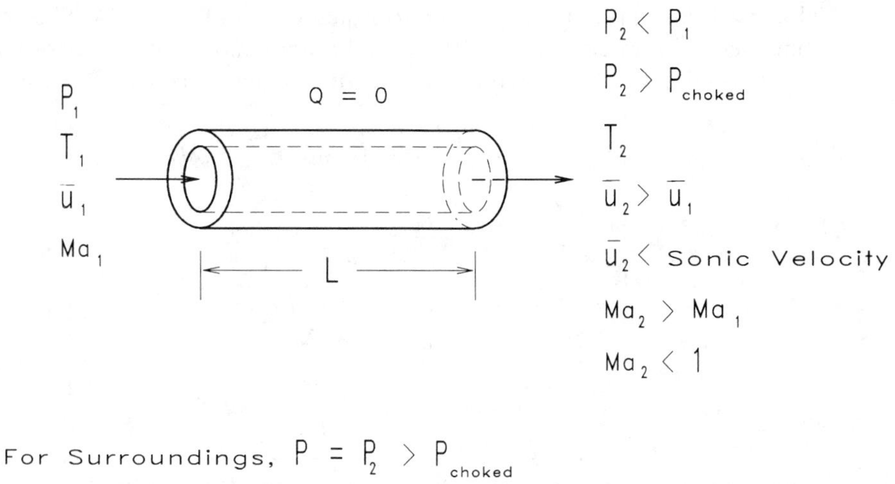

Figure 4-11 Adiabatic, nonchoked flow of gas through a pipe. The gas temperature might increase or decrease, depending on the magnitude of the frictional losses.

is valid for gases, and

$$dF = \frac{2f\bar{u}^2 dL}{g_c d}$$

from Equation 4-23, assuming constant f, and

$$\delta W_s = 0$$

since no mechanical linkages are present. An important part of the frictional loss term is the assumption of a constant Fanning friction factor, f, across the length of the pipe. This assumption is only valid at high Reynolds number.

A total energy balance is useful for describing the temperature changes within the flowing gas. For this open, steady flow process it is given by

$$dh + \frac{\bar{u}d\bar{u}}{\alpha g_c} + \frac{g}{g_c}dz = \delta q - \frac{\delta W_s}{m} \tag{4-45}$$

where h is the enthalpy of the gas and q is the heat. The following assumptions are invoked.

$dh = C_p dT$ for an ideal gas,

$\frac{g}{g_c}dz \approx 0$ is valid for gases,

$\delta q = 0$ since the pipe is adiabatic,

$\delta W_s = 0$ since no mechanical linkages are present.

The above assumptions are applied to Equations 4-45 and 4-44. The equations are combined, integrated (between the initial point denoted by subscript o and any arbitrary final point), and manipulated to yield, after considerable effort,[3]

$$\frac{T_2}{T_1} = \frac{Y_1}{Y_2} \text{ where } Y_i = 1 + \frac{\gamma - 1}{2} Ma_i^2 \tag{4-46}$$

$$\frac{P_2}{P_1} = \frac{Ma_1}{Ma_2}\sqrt{\frac{Y_1}{Y_2}} \tag{4-47}$$

$$\frac{\rho_2}{\rho_1} = \frac{Ma_1}{Ma_2}\sqrt{\frac{Y_2}{Y_1}} \tag{4-48}$$

$$G = \rho\bar{u} = Ma_1 P_1 \sqrt{\frac{\gamma g_c M}{R_g T_1}} = Ma_2 P_2 \sqrt{\frac{\gamma g_c M}{R_g T_2}} \tag{4-49}$$

where G is the mass flux with units of mass/(area time), and

$$\underbrace{\frac{\gamma+1}{2}\ln\left(\frac{Ma_2^2 Y_1}{Ma_1^2 Y_2}\right)}_{\text{kinetic energy}} - \underbrace{\left(\frac{1}{Ma_1^2} - \frac{1}{Ma_2^2}\right)}_{\text{compressibility}} + \underbrace{\gamma\left(\frac{4fL}{d}\right)}_{\text{pipe friction}} = 0 \tag{4-50}$$

Equation 4-50 relates the Mach numbers to the frictional losses in the pipe. The various energy contributions are identified. The compressibility term accounts for the change in velocity due to the expansion of the gas.

Equations 4-49 and 4-50 are converted to a more convenient and useful form by replacing the Mach numbers with temperatures and pressures using Equations 4-46 through 4-48.

$$\frac{\gamma+1}{\gamma}\ln\frac{P_1 T_2}{P_2 T_1} - \frac{\gamma-1}{2\gamma}\left(\frac{P_1^2 T_2^2 - P_2^2 T_1^2}{T_2 - T_1}\right)\left(\frac{1}{P_1^2 T_2} - \frac{1}{P_2^2 T_1}\right) + \frac{4fL}{d} = 0 \tag{4-51}$$

$$G = \sqrt{\frac{2 g_c M}{R_g} \frac{\gamma}{\gamma - 1} \frac{T_2 - T_1}{(T_1/P_1)^2 - (T_2/P_2)^2}} \tag{4-52}$$

For most problems the pipe length (L), inside diameter (d), upstream temperature (T_1) and pressure (P_1), and downstream pressure (P_2) are known. To compute the mass flux, G, the procedure is as follows.

1. Determine pipe roughness, ε from Table 4-1. Compute ε/d.
2. Determine the Fanning friction factor, f, from Equation 4-27. This assumes fully developed turbulent flow at high Reynolds numbers. This assumption can be checked later, but is normally valid.
3. Determine T_2 from Equation 4-51.
4. Compute the total mass flux, G, from Equation 4-52.

[3]Octave Levenspiel, *Engineering Flow and Heat Exchange* (New York: Plenum Press, 1986), p. 43.

For long pipes, or for large pressure differences across the pipe, the velocity of the gas can approach the sonic velocity. This case is shown in Figure 4-12. At the sonic velocity the flow will be choked. The gas velocity will remain at the sonic velocity, temperature, and pressure for the remainder of the pipe. For choked flow, Equations 4-46 through 4-50 are simplified by setting $Ma_2 = 1.0$. The results are

$$\frac{T_{\text{choked}}}{T_1} = \frac{2Y_1}{\gamma+1} \tag{4-53}$$

$$\frac{P_{\text{choked}}}{P_1} = Ma_1\sqrt{\frac{2Y_1}{\gamma+1}} \tag{4-54}$$

$$\frac{\rho_{\text{choked}}}{\rho_1} = Ma_1\sqrt{\frac{\gamma+1}{2Y_1}} \tag{4-55}$$

$$G_{\text{choked}} = \rho\bar{u} = Ma_1 P_1\sqrt{\frac{\gamma g_c M}{R_g T_1}} = P_{\text{choked}}\sqrt{\frac{\gamma g_c M}{R_g T_{\text{choked}}}} \tag{4-56}$$

$$\frac{\gamma+1}{2}\ln\left[\frac{2Y_1}{(\gamma+1)Ma_1^2}\right] - \left(\frac{1}{Ma_1^2} - 1\right) + \gamma\left(\frac{4fL}{d}\right) = 0 \tag{4-57}$$

Choked flow occurs if the downstream pressure is less than P_{choked}. This is checked using Equation 4-54.

For most problems involving choked, adiabatic flows, the pipe length (L), inside diameter (d), and upstream pressure (P_1) and temperature (T_1) are known. To compute the mass flux, G, the procedure is as follows.

1. Determine the Fanning friction factor, f, using Equation 4-27. This assumes fully developed turbulent flow at high Reynolds number. This assumption can be checked later, but is normally valid.
2. Determine Ma_1 from Equation 4-57.
3. Determine the mass flux, G_{choked}, from Equation 4-56.
4. Determine P_{choked} from Equation 5-54 to confirm operation at choked conditions.

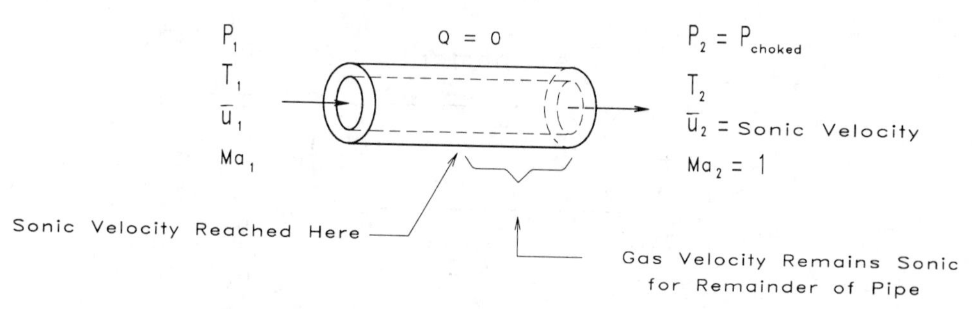

Figure 4-12 Adiabatic, choked flow of gas through a pipe. The maximum velocity reached is the sonic velocity of the gas.

Isothermal Flows

Isothermal flow of gas in a pipe with friction is shown in Figure 4-13. For this case the gas velocity is assumed to be well below the sonic velocity of the gas. A pressure gradient across the pipe provides the driving force for the gas transport. As the gas expands through the pressure gradient the velocity must increase to maintain the same mass flowrate. The pressure at the end of the pipe is equal to the pressure of the surroundings. The temperature is constant across the entire pipe length.

Isothermal flow is represented by the mechanical energy balance in the form shown in Equation 4-44. The following assumptions are valid for this case.

$$\frac{g}{g_c}\,dz \approx 0$$

is valid for gases, and

$$dF = \frac{2f\bar{u}^2 dL}{g_c d}$$

from Equation 4-23, assuming constant f, and

$$\delta W_s = 0$$

since no mechanical linkages are present. A total energy balance is not required since the temperature is constant.

Applying the above assumptions to Equation 4-44, and, after considerable manipulation[4]

$$T_2 = T_1 \tag{4-58}$$

$$\frac{P_2}{P_1} = \frac{Ma_1}{Ma_2} \tag{4-59}$$

$$\frac{\rho_2}{\rho_1} = \frac{Ma_1}{Ma_2} \tag{4-60}$$

$$G = \rho\bar{u} = Ma_1 P_1 \sqrt{\frac{\gamma g_c M}{R_g T}} \tag{4-61}$$

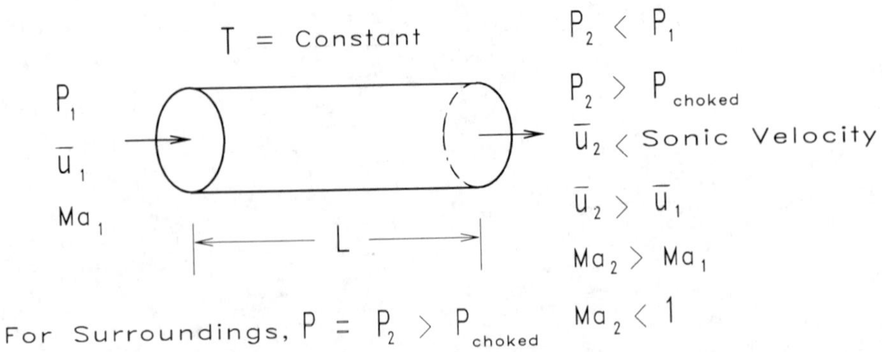

Figure 4-13 Isothermal, nonchoked flow of gas through a pipe.

[4]Levenspiel, *Engineering Flow*, p. 46.

where G is the mass flux with units of mass/(area time), and,

$$2 \ln \frac{Ma_2}{Ma_1} - \frac{1}{\gamma}\left(\frac{1}{Ma_1^2} - \frac{1}{Ma_2^2}\right) + \frac{4fL}{d} = 0 \tag{4-62}$$

kinetic energy — compressibility — pipe friction

The various energy terms in Equation 4-62 have been identified.

A more convenient form of Equation 4-62 is in terms of pressure instead of Mach numbers. This form is achieved by using Equations 4-58 through 4-60. The result is

$$2 \ln \frac{P_1}{P_2} - \frac{g_c M}{G^2 R_g T}(P_1^2 - P_2^2) + \frac{4fL}{d} = 0 \tag{4-63}$$

A typical problem is to determine the mass flux, G, given the pipe length (L), inside diameter (d), and upstream and downstream pressures (P_1 and P_2). The procedure is as follows.

1. Determine the Fanning friction factor, f, using Equation 4-27. This assumes fully developed turbulent flow at high Reynolds number. This assumption can be checked later, but is usually valid.
3. Compute the mass flux, G, from Equation 4-63.

Levenspiel[5] has shown that the maximum velocity possible during the isothermal flow of gas in a pipe is not the sonic velocity as in the adiabatic case. In terms of the Mach number, the maximum velocity is

$$Ma_{\text{choked}} = \frac{1}{\sqrt{\gamma}} \tag{4-64}$$

This result is shown by starting with the mechanical energy balance and rearranging it into the following form.

$$-\frac{dP}{dL} = \frac{2fG^2}{g_c \rho d}\left[\frac{1}{1-(\bar{u}^2\rho/g_c P)}\right] = \frac{2fG^2}{g_c \rho d}\left[\frac{1}{1-\gamma Ma^2}\right] \tag{4-65}$$

The quantity $-(dP/dL) \rightarrow \infty$ when $Ma \rightarrow 1/\sqrt{\gamma}$. Thus, for choked flow in an isothermal pipe, as shown in Figure 4-14, the following equations apply.

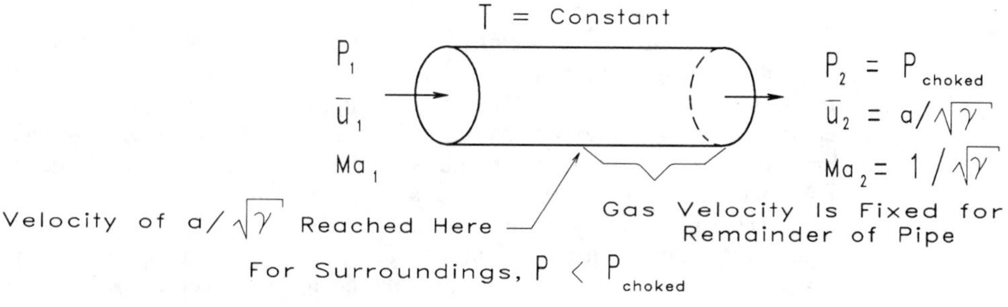

Figure 4-14 Isothermal, choked flow of gas through a pipe. The maximum velocity reached is $a/\sqrt{\gamma}$.

[5]Levenspiel, *Engineering Flow,* p. 46.

$$T_{\text{choked}} = T_1 \tag{4-66}$$

$$\frac{P_{\text{choked}}}{P_1} = Ma_1\sqrt{\gamma} \tag{4-67}$$

$$\frac{\rho_{\text{choked}}}{\rho_1} = Ma_1\sqrt{\gamma} \tag{4-68}$$

$$\frac{\bar{u}_{\text{choked}}}{\bar{u}_1} = \frac{1}{Ma_1\sqrt{\gamma}} \tag{4-69}$$

$$G_{\text{choked}} = \rho\bar{u} = \rho_1\bar{u}_1 = Ma_1 P_1\sqrt{\frac{\gamma g_c M}{R_g T}} = P_{\text{choked}}\sqrt{\frac{g_c M}{R_g T}} \tag{4-70}$$

where G_{choked} is the mass flux with units of mass/(area time), and

$$\ln\left(\frac{1}{\gamma Ma_1^2}\right) - \left(\frac{1}{\gamma Ma_1^2} - 1\right) + \frac{4fL}{d} = 0 \tag{4-71}$$

For most typical problems the pipe length (L), inside diameter (d), upstream pressure (P_1), and temperature (T) are known. The mass flux, G, is determined using the following procedure.

1. Determine the Fanning friction factor using Equation 4-27. This assumes fully developed turbulent flow at high Reynolds number. This assumption can be checked later, but is usually valid.
3. Determine Ma_1 from Equation 4-71.
4. Determine the mass flux, G, from Equation 4-70.

Example 4-5

The vapor space above liquid ethylene oxide (EO) in storage tanks must be purged of oxygen and then padded with 81-psig nitrogen to prevent explosion. The nitrogen in a particular facility is supplied from a 200-psig source. It is regulated to 81-psig and supplied to the storage vessel through 33 feet of commercial steel pipe with and ID of 1.049 inches.

In the event of a failure of the nitrogen regulator, the vessel will be exposed to the full 200-psig pressure from the nitrogen source. This will exceed the pressure rating of the storage vessel. To prevent rupture of the storage vessel it must be equipped with a relief device to vent this nitrogen. Determine the required minimum mass flow rate of nitrogen through the relief device to prevent the pressure from rising within the tank in the event of a regulator failure.

Determine the mass flow rate assuming (a) an orifice with a throat diameter equal to the pipe diameter, (b) an adiabatic pipe, and (c) an isothermal pipe. Decide which result most closely corresponds to the real situation. Which mass flow rate should be used?

Solution a. The maximum flowrate through the orifice occurs under choked conditions.

The area of the pipe is

$$A = \frac{\pi d^2}{4} = \frac{(3.14)(1.049 \text{ in})^2(1 \text{ ft}^2/144 \text{ in}^2)}{4}$$

$$= 6.00 \times 10^{-3} \text{ ft}^2$$

The absolute pressure of the nitrogen source is

$$P_o = 200 + 14.7 = 214.7 \text{ psia} = 3.09 \times 10^4 \text{ lb}_f/\text{ft}^2$$

The choked pressure from Equation 4-39 is, for a diatomic gas,

$$P_{\text{choked}} = (0.528)(214.7 \text{ psia}) = 113.4 \text{ psia}$$

$$= 1.63 \times 10^4 \text{ lb}_f/\text{ft}^2$$

Choked flow can be expected since the system will be venting to atmospheric conditions. Equation 4-40 provides the maximum mass flowrate. For nitrogen, $\gamma = 1.4$ and

$$\left(\frac{2}{\gamma+1}\right)^{(\gamma+1)/(\gamma-1)} = \left(\frac{2}{2.4}\right)^{2.4/0.4} = 0.335$$

The molecular weight of nitrogen is 28 lb_m/lb-mole. Without any additional information, assume a unit discharge coefficient, $C_o = 1.0$. Thus,

$$Q_m = (1.0)(6.00 \times 10^{-3} \text{ ft}^2)(3.09 \times 10^4 \text{ lb}_f/\text{ft}^2)$$

$$\times \sqrt{\frac{(1.4)(32.17 \text{ ft lb}_m/\text{lb}_f\text{s}^2)(28 \text{ lb}_m/\text{lb-mole})}{(1545 \text{ ft lb}_f/\text{lb-mole}°\text{R})(540°\text{R})}(0.335)}$$

$$= (185 \text{ lb}_f)\sqrt{5.06 \times 10^{-4} \text{ lb}_m^2/\text{lb}_f^2 \text{ s}^2}$$

$$\boxed{Q_m = 4.16 \text{ lb}_m/\text{s}}$$

b. Assume adiabatic, choked flow conditions. For commercial steel pipe, from Table 4-1, $\varepsilon = 0.046$ mm. The diameter of the pipe in mm is (1.049 in)(25.4 mm/in) = 26.6 mm. Thus,

$$\frac{\varepsilon}{d} = \frac{0.046 \text{ mm}}{26.6 \text{ mm}} = 0.00173$$

From Equation 4-27,

$$\frac{1}{\sqrt{f}} = 4 \log\left(3.7\frac{d}{\varepsilon}\right)$$

$$= 4 \log(3.7/0.00173) = 13.32$$

$$\sqrt{f} = 0.0751$$

$$f = 0.00564$$

For nitrogen, $\gamma = 1.4$.

The upstream Mach number is determined from Equation 4-57.

$$\frac{\gamma+1}{2}\ln\left[\frac{2Y_1}{(\gamma+1)Ma_1^2}\right] - \left(\frac{1}{Ma_1^2} - 1\right) + \gamma\left(\frac{4fL}{d}\right) = 0$$

with Y_1 given by Equation 4-46. Substituting the numbers provided,

$$\frac{1.4+1}{2}\ln\left[\frac{2 + (1.4-1)Ma^2}{(1.4+1)Ma^2}\right] - \left(\frac{1}{Ma^2} - 1\right) + 1.4\left[\frac{(4)\,(0.00564)\,(33\text{ ft})}{(1.049\text{ in})\,(1\text{ ft}/12\text{ in})}\right] = 0$$

$$1.2\ln\left(\frac{2 + 0.4\,Ma^2}{2.4Ma^2}\right) - \left(\frac{1}{Ma^2} - 1\right) + 11.92 = 0$$

This equation is solved by trial and error for the value of Ma. The results are tabulated below.

Guessed Ma	Value of LHS of equation
0.20	−8.43
0.25	0.043

This last value looks very close. Then from Equation 4-46,

$$Y_1 = 1 + \frac{\gamma-1}{2}Ma^2 = 1 + \frac{1.4-1}{2}(0.25)^2 = 1.012$$

and from Equations 4-53 and 4-54

$$\frac{T_{\text{choked}}}{T_1} = \frac{2Y_1}{\gamma+1} = \frac{2(1.012)}{1.4 + 1} = 0.843$$

$$T_{\text{choked}} = (0.843)\,(80+460)°\text{R} = 455°\text{R}$$

$$\frac{P_{\text{choked}}}{P_1} = Ma\sqrt{\frac{2Y_1}{\gamma+1}} = (0.25)\sqrt{0.843} = 0.230$$

$$P_{\text{choked}} = (0.230)\,(214.7\text{ psia}) = 49.4\text{ psia} = 7.11 \times 10^3\text{ lb}_\text{f}/\text{ft}^2$$

The pipe outlet pressure must be less than 49.4 psia to insure choked flow. The mass flux is computed using Equation 4-56,

$$G_{\text{choked}} = P_{\text{choked}}\sqrt{\frac{\gamma g_c M}{R_g T_{\text{choked}}}}$$

$$= (7.11 \times 10^3\text{ lb}_\text{f}/\text{ft}^2) \times \sqrt{\frac{(1.4)\,(32.17\text{ ft lb}_\text{m}/\text{lb}_\text{f}\text{ s}^2)\,(28\text{ lb}_\text{m}/\text{lb-mole})}{(1545\text{ ft lb}_\text{f}/\text{lb-mole}°\text{R})\,(455°\text{R})}}$$

$$= 7.11 \times 10^3\text{ lb}_\text{f}/\text{ft}^2\sqrt{1.79 \times 10^{-3}\text{ lb}_\text{m}^2/\text{lb}_\text{f}^2\text{ s}^2} = 301\text{ lb}_\text{m}/\text{ft}^2\text{ s}$$

$$Q_\text{m} = GA = (301\text{ lb}_\text{m}/\text{ft}^2\text{ s})\,(6.00 \times 10^{-3}\text{ ft}^2)$$

$$\boxed{= 1.81\text{ lb}_\text{m}/\text{s}}$$

c. For the isothermal case, the upstream Mach number is given by Equation 4-71. Substituting the numbers provided,

$$\ln\left[\frac{1}{1.4Ma^2}\right] - \left(\frac{1}{1.4Ma^2} - 1\right) + 8.52 = 0$$

The solution is found by trial and error.

Guessed *Ma*	Value of *LHS* of equation
0.25	0.526
0.24	−0.362
0.245	0.097
0.244	0.005 ← Final result

The choked pressure is, from Equation 4-67.

$$P_{\text{choked}} = P_1 Ma_1 \sqrt{\gamma} = (214.7 \text{ lb}_f/\text{in}^2)(0.244)\sqrt{1.4} = 62.0 \text{ psia} = 8.93 \times 10^3 \text{ lb}_f/\text{ft}^2$$

The mass flow rate is computed using Equation 4-70.

$$G_{\text{choked}} = P_{\text{choked}} \sqrt{\frac{g_c M}{R_g T}} = 8.93 \times 10^3 \text{ lb}_f/\text{ft}^2$$

$$\times \sqrt{\frac{(32.17 \text{ ft lb}_m/\text{lb}_f \text{ s}^2)(28 \text{ lb}_m/\text{lb-mole})}{(1545 \text{ ft lb}_f/\text{lb-mole}^\circ\text{R})(540^\circ\text{R})}}$$

$$= 8.93 \times 10^3 \text{ lb}_f/\text{ft}^2 \sqrt{1.08 \times 10^{-3} \text{ lb}_m^2/\text{lb}_f^2 \text{ s}^2} = 293 \text{ lb}_m/\text{ft}^2 \text{ s}$$

$$Q_m = G_{\text{choked}} A = (293 \text{ lb}_m/\text{ft}^2 \text{ s})(6.00 \times 10^{-3} \text{ ft}^2)$$

$$\boxed{= 1.76 \text{ lb}_m/\text{s}}$$

The results are summarized in the following table,

Case	P_{choked} (psia)	Q_m (lb_m/s)
Orifice	113.4	4.16
Adiabatic pipe	49.4	1.81
Isothermal pipe	62.0	1.76

A standard procedure for these types of problems is to represent the discharge through the pipe as an orifice. The results show that this approach results in a large result for this case. The orifice method will always produce a larger value than the adiabatic pipe method, insuring a conservative safety design. The orifice calculation, however, is easier to apply, requiring only the pipe diameter and upstream supply pressure and temperature. The configurational details of the piping are not required, as in the adiabatic or isothermal pipe methods.

Also note that the choked pressures computed differ for each case, with a substantial difference between the orifice and adiabatic/isothermal cases. A choking design based on an orifice calculation might not be choked in reality due to high downstream pressures.

Finally, note that the adiabatic and isothermal pipe methods produce results that are reasonably close. For most real situations the heat transfer characteristics cannot be easily determined. Thus, the adiabatic pipe method is the method of choice; it will always produce the larger number for a conservative safety design.

4-6 FLASHING LIQUIDS

Liquids stored under pressure above their normal boiling point temperature present substantial problems due to flashing. If the tank, pipe, or other containment device develops a leak, the liquid will partially flash into vapor, sometimes explosively.

Flashing occurs so rapidly that the process is assumed to be adiabatic. The excess energy contained in the superheated liquid vaporizes the liquid and lowers the temperature to the new boiling point. If m is the mass of original liquid, C_p the heat capacity of the liquid (energy/mass deg), T_o the temperature of the liquid prior to depressurization, and T_b the depressurized boiling point of the liquid, then the excess energy contained in the superheated liquid is given by,

$$Q = mC_p(T_o - T_b) \tag{4-72}$$

This energy vaporizes the liquid. If ΔH_v is the heat of vaporization of the liquid, the mass of liquid vaporized, m_v, is given by

$$m_v = \frac{Q}{\Delta H_v} = \frac{mC_p(T_o - T_b)}{\Delta H_v} \tag{4-73}$$

The fraction of the liquid vaporized is

$$\boxed{f_v = \frac{m_v}{m} = \frac{C_p(T_o - T_b)}{\Delta H_v}} \tag{4-74}$$

Equation 4-74 assumes constant physical properties over the temperature range T_o to T_b. A more general expression without this assumption is derived as follows.

The change in liquid mass, m, due to a change in temperature, T is given by

$$dm = \frac{mC_p}{\Delta H_v}\,dT \tag{4-75}$$

Equation 4-75 is integrated between the initial temperature T_o (with liquid mass m) and the final boiling point temperature T_b (with liquid mass $m - m_v$),

$$\int_m^{m-m_v} \frac{dm}{m} = \int_{T_o}^{T_b} \frac{C_p}{\Delta H_v}\,dT \tag{4-76}$$

$$\ln\left(\frac{m - m_v}{m}\right) = -\frac{\overline{C_p}(T_o - T_b)}{\overline{\Delta H_v}} \tag{4-77}$$

where $\overline{C_p}$ and $\overline{\Delta H_v}$ are the mean heat capacity and mean latent heat of vaporization, respectively, over the temperature range T_o to T_b. Solving for the fraction of the liquid vaporized, $f_v = m_v/m$,

$$\boxed{f_v = 1 - \exp[-\overline{C}_p(T_o - T_b)/\overline{\Delta H_v}]} \tag{4-78}$$

Example 4-6

One lb_m of saturated liquid water is contained in a vessel at 350°F. The vessel ruptures and the pressure is reduced to 1 atm. Compute the fraction of material vaporized using (a) the steam tables, (b) Equation 4-74, and (c) Equation 4-78.

Solution **a.** The initial state is saturated steam at T_o = 350°F. From the steam tables,

$$P = 134.6 \text{ psia}$$

$$H = 321.6 \text{ BTU/lb}_m$$

The final temperature is the boiling point at 1 atm, or 212°F. At this temperature, and saturated conditions,

$$H_{vapor} = 1150.4 \text{ BTU/lb}_m$$

$$H_{liquid} = 180.07 \text{ BTU/lb}_m$$

Since the process occurs adiabatically, $H_{final} = H_{initial}$ and the fraction of vapor (or quality) is computed from,

$$H_{final} = H_{liquid} + f_v (H_{vapor} - H_{liquid})$$

$$321.6 = 180.07 + f_v(1150.4 - 180.07)$$

$$\boxed{f_v = 0.1459}$$

14.59% of the mass of the original liquid is vaporized.

b. For liquid water at 212°F,

$$C_p = 1.01 \text{ BTU/lb}_m \text{ °F}$$

$$\Delta H_v = 970.3 \text{ BTU/lb}_m$$

From Equation 4-74

$$f_v = \frac{C_p(T_o - T_b)}{\Delta H_v} = \frac{(1.01 \text{ BTU/lb}_m \text{ °F})(350 - 212)\text{°F}}{970.3 \text{ BTU/lb}_m}$$

$$\boxed{f_v = 0.1436}$$

c. The mean properties for liquid water between T_o and T_b are

$$\overline{C_p} = 1.04 \text{ BTU/lb}_m \text{ °F}$$

$$\overline{\Delta H_v} = 920.7 \text{ BTU/lb}_m$$

Substituting into Equation 4-78,

$$f_v = 1 - \exp[-\overline{C}_p(T_o - T_b) / \overline{\Delta H_v}]$$

$$= 1 - \exp[-(1.04 \text{ BTU/lb}_m \text{ °F})(350 - 212)\text{°F}/(920.7 \text{ BTU/lb}_m)]$$

$$f_v = 1 - 0.8557$$

$$\boxed{f_v = 0.1443}$$

Both expressions work about as well when compared to the actual value from the steam table.

For flashing liquids composed of many miscible substances, the flash calculation is complicated considerably, since the more volatile components will flash preferentially. Procedures are available to solve this problem.[6]

Flashing liquids escaping through holes and pipes require very special consideration since two-phase flow conditions may be present. Several special cases need consideration.[7] If the fluid path length of the release is very short (through a hole in a thin-walled container), nonequilibrium conditions exist, and the liquid does not have time to flash within the hole; the fluid flashes external to the hole. The equations describing incompressible fluid flow through holes apply (see Section 4-1).

If the fluid path length through the release is greater than 10 cm (through a pipe or thick-walled container), equilibrium flashing conditions are achieved and the flow is choked. A good approximation is to assume a choked pressure equal to the saturation vapor pressure of the flashing liquid. The result will only be valid for liquids stored at a pressure higher than the saturation vapor pressure. With this assumption the mass flow rate is given by

$$\boxed{Q_m = AC_o\sqrt{2\rho_f g_c(P - P^{sat})}} \tag{4-79}$$

where

A is the area of the release,

C_o is the discharge coefficient (unitless)

ρ_f is the density of the liquid (mass/volume),

P is the pressure within the tank, and,

P^{sat} is the saturation vapor pressure of the flashing liquid at ambient temperature.

Example 4-7

Liquid ammonia is stored in a tank at 24°C and a pressure of 1.4×10^6 Pa. A leak of diameter 0.0945 m forms in the tank, allowing the flashing ammonia to escape. The saturation vapor pressure of liquid ammonia at this temperature is 0.968×10^6 Pa and its density is 603 kg/m^3. Determine the mass flow rate through the leak. Equilibrium flashing conditions can be assumed.

Solution Equation 4-79 applies for the case of equilibrium flashing conditions. Assume a discharge coefficient of 0.61.

$$Q_m = AC_o\sqrt{2\rho_f g_c(P - P^{sat})}$$

$$= (0.61)\frac{(3.14)(0.0945\text{ m})^2}{4}$$

$$\times \sqrt{2(603\text{ kg/m}^3)[1\text{ (kg m/s}^2)/\text{N}](1.4 \times 10^6 - 0.968 \times 10^6)(\text{N/m}^2)}$$

$$\boxed{Q_m = 97.6\text{ kg/s}}$$

[6]J. M. Smith and H. C. Van Ness, *Introduction to Chemical Engineering Thermodynamics,* 4th Ed. (New York: McGraw-Hill Book Company, 1987), p. 314.

[7]Hans K. Fauske, "Flashing Flows or: Some Practical Guidelines for Emergency Releases," *Plant/Operations Progress,* July, 1985, p. 133.

For liquids stored at their saturation vapor pressure, $P = P^{sat}$, Equation 4-79 is no longer valid. For this case the choked, two-phase mass flow rate is given by[8]

$$Q_m = A\sqrt{-\frac{g_c}{(dv/dP)}} \tag{4-80}$$

where v is the specific volume with units of (volume/mass). The two-phase specific volume is given by

$$v = v_{fg} f_v + v_f \tag{4-81}$$

where

v_{fg} is the difference in specific volume between vapor and liquid,
v_f is the liquid specific volume, and
f_v is the mass fraction of vapor.

Differentiating Equation 4-81 with respect to pressure,

$$\frac{dv}{dP} = v_{fg}\frac{df_v}{dP} \tag{4-82}$$

But, from Equation 4-74,

$$df_v = -\frac{C_p}{\Delta H_v}dT \tag{4-83}$$

and, from the Clausius-Clapyron equation, at saturation,

$$\frac{dP}{dT} = \frac{\Delta H_v}{T v_{fg}} \tag{4-84}$$

Substituting Equations 4-84 and 4-83 into Equation 4-82 yields,

$$\frac{dv}{dP} = -\frac{v_{fg}^2}{\Delta H_v^2}C_p T \tag{4-85}$$

The mass flow rate is determined by combining Equation 4-85, with Equation 4-80.

$$Q_m = \frac{\Delta H_v A}{v_{fg}}\sqrt{\frac{g_c}{C_p T}} \tag{4-86}$$

Small droplets of liquid also form in a jet of flashing vapor. These aerosol droplets are readily entrained by the wind and transported away from the release site. The assumption that the quantity of droplets formed is equal to the amount of material flashed is frequently made.[9]

[8]Hans K. Fauske and Michael Epstein, "Source Term Considerations in Connection with Chemical Accidents and Vapor Cloud Modeling," *International Conference on Vapor Cloud Modeling* (New York: American Institute of Chemical Engineers, 1987), p. 251.

[9]Trevor A. Kletz, "Unconfined Vapor Cloud Explosions," *Eleventh Loss Prevention Symposium* (New York: American Institute of Chemical Engineers, 1977).

Example 4-8

Propylene is stored at 25°C in a tank at its saturation pressure. A 1-cm diameter hole develops in the tank. Estimate the mass flow rate through the hole. At these conditions, for propylene.

$\Delta H_v = 3.34 \times 10^5$ J/kg
$v_{fg} = 0.042$ m^3/kg
$P^{sat} = 1.15 \times 10^6$ Pa
$C_p = 2.18 \times 10^3$ J/kg K

Equation 4-86 applies to this case. The area of the leak is

$$A = \frac{\pi d^2}{4} = \frac{(3.14)(1 \times 10^{-2}\ \text{m})^2}{4} = 7.85 \times 10^{-5}\ \text{m}^2$$

Using Equation 4-86,

$$Q_m = \frac{\Delta H_v A}{v_{fg}} \sqrt{\frac{g_c}{C_p T}}$$

$$= (3.34 \times 10^5\ \text{J/kg})(1\ \text{N m/J}) \frac{(7.85 \times 10^{-5}\ \text{m}^2)}{(0.042\ \text{m}^3/\text{kg})}$$

$$\times \sqrt{\frac{1.0\ (\text{kg m/s}^2)/\text{N}}{(2.18 \times 10^3\ \text{J/kg K})(298\ \text{K})(1\ \text{N m/J})}}$$

$$\boxed{Q_m = 0.774\ \text{kg/s}}$$

4-7 LIQUID POOL EVAPORATION OR BOILING

The case for evaporation of volatile from a pool of liquid has already been considered in Chapter 3. The total mass flow rate from the evaporating pool is given by

$$\boxed{Q_m = \frac{MKAP^{sat}}{R_g T_L}} \qquad (3\text{-}12)$$

where

Q_m is the mass vaporization rate (mass/time),
M is the molecular weight of the pure material,
K is the mass transfer coefficient (length/time),
A is the area of exposure,
P^{sat} is the saturation vapor pressure of the liquid,
R_g is the ideal gas constant, and
T_L is the temperature of the liquid.

For liquids boiling from a pool, the boiling rate is limited by the heat transfer from the surroundings to the liquid in the pool. Heat is transferred (1) from the ground by conduction, (2) from the air by conduction and convection, and, (3) by radiation from the sun and/or adjacent sources such as a fire.

SUGGESTED READING

Flow of Liquid through Holes

FRANK P. LEES, *Loss Prevention in the Process Industries* (London: Butterworths, 1986), p. 417.

ALAN S. FOUST, LEONARD A. WENZEL, CURTIS W. CLUMP, LOUIS MAUS, and L. BRYCE ANDERSON, *Principles of Unit Operations* (New York: John Wiley and Sons, 1980), p. 560.

Flow of Liquid through Pipes

OCTAVE LEVENSPIEL, *Engineering Flow and Heat Exchange* (New York: Plenum Press, 1984), Chapter 2.

WARREN L. MCCABE, JULIAN C. SMITH, and PETER HARRIOTT, *Unit Operations of Chemical Engineering* (New York: McGraw-Hill Book Company, 1985), Chapter 5.

Flow of Vapor through Holes

LEES, *Loss Prevention,* p. 417-420.

LEVENSPIEL, *Engineering Flow,* pp. 48-51.

Flow of Vapor through Pipes

LEVENSPIEL, *Engineering Flow,* Chapter 3.

Flashing Liquids

LEES, *Loss Prevention,* p. 426.

STEVEN R. HANNA and PETER J. DRIVAS, *Guidelines for Use of Vapor Dispersion Models* (New York: American Institute of Chemical Engineers, 1987), pp. 24-32.

Liquid Pool Evaporation and Boiling

HANNA and DRIVAS, *Guidelines,* pp. 32-35.

PROBLEMS

4-1. A 0.20 inch hole develops in a pipeline containing toluene. The pressure in the pipeline at the point of the leak is 100 psig. Determine the leakage rate. The specific gravity of toluene is 0.866.

4-2. A 100-foot long horizontal pipeline transporting benzene develops a leak 43 feet from the high pressure end. The diameter of the leak is estimated to be 0.1 inch. At the time the upstream pressure in the pipeline is 50-psig and the downstream pressure is 40 psig. Estimate the mass flow rate of benzene through the leak. The specific gravity of benzene is 0.8794.

4-3. The TLV-TWA for hydrogen sulfide gas is 10 ppm. Hydrogen sulfide gas is stored in a tank at 100 psig and 80°F. Estimate the diameter of a hole in the tank leading to a local hydrogen sulfide concentration equal to the TLV. The local ventilation rate is 2000 ft^3/min and is deemed average. The ambient pressure is 1 atm.

4-4. A tank contains pressurized gas. Develop an equation describing the gas pressure as a function of time if the tank develops a leak. Assume choked flow and a constant tank gas temperature of T_o.

4-5. Show that for incompressible flow in a horizontal pipe of constant diameter and without fittings or valves that the pressure is a linear function of pipe length. What other assumptions are required for this result? Is this result valid for nonhorizontal pipes? How will the presence of fittings, valves, and other hardware affect this result?

4-6. Show that for incompressible flow in a horizontal pipe of constant diameter the pressure, P, at any point is given by

$$P = P_1 + \frac{\left(\sum L_{\text{equiv}}\right)_{\text{at point in question}}}{\left(\sum L_{\text{equiv}}\right)_{\text{total across pipe}}} \Delta P$$

where P_1 is the upstream pressure in the pipe, and ΔP is the total pressure drop across the pipe.

4-7. Water is pumped through a 1 inch schedule 40 pipe (ID = 1.049 in) at 400 gallons per hour. If the pressure at one point in the pipe is 103 psig, and a small leak develops 22 feet downstream, compute the fluid pressure at the leak. The pipe section is horizontal and without fittings or valves. For water at these conditions the viscosity is 1.0 CP and the density is 62.4 lb_m/ft^3.

4-8. If a globe valve is added to the pipe section of Problem 4-7, compute the pressure assuming the valve is wide open.

4-9. A 31.5% hydrochloric acid solution is pumped from one storage tank to another. The power input to the pump is 2 kw and is 50% efficient. The pipe is plastic PVC pipe with an ID of 50 mm. At a certain time the liquid level in the first tank is 4.1 m above the pipe outlet. Due to an accident, the pipe is severed between the pump and the second tank, at a point 2.1 m below the pipe outlet of the first tank. This point is 27 m in equivalent pipe length from the first tank. Compute the flow rate, in kg/s, from the leak. The viscosity of the solution is 1.8×10^{-3} kg/m s and the density is 1600 kg/m^3.

4-10. The morning inspection of the tank farm finds a leak in the turpentine tank. The leak is repaired. An investigation finds that the leak was 0.1 inch in diameter and 7 feet above the tank bottom. Records show that the turpentine level in the tank was 17.3 feet before the leak occurred and 13.0 feet after the leak was repaired. The tank diameter is 15 feet. Determine (a) the total amount of turpentine spilled, (b) the maximum spill rate, and (c) the total time the leak was active. The density of turpentine at these conditions is 55 lb/ft^3.

4-11. Compute the pressure in the pipe at the location shown on Figure P4-11. The flow rate through the pipe is 10,000 liters/hour. The pipe is 50 mm ID commercial steel pipe. The liquid in the pipe is crude oil with a density of 928 kg/m^3 and a viscosity of 0.004 kg/m s. The tank is vented to the atmosphere.

4-12. A tank with a drain pipe is shown in Figure P4-12. The tank contains crude oil and there is concern that the drain pipe might shear off below the tank, allowing the tank contents to leak out. (a) If the drain pipe shears 2 meters below the tank, and the oil level is 7 meters at the time, estimate the initial mass flow rate of material out of the drain pipe. (b) If the pipe shears off at the tank bottom, leaving a 50 mm hole, estimate the initial mass flow rate. The crude oil has a density of 928 kg/m^3 and a viscosity of 0.004 kg/m s.

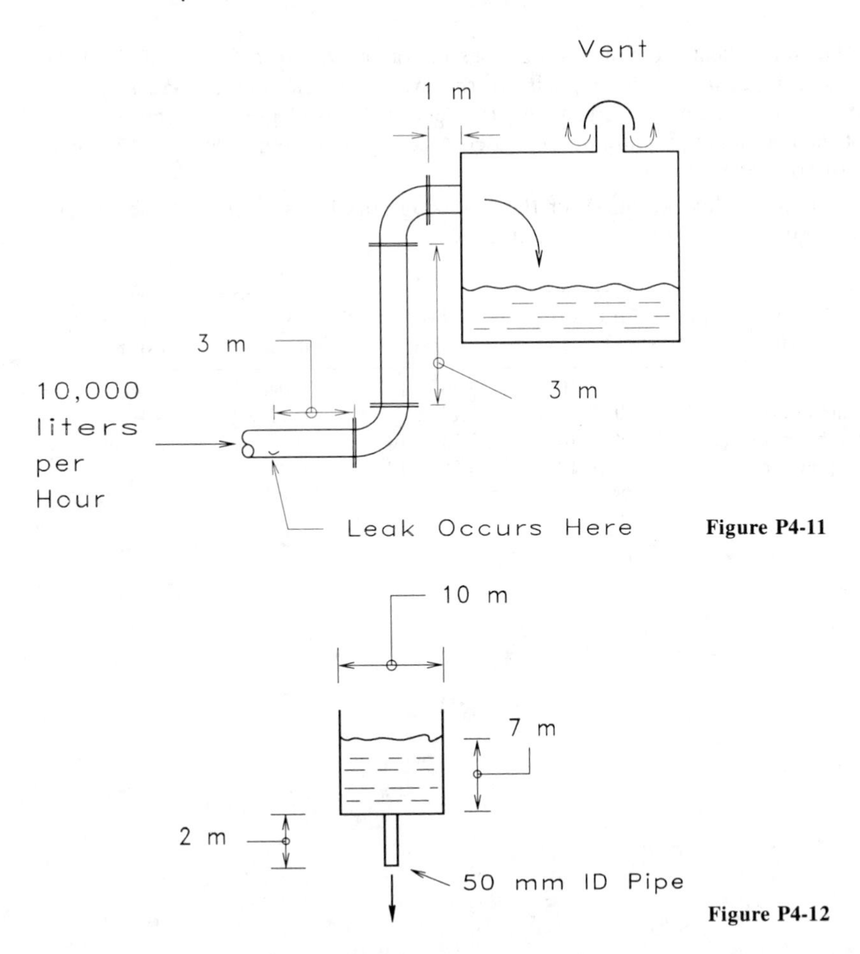

Figure P4-11

Figure P4-12

4-13. A cylinder in the laboratory contains nitrogen at 2200 psia. If the cylinder falls and the valve is sheared off, estimate the initial mass flow rate of nitrogen from the tank. Assume a hole diameter of 1/2 inch. What is the force created by the jet of nitrogen?

4-14. A laboratory apparatus uses nitrogen at 250 psig. The nitrogen is supplied from a cylinder, through a regulator, and to the apparatus via 15 feet of 1/4-inch ID drawn copper tubing. If the tubing separates from the apparatus, estimate the flow of nitrogen from the tubing. The nitrogen in the tank is at 75°F.

4-15. Steam is supplied to the heating coils of a reactor vessel at 125 psig, saturated. The coils are 1/2-inch Schedule 80 pipe (ID = 0.546 inch). The steam is supplied from a main header through similiar pipe with an equivalent length of 53 feet. The heating coils consist of 20 feet of the pipe wound in a coil within the reactor.

If the heating coil pipe shears accidently, the reactor vessel will be exposed to the full 125 psig pressure of the steam, exceeding the vessel's pressure rating. As a result, the reactor must be equipped with a relief system to discharge the steam in the event of a coil shear. Compute the maximum mass flow rate of steam from the sheared coils using two approaches:

a. Assuming the leak in the coil is represented by an orifice.

b. Assuming adiabatic flow through the pipe.

4-16. A home hot water heater contains 40 gallons of water. Due to a failure of the heat control, heat is continuously applied to the water in the tank, increasing the temperature and pressure. Unfortunately, the relief valve is clogged and the pressure rises past the maximum pressure of the vessel. At 250 psig the tank ruptures. Estimate the quantity of water flashed.

4-17. Calculate the mass flux (kg/m^2 s) for the following tank leaks given that the storage pressure is equal to the vapor pressure at 25°C.

Toxic material	Pressure Pa	Heat of vap. J/kg	v_{fg} m^3/kg	Heat cap. J/kg K
a. Propane	0.95×10^6	3.33×10^5	0.048	2.23×10^3
b. Ammonia	1×10^6	1.17×10^6	0.127	4.49×10^3
c. Methyl chloride	0.56×10^6	3.75×10^5	0.077	1.5×10^3
d. Sulphur dioxide	0.39×10^6	3.56×10^5	0.09	1.36×10^3

Toxic Release and Dispersion Models

5

During an accident, process equipment can release toxic materials very quickly and in significant enough quantities to spread in dangerous clouds throughout a plant site and the local community. A few examples are

- Explosive rupture of a process vessel due to excessive pressure caused by a runaway reaction.
- Rupture of a pipeline containing toxic materials at high pressure.
- Rupture of a tank containing toxic material stored above its atmospheric boiling point.
- Rupture of a train or truck transportation tank following an accident.

Serious accidents (such as Bhopal) emphasize the importance of emergency planning and for designing plants to minimize the occurrence and consequences of a toxic release. Toxic release models are routinely used to estimate the effects of a release on the plant and community environments.

An excellent safety program strives to identify problems before they occur. Chemical engineers must understand all aspects of toxic release to prevent the existence of release situations and to reduce the impact of a release if one occurs. This requires a toxic release model.

There are three steps in utilizing a toxic release model.

1. Identify the design basis. What process situations can lead to a release, and which situation is the worst?
2. Develop a source model to describe how materials are released and the rate of release.

3. Use a dispersion model to describe how materials spread throughout the adjacent areas.

The main emphasis of the toxic release model is to provide a tool useful for release mitigation. The source and dispersion models predict the area affected and the concentration of vapor throughout. The design basis is valuable for eliminating situations that could result in a release.

Various options are available based on the predictions of the toxic release model. To name a few, these are (1) develop an emergency response plan with the surrounding community, (2) develop engineering modifications to eliminate the source of the release, (3) enclose the potential release and add appropriate vent scrubbers or other vapor removal equipment, (4) reduce inventories of hazardous materials to reduce the quantity released, and (5) add area monitors to detect incipient leaks and provide block valves and engineering controls to eliminate hazardous levels of spills and leaks. These options are discussed in more detail in Section 5-7 on release mitigation.

5-1 DESIGN BASIS

The design basis describes the various scenarios leading to toxic release; it looks for what can go wrong. For any reasonably complex chemical facility, thousands of release scenarios are possible; it is not practicable to elucidate every scenario. Most toxic release studies strive to determine the largest practicable release and the largest potential release. The largest practicable release considers releases having a reasonable chance for occurrence. This includes pipe ruptures, holes in storage tanks and process vessels, ground spills, and so forth. The largest potential release is a catastrophic situation resulting in release of the largest quantity of material. This includes complete spillage of tank contents, rupture of large bore piping, explosive rupture of reactors, and so forth. Table 5-1 contains examples of largest practicable and largest potential releases.

Development of a proper design basis requires skill, experience, and considerable knowledge of the process. Hazards identification procedures (discussed in Chapter 10) are very helpful.

The completed design basis describes (1) what went wrong, (2) the state of the toxic material released (solid, liquid, or vapor), and (3) the mechanism of release (ruptured pipe, hole in storage vessel, and so on).

Example 5-1

Water is treated at a swimming pool using a 100-lb bottle of chlorine. The chlorine is fed from the bottle through a 1/4-in line to the water treatment facility. A relief valve on the tank prevents excessive pressure from rupturing the tank. Chlorine is stored in the bottle as a liquid under pressure and will boil when the pressure is reduced. Identify the release scenarios.

Solution Scenario 1: The bottle of chlorine ruptures, possibly from dropping the tank while unloading from a truck. The entire contents is spilled, with a fraction flashing immediately into vapor and the remaining liquid forming a boiling pool on the ground.

Scenario 2: A hole forms in the tank either because of mechanical rupture or corrosion. A jet of flashing chlorine and a boiling pool of liquid chlorine forms.
Scenario 3: The relief valve fails open, forming a jet and pool of boiling chlorine.
Scenario 4: The feed line to the treatment plant fails with a jet and pool of boiling chlorine forming.
Scenario 5: A fire develops around the chlorine tank, heating the tank until the relief valve opens.
Scenario 6: A fire develops around the chlorine tank, but the relief valve fails closed. The tank pressure builds until it ruptures, spilling the entire tank contents explosively.
The largest practicable release could be either scenarios 2, 3 or 4, depending on the rate of material release computed using an appropriate source model. The largest potential release is scenario 6, releasing the entire tank contents almost immediately.

5-2 SOURCE MODELS

Source models were covered in detail in Chapter 4. The purpose of the source model is to determine (1) the form of material released, solid, liquid or vapor, (2) the total quantity of material released and, and (3) the rate at which it is released. This information is required for any quantitative dispersion model study.

5-3 DISPERSION MODELS

Dispersion models describe the airborne transport of toxic materials away from the accident site and into the plant and community. After a release, the airborne toxic is carried away by the wind in a characteristic plume as shown in Figure 5-1 or a puff, shown in Figure 5-2. The maximum concentration of toxic material occurs at the release point (which may not be at ground level). Concentrations downwind are less, due to turbulent mixing and dispersion of the toxic substance with air.

TABLE 5-1 EXAMPLES OF LARGEST PRACTICABLE AND LARGEST POTENTIAL RELEASES.*

Largest practicable release: large release with a reasonable chance to occur.
Rupture of small bore piping, 1-inch maximum. Partial flange gasket blowout of large diameter piping (for example, 50% blowout of a 2-inch line resulting in an equivalent hole diameter of 1-inch). Failure of a 3/4-inch fusible plug on a 1-ton cylinder. Generally limited release duration (15 minutes typical based on time required for operator intervention to stop the leak).
Largest potential release: catastrophic release of maximum amount of material.
Rupture of a 2 or 3 inch liquid line. Tank truck rupture on a highway (3 or 4 inch hole size assumed, typical). Typically, entire source vessel inventory spilled.

*Selected from Steven R. Hanna and Peter J. Drivas, *Guidelines for the Use of Vapor Cloud Dispersion Models* (New York: American Institute of Chemical Engineers, 1987), p. 145.

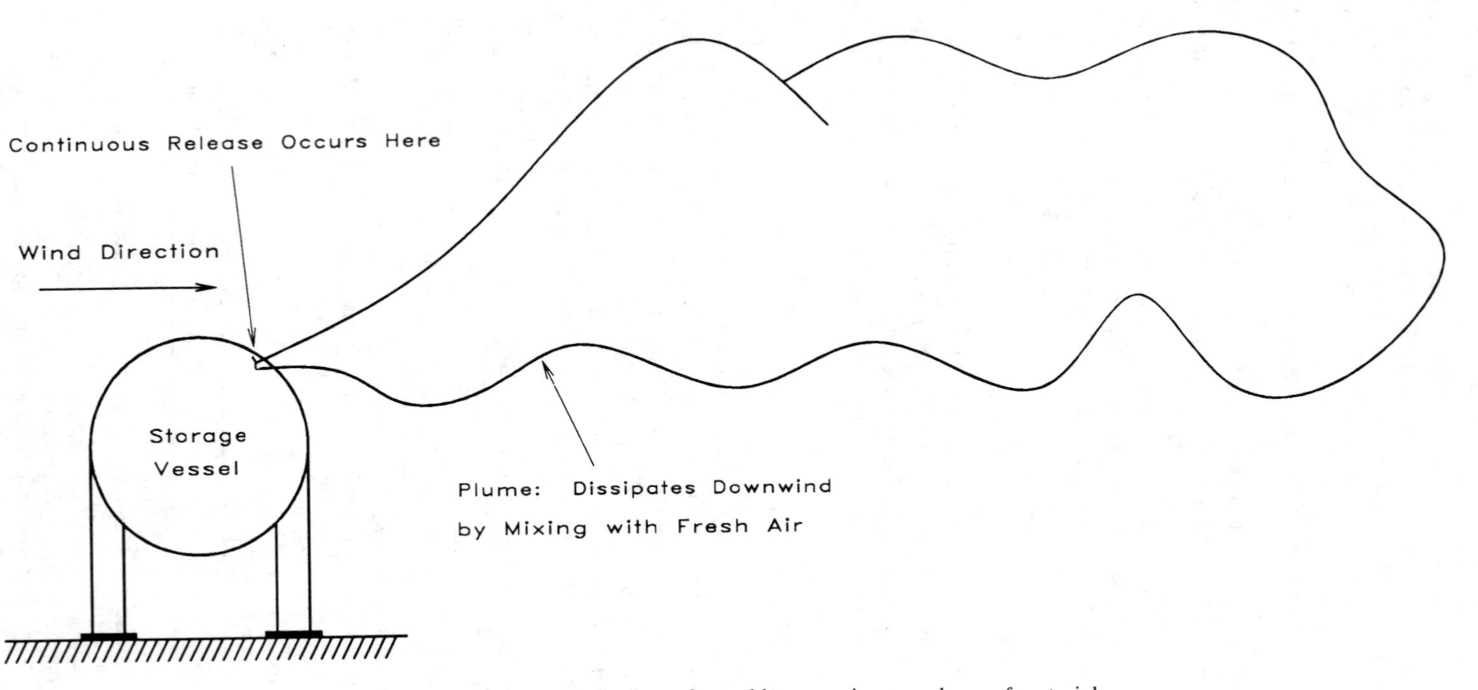

Figure 5-1 Characteristic plume formed by a continuous release of material.

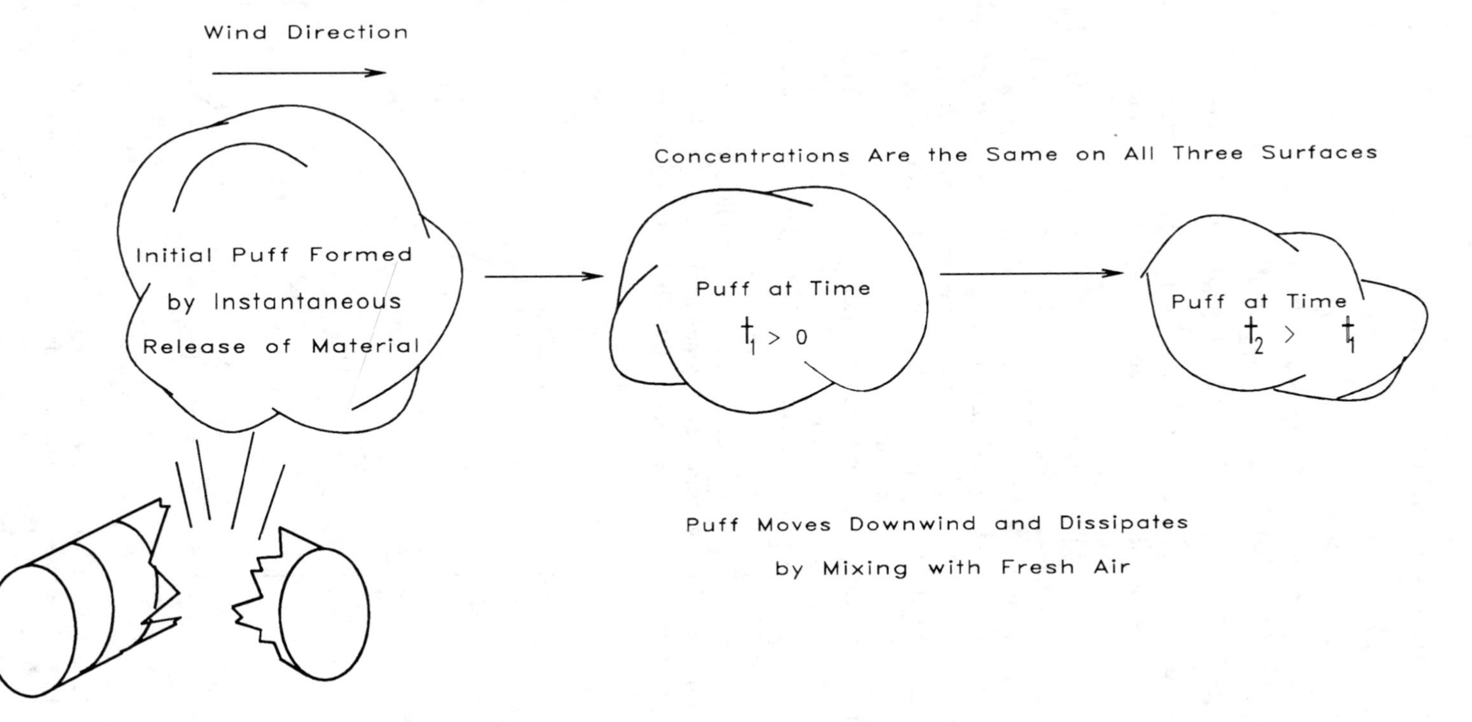

Figure 5-2 Puff formed by near instantaneous release of material.

A wide variety of parameters affect atmospheric dispersion of toxic materials.

- Wind speed
- Atmospheric stability
- Ground conditions, buildings, water, trees
- Height of the release above ground level
- Momentum and buoyancy of the initial material released

As the wind speed increases, the plume in Figure 5-1 becomes longer and narrower; the substance is carried downwind faster but is diluted faster by a larger quantity of air.

Atmospheric stability relates to vertical mixing of the air. During the day the air temperature decreases rapidly with height, encouraging vertical motions. At night the temperature decrease is less, resulting in less vertical motion. Temperature profiles for day and night situations are shown in Figure 5-3. Sometimes an inversion will occur. During an inversion, the temperature increases with height, resulting in minimal vertical motion. This most often occurs at night as the ground cools rapidly due to thermal radiation.

Ground conditions affect the mechanical mixing at the surface and the wind profile with height. Trees and buildings increase mixing while lakes and open areas decrease it. Figure 5-4 shows the change in wind speed versus height for a variety of surface conditions.

The release height significantly affects ground level concentrations. As the release height increases, ground level concentrations are reduced since the plume must disperse a greater distance vertically. This is shown in Figure 5-5.

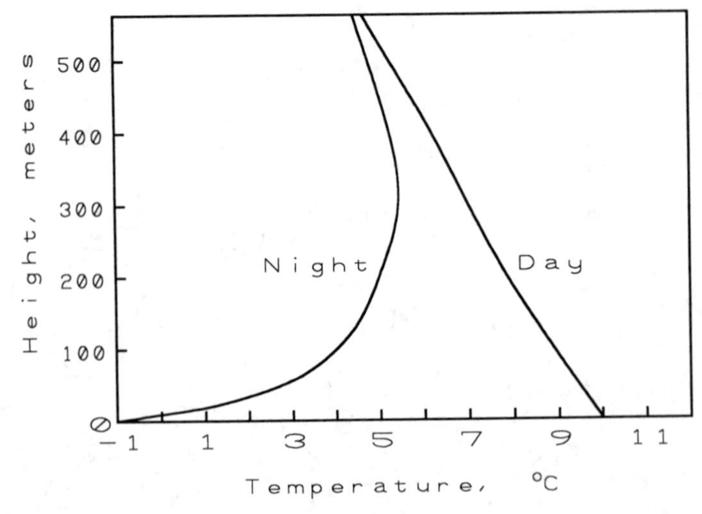

Figure 5-3 Air temperature as a function of altitude for day and night conditions. The temperature gradient affects the vertical air motion. (Adapted from D. Bruce Turner, *Workbook of Atmospheric Dispersion Estimates,* U.S. Department of Health, Education and Welfare, Cincinnati, OH, 1970, p. 1.)

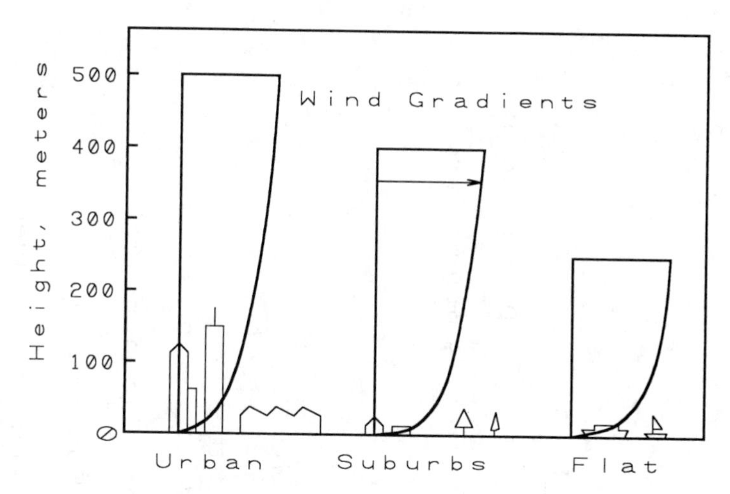

Figure 5-4 Effect of ground conditions on vertical wind gradient. (Adapted from D. Bruce Turner, *Workbook of Atmospheric Dispersion Estimates,* U.S. Department of Health, Education and Welfare, Cincinnati, OH, 1970, p. 2.)

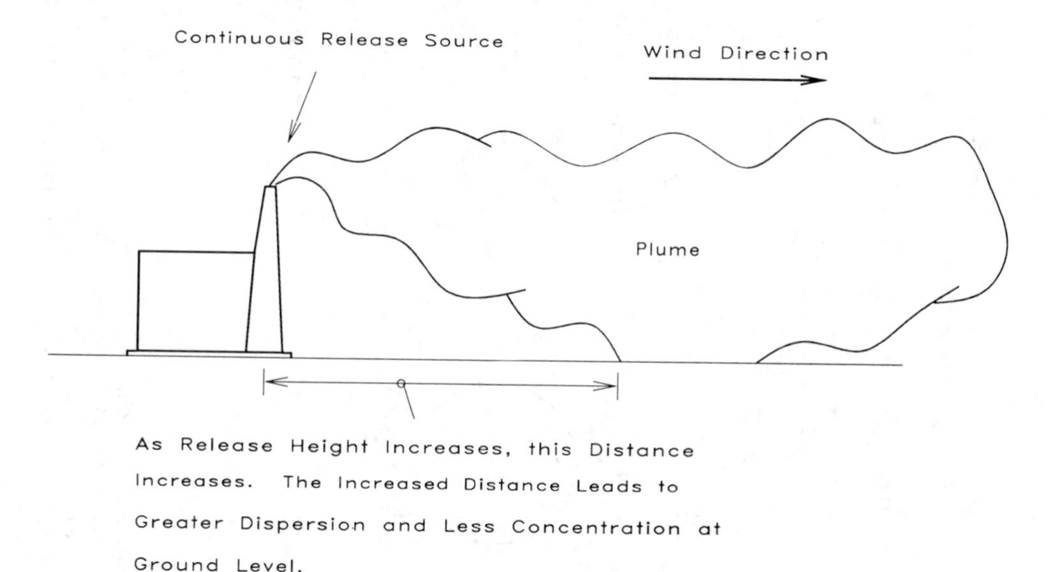

Figure 5-5 Increased release height decreases the ground concentration.

The buoyancy and momentum of the material released changes the "effective" height of the release. Figure 5-6 demonstrates these effects. After the initial momentum and buoyancy has dissipated, ambient turbulent mixing becomes the dominant effect.

Two types of vapor cloud dispersion models are commonly used: the plume and puff models. The plume model describes the steady-state concentration of material released from a continuous source. The puff model describes the temporal

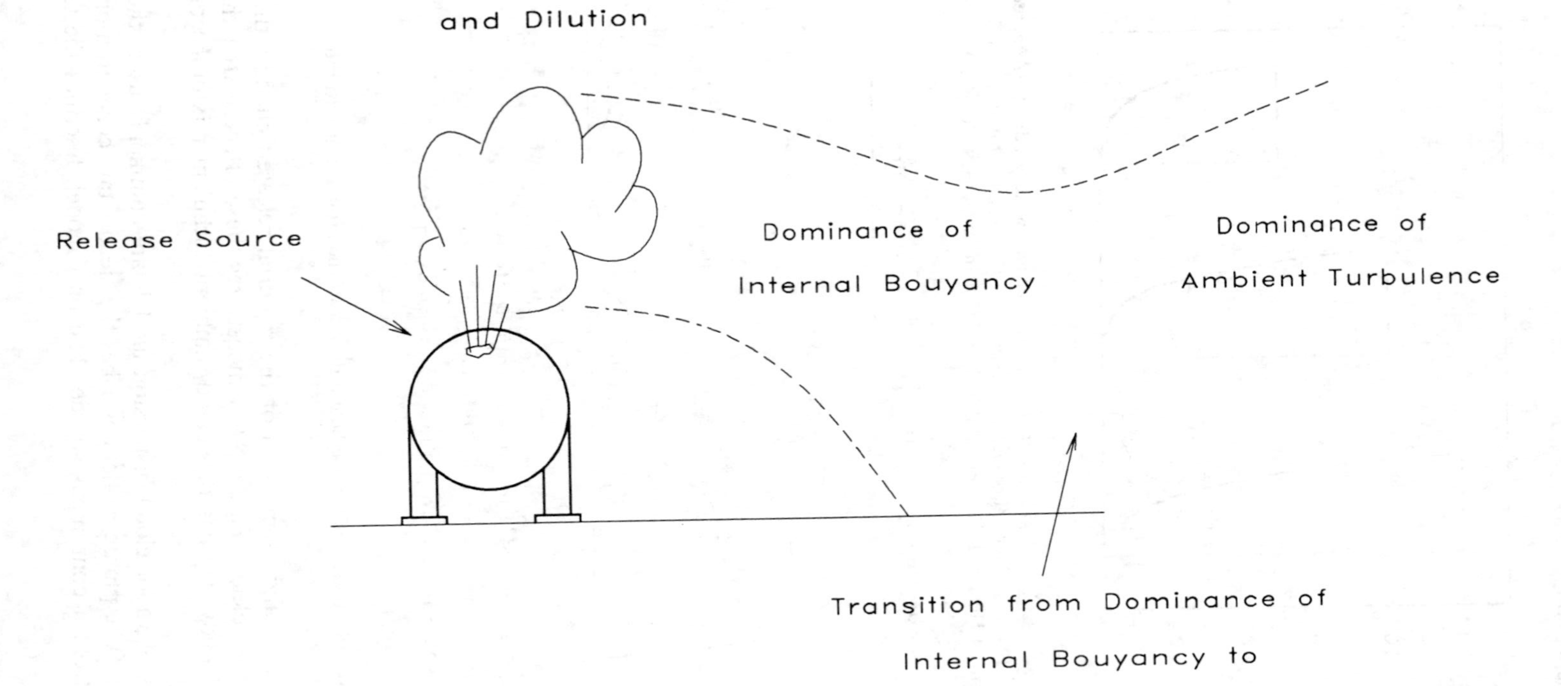

Figure 5-6 The initial acceleration and buoyancy of the released material affects the plume character. The dispersion models discussed in this chapter represent only ambient turbulence. (Adapted from Steven R. Hanna and Peter J. Drivas, *Guidelines for Use of Vapor Cloud Dispersion Models,* The American Institute of Chemical Engineers, New York, 1987, p. 6.)

concentration of material from a single release of a fixed amount of material. The distinction between the two models is shown graphically in Figures 5-1 and 5-2. For the plume model, a typical example is the continuous release of gases from a smokestack. A steady-state plume is formed downwind from the smokestack. For the puff model, a typical example is the sudden release of a fixed amount of material due to the rupture of a storage vessel. A large vapor cloud is formed that moves away from the rupture point.

The puff model can be used to describe a plume; a plume is simply the release of continuous puffs. If, however, steady-state plume information is all that is required, the plume model is recommended since it is easier to use. For studies involving dynamic plumes (for instance the effect on a plume due to a change in wind direction), the puff model must be used.

Consider the instantaneous release of a fixed mass of material, Q_m^*, into an infinite expanse of air (a ground surface will be added later). The coordinate system is fixed at the source. Assuming no reaction or molecular diffusion, the concentration, C, of material due to this release is given by the advection equation

$$\frac{\partial C}{\partial t} + \frac{\partial}{\partial x_j}(u_j C) = 0 \tag{5-1}$$

where u_j is the velocity of the air and the subscript j represents the summation over all coordinate directions, x, y and z. If the velocity, u_j, in Equation 5-1 is set equal to the average wind velocity and the equation is solved, one would find that the material disperses much faster than predicted. This is due to turbulence in the velocity field. If one were able to specify the wind velocity exactly with time and position, including the effects due to turbulence, Equation 5-1 would predict the correct concentration. Unfortunately, no models are currently available to adequately describe turbulence. As a result, an approximation is used. Let the velocity be represented by an average (or mean) and stochastic quantity,

$$u_j = \langle u_j \rangle + u_j' \tag{5-2}$$

where $\langle u_j \rangle$ is the average velocity and u_j' is the stochastic fluctuation due to turbulence. It follows that the concentration, C, will also fluctuate as a result of the velocity field, so,

$$C = \langle C \rangle + C' \tag{5-3}$$

where $\langle C \rangle$ is the mean concentration and C' is the stochastic fluctuation. Since the fluctuations in both C and u_j are around the average or mean values, it follows that,

$$\begin{aligned} \langle u_j' \rangle &= 0 \\ \langle C' \rangle &= 0 \end{aligned} \tag{5-4}$$

Substituting Equations 5-2 and 5-3 into Equation 5-1 and averaging the result over time, yields

$$\frac{\partial \langle C \rangle}{\partial t} + \frac{\partial}{\partial x_j}(\langle u_j \rangle \langle C \rangle) + \frac{\partial}{\partial x_j}\langle u_j' C' \rangle = 0 \tag{5-5}$$

The terms $\langle u_j \rangle C'$ and $u_j' \langle C \rangle$ are zero when averaged ($\langle \langle u_j \rangle C' \rangle = \langle u_j \rangle \langle C' \rangle = 0$), but the turbulent flux term $\langle u_j' C' \rangle$ is not necessarily zero and remains in the equation.

An additional equation is required to describe the turbulent flux. The usual approach is to define an eddy diffusivity, K_j (with units of area/time), such that

$$\langle u_j' C' \rangle = -K_j \frac{\partial \langle C \rangle}{\partial x_j} \tag{5-6}$$

Substituting Equation 5-6 into Equation 5-5 yields,

$$\frac{\partial \langle C \rangle}{\partial t} + \frac{\partial}{\partial x_j}(\langle u_j \rangle \langle C \rangle) = \frac{\partial}{\partial x_j}\left(K_j \frac{\partial \langle C \rangle}{\partial x_j}\right) \tag{5-7}$$

If the atmosphere is assumed to be incompressible,

$$\frac{\partial \langle u_j \rangle}{\partial x_j} = 0 \tag{5-8}$$

and Equation 5-7 becomes

$$\boxed{\frac{\partial \langle C \rangle}{\partial t} + \langle u_j \rangle \frac{\partial \langle C \rangle}{\partial x_j} = \frac{\partial}{\partial x_j}\left(K_j \frac{\partial \langle C \rangle}{\partial x_j}\right)} \tag{5-9}$$

Equation 5-9, together with appropriate boundary and initial conditions, forms the fundamental basis for dispersion modeling. This equation will be solved for a variety of cases.

The coordinate system used for the dispersion models is shown on Figures 5-7 and 5-8. The x-axis is the centerline directly downwind from the release point and is rotated for different wind directions. The y-axis is the distance off of the centerline and the z-axis is the elevation above the release point. The point $(x, y, z) = (0, 0, 0)$ is at the release point. The coordinates $(x, y, 0)$ are level with the release point, and the coordinates $(x, 0, 0)$ are along the centerline, or x-axis.

Case 1: Steady-state, Continuous Point Release with No Wind

The applicable conditions are

- Constant mass release rate, Q_m = constant,
- No wind, $\langle u_j \rangle = 0$,
- Steady state, $\dfrac{\partial \langle C \rangle}{\partial t} = 0$, and
- Constant eddy diffusivity, $K_j = K^*$ in all directions.

For this case, Equation 5-9 reduces to the form,

$$\frac{\partial^2 \langle C \rangle}{\partial x^2} + \frac{\partial^2 \langle C \rangle}{\partial y^2} + \frac{\partial^2 \langle C \rangle}{\partial z^2} = 0 \tag{5-10}$$

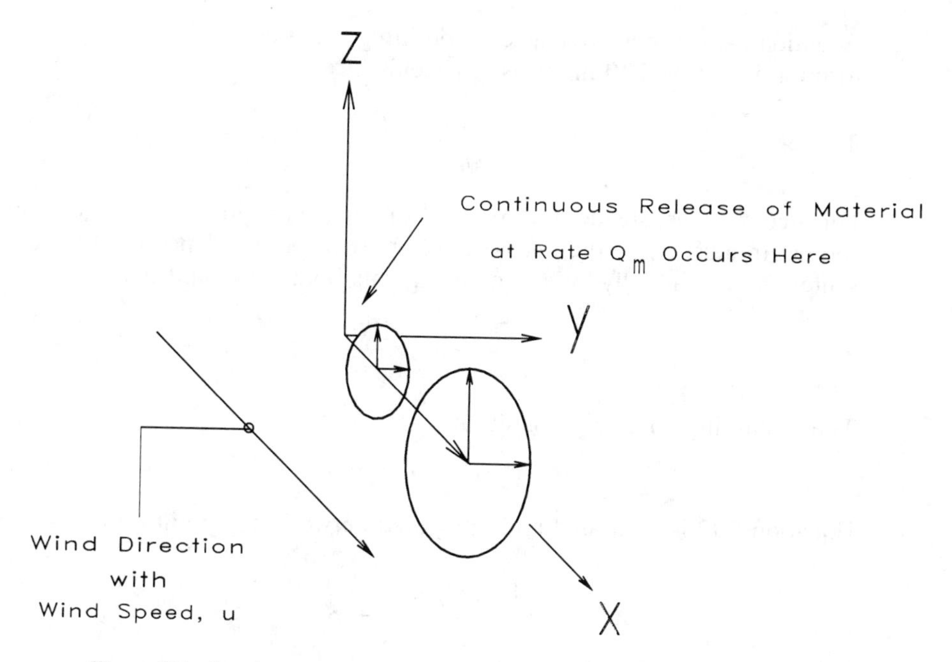

Figure 5-7 Steady-state, continuous point source release with wind. Note coordinate system: x is downwind direction, y is off-wind direction, and z is vertical direction.

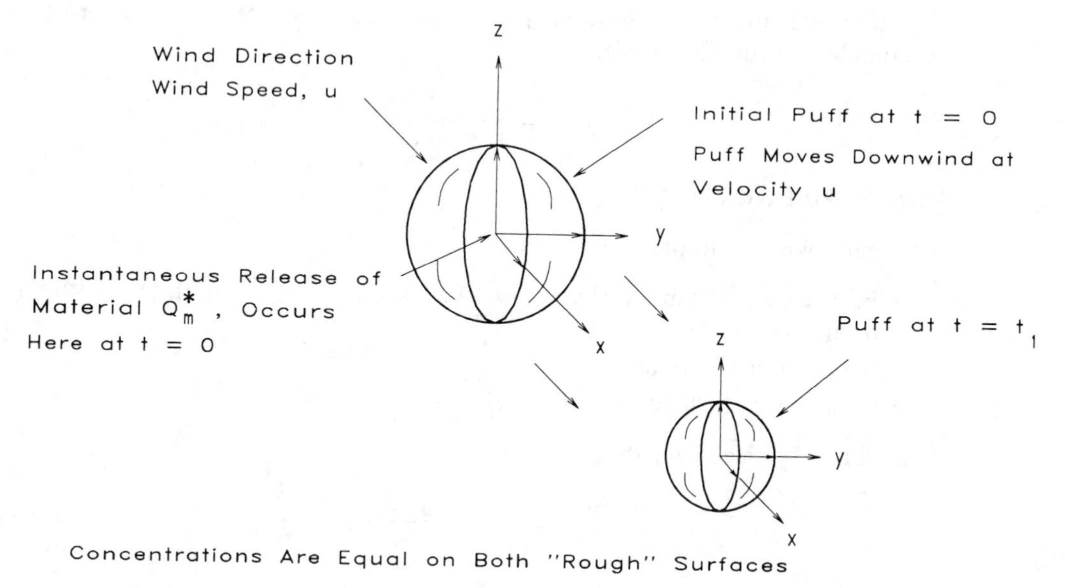

Figure 5-8 Puff with wind. After the initial, instantaneous release, the puff moves with the wind.

Equation 5-10 is more tractable by defining a radius as $r^2 = x^2 + y^2 + z^2$. Transforming Equation 5-10 in terms of r yields

$$\frac{d}{dr}\left(r^2 \frac{d\langle C\rangle}{dr}\right) = 0 \tag{5-11}$$

For a continuous, steady state release, the concentration flux at any point, r, from the origin must equal the release rate, Q_m (with units of mass/time). This is represented mathematically by the following flux boundary condition.

$$-4\pi r^2 K^* \frac{d\langle C\rangle}{dr} = Q_m \tag{5-12}$$

The remaining boundary condition is

$$\text{As } r \to \infty, \langle C\rangle \to 0 \tag{5-13}$$

Equation 5-12 is separated and integrated between any point r and $r = \infty$.

$$\int_{\langle C\rangle}^{0} d\langle C\rangle = -\frac{Q_m}{4\pi K^*} \int_r^{\infty} \frac{dr}{r^2} \tag{5-14}$$

Solving Equation 5-14 for $\langle C\rangle$ yields,

$$\langle C\rangle(r) = \frac{Q_m}{4\pi K^* r} \tag{5-15}$$

It is easy to verify by substitution that Equation 5-15 is also a solution to Equation 5-11 and thus a solution to this case. Equation 5-15 is transformed to rectangular coordinates to yield,

$$\langle C\rangle(x, y, z) = \frac{Q_m}{4\pi K^* \sqrt{x^2 + y^2 + z^2}} \tag{5-16}$$

Case 2: Puff with No Wind

The applicable conditions are

- Puff release, instantaneous release of a fixed mass of material, Q_m^* (with units of mass),
- No wind, $\langle u_j\rangle = 0$, and
- Constant eddy diffusivity, $K_j = K^*$, in all directions.

Equation 5-9 reduces, for this case, to

$$\frac{1}{K^*}\frac{\partial\langle C\rangle}{\partial t} = \frac{\partial^2\langle C\rangle}{\partial x^2} + \frac{\partial^2\langle C\rangle}{\partial y^2} + \frac{\partial^2\langle C\rangle}{\partial z^2} \tag{5-17}$$

The initial condition required to solve Equation 5-17 is

$$\langle C\rangle(x, y, z, t) = 0 \text{ at } t = 0 \tag{5-18}$$

The solution to Equation 5-17 in spherical coordinates[1] is

$$\langle C \rangle (r,t) = \frac{Q_m^*}{8(\pi K^* t)^{3/2}} \exp\left(-\frac{r^2}{4K^* t}\right) \tag{5-19}$$

and in rectangular coordinates is

$$\langle C \rangle (x,y,z,t) = \frac{Q_m^*}{8(\pi K^* t)^{3/2}} \exp\left[-\frac{(x^2 + y^2 + z^2)}{4K^* t}\right] \tag{5-20}$$

Case 3: Non Steady-state, Continuous Point Release with No Wind

The applicable conditions are

- Constant mass release rate, Q_m = constant,
- No wind, $\langle u_j \rangle = 0$, and
- Constant eddy diffusivity, $K_j = K^*$ in all directions.

For this case, Equation 5-9 reduces to Equation 5-17 with initial condition, Equation 5-18, and boundary condition, Equation 5-13. The solution is found by integrating the instantaneous solution, Equation 5-19 or 5-20 with respect to time. The result in spherical coordinates[2] is

$$\langle C \rangle (r,t) = \frac{Q_m}{4\pi K^* r} \operatorname{erfc}\left(\frac{r}{2\sqrt{K^* t}}\right) \tag{5-21}$$

and in rectangular coordinates is

$$\langle C \rangle (x,y,z,t) = \frac{Q_m}{4\pi K^* \sqrt{x^2 + y^2 + z^2}} \operatorname{erfc}\left(\frac{\sqrt{x^2 + y^2 + z^2}}{2\sqrt{K^* t}}\right) \tag{5-22}$$

As $t \to \infty$, Equations 5-21 and 5-22 reduce to the corresponding steady state solutions, Equations 5-15 and 5-16.

Case 4: Steady-state, Continuous Point Source Release with Wind

This case is shown in Figure 5-7. The applicable conditions are

- Continuous release, Q_m = constant,
- Wind blowing in x direction only, $\langle u_j \rangle = \langle u_x \rangle = u$ = constant, and
- Constant eddy diffusivity, $K_j = K^*$, in all directions.

For this case, Equation 5-9 reduces to

$$\frac{u}{K^*} \frac{\partial \langle C \rangle}{\partial x} = \frac{\partial^2 \langle C \rangle}{\partial x^2} + \frac{\partial^2 \langle C \rangle}{\partial y^2} + \frac{\partial^2 \langle C \rangle}{\partial z^2} \tag{5-23}$$

[1] H. S. Carslaw and J. C. Jaeger, *Conduction of Heat in Solids* (London: Oxford University Press, 1959), p. 256.

[2] Carslaw and Jaeger, *Conduction of Heat*, p. 261.

Equation 5-23 is solved together with boundary conditions, Equation 5-12 and 5-13. The solution for the average concentration at any point[3] is

$$\langle C\rangle(x,y,z) = \frac{Q_m}{4\pi K^* \sqrt{x^2+y^2+z^2}} \exp\left(-\frac{u}{2K^*}\left[\sqrt{x^2+y^2+z^2} - x\right]\right) \quad (5\text{-}24)$$

If a slender plume is assumed (the plume is long and slender and is not far removed from the x-axis),

$$y^2 + z^2 \ll x^2 \quad (5\text{-}25)$$

and, using $\sqrt{1+a} \approx 1 + a/2$, Equation 5-24 is simplified to

$$\langle C\rangle(x,y,z) = \frac{Q_m}{4\pi K^* x} \exp\left[-\frac{u}{4K^* x}(y^2+z^2)\right] \quad (5\text{-}26)$$

Along the centerline of this plume, $y = z = 0$ and

$$\langle C\rangle(x) = \frac{Q_m}{4\pi K^* x} \quad (5\text{-}27)$$

Case 5: Puff with No Wind. Eddy Diffusivity a Function of Direction

This is the same as Case 2, but with eddy diffusivity a function of direction. The applicable conditions are

- Puff release, $Q_m^* =$ constant,
- No wind, $\langle u_j\rangle = 0$, and
- Each coordinate direction has a different, but constant eddy diffusivity, K_x, K_y and K_z.

Equation 5-9 reduces to the following equation for this case.

$$\frac{\partial\langle C\rangle}{\partial t} = K_x\frac{\partial^2\langle C\rangle}{\partial x^2} + K_y\frac{\partial^2\langle C\rangle}{\partial y^2} + K_z\frac{\partial^2\langle C\rangle}{\partial z^2} \quad (5\text{-}28)$$

The solution is[4]

$$\langle C\rangle(x,y,z,t) = \frac{Q_m^*}{8(\pi t)^{3/2}\sqrt{K_x K_y K_z}} \exp\left[-\frac{1}{4t}\left(\frac{x^2}{K_x} + \frac{y^2}{K_y} + \frac{z^2}{K_z}\right)\right] \quad (5\text{-}29)$$

Case 6: Steady-state, Continuous Point Source Release with Wind. Eddy Diffusivity a Function of Direction

This is the same as Case 4, but with eddy diffusivity a function of direction. The applicable conditions are

[3] Carslaw and Jaeger, *Conduction of Heat,* p. 267.

[4] Frank P. Lees, *Loss Prevention in the Process Industries* (London: Butterworths, 1986), p. 440.

- Continuous release, Q_m = constant,
- Steady-state, $\frac{\partial \langle C \rangle}{\partial t} = 0$,
- Wind blowing in x direction only, $\langle u_j \rangle = \langle u_x \rangle = u$ = constant,
- Each coordinate direction has a different but constant eddy diffusivity, K_x, K_y, and K_z, and
- Slender plume approximation, Equation 5-25.

Equation 5-9 reduces to

$$u\frac{\partial \langle C \rangle}{\partial x} = K_x\frac{\partial^2 \langle C \rangle}{\partial x^2} + K_y\frac{\partial^2 \langle C \rangle}{\partial y^2} + K_z\frac{\partial^2 \langle C \rangle}{\partial z^2} \tag{5-30}$$

The solution is[5]

$$\langle C \rangle (x, y, z) = \frac{Q_m}{4\pi x\sqrt{K_x K_y}} \exp\left[-\frac{u}{4x}\left(\frac{y^2}{K_y} + \frac{z^2}{K_z}\right)\right] \tag{5-31}$$

Along the centerline of this plume, $y = z = 0$ and the average concentration is given by

$$\langle C \rangle (x) = \frac{Q_m}{4\pi x\sqrt{K_y K_z}} \tag{5-32}$$

Case 7: Puff with Wind

This is the same as Case 5, but with wind. Figure 5-8 shows the geometry. The applicable conditions are

- Puff release, Q_m^* = constant,
- Wind blowing in x direction only, $\langle u_j \rangle = \langle u_x \rangle = u$ = constant, and
- Each coordinate direction has a different, but constant eddy diffusivity, K_x, K_y, and K_z.

The solution to this problem is found by a simple transformation of coordinates. The solution to Case 5 represents a puff fixed around the release point. If the puff moves with the wind along the x-axis, the solution to this case is found by replacing the existing coordinate x by a new coordinate system, $x - ut$, that moves with the wind velocity. The variable t is the time since the release of the puff, and u is the wind velocity. The solution is simply Equation 5-29, transformed into this new coordinate system.

$$\langle C \rangle (x, y, z, t) = \frac{Q_m^*}{8(\pi t)^{3/2}\sqrt{K_x K_y K_z}} \exp\left\{-\frac{1}{4t}\left[\frac{(x - ut)^2}{K_x} + \frac{y^2}{K_y} + \frac{z^2}{K_z}\right]\right\} \tag{5-33}$$

[5]Lees, *Loss Prevention*, p. 440.

Case 8: Puff with No Wind with Source on Ground

This is the same as Case 5, but with the source on the ground. The ground represents an impervious boundary. As a result, the concentration is twice the concentration as for Case 5. The solution is 2 times Equation 5-29.

$$\langle C\rangle(x,y,z,t) = \frac{Q_m^*}{4(\pi t)^{3/2}\sqrt{K_x K_y K_z}} \exp\left[-\frac{1}{4t}\left(\frac{x^2}{K_x} + \frac{y^2}{K_y} + \frac{z^2}{K_z}\right)\right] \tag{5-34}$$

Case 9: Steady-state Plume with Source on Ground

This is the same as Case 6, but with the release source on the ground, as shown in Figure 5-9. The ground represents an impervious boundary. As a result, the concentration is twice the concentration of Case 6. The solution is 2 times Equation 5-31.

$$\langle C\rangle(x,y,z) = \frac{Q_m}{2\pi x\sqrt{K_x K_y}} \exp\left[-\frac{u}{4x}\left(\frac{y^2}{K_y} + \frac{z^2}{K_z}\right)\right] \tag{5-35}$$

Case 10: Continuous, Steady-state Source. Source at Height H_r above the Ground.

For this case the ground acts as an impervious boundary at a distance H from the source. The solution is[6]

$$\begin{aligned}\langle C\rangle(x,y,z) &= \frac{Q_m}{4\pi x\sqrt{K_y K_z}} \exp\left(-\frac{uy^2}{4K_y x}\right) \\ &\times \left\{\exp\left[-\frac{u}{4K_z x}(z - H_r)^2\right] + \exp\left[-\frac{u}{4K_z x}(z + H_r)^2\right]\right\}\end{aligned} \tag{5-36}$$

If $H_r = 0$, Equation 5-36 reduces to Equation 5-35 for a source on the ground.

5-4 PASQUILL-GIFFORD MODEL

Cases 1 through 10 above all depend on the specification of a value for the eddy diffusivity, K_j. In general, K_j changes with position, time, wind velocity, and prevailing weather conditions. While the eddy diffusivity approach is useful theoretically, it is not convenient experimentally and does not provide a useful framework for correlation.

Sutton[7] solved this difficulty by proposing the following definition for a *dispersion coefficient.*

$$\sigma_x^2 = \frac{1}{2}\langle C\rangle^2(ut)^{2-n} \tag{5-37}$$

with similiar relations given for σ_y and σ_z. The dispersion coefficients, σ_x, σ_y, and σ_z represent the standard deviations of the concentration in the downwind, crosswind,

[6] Lees, *Loss Prevention,* p. 441.

[7] O. G. Sutton, *Micrometeorology* (New York: McGraw-Hill Book Company, 1953), p. 286.

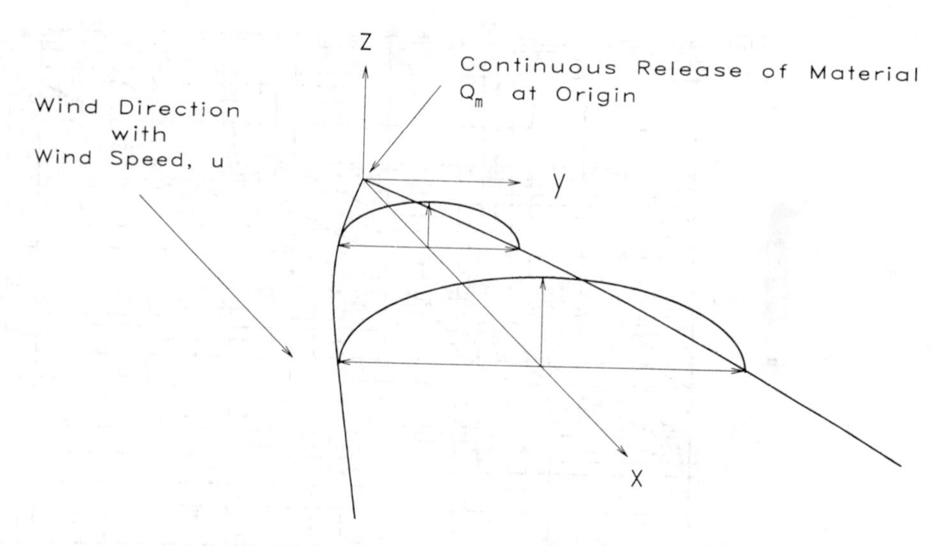

Figure 5-9 Steady-state plume with source at ground level. The concentration is twice the concentration of a plume without the ground.

and vertical (x, y, z) directions, respectively. Values for the dispersion coefficients are much easier to obtain experimentally than eddy diffusivities.

The dispersion coefficients are a function of atmospheric conditions and the distance downwind from the release. The atmospheric conditions are classified according to six different stability classes shown in Table 5-2. The stability classes depend on wind speed and quantity of sunlight. During the day, increased wind speed results in greater atmospheric stability, while at night the reverse is true. This is due to a change in vertical temperature profiles from day to night.

The dispersion coefficients, σ_y and σ_z for a continuous source were developed by Gifford[8] and are given in Figures 5-10 and 5-11, with the corresponding correla-

TABLE 5-2 ATMOSPHERIC STABILITY CLASSES FOR USE WITH THE PASQUILL-GIFFORD DISPERSION MODEL

Wind speed (m/s)	Day radiation intensity			Night cloud cover	
	strong	medium	slight	cloudy	calm & clear
< 2	A	A-B	B		
2-3	A-B	B	C	E	F
3-5	B	B-C	C	D	E
5-6	C	C-D	D	D	D
> 6	C	D	D	D	D

Stability classes for puff model:
A, B: unstable
C, D: neutral
E, F: stable

[8]F. A. Gifford, "Use of Routine Meteorological Observations for Estimating Atmospheric Dispersion," *Nuclear Safety*, Vol. 2, No. 4, (1961), p. 47.

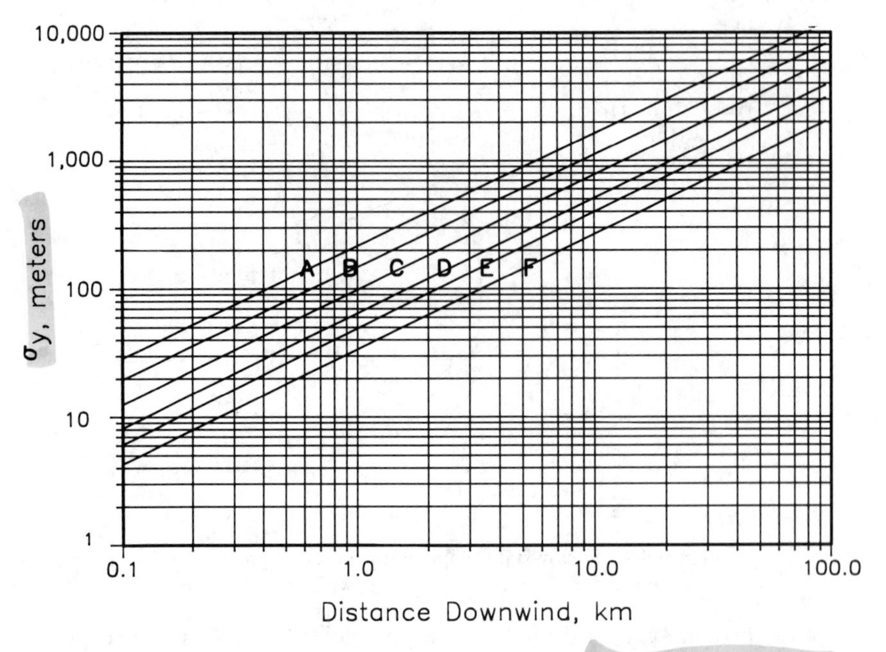

Figure 5-10 Horizontal dispersion coefficient for Pasquill-Gifford plume model. The dispersion coefficient is a function of distance downwind and the atmospheric stability class.

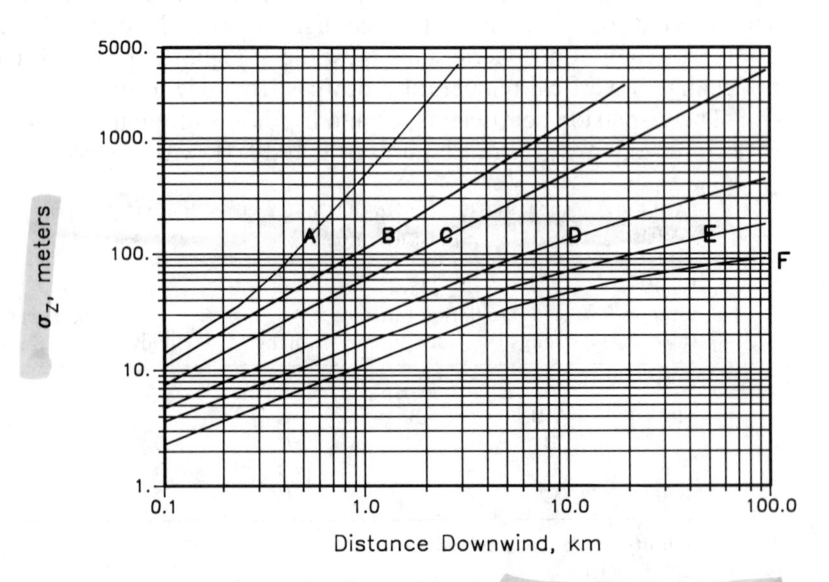

Figure 5-11 Vertical dispersion coefficient for Pasquill-Gifford plume model. The dispersion coefficient is a function of distance downwind and the atmospheric stability class.

TABLE 5-3 EQUATIONS AND DATA FOR PASQUILL-GIFFORD DISPERSION COEFFICIENTS[1]

Equations for continuous plumes

Stability class	σ_y (m)
A	$\sigma_y = 0.493x^{0.88}$
B	$\sigma_y = 0.337x^{0.88}$
C	$\sigma_y = 0.195x^{0.90}$
D	$\sigma_y = 0.128x^{0.90}$
E	$\sigma_y = 0.091x^{0.91}$
F	$\sigma_y = 0.067x^{0.90}$

Stability class	x (m)	σ_z (m)
A	100 - 300	$\sigma_z = 0.087x^{1.10}$
	300 - 3000	$\log_{10}\sigma_z = -1.67 + 0.902 \log_{10}x + 0.181(\log_{10}x)^2$
B	100 - 500	$\sigma_z = 0.135x^{0.95}$
	500 - 2×10^4	$\log_{10}\sigma_z = -1.25 + 1.09 \log_{10}x + 0.0018(\log_{10}x)^2$
C	100 - 10^5	$\sigma_z = 0.112x^{0.91}$
D	100 - 500	$\sigma_z = 0.093x^{0.85}$
	500 - 10^5	$\log_{10}\sigma_z = -1.22 + 1.08 \log_{10}x - 0.061(\log_{10}x)^2$
E	100 - 500	$\sigma_z = 0.082x^{0.82}$
	500 - 10^5	$\log_{10}\sigma_z = -1.19 + 1.04 \log_{10}x - 0.070(\log_{10}x)^2$
F	100 - 500	$\sigma_z = 0.057x^{0.80}$
	500 - 10^5	$\log_{10}\sigma_z = -1.91 + 1.37 \log_{10}x - 0.119(\log_{10}x)^2$

Data for puff releases

Stability condition	$x = 100$ m		$x = 4000$ m	
	σ_y (m)	σ_z (m)	σ_y (m)	σ_z (m)
Unstable	10	15	300	220
Neutral	4	3.8	120	50
Very stable	1.3	0.75	35	7

[1]Frank P. Lees, *Loss Prevention in the Process Industries* (London: Butterworths, 1986), p. 443.

tions given in Table 5-3. Values for σ_x are not provided since it is reasonable to assume $\sigma_x = \sigma_y$. The dispersion coefficients σ_y and σ_z for a puff release are given in Figures 5-12 and 5-13. The puff dispersion coefficients are based on limited data (shown in Table 5-3) and should not be considered precise.

The equations for Cases 1 through 10 were rederived by Pasquill[9] using relations of the form of Equation 5-37. These equations, along with the correlations for the dispersion coefficients are known as the *Pasquill-Gifford model*.

[9]F. Pasquill, *Atmospheric Diffusion* (London: Van Nostrand, 1962).

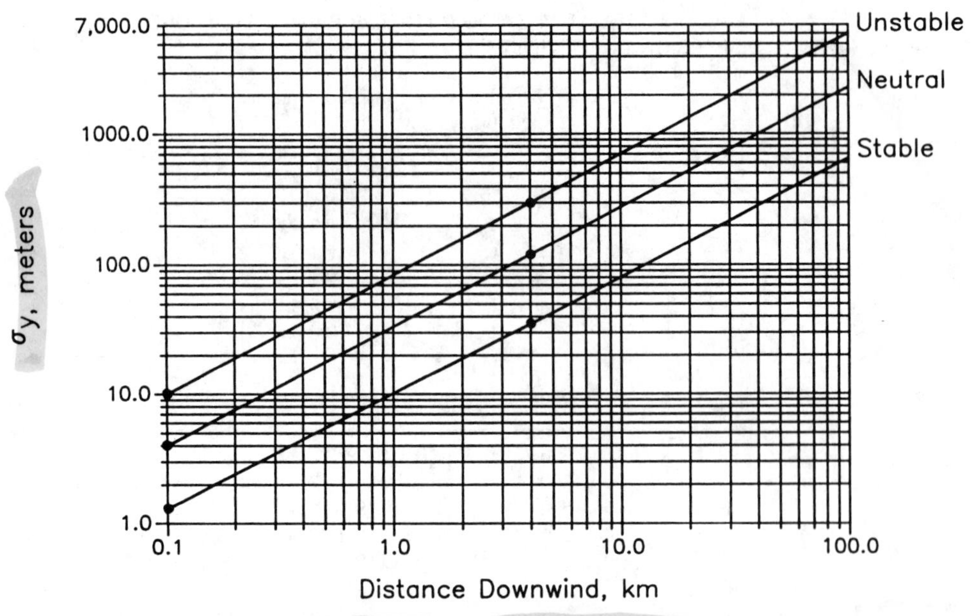

Figure 5-12 Horizontal dispersion coefficient for puff model. This data is based only on the data points shown and should not be considered reliable at other distances.

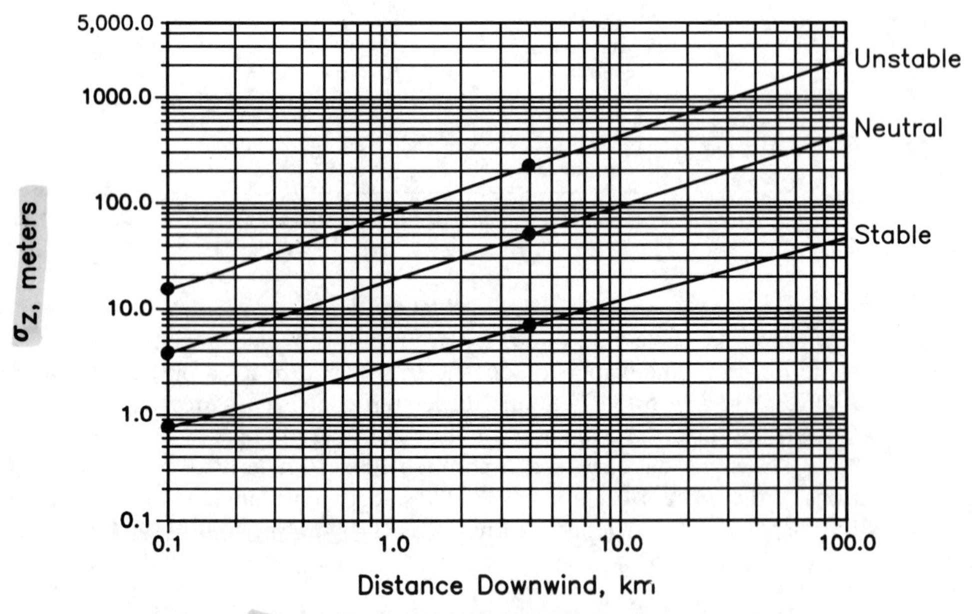

Figure 5-13 Vertical dispersion coefficient for puff model. This data is based only on the data points shown and should not be considered reliable at other distances.

Case 1: Puff. Instantaneous Point Source at Ground Level, Coordinates Fixed at Release Point. Constant Wind in *x* Direction Only with Constant Velocity *u*.

This case is identical to Case 7. The solution has a form similiar to Equation 5-33.

$$\langle C\rangle(x,y,z,t) = \frac{Q^*_{\mathrm{m}}}{\sqrt{2}\pi^{3/2}\sigma_x\sigma_y\sigma_z}\exp\left\{-\frac{1}{2}\left[\left(\frac{x-ut}{\sigma_x}\right)^2 + \frac{y^2}{\sigma_y^2} + \frac{z^2}{\sigma_z^2}\right]\right\} \tag{5-38}$$

The ground level concentration is given at $z = 0$.

$$\langle C\rangle(x,y,0,t) = \frac{Q^*_{\mathrm{m}}}{\sqrt{2}\pi^{3/2}\sigma_x\sigma_y\sigma_z}\exp\left\{-\frac{1}{2}\left[\left(\frac{x-ut}{\sigma_x}\right)^2 + \frac{y^2}{\sigma_y^2}\right]\right\} \tag{5-39}$$

The ground level concentration along the x-axis is given at $y = z = 0$.

$$\langle C\rangle(x,0,0,t) = \frac{Q^*_{\mathrm{m}}}{\sqrt{2}\pi^{3/2}\sigma_x\sigma_y\sigma_z}\exp\left[-\frac{1}{2}\left(\frac{x-ut}{\sigma_x}\right)^2\right] \tag{5-40}$$

The center of the cloud is found at coordinates $(ut, 0, 0)$. The concentration at the center of this moving cloud is given by

$$\langle C\rangle(ut,0,0,t) = \frac{Q^*_{\mathrm{m}}}{\sqrt{2}\pi^{3/2}\sigma_x\sigma_y\sigma_z} \tag{5-41}$$

The total integrated dose, D_{tid} received by an individual standing at fixed coordinates (x, y, z) is the time integral of the concentration.

$$D_{\mathrm{tid}}(x,y,z) = \int_0^\infty \langle C\rangle(x,y,z,t)\,dt \tag{5-42}$$

The total integrated dose at ground level is found by integrating Equation 5-39 according to Equation 5-42. The result is

$$D_{\mathrm{tid}}(x,y,0) = \frac{Q^*_{\mathrm{m}}}{\pi\sigma_y\sigma_z u}\exp\left(-\frac{1}{2}\frac{y^2}{\sigma_y^2}\right) \tag{5-43}$$

The total integrated dose along the x-axis on the ground is

$$D_{\mathrm{tid}}(x,0,0) = \frac{Q^*_{\mathrm{m}}}{\pi\sigma_y\sigma_z u} \tag{5-44}$$

Frequently the cloud boundary defined by a fixed concentration is required. The line connecting points of equal concentration around the cloud boundary is called an isopleth. For a specified concentration, $\langle C\rangle^*$, the isopleths at ground level are determined by dividing the equation for the centerline concentration, Equation 5-40, by the equation for the general ground level concentration, Equation 5-39. This equation is solved directly for y.

$$y = \sigma_y\sqrt{2\ln\left(\frac{\langle C\rangle(x,0,0,t)}{\langle C\rangle(x,y,0,t)}\right)} \tag{5-45}$$

The procedure is

1. Specify $\langle C\rangle^*$, u, and t.
2. Determine the concentrations, $\langle C\rangle(x, 0, 0, t)$, along the x-axis using Equation 5-40. Define the boundary of the cloud along the x-axis.
3. Set $\langle C\rangle(x, y, 0, t) = \langle C\rangle^*$ in Equation 5-45 and determine the values of y at each centerline point determined in step 2.

The procedure is repeated for each value of t required.

Case 2: Plume. Continuous, Steady-state, Source at Ground Level, Wind Moving in *x* Direction at Constant Velocity, *u*.

This case is identical to Case 9. The solution has a form similiar to Equation 5-35.

$$\langle C\rangle(x, y, z) = \frac{Q_m}{\pi\sigma_y\sigma_z u}\exp\left[-\frac{1}{2}\left(\frac{y^2}{\sigma_y^2} + \frac{z^2}{\sigma_z^2}\right)\right] \tag{5-46}$$

The ground concentration is given at $z = 0$.

$$\langle C\rangle(x, y, 0) = \frac{Q_m}{\pi\sigma_y\sigma_z u}\exp\left[-\frac{1}{2}\left(\frac{y}{\sigma_y}\right)^2\right] \tag{5-47}$$

The concentration along the centerline of the plume directly downwind is given at $y = z = 0$.

$$\langle C\rangle(x, 0, 0) = \frac{Q_m}{\pi\sigma_y\sigma_z u} \tag{5-48}$$

The isopleths are found using a procedure identical to the isopleth procedure used for Case I.

For continuous ground level releases the maximum concentration occurs at the release point.

Case 3: Plume. Continuous, Steady-state Source at Height H_r above Ground Level, Wind Moving in *x* Direction at Constant Velocity, *u*.

This is identical to Case 10. The solution has a form similiar to Equation 5-36.

$$\langle C\rangle(x, y, z) = \frac{Q_m}{2\pi\sigma_y\sigma_z u}\exp\left[-\frac{1}{2}\left(\frac{y}{\sigma_y}\right)^2\right] \times\left\{\exp\left[-\frac{1}{2}\left(\frac{z - H_r}{\sigma_z}\right)^2\right] + \exp\left[-\frac{1}{2}\left(\frac{z + H_r}{\sigma_z}\right)^2\right]\right\} \tag{5-49}$$

The ground level concentration is found by setting $z = 0$.

$$\langle C\rangle(x, y, 0) = \frac{Q_m}{\pi\sigma_y\sigma_z u}\exp\left[-\frac{1}{2}\left(\frac{y}{\sigma_y}\right)^2 - \frac{1}{2}\left(\frac{H_r}{\sigma_z}\right)^2\right] \tag{5-50}$$

The ground centerline concentrations are found by setting $y = z = 0$.

$$\langle C \rangle (x, 0, 0) = \frac{Q_{\text{m}}}{\pi \sigma_y \sigma_z u} \exp\left[-\frac{1}{2} \left(\frac{H_{\text{r}}}{\sigma_z} \right)^2 \right] \tag{5-51}$$

The maximum ground level concentration along the x-axis, $\langle C \rangle_{\text{max}}$, is found using

$$\boxed{\langle C \rangle_{\text{max}} = \frac{2Q_{\text{m}}}{e\pi u H_{\text{r}}^2} \left(\frac{\sigma_z}{\sigma_y} \right)} \tag{5-52}$$

The distance downwind at which the maximum ground level concentration occurs is found from

$$\boxed{\sigma_z = \frac{H_{\text{r}}}{\sqrt{2}}} \tag{5-53}$$

The procedure for finding the maximum concentration and the downwind distance is to use Equation 5-53 to determine the distance followed by Equation 5-52 to determine the maximum concentration.

Case 4: Puff. Instantaneous Point Source at Height H_r above Ground Level. Coordinate System on Ground Moves with Puff.

For this case the center of the puff is found at $x = ut$. The average concentration is given by

$$\boxed{\begin{aligned} \langle C \rangle (x, y, z, t) &= \frac{Q_{\text{m}}^*}{(2\pi)^{3/2} \sigma_x \sigma_y \sigma_z} \exp\left[-\frac{1}{2} \left(\frac{y}{\sigma_y} \right)^2 \right] \\ &\times \left\{ \exp\left[-\frac{1}{2} \left(\frac{z - H_{\text{r}}}{\sigma_z} \right)^2 \right] + \exp\left[-\frac{1}{2} \left(\frac{z + H_{\text{r}}}{\sigma_z} \right)^2 \right] \right\} \end{aligned}} \tag{5-54}$$

The time dependence is achieved through the dispersion coefficients, since their values change as the puff moves downwind from the release point. If wind is absent ($u = 0$), Equation 5-54 will not predict the correct result.

At ground level, $z = 0$, and the concentration is computed using

$$\langle C \rangle (x, y, 0, t) = \frac{Q_{\text{m}}^*}{\sqrt{2}\pi^{3/2} \sigma_x \sigma_y \sigma_z} \exp\left[-\frac{1}{2} \left(\frac{y}{\sigma_y} \right)^2 - \frac{1}{2} \left(\frac{H_{\text{r}}}{\sigma_z} \right)^2 \right] \tag{5-55}$$

The concentration along the ground at the centerline is given at $y = z = 0$,

$$\langle C \rangle (x, 0, 0, t) = \frac{Q_{\text{m}}^*}{\sqrt{2}\pi^{3/2} \sigma_x \sigma_y \sigma_z} \exp\left[-\frac{1}{2} \left(\frac{H_{\text{r}}}{\sigma_z} \right)^2 \right] \tag{5-56}$$

The total integrated dose at ground level is found by application of Equation 5-42 to Equation 5-55. The result is

$$\boxed{D_{\text{tid}} (x, y, 0) = \frac{Q_{\text{m}}^*}{\pi \sigma_y \sigma_z u} \exp\left[-\frac{1}{2} \left(\frac{y}{\sigma_y} \right)^2 - \frac{1}{2} \left(\frac{H_{\text{r}}}{\sigma_z} \right)^2 \right]} \tag{5-57}$$

Case 5: Puff. Instantaneous Point Source at Height H_r above Ground Level. Coordinate System Fixed on Ground at Release Point.

For this case, the result is obtained using a transformation of coordinates similiar to the transformation used for Case 7. The result is

$$\langle C \rangle (x, y, z, t) = \text{(Puff equations with moving coordinate system, Equations 5-54 through 5-56)} \times \exp\left[-\frac{1}{2}\left(\frac{x - ut}{\sigma_x}\right)^2\right] \tag{5-58}$$

where t is the time since the release of the puff.

Comparison of the Plume and Puff Models

The plume model describes the steady state behavior of material ejected from a continuous source. The puff model describes the behavior of a single, instantaneous release of material. The puff model is not steady-state and follows the cloud of material as it moves with the wind. As a result, only the puff model is capable of providing a time dependence for the release.

The puff model is also used for continuous releases by representing the release as a succession of puffs. For leaks from pipes and vessels, if t_p is the time to form one puff, then the number of puffs formed, n, is given by

$$n = \frac{t}{t_p} \tag{5-59}$$

where t is the duration of the spill. The time to form one puff, t_p, is determined by defining an effective leak height, H_{eff}. Then,

$$t_p = \frac{H_{eff}}{u} \tag{5-60}$$

where u is the wind speed. Empirical results show that the best H_{eff} to use is

$$H_{eff} = \text{(height of leak)} \times 1.5 \tag{5-61}$$

For a continuous leak,

$$Q_m^* = Q_m t_p \tag{5-62}$$

and for an instantaneous release divided into a number of smaller puffs,

$$Q_m^* = (Q_m^*)_{total} / n \tag{5-63}$$

where $(Q_m^*)_{total}$ is the total release amount.

This approach works for liquid spills, but not for vapor releases. For vapor releases a single puff is suggested.

The puff model is also used to represent changes in wind speed and direction.

Example 5-2

On an overcast day, a stack with an effective height of 60 meters is releasing sulfur dioxide at the rate of 80 grams per second. The wind speed is 6 meters per second.

Determine

a. The mean concentration of SO_2 on the ground 500 meters downwind.

b. The mean concentration on the ground 500 meters downwind and 50 meters crosswind.

c. The location and value of the maximum mean concentration on ground level directly downwind.

Solution **a.** This is a continuous release. The ground concentration directly downwind is given by Equation 5-51.

$$\langle C \rangle (x, 0, 0) = \frac{Q_m}{\pi \sigma_y \sigma_z u} \exp\left[-\frac{1}{2}\left(\frac{H_r}{\sigma_z}\right)^2\right] \tag{5-51}$$

From Table 5-2, the stability class is D. The dispersion coefficients are obtained from Figures 5-10 and 5-11. The resulting values are $\sigma_y = 36$ meters and $\sigma_z = 18.5$ meters. Substituting into Equation 5-51

$$\langle C \rangle (500 \text{ m}, 0, 0) = \frac{80 \text{ gm/s}}{(3.14)(36 \text{ m})(18.5 \text{ m})(6 \text{ m/s})} \exp\left[-\frac{1}{2}\left(\frac{60 \text{ m}}{18.5 \text{ m}}\right)^2\right]$$

$$= 3.31 \times 10^{-5} \text{ gm/m}^3$$

b. The mean concentration 50 meters crosswind is found using Equation 5-50 and setting $y = 50$. The results from part a are applied directly,

$$\langle C \rangle (500 \text{ m}, 50 \text{ m}, 0) = \langle C \rangle (500 \text{ m}, 0, 0) \exp\left[-\frac{1}{2}\left(\frac{y}{\sigma_y}\right)^2\right]$$

$$= (3.31 \times 10^{-5} \text{ gm/m}^3) \exp\left[-\frac{1}{2}\left(\frac{50 \text{ m}}{36 \text{ m}}\right)^2\right]$$

$$= 1.26 \times 10^{-5} \text{ gm/m}^3$$

c. The location of the maximum concentration is found from Equation 5-53,

$$\sigma_z = \frac{H_r}{\sqrt{2}} = \frac{60 \text{ m}}{\sqrt{2}} = 42.4 \text{ m}$$

From Figure 5-11, the dispersion coefficient has this value at $x = 1500$ meters. At $x = 1500$ meters, from Figure 5-10, $\sigma_y = 100$ m. The maximum concentration is determined using Equation 5-52.

$$\langle C \rangle_{max} = \frac{2Q_m}{e\pi u H_r^2}\left(\frac{\sigma_z}{\sigma_y}\right)$$

$$= \frac{(2)(80 \text{ gm/s})}{(2.72)(3.14)(6 \text{ m/s})(60 \text{ m})^2}\left(\frac{42.4 \text{ m}}{100 \text{ m}}\right) \tag{5-52}$$

$$= 3.68 \times 10^{-4} \text{ gm/m}^3$$

Example 5-3

Chlorine is used in a particular chemical process. A source model study indicates that for a particular accident scenario 1.0 kg of chlorine will be released instantaneously. The release will occur at ground level.

A residential area is 500 m away from the chlorine source. Determine

a. The time required for the center of the cloud to reach the residential area. Assume a wind speed of 2 m/s.

b. The maximum concentration of chlorine in the residential area. Compare this with a TLV for chlorine of 0.5 ppm. What stability conditions and wind speed produces the maximum concentration?

c. Determine the distance the cloud must travel to disperse the cloud to a maximum concentration below the TLV. Use the conditions of Part b.

d. Determine the size of the cloud, based on the TLV, at a point 5 km directly downwind on the ground. Assume the conditions of Part b.

Solution **a.** For a distance of 500 m and a wind speed of 2 m/s, the time required for the center of the cloud to reach the residential area is

$$t = \frac{x}{u} = \frac{500 \text{ m}}{2 \text{ m/s}} = 250 \text{ s} = 4.2 \text{ min}$$

This leaves very little time for emergency warning.

b. The maximum concentration will occur at the center of the cloud directly downwind from the release. The concentration is given by Equation 5-41.

$$\langle C \rangle (ut, 0, 0, t) = \frac{Q_m^*}{\sqrt{2}\pi^{3/2}\sigma_x\sigma_y\sigma_z} \tag{5-41}$$

The stability conditions are selected to maximize $\langle C \rangle$ in Equation 5-41. This requires dispersion coefficients of minimum value. From Figures 5-12 and 5-13, this occurs under stable conditions. From Table 5-2, this will occur at night with a 2-3 m/s wind. Assume a slow moving cloud of 2 m/s. From Figures 5-12 and 5-13, at 500 m, $\sigma_y = 5.2$ m and $\sigma_z = 2.2$ m. Also assume $\sigma_x = \sigma_y$. From Equation 5-41,

$$\langle C \rangle = \frac{1.0 \text{ kg}}{\sqrt{2}(3.14)^{3/2}(5.2 \text{ m})^2(2.2 \text{ m})} = 2.14 \times 10^{-3} \text{ kg/m}^3 = 2{,}140 \text{ mg/m}^3$$

This is converted to ppm using Equation 2-6. Assuming a pressure of 1 atm and a temperature to 298°K, the concentration in ppm is 737 ppm. This is much higher than the TLV of 0.5 ppm. Any individuals within the immediate residential area, and any personnel within the plant will be excessively exposed if they are outside and downwind from the source.

c. From Table 2-8, the TLV of 0.5 ppm is 1.45 mg/m^3 or 1.45×10^{-6} kg/m^3. The concentration at the center of the cloud is given by Equation 5-41. Substituting the known values,

$$1.45 \times 10^{-6} \text{ kg/m}^3 = \frac{1.0 \text{ kg}}{\sqrt{2}(3.14)^{3/2}\sigma_y^2\sigma_z}$$

$$\sigma_y^2\sigma_z = 8.76 \times 10^4 \text{ m}^3$$

This equation is satisfied at the correct distance from the release point. A trial and error procedure is required. The procedure is

1. Select a distance, x.
2. Determine σ_x, σ_y, and σ_z using Figures 5-12 and 5-13.
3. Check if dispersion coefficients satisfy above equation.

The procedure is continued until the equation is satisfied. This produces the following results,

Guessed distance (km)	σ_y	σ_z	$\sigma_y^2\sigma_z$
1	10.0	3.2	3.2×10^2
10	80	12.0	8.07×10^4
11	88	13.0	1.01×10^5

The distance is interpolated to about 10.3 km. This is quite a substantial distance considering that only 1.0 kg of chlorine is released.

d. The downwind centerline concentration is given by Equation 5-40.

$$\langle C \rangle (x, 0, 0, t) = \frac{Q_m^*}{\sqrt{2}\pi^{3/2}\sigma_x\sigma_y\sigma_z} \exp\left[-\frac{1}{2}\left(\frac{x - ut}{\sigma_x}\right)^2\right] \tag{5-40}$$

The time required for the center of the plume to arrive is

$$t = \frac{x}{u} = \frac{5000 \text{ m}}{2 \text{ m/s}} = 2500 \text{ s}$$

At a downwind distance of 5 km, from Figures 5-12 and 5-13,

$$\sigma_y = \sigma_x = 44 \text{ m and } \sigma_z = 8 \text{ m}$$

Substituting the numbers provided,

$$1.45 \times 10^{-6} \text{ kg/m}^3 = \frac{1.0 \text{ kg}}{\sqrt{2}\pi^{3/2}(44 \text{ m})^2(8 \text{ m})} \exp\left[-\frac{1}{2}\left(\frac{x - 5000}{44 \text{ m}}\right)^2\right]$$

where x has units of meters. Rearranging and combining leads to a quadratic equation,

$$x^2 - 10{,}000x + 2.499328 \times 10^7 = 0$$

$$x = 5000 \pm 82 \text{ m}$$

The cloud is 164 meters wide at this point, based on the TLV concentration. At 2 m/s, it will take approximately,

$$\frac{164 \text{ m}}{2 \text{ m/s}} = 82 \text{ s}$$

to pass.

An appropriate emergency procedure would be to alert residents to stay indoors with the windows closed and ventilation off until the cloud passes. An effort by the plant to reduce the quantity of chlorine released is also indicated.

5-5 EFFECT OF RELEASE MOMENTUM AND BUOYANCY

Figure 5-6 indicates that the release characteristics of a puff or plume are dependent on the initial release momentum and buoyancy. The initial momentum and buoyancy will change the effective height of release. A release that occurs at ground level but in an upward spouting jet of vaporizing liquid will have a greater

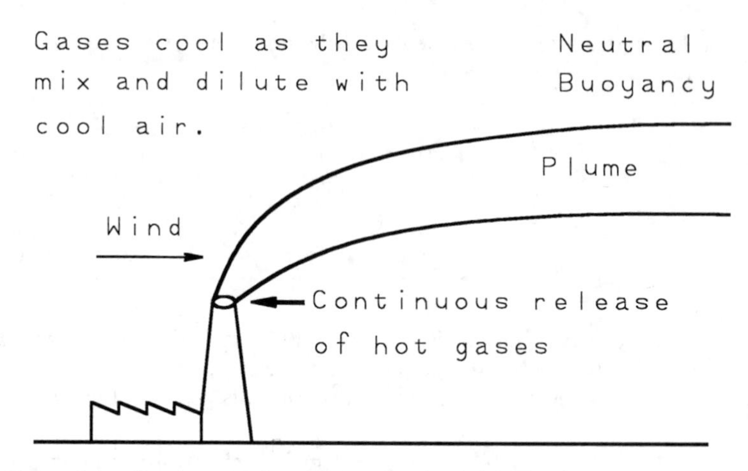

Figure 5-14 Smokestack plume demonstrating initial buoyant rise of hot gases.

"effective" height than a release without a jet. Similiarly, a release of vapor at a temperature higher than the ambient air temperature will rise due to buoyancy effects, increasing the "effective" height of the release.

Both of these effects are demonstrated by the traditional smokestack release shown in Figure 5-14. The material released from the smokestack contains momentum, based on its upward velocity within the stack pipe, and it is also buoyant, since its temperature is higher than the ambient temperature. Thus, the material continues to rise after its release from the stack. The upward rise is slowed and eventually stopped as the released material cools and the momentum is dissipated.

For smokestack releases, Turner[10] suggests using the empirical Holland formula to compute the additional height due to the buoyancy and momentum of the release,

$$\Delta H_{\mathrm{r}} = \frac{\bar{u}_{s}d}{\bar{u}}\left[1.5 + 2.68 \times 10^{-3}Pd\left(\frac{T_{\mathrm{s}}-T_{\mathrm{a}}}{T_{\mathrm{s}}}\right)\right] \tag{5-64}$$

where

ΔH_{r} is the correction to the release height, H_{r}
$\bar{u}_{\mathrm{s}}$ is the stack gas exit velocity, in m/s
d is the inside stack diameter, in m
$\bar{u}$ is the wind speed, in m/s
P is the atmospheric pressure, in mb
T_{s} is the stack gas temperature, in °K
T_{a} is the air temperature, in °K

For heavier than air vapors, if the material is released above ground level, the material will initially fall towards the ground until it disperses enough to reduce the cloud density.

[10] D. Bruce Turner, *Workbook of Atmospheric Dispersion Estimates* (Cincinnati: U.S. Department of Health, Education and Welfare, 1970), p. 31.

5-6 EFFECT OF BUILDINGS AND STRUCTURES

Buildings and structures provide barriers to vapor clouds and ground releases. The behavior of vapor clouds moving around buildings and structures is not well understood.

5-7 RELEASE MITIGATION

The purpose of the toxic release model is to provide a tool for performing release mitigation. Release mitigation is defined as "lessening the risk of a release incident by acting on the source (at the point of release) either (1) in a preventive way by reducing the liklihood of an event which could generate a hazardous vapor cloud or (2) in a protective way by reducing the magnitude of the release and/or the exposure of local persons or property."[11]

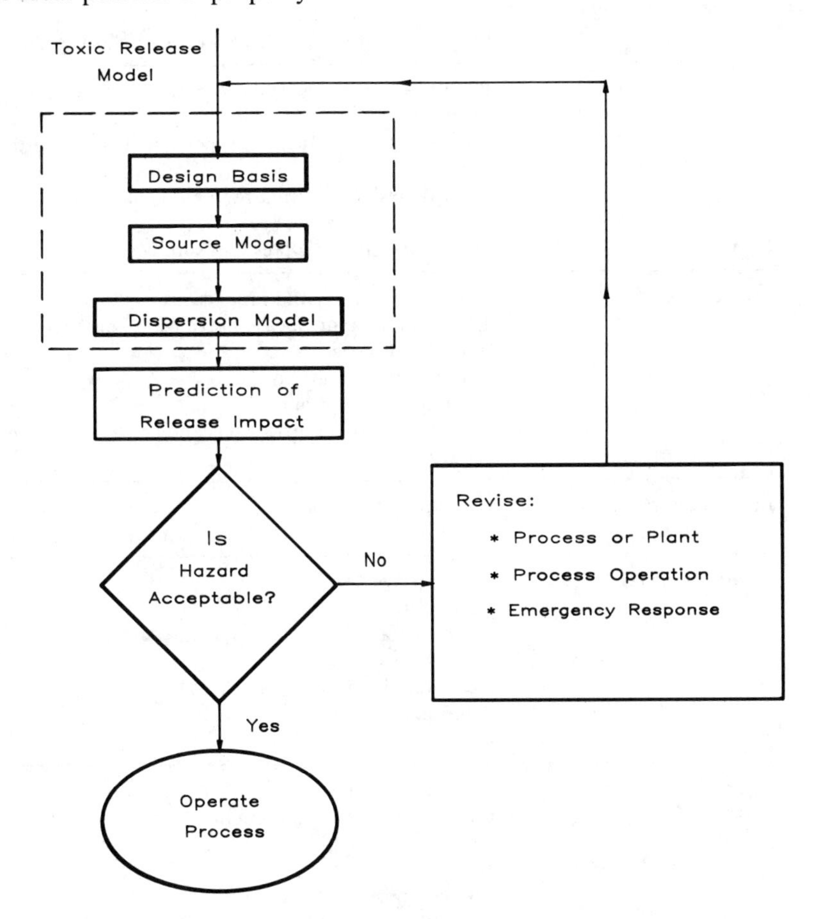

Figure 5-15 The release mitigation procedure.

[11]Richard W. Prugh and Robert W. Johnson, *Guidelines for Vapor Release Mitigation* (New York: American Institute of Chemical Engineers, 1988), p. 2.

The release mitigation design procedure is shown in Figure 5-15. Once the toxic release model is completed, it is used to predict the impact of the release. This includes the area and number of people affected and the manner in which they are affected. At this point a decision is made whether the hazards are acceptable. If the hazards are acceptable, the process is operated. If the hazards are unacceptable, a change is made to reduce the hazard. This includes changing the process, the operation of the process, or invoking an improved emergency procedure. A new

TABLE 5-4 RELEASE MITIGATION APPROACHES[1]

Major area	Examples
Inherent Safety	Inventory reduction: Less chemicals inventoried or less in process vessels. Chemical substitution: Substitute a less hazardous chemical for one more hazardous. Process attentuation: Use lower temperatures and pressures.
Engineering Design	Plant physical integrity: Use better seals or materials of construction. Process integrity: Insure proper operating conditions and material purity. Process design features for emergency control: Emergency relief systems. Spill containment: Dikes and spill vessels.
Management	Operating policies and procedures Training for vapor release prevention and control. Audits and inspections Equipment testing Maintenance program Management of modifications and changes to prevent new hazards. Security
Early Vapor Detection and Warning	Detection by sensors Detection by personnel
Countermeasures	Water sprays Water curtains Steam curtains Air curtains Deliberate ignition of explosive cloud. Dilution Foams
Emergency Response	On-site communications Emergency shutdown equipment and procedures. Site evacuation Safe havens Personal protective equipment Medical treatment. On-site emergency plans, procedures, training and drills.

[1]Richard W. Prugh and Robert W. Johnson, *Guidelines for Vapor Release Mitigation* (New York: American Institute of Chemical Engineers, 1988).

toxic release model is developed for the process incorporating the changes and the release impact is again assessed. The procedure is continued until the hazards are reduced to acceptable levels.

The best method for preventing a release situation is to prevent the accident leading to the release in the first place. However, engineers must be prepared in the event of an accident. Release mitigation involves (1) detecting the release as quickly as possible, (2) stopping the release as quickly as possible, and (3) invoking a mitigation procedure to reduce the impact of the release on the surroundings. Once a release is in vapor form, the resulting cloud is nearly impossible to control. Thus, an emergency procedure must strive to reduce the amount of vapor formed.

Table 5-4 provides additional methods and detail on release mitigation techniques.

SUGGESTED READING

Vapor Cloud Modeling

STEVEN R. HANNA and PETER J. DRIVAS, *Vapor Cloud Dispersion Models* (New York: American Institute of Chemical Engineers, 1988).

FRANK P. LEES, *Loss Prevention in the Process Industries* (London: Butterworths, 1986), Chapter 15.

JOHN H. SEINFELD, *Atmospheric Chemistry and Physics of Air Pollution* (New York: John Wiley and Sons, 1986), Chapters 12, 13, and 14.

D. BRUCE TURNER, *Workbook of Atmospheric Dispersion Estimates* (Cincinnati: U.S. Department of Health , Education and Welfare, 1970).

Release Mitigation

RICHARD W. PRUGH and ROBERT W. JOHNSON, *Guidelines for Vapor Release Mitigation* (New York: American Institute of Chemical Engineers, 1988).

PROBLEMS

5-1. A backyard barbeque grill contains a 20-lb tank of propane. The propane leaves the tank through a valve and regulator and is fed through a 1/2-in rubber hose to a dual valve assembly. After the valves, the propane flows through a dual set of ejectors where it is mixed with air. The propane-air mixture then arrives at the burner assembly where it is burned. Describe the possible propane release scenarios for this equipment. Identify the largest practicable and largest potential release.

5-2. Contaminated toluene is fed to a water wash system shown in Figure P5-2. The toluene is pumped from a 50-gallon drum into a countercurrent centrifugal extractor. The extractor separates the water from the toluene by centrifugal force acting on the difference in densities. The contaminated toluene enters the extractor at the periphery and flows to the center. The water enters the center of the extractor and flows to the periphery. The washed toluene and contaminated water flow into 50-gallon drums. Determine a number of release scenarios for this equipment. See if you can identify the largest practicable and largest potential releases.

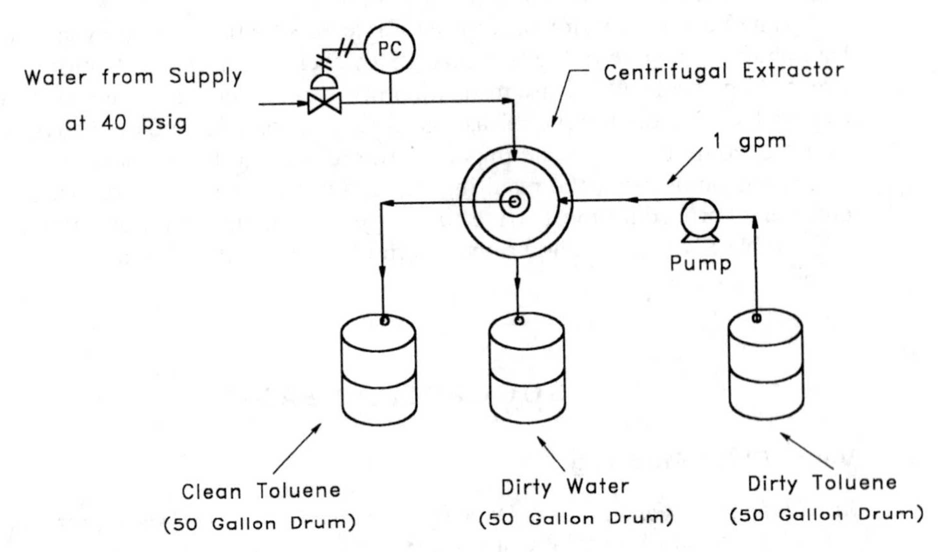

Figure P5-2 Toluene water wash process.

5-3. A burning dump emits an estimated 3 gm/s of oxides of nitrogen. What is the average concentration of oxides of nitrogen from this source directly downwind at a distance of 3 km on an overcast night with a wind speed of 7 m/s? Assume this dump to be a point ground-level source.

5-4. A trash incinerator has an effective stack height of 100 m. On a sunny day with a 2 m/s wind, the concentration of sulfur dioxide 200 m directly downwind is measured at 5.0×10^{-5} gm/m^3. Estimate the mass release rate, in gm/sec, of sulfur dioxide from this stack. Also estimate the maximum sulfur dioxide concentration expected on the ground and its location downwind from the stack.

5-5. You have been suddenly enveloped by a plume of toxic material from a nearby chemical plant. Which way should you run with respect to the wind to minimize your exposure?

5-6. An air sampling station is located at an azimuth of 203° from a cement plant at a distance of 1500 meters. The cement plant releases fine particulates (less than 15 microns diameter) at the rate of 750 pounds per hour from a 30-meter stack. What is the concentration of particulates at the air sampling station when the wind is from 30° at 3 m/s on a clear day in the late Fall at 4:00 P.M.?

5-7. A storage tank containing acrolein (TLV-STEL = 0.3 ppm) is located 1500 meters from a residential area. Estimate the amount of acrolein that must be instantaneously released at ground level to produce a concentration at the boundary of the residential area equal to the TLV-STEL.

5-8. Consider again Problem 5-7, but assume a continuous release at ground level. What is the release rate required to produce an average concentration at the boundary to the residential area equal to the TLV-TWA of 0.1 ppm? Explain why the TLV-TWA might be appropriate for this case rather than the TLV-STEL.

5-9. The concentration of vinyl chloride 2-km downwind from a continuous release 25 m high is 1.6 mg/m^3. It is a sunny day and the wind is 18 km/hr. Determine the average concentration 0.1 km perpendicular to the plume 2 km downwind.

5-10. Diborane is used in silicon chip manufacture. One facility uses a 500-lb bottle. If the entire bottle is released continuously during a 20-minute period, determine the location of the 5 mg/m^3 ground level isopleth. It is a clear, sunny day with a 5 mph wind. Assume the release is at ground level.

5-11. Reconsider Problem 5-10. Assume now that the bottle ruptures and the entire contents of diborane is released instantaneously. Determine, at 15-minutes after the release,

a. The location of the vapor cloud.
b. The location of the 5 mg/m^3 isopleth.
c. The concentration at the center of the cloud.
d. The total dosage received by an individual standing on the downwind axis at the 15-minute downwind location.
e. How far and long will the cloud need to travel to reduce the maximum concentration to 5 mg/m^3?

5-12. An 800-pound tank of chlorine is stored at a water treatment plant. A study of the release scenarios indicate that the entire tank contents could be released as vapor in a period of 10 minutes. For chlorine gas, evacuation of the population must occur for areas where the vapor concentration exceeds 7.3 mg/m^3. Without any additional information, estimate the distance downwind that must be evacuated.

5-13. A reactor in a pesticide plant contains 1000 lb of a liquid mixture of 50% by weight liquid methyl isocyanate (MIC). The liquid is near its boiling point. A study of various release scenarios indicates that a rupture of the reactor will spill the liquid into a boiling pool on the ground. The boiling rate has been estimated to be 20 lb per minute of MIC. Evacuation of the population must occur in areas where the vapor concentration exceeds 4.7 mg/m^3. If the wind speed is 3.4 mph on a clear night, estimate the area downwind that must be evacuated.

5-14. A chemical plant has 10,000 pounds of solid acrylamide stored in a large bin. About 20% by weight of the solid has a particle size less than 10 microns. A scenario study indicates that all of the fine particles could be airborne in a period of 10 minutes. If evacuation must occur in areas where the particle concentration exceeds 110 mg/m^3, estimate the area that must be evacuated.

5-15. You have been appointed emergency coordinator for the community of Smallville, shown on Figure P5-15.

ABC Chemical Company is shown on the map. They report the following chemicals and amounts:

Hydrogen chloride:	100 pounds
Sulfuric acid:	100 gallons

You are required to develop an emergency plan for the community.

a. Determine which chemical presents the greater hazard to the community.
b. Assuming all of the chemical is released during a 10-minute period, determine the distance downwind that must be evacuated.
c. Identify locations that might be affected by a release incident at the plant, or might contribute to the incident due to its proximity to the plant.
d. Determine transportation routes that will be used to transport hazardous materials into or out of the facility. Identify any "high risk" intersections where accidents might occur.

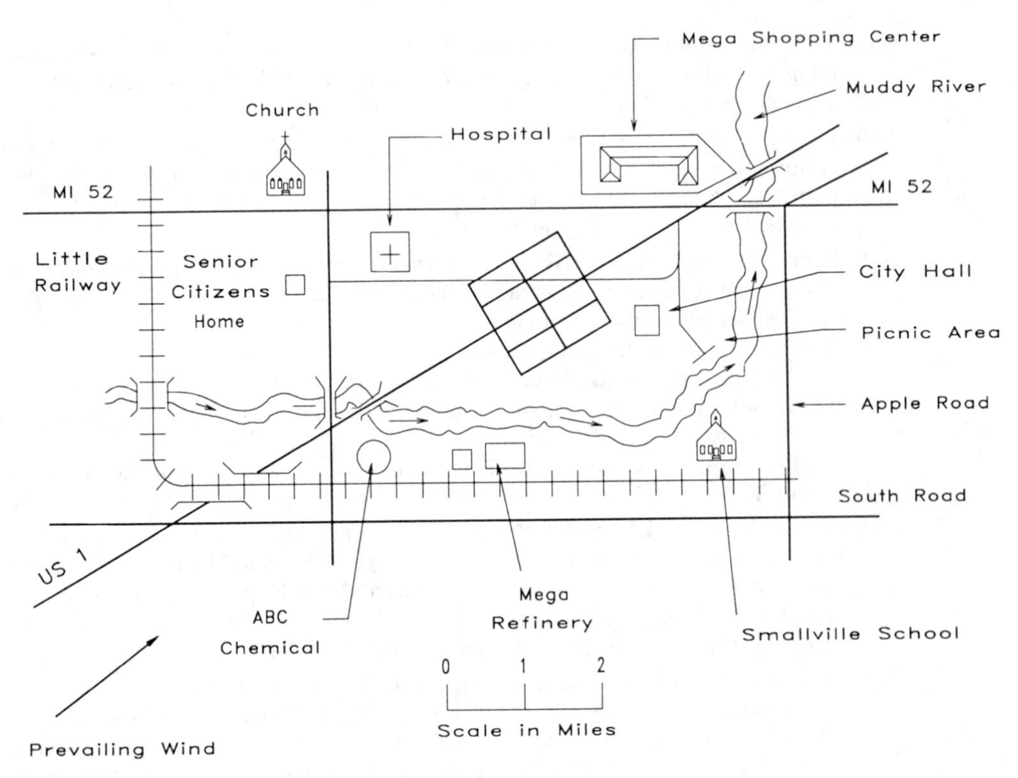

Figure P5-15 Map of Smallville.

e. Determine the vulnerable zone along the transportation routes identified in Part d. Use a distance of 1/2 mile on either side of the route, unless a smaller distance is indicated by Part b.
f. Identify any special concerns (schools, nursing homes, shopping centers, and the like) that appear in the transportation route vulnerable zone.
g. Determine evacuation routes for the areas surrounding the plant.
h. Determine alternate traffic routes around the potential hazard.
i. Determine the manpower required to support the needs of Parts g and h.
j. Identify the means required to warn the area, and describe the content of an example warning message that could be used in an emergency at the facility.
k. Estimate the potential number of people evacuated during an emergency. Determine how these people are to be moved and where they might be evacuated to.
l. What other concerns might be important during a chemical emergency?

5-16. Derive Equation 5-43.

5-17. One response to a short-term release is to warn people to stay in their homes or offices with the windows closed and the ventilation off.

An average house, with the windows closed, exchanges air with the surroundings equal to three times the volume of the house per hour (although wide variations are expected).

a. Derive an equation for the concentration of chemical vapor within the house based on a parameter, N_t, equal to the number of volume exchanges per hour. Assume well-mixed behavior for the air, an initial zero concentration of vapor within the house, and a constant external concentration during the exposure period.

b. A vapor cloud with a maximum concentration of 20 ppm is moving through a community. Determine the time before the vapor concentration within an average house reaches 10 ppm.

c. If the wind is blowing at 2 mph and the plant is 1 mile upwind from the community, what is the maximum time available to the plant personnel to stop or reduce the release to insure that the concentrations within the homes do not exceed the 10 ppm value?

5-18. A supply line (ID = 0.493 inches) containing chlorine gas is piped from a regulated supply at 50-psig. If the supply line ruptures, estimate the distance the plume must travel to reduce the concentration to 7.3 mg/m^3. Assume an overcast day with a 15 mph wind, and a temperature of 80°F. The release is near ground level.

5-19. A tank has ruptured and a pool of benzene has formed. The pool is approximately rectangular with dimensions of 20 feet by 30 feet. Estimate the evaporation rate and the distance affected downwind. Define the plume boundary using the TLV-TWA of 10 ppm. It is an overcast day with a 9 mph wind. The temperature is 90°F.

Fires and Explosions

6

Chemicals present a very substantial hazard due to fires and explosions. The combustion of one gallon of toluene can destroy an ordinary chemistry laboratory in minutes; persons present may be killed. The potential consequences of fires and explosions in pilot plants and plant environments are even greater.

The three most common chemical plant accidents are fires, explosions, and toxic releases, in that order (see Chapter 1). Organic solvents are the most common source of fires and explosions in the chemical industry.

Yearly losses due to fires and explosions are substantial.[1] Property losses for explosions in the United States are estimated at over \$150 million (1979 dollars). Additional losses due to business interruptions are estimated to exceed \$150 million annually. To prevent accidents due to fires and explosions, engineers must be familiar with

- the fire and explosion properties of materials,
- the nature of the fire and explosion process, and
- procedures to reduce fire and explosion hazards.

This chapter covers the first two topics, emphasizing definitions and calculation methods for estimating the magnitude and consequences of fires and explosions. Chapter 7 discusses procedures to reduce fire and explosion hazards.

[1]Frank T. Bodurtha, *Industrial Explosion Prevention and Protection* (New York: McGraw-Hill Book Company, 1980), p. 1.

6-1 THE FIRE TRIANGLE

The essential elements for combustion are fuel, oxidizer, and an ignition source. These elements are illustrated by the fire triangle shown in Figure 6-1.

Fire, or burning, is the rapid, exothermic oxidation of an ignited fuel. The fuel can be in solid, liquid or vapor form, but vapor and liquid fuels are generally easier to ignite. The combustion always occurs in the vapor phase; liquids are volatized and solids are decomposed into vapor prior to combustion.

When fuel, oxidizer, and an ignition source are present at the necessary levels, burning will occur. This means a fire will *not* occur if (1) fuel is not present or is not present in sufficient quantities, (2) oxidizer is not present or is not present in sufficient quantities, and (3) the ignition source is not energetic enough to initiate the fire.

Two common examples of the three components of the fire triangle are wood, air, and a match; or gasoline, air, and a spark. However, other, less obvious combinations of chemicals can lead to fires and explosions. Various fuels, oxidizers and ignition sources common in the chemical industry are

FUELS

Liquids	Gasoline, acetone, ether, pentane.
Solids	Plastics, wood dust, fibers, metal particles.
Gases	Acetylene, propane, carbon monoxide, hydrogen.

OXIDIZERS

Gases	Oxygen, fluorine, chlorine.
Liquids	Hydrogen peroxide, nitric acid, perchloric acid.
Solids	Metal peroxides, ammonium nitrite.

IGNITION SOURCES

Sparks, flames, static electricity, heat.

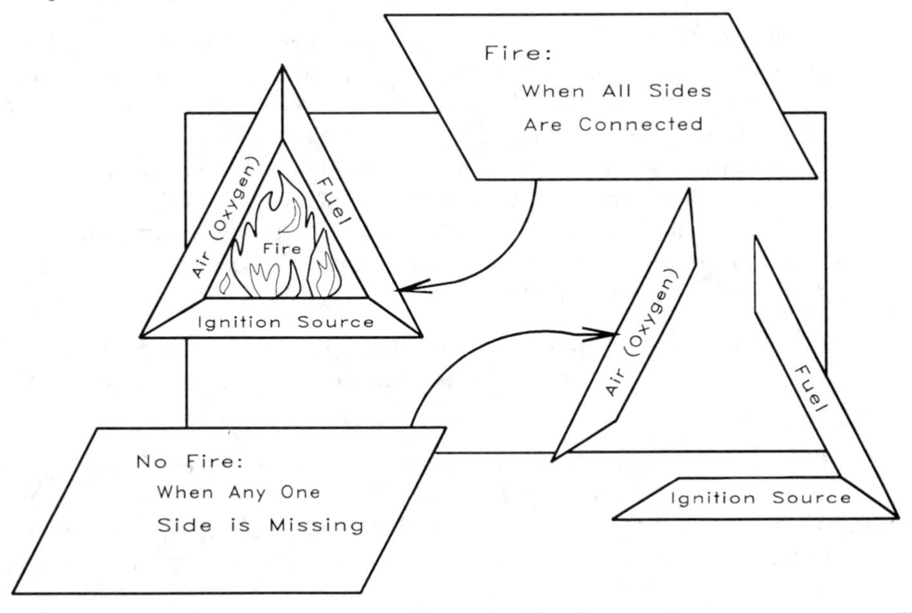

Figure 6-1 The fire triangle.

6-2 DISTINCTION BETWEEN FIRES AND EXPLOSIONS

The major distinction between fires and explosions is the rate of energy release. Fires release energy slowly, while explosions release energy very rapidly, typically on the order of microseconds. Fires can also result from explosions and explosions can result from fires.

A good example of how the energy release rate affects the consequences of an accident is a standard automobile tire. The compressed air within the tire contains energy. If the energy is released slowly through the nozzle, the tire is harmlessly deflated. If the tire ruptures suddenly and all of the energy within the compressed tire releases rapidly, the result is a dangerous explosion.

6-3 DEFINITIONS

Some of the commonly used definitions related to fires and explosions are given below.

Combustion or Fire: Combustion or fire is a chemical reaction in which a substance combines with an oxidant and releases energy. Part of the energy released is used to sustain the reaction.

Ignition: Ignition of a flammable mixture may be caused by a flammable mixture coming in contact with a source of ignition with sufficient energy or the gas reaching a temperature high enough to cause the gas to autoignite.

Autoignition temperature (AIT): A fixed temperature above which a flammable mixture is capable of extracting enough energy from the environment to self-ignite.

Flash Point (FP): The flash point of a liquid is the lowest temperature at which it gives off enough vapor to form an ignitable mixture with air. At the flash point, the vapor will burn, but only briefly; inadequate vapor is produced to maintain combustion. The flash point generally increases with increasing pressure.

There are several different experimental methods used to determine flash points. Each method produces a somewhat different value. The two most commonly used methods are open cup and closed cup, depending on the physical configuration of the experimental equipment. The open cup flash point is a few degrees higher then the closed cup.

Fire Point: The fire point is the lowest temperature at which a vapor above a liquid will continue to burn once ignited; the fire point temperature is higher than the flash point.

Flammability Limits (LFL and UFL): Vapor-air mixtures will only ignite and burn over a well-specified range of compositions. The mixture will not burn when the composition is lower than the lower flammable limit (LFL); the mixture is too lean for combustion. The mixture is also not combustible when the composition is too rich; that is, when it is above the upper flammable limit (UFL). A mixture is flammable only when the composition is between the LFL and the UFL. Commonly used units are volume percent fuel (percent of fuel plus air).

Lower explosive limit (LEL) and upper explosive limit (UEL) are used interchangeably with LFL and UFL.

Explosion: An explosion is a rapid expansion of gases resulting in a rapidly moving pressure or shock wave. The expansion can be mechanical (via the sudden rupture of a pressurized vessel) or it can be the result of a rapid chemical reaction. Explosion damage is caused by the pressure or shock wave.

Mechanical explosion: An explosion due to the sudden failure of a vessel containing high pressure, nonreactive gas.

Deflagration: An explosion with a resulting shock wave moving at a speed less then the speed of sound in the unreacted medium.

Detonation: An explosion with a resulting shock wave moving at a speed greater than the speed of sound in the unreacted medium.

Confined explosion: An explosion occurring within a vessel or a building. These are most common and usually result in injury to the building inhabitants and extensive damage.

Unconfined explosion: Unconfined explosions occur in the open. This type of explosion is usually the result of a flammable gas spill. The gas is dispersed and mixed with air until it comes in contact with an ignition source. Unconfined explosions are rarer than confined explosions since the explosive material is frequently diluted below the LFL by wind dispersion. These explosions are very destructive since large quantities of gas and large areas are frequently involved.

Boiling liquid expanding vapor explosion (BLEVE): A BLEVE occurs if a vessel ruptures which contains a liquid at a temperature above its atmospheric-pressure boiling point. The subsequent BLEVE is the explosive vaporization of a large fraction of the vessel contents; possibly followed by combustion or explosion of the vaporized cloud if it is combustible. This type of explosion occurs when an external fire heats the contents of a tank of volatile material. As the tank contents heat, the vapor pressure of the liquid within the tank increases and the tank's structural integrity is reduced due to the heating. If the tank ruptures the hot liquid volatilizes explosively.

Dust explosion: This explosion results from the rapid combustion of fine solid particles. Many solid materials (including common metals such as iron and aluminum) become very flammable when reduced to a fine powder.

Shock wave: A pressure wave moving through a gas. A shock wave in open air is followed by a strong wind; the combined shock wave and wind is called a blast wave. The pressure increase in the shock wave is so rapid that the process is mostly adiabatic.

Overpressure: The pressure on an object as a result of an impacting shock wave.

6-4 FLAMMABILITY CHARACTERISTICS OF LIQUIDS AND VAPORS

Flammability characteristics of some important organic chemicals (liquids and gases) are provided in Table 6-1.

Liquids

The flash point (FP) is one of the major physical properties used to determine the fire and explosion hazards of liquids. Flash points for pure components are easily determined experimentally. Table 6-1 lists flash points for a number of substances.

Flash points can be estimated for multicomponent mixtures if only one component is flammable and if the flash point of the flammable is known. In this case the flash point temperature is estimated by determining the temperature at which the vapor pressure of the flammable in the mixture is equal to the pure component vapor pressure at its flash point. Experimentally determined flash points are recommended for multicomponent mixtures with more than one flammable component.

Example 6-1

Methanol has a flash point of 54°F and its vapor pressure at this temperature is 62 mm Hg. What is the flash point of a solution containing 75% methanol and 25% water by weight?

Solution The mole fractions of each component are needed to apply Raoult's Law. Assuming a basis of 100 pounds of solution

	Pounds	Molecular weight	Moles	Mole fraction
Water	25	18	1.39	0.37
Methanol	75	32	2.34	0.63
			3.73	1.00

Raoult's law is used to compute the vapor pressure (p^{sat}) of pure methanol, based on the partial pressure required to flash.

$$p = xP^{sat}$$

$$P^{sat} = p / x = 62/0.63 = 98.4 \text{ mm Hg}$$

Using a graph of the vapor pressure versus temperature, shown on Figure 6-2, the flash point of the solution is 20.5°C or 68.9°F.

Vapors

Flammable limits for vapors are determined experimentally in a specially designed closed vessel apparatus shown in Figure 6-5. Vapor-air mixtures of known concen-

TABLE 6-1 FLAMMABILITY CHARACTERISTICS OF LIQUIDS AND GASES[1]

Compound	Flash point (°F)	LFL % in air	UFL % in air	Autoignition temperature (°F)
Acetone	0.0*	2.5	13	1000
Acetylene	Gas	2.5	100	
Acrolein	−14.8	2.8	31	
Acrylonitrile	32	3.0	17	
Aniline	158	1.3	11	
Benzene	12.0**	1.3	7.9	1044
Butane	−76	1.6	8.4	761
Carbon monoxide	Gas	12.5	74	
Chlorobenzene	85**	1.3	9.6	1180
Cyclohexane	−1**	1.3	8	473
Diborane	Gas	0.8	88	
Dioxane	53.6	2.0	22	
Ethane	−211	3.0	12.5	959
Ethyl alcohol	55	3.3	19	793
Ethylene	Gas	2.7	36.0	914
Ethylene oxide	−20*	3.0	100	800
Ethyl ether	−49.0**	1.9	36.0	180
Formaldehyde		7.0	73	
Gasoline	−45.4	1.4	7.6	
Heptane	24.8	1.1	6.7	
Hexane	−15	1.1	7.5	500
Hydrogen	Gas	4.0	75	1075
Isopropyl alcohol	53*	2.0	12	850
Isopropyl ether	0	1.4	7.9	830
Methane	−306	5	15	1000
Methyl acetate	15	3.1	16	935
Methyl alcohol	54*	6	36	867
Methyl chloride	32	8.1	17.4	1170
Methyl ethyl ketone	24*	1.4	11.4	960
Methyl isobutyl ketone	73	1.2	8.0	860
Methyl methacrylate	50*	1.7	8.2	790
Methyl propyl ketone	45	1.5	8.2	941
Naptha	−57	1.2	6.0	550
Octane	55.4	1.0	6.5	
Pentane	−40	1.51	7.8	588
Phenol	174	1.8	8.6	
Propane	Gas	2.1	9.5	
Propylene	−162	2.0	11.1	927
Propylene dichloride	61	3.4	14.5	1035
Propylene oxide	−35	2.3	36	869
Styrene	87**	1.1	7.0	914
Toluene	40	1.2	7.1	997

* Open cup flash point

**Closed cup flash point

[1]Martha W. Windholtz, ed., *The Merck Index: An Encyclopedia of Chemicals, Drugs, and Biologicals,* 10th ed. (Rahway, NJ: Merck and Company, 1983), p. 1124; Gressner G. Hawley, ed., *The Condensed Chemical Dictionary,* 10th ed. (New York: Van Nostrand Reinhold, 1981), pp 860-861; and Richard A. Wadden and Peter A. Scheff, *Engineering Design for the Control of Workplace Hazards* (New York: McGraw-Hill Book Company, 1987), pp. 146-156.

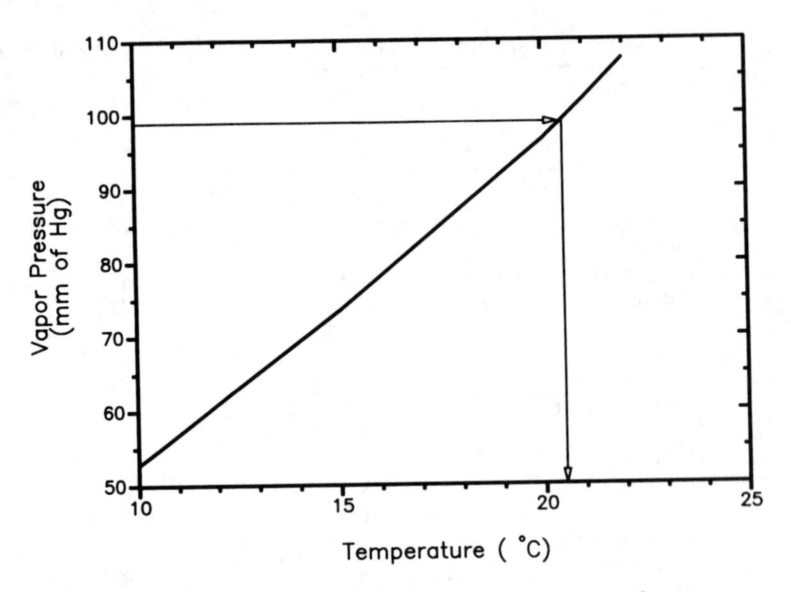

Figure 6-2 Saturation vapor pressure for methanol.

tration are added and then ignited. The maximum explosion pressure is measured. This test is repeated with different concentrations to establish the range of flammability for the specific gas. Figure 6-3 shows the results of one such experimental run; this particular substance has an LFL of 2.2 per cent and a UFL of 7.8 per cent.

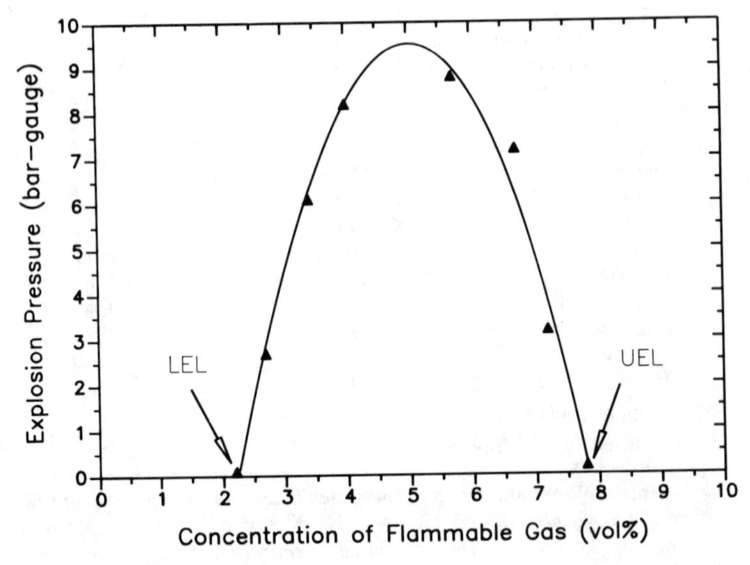

Figure 6-3 Flammability limits for a typical vapor.

Vapor Mixtures[2]

Frequently LFLs and UFLs for mixtures are needed. These mixture limits are computed using the Le Chatelier equation[3]

$$LFL_{mix} = \frac{1}{\sum_{i=1}^{n} \frac{y_i}{LFL_i}} \tag{6-1}$$

where

LFL_i is the lower flammable limit for component i in volume % of component i in fuel and air,

y_i is the mole fraction of component i on a combustible basis, and

n is the number of combustible species.

Similiarly,

$$UFL_{mix} = \frac{1}{\sum_{i=1}^{n} \frac{y_i}{UFL_i}} \tag{6-2}$$

where

UFL_i is the upper flammable limit for component i in volume % of component i in fuel and air.

The Le Chatelier's equation is an empirically derived equation which is not universally applicable. The limitations are covered in the literature.[4]

Example 6-2

What is the LFL and UFL of a gas mixture composed of 0.8% hexane, 2.0% methane, and 0.5% ethylene by volume?

Solution The mole fractions on a fuel only basis are calculated below. The LFL and UFL data are obtained from Table 6-1.

	Volume %	Mole fraction on combustible basis	LFL_i (vol. %)	UFL_i (vol. %)
Hexane	0.8	0.24	1.1	7.5
Methane	2.0	0.61	5.0	15
Ethylene	0.5	0.15	2.7	36.0
Total combustibles	3.3			
Air	96.7			

[2]Bodurtha, *Industrial Explosion Prevention and Protection,* pp. 7-24.

[3]H. Le Chatelier, "Estimation of Firedamp by Flammability Limits," *Ann. Mines,* Vol. 19, ser. 8, 1891, pp. 388-395.

[4]*U.S. Bureau of Mines Bulletin* 503 (1952), p. 6.

Equation 6-1 is used to determine the LFL of the mixture.

$$LFL_{\text{mix}} = \frac{1}{\sum_{i=1}^{n} \frac{y_i}{LFL_i}}$$

$$= \frac{1}{\frac{0.24}{1.1} + \frac{0.61}{5.0} + \frac{0.15}{2.7}}$$

$$= 1/0.396 = 2.53\% \text{ by volume total combustibles}$$

Equation 6-2 is used to determine the UFL of the mixture,

$$UFL_{\text{mix}} = \frac{1}{\sum_{i=1}^{n} \frac{y_i}{\text{UFL}_i}}$$

$$= \frac{1}{\frac{0.24}{7.5} + \frac{0.61}{15} + \frac{0.15}{36.0}}$$

$$= 13.0\% \text{ by volume total combustibles}$$

Since the above mixture contains 3.3% total combustibles, it is flammable.

Flammability Limit Dependence on Temperature

In general, the flammability range increases with temperature.[5] The following empirically derived equations are available for vapors.

$$LFL_T = LFL_{25}\,[1 - 0.75(T - 25)\,/\,\Delta H_c] \tag{6-3}$$

$$UFL_T = UFL_{25}\,[1 + 0.75(T - 25)\,/\,\Delta H_c] \tag{6-4}$$

where

ΔH_c is the net heat of combustion, (kcal/mole)
T is the temperature, °C.

Flammability Limit Dependence on Pressure

Pressure has little affect on the LFL except at very low pressures (< 50 mm Hg absolute), where flames do not propagate.

[5]M. G. Zabetakis, S. Lambiris, and G. S. Scott, "Flame Temperatures of Limit Mixtures," *Seventh Symposium on Combustion* (London, Butterworths, 1959), p. 484.

The UFL increases significantly as the pressure is increased, broadening the flammability range. An empirical expression for the UFL for vapors as a function of pressure is available.[6]

$$UFL_p = UFL + 20.6 (\log P + 1) \tag{6-5}$$

where

P is the pressure, (mega pascals absolute),

UFL is the upper flammable limit, (volume % of fuel plus air at 1 atm).

Example 6-3

If the UFL for a substance is 11.0% by volume at 0.0 MPa gauge, what is the UFL at 6.2 MPa gauge?

Solution The absolute pressure is $P = 6.2 + 0.101 = 6.301$ MPa. The UFL is determined using Equation 6-5.

$$UFL_P = UFL + 20.6 (\log P + 1)$$

$$UFL_P = 11.0 + 20.6 (\log 6.301 + 1)$$

$$UFL_P = 48 \text{ volume \% fuel in air}$$

Estimating Flammability Limits

For some situations it may be necessary to estimate the flammability limits without experimental data. Flammability limits are easily measured; experimental determination is always recommended.

Jones[7] found that for many hydrocarbon vapors the LFL and UFL are a function of the stoichiometric concentration (C_{st}) of fuel

$$LFL = 0.55\, C_{st} \tag{6-6}$$

$$UFL = 3.50\, C_{st} \tag{6-7}$$

where C_{st} is volume % fuel in fuel plus air.

The stoichiometric concentration for most organics is determined using the general combustion reaction

$$C_mH_xO_y + z\, O_2 \rightarrow m\, CO_2 + x/2\, H_2O \tag{6-8}$$

It follows from the stoichiometry that

$$z = m + x/4 - y/2$$

where z has units of moles O_2/mole fuel.

[6]M. G. Zabetakis, "Fire and Explosion Hazards at Temperature and Pressure Extremes," *AIChE-Inst. Chem. Engr. Symp., Ser. 2, Chem. Engr. Extreme Cond. Proc. Symp.*, 1965, pp. 99-104.

[7]G. W. Jones, "Inflammation Limits and Their Practical Application in Hazardous Industrial Operations," *Chem. Rev.*, Vol. 22, No. 1, (1938), pp. 1-26.

Additional stoichiometric and unit changes are required to determine C_{st} as a function of z.

$$C_{st} = \frac{\text{Moles Fuel}}{\text{Moles Fuel} + \text{Moles Air}} \times 100$$

$$= \frac{100}{1 + \left(\dfrac{\text{Moles Air}}{\text{Moles Fuel}}\right)}$$

$$= \frac{100}{1 + \left(\dfrac{1}{0.21}\right)\left(\dfrac{\text{Moles } O_2}{\text{Moles Fuel}}\right)}$$

$$= \frac{100}{1 + \left(\dfrac{z}{0.21}\right)}$$

Substituting z and applying Equations 6-6 and 6-7 yields

$$LFL = \frac{0.55\ (100)}{4.76m + 1.19x - 2.38y + 1} \tag{6-9}$$

$$UFL = \frac{3.50\ (100)}{4.76m + 1.19x - 2.38y + 1} \tag{6-10}$$

Example 6-4

Estimate the LFL and UFL for hexane and compare the calculated limits to the actual values determined experimentally.

Solution The stoichiometry is

$$C_6H_{14} + z\ O_2 \rightarrow m\ CO_2 + \frac{x}{2} H_2O$$

and z, m, x, and y are found by balancing this chemical reaction using the definitions in Equation 6-8.

$m = 6$

$x = 14$

$y = 0$

The LFL and UFL are determined using Equations 6-9 and 6-10.

$LFL = 0.55(100)/[4.76\ (6) + 1.19\ (14) + 1]$

$LFL = 1.19$ volume % versus 1.1 volume % actual

$UFL = 3.5\ (100)/[4.76\ (6) + 1.19\ (14) + 1]$

$UFL = 7.57$ volume % versus 7.5 volume % actual

6-5 MINIMUM OXYGEN CONCENTRATION (MOC) AND INERTING

The LFL is based on fuel in air. However, oxygen is the key ingredient and there is a minimum oxygen concentration required to propagate a flame. This is an especially useful result, because explosions and fires are preventable by reducing the oxygen concentration regardless of the concentration of the fuel. This concept is the basis for a common procedure called inerting (see Chapter 7).

Below the MOC, the reaction cannot generate enough energy to heat the entire mixture of gases (including the inerts) to the extent required for the self propagation of the flame.

The MOC has units of per cent oxygen in air plus fuel. If experimental data are not available, the MOC is estimated using the stoichiometry of the combustion reaction and the LFL. This procedure works for many hydrocarbons.

Example 6-5

Estimate the MOC for butane (C_4H_{10}).

Solution The stoichiometry for this reaction is

$$C_4H_{10} + 6.5\ O_2 \rightarrow 4\ CO_2 + 5\ H_2O$$

The LFL for butane (from Table 6-1) is 1.6% by volume. From the stoichiometry,

$$MOC = \left(\frac{\text{Moles Fuel}}{\text{Moles Fuel + Moles Air}}\right)\left(\frac{\text{Moles } O_2}{\text{Moles Fuel}}\right) = \text{LFL}\left(\frac{\text{Moles } O_2}{\text{Moles Fuel}}\right)$$

By substitution

$$MOC = \left(1.6\ \frac{\text{Moles Fuel}}{\text{Moles Fuel + Moles Air}}\right)\left(\frac{6.5 \text{ Moles } O_2}{1.0 \text{ Moles Fuel}}\right)$$

$$MOC = 10.4 \text{ volume } \%\ O_2$$

The combustion of butane is preventable by adding nitrogen, carbon dioxide or even water vapor until the oxygen concentration is below 10.4 %. The addition of water, however, is not recommended because any condition which condenses water would move the oxygen concentration back into the flammable region.

6-6 IGNITION ENERGY[8]

The minimum ignition energy (MIE) is the minimum energy input required to initiate combustion. All flammables (including dusts) have minimum ignition energies. The MIE depends on the specific chemical or mixture, the concentration, pressure, and temperature. A few MIEs are given in Table 6-2.

Experimental data indicates that

- The MIE decreases with an increase in pressure,
- The MIE of dusts are, in general, at energy levels comparable to combustible gases, and
- An increase in the nitrogen concentration increases the MIEs.

[8] P. Field, *Dust Explosions* (Amsterdam: Elsevier, 1982).

TABLE 6-2 MINIMUM IGNITION ENERGIES[1]

Combustible	Pressure (atm)	Minimum energy (mJ)
Methane	1	0.29
Propane	1	0.26
Heptane	1	0.25
Hydrogen	1	0.03
Propane (mole %) $[O_2/(O_2 + N_2)][100]$		
1.0	1	0.004
0.5	1	0.012
0.21	1	0.150
1.0	0.5	0.01
Cornstarch dust		0.3
Iron dust		0.12

[1]M. G. Zabetakis, "Flammability Characteristics of Combustible Gases and Vapors," *U.S. Bureau of Mines Bulletin* 627 (USNT AD 701, 576, 1965).

Many hydrocarbons have MIEs of about 0.25 mJ. This is low when compared to sources of ignition. For example, a static discharge of 22 mJ is initiated by walking across a rug, and an ordinary spark plug has a discharge energy of 25 mJ. Electrostatic discharges, as a result of fluid flow, also have energy levels exceeding the MIEs of flammables and can provide an ignition source, contributing to plant explosions (see Chapter 7).

6-7 AUTOIGNITION

The autoignition temperature (AIT) of a vapor, sometimes called the spontaneous ignition temperature (SIT), is the temperature at which the vapor ignites spontaneously from the energy of the environment. The autoignition temperature is a function of the concentration of vapor, volume of vapor, pressure of the system, presence of catalytic material, and flow conditions. It is essential to experimentally determine AITs at conditions as close as possible to process conditions.

Composition affects the AIT; rich or lean mixtures have higher AITs. Larger system volumes decrease the AITs; an increase in pressure decreases AITs; and increases in oxygen concentration decrease the AITs. This strong dependence on conditions illustrates the importance for exercising caution when using AIT data.

AIT data are provided in Table 6-1.

6-8 AUTOOXIDATION

Autooxidation is the process of slow oxidation with accompanying evolution of heat, sometimes leading to autoignition if the energy is not removed from the system. Liquids with relatively low volatility are particularly susceptible to this problem. Liquids with high volatility are less susceptible to autoignition because they self cool as a result of evaporation.

Many fires are initiated as a result of autooxidation, referred to as spontaneous combustion. Some examples of autooxidation with a potential for spontaneous combustion include

- Oils on a rag in a warm storage area,
- Insulation on a steam pipe saturated with certain polymers, and
- Filter aid saturated with certain polymers. Cases have been recorded where ten year old filter aid residues were ignited when the land-filled material was bulldozed, allowing autooxidation and eventual autoignition.

These examples illustrate why special precautions must be taken to prevent fires due to autooxidation and autoignition.

6-9 ADIABATIC COMPRESSION

An additional means of ignition is adiabatic compression. For example, gasoline and air in an automobile cylinder will ignite if the vapors are compressed to an adiabatic temperature which exceeds the autoignition temperature. This is the cause of preignition knock in engines which are running too hot and too lean. It is also the reason why some over-heated engines continue to run after the ignition is turned off.

Several large accidents were caused by flammable vapors being sucked into the intake of air compressors; subsequent compression resulted in autoignition. A compressor is particularly susceptible to autoignition if it has a fouled after-cooler. Safeguards must be included in the process design to prevent undesirable fires due to adiabatic compression.

The adiabatic temperature rise for an ideal gas is computed from the thermodynamic adiabatic compression equation

$$T_f = T_i \left(\frac{P_f}{P_i}\right)^{(\gamma-1)/\gamma} \tag{6-11}$$

where

T_f is the final absolute temperature,

T_i is the initial absolute temperature,

P_f is the final absolute pressure,

P_i is the initial absolute pressure, and

$\gamma = C_p / C_v$.

The potential consequences of adiabatic temperature increases within a chemical plant are illustrated in the following two examples.

Example 6-6

What is the final temperature after compressing air over liquid hexane from 14.7 psia to 500 psia if the initial temperature is 100°F? The AIT of hexane is 500°F (Table 6-1) and γ for air is 1.4.

Solution From Equation 6-11

$$T_f = (560)\left(\frac{500}{14.7}\right)^{(0.4/1.4)}$$

$$T_f = 1533°R = 1073°F$$

This exceeds the AIT for hexane, resulting in an explosion.

Example 6-7

The lubricating oil in piston type compressors is always found in minute amounts in the cylinder bore; compressor operations must always be maintained well below the AIT of the oil to prevent explosion.

A particular lubricating oil has an AIT of 400°C. Compute the compression ratio required to raise the temperature of air to the AIT of this oil. Assume an initial air temperature of 25°C and 1 atm.

Solution Equation 6-11 applies. Solving for the compression ratio,

$$\left(\frac{P_f}{P_i}\right) = \left(\frac{T_f}{T_i}\right)^{\gamma/(\gamma-1)}$$

$$= \left(\frac{400 + 273}{25 + 273}\right)^{1.4/0.4}$$

$$= 17.3$$

This represents an output pressure of only (17.3) (14.7 psia) = 254 psia. The actual compression ratio or pressure should be kept well below this.

These examples illustrate the importance of careful design, careful monitoring of conditions, and the need for periodic preventative maintenance programs when working with flammable gases and compressors. This is especially important today, because high pressure process conditions are becoming more common in modern chemical plants.

6-10 IGNITION SOURCES[9]

As illustrated by the fire triangle, fires and explosions are preventable by eliminating ignition sources. Various ignition sources were tabulated for over 25,000 fires by the Factory Mutual Engineering Corporation and are summarized in Table 6-3. The sources of ignition are numerous; consequently it is impossible to identify and eliminate them all. The main reason for rendering a flammable liquid inert, for example, is to prevent a fire or explosion by ignition from an unidentified source. Although all sources of ignition are not likely to be identified, engineers must still continue to identify and eliminate them.

Some very special situations might occur in a process facility where it is impossible to avoid flammable mixtures. In these cases a very thorough safety analysis is required to eliminate all possible ignition sources in each of the units where flammable gases are present.

[9]*Accident Prevention Manual for Industrial Operations* (Chicago: National Safety Council, 1974).

TABLE 6-3 IGNITION SOURCES OF MAJOR FIRES[1]

Electrical (wiring of motors)	23%
Smoking	18%
Friction (bearings or broken parts)	10%
Overheated materials (abnormally high temperatures)	8%
Hot surfaces (heat from boilers, lamps, etc.)	7%
Burner flames (improper use of torches, etc.)	7%
Combustion sparks (sparks and embers)	5%
Spontaneous ignition (rubbish, etc.)	4%
Cutting and welding (sparks, arcs, heat, etc.)	4%
Exposure (fires jumping into new areas)	3%
Incendiarism (fires maliciously set)	3%
Mechanical sparks (grinders, crushers, etc.)	2%
Molten substances (hot spills)	2%
Chemical action (processes not in control)	1%
Static sparks (release of accumulated energy)	1%
Lightning (where lightning rods are not used)	1%
Miscellaneous	1%

[1]*Accident Prevention Manual for Industrial Operations* (Chicago: National Safety Council, 1974).

The elimination of the ignition sources with the greatest probability of occurrence (see Table 6-3) should be given the greatest attention. Combinations of sources must also be investigated. The goal is to eliminate or minimize ignition sources since the probability of a fire or explosion increases rapidly as the number of ignition sources is increased. The effort required increases significantly as the size of the plant increases; potential ignition sources may be in the thousands.

6-11 SPRAYS AND MISTS[10]

Static electricity is generated when mists of sprays pass through orifices. A charge may accumulate and discharge in a spark. If flammable vapors are present, a fire or explosion will occur.

Mists and sprays also affect flammability limits.[11] For suspensions with drop diameters less than 0.01 mm, the lower flammability limit is virtually the same as the substance in vapor form. This is true even at low temperatures where the liquid is nonvolatile and no vapor is present. Mists of this type are formed by condensation.

For mechanically formed mists with drop diameters between 0.01 mm and 0.2 mm the LFL decreases as the drop diameter increases. In experiments with larger drop diameters, the lower flammability limit was less than one-tenth of the normal LFL. This is important when inerting in the presence of mists.

When sprays have drop diameters between 0.6 and 1.5 mm, flame propagation is impossible. In this situation, however, the presence of small drops and/or disturbances which shatter the larger drops, may create a hazardous condition.

[10]Frank P. Lees, *Loss Prevention in the Process Industries* (Boston: Butterworths, 1980).

[11]J. H. Borgoyne, "The Flammability of Mists and Sprays," *Chemical Process Hazards,* Vol 2, (1965), p. 1.

6-12 EXPLOSIONS

Explosion behavior depends on a large number of parameters. A summary of the more important parameters is shown in Table 6-4.

Explosion behavior is very difficult to characterize. Many approaches to the problem have been undertaken, including theoretical, semiempirical, and empirical studies. Despite these efforts, explosion behavior is still not completely understood. The practicing engineer, therefore, should use extrapolated results cautiously and provide a suitable "margin of safety" in all designs.

Detonation and Deflagration

Explosions are either detonations or deflagrations; the difference depends on the speed of the shock wave emanating from the explosion.

Suppose a combustible mixture is placed within a long pipe. A small spark, flame, or other ignition source initiates the reaction at one end of the pipe. After ignition, a flame or reaction front moves down the pipe.

In front of the flame front is a pressure or shock wave as shown in Figure 6-4. If the pressure wave moves faster than the speed of sound in the unreacted medium the explosion is a detonation; if it moves at a speed less than the speed of sound it is a deflagration.

The combustion of a mixture of gasoline and air within an internal combustion engine is a deflagration, even though the complete combustion process occurs in approximately 1/300 of a second. A detonation occurs much faster, within approximately 1/10,000 of a second.

The shock wave is due to the expansion of gases by the reaction. This is caused by either stoichiometric effects (change in the number of moles) or thermal expansion effects.

The relationship between the pressure-shock and reaction fronts determines whether a deflagration or detonation will occur. For deflagrations, the pressure increase is typically several atmospheres. For detonations the pressure rise is typically ten times higher, or more.

If the reaction or flame front propagation depends upon molecular or turbulent diffusion then the energy release rate is mass transfer limited. The result will

TABLE 6-4 PARAMETERS SIGNIFICANTLY AFFECTING THE BEHAVIOR OF EXPLOSIONS

1. Ambient temperature
2. Ambient pressure.
3. Composition of explosive material.
4. Physical properties of explosive material.
5. Nature of ignition source: type, energy, and duration.
6. Geometry of surroundings: confined or unconfined.
7. Amount of combustible material.
8. Turbulence of combustible material.
9. Time before ignition.
10. Rate at which combustible is released.

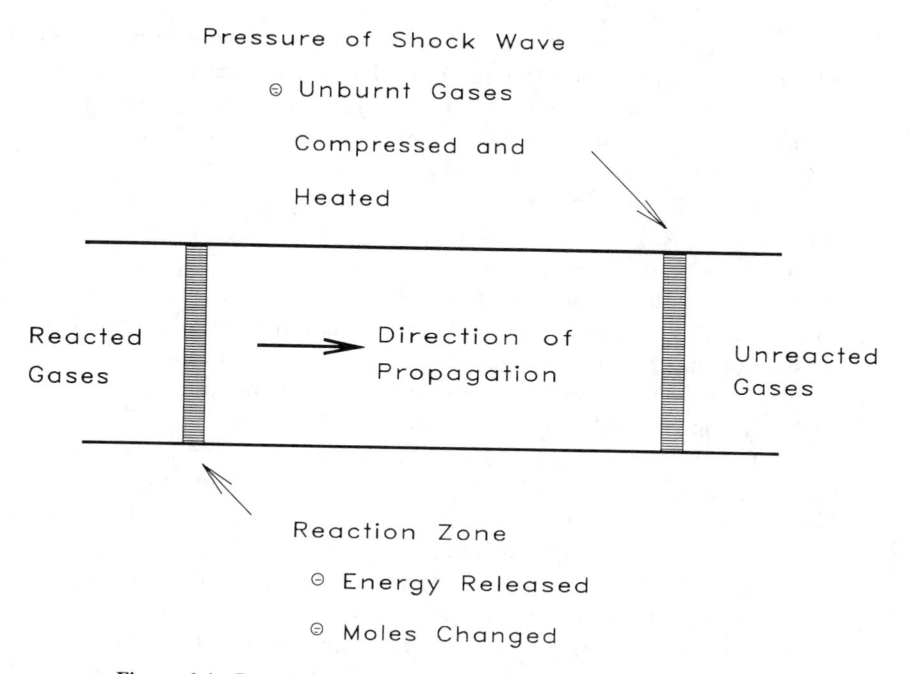

Figure 6-4 Reaction and pressure fronts propagating through a pipe.

be a relatively slow flame and pressure front movement. This is typical of deflagrations with fronts moving slower than the speed of sound.

There are several mechanisms leading to explosive detonation. The essential ingredient is that the energy must be released in a very short time within a very small volume to produce a significant initial pressure or shock wave. Two mechanisms have been proposed to describe such an event. In the first mechanism, called the thermal mechanism, the gas temperature increases by reaction, leading to self-acceleration of the reaction rate. In the second mechanism, called the chain branching mechanism, reactive free radicals or "centers" are rapidly increased in number by an elementary reaction. Typically a situation occurs where one free radical participates in a reaction that produces two free radicals. Both of these mechanisms can account for explosive behavior. In reality both are bound to occur.

A deflagration can also evolve into a detonation. This is particularly common in pipes but unlikely in vessels or open spaces. In a piping system, energy from a deflagration can feed forward to the pressure wave, resulting in an increase in the adiabatic pressure rise. The pressure builds and results in a full detonation.

Confined Explosions

A confined explosion occurs in a confined space, such as, a vessel or a building. The two most common confined explosion scenarios involve explosive vapors and explosive dusts. Empirical studies have shown that the nature of the explosion is a function of several experimentally determined characterisitics. These characteristics are dependent on the explosive material used and include flammability or explosive limits, the rate of pressure rise after the flammable mixture is ignited, and the maximum pressure after ignition. These characteristics are determined using two similiar laboratory devices, shown in Figures 6-5 and 6-8.

Explosion apparatus for vapors. The apparatus used to determine the explosive nature of vapors is shown in Figure 6-5. The test procedure includes (1) evacuating the vessel, (2) adjusting the temperature, (3) metering in the gases to obtain the proper mixture, (4) igniting the gas by a spark, and (5) measuring the pressure as a function of time.

After ignition the pressure wave moves outward within the vessel until it collides with the wall; the reaction is terminated at the wall. The pressure within the vessel is measured by a transducer located on the external wall. A typical pressure versus time plot is shown in Figure 6-6. Experiments of this type usually result in a deflagration with a few atmospheres of pressure rise.

The rate of pressure rise is indicative of the flame front propagation rate and thus the magnitude of the explosion. The pressure rate or slope is computed at the inflection point of the pressure curve as shown in Figure 6-6. The experiment is repeated at different concentrations. The pressure rate and maximum pressure for each run are plotted versus concentration as shown in Figure 6-7. The maximum pressure and maximum rate of pressure rise are determined. Typically, the maximum pressure and pressure rates occur somewhere within the range of flammability (but not necessarily at the same concentration). Using this relatively simple set of experiments, the explosive characteristics are completely established; in this example the flammability limits are between 2% and 8%, the maximum pressure is 7.4 bar, and the maximum rate of pressure rise is 360 bar/s.

Explosion apparatus for dusts. The experimental apparatus used to characterize the explosive nature of dusts is shown in Figure 6-8. The device is similiar to the

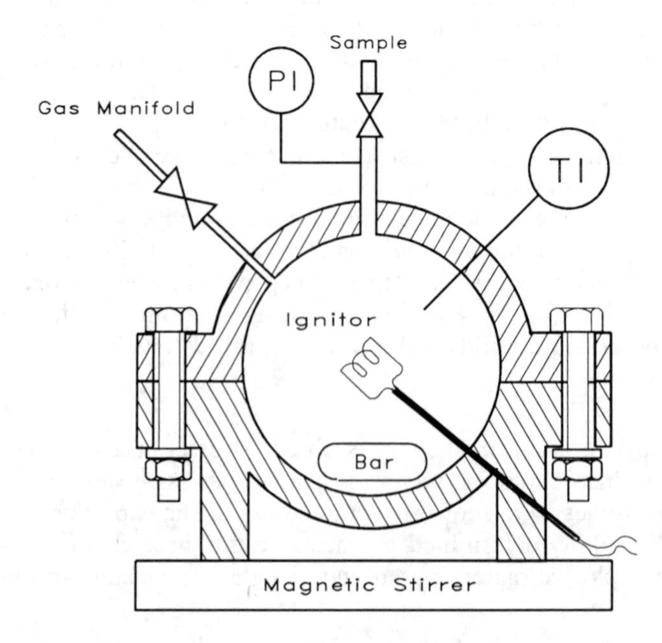

Figure 6-5 Test apparatus for acquiring vapor explosion data.

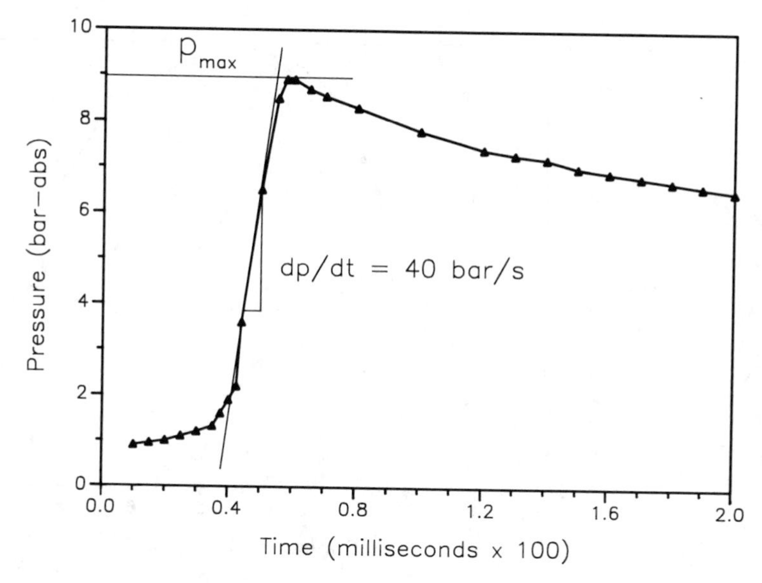

Figure 6-6 Typical pressure versus time data obtained from explosivity apparatus shown in Figure 6-5.

vapor explosion apparatus, with the exception of a larger volume and the addition of a sample container and a dust distribution ring. The distribution ring insures proper mixing of the dust prior to ignition.

The experimental procedure is as follows. The dust sample is placed in the sample container. The computer system opens the solenoid valve and the dust is driven by air pressure from the sample container through the distribution ring and into the dust sphere. After a delay of several milliseconds to insure proper mixing and distribution of the dust, the ignitor is discharged. The computer measures the pressure as a function of time using high and low speed pressure transducers. The air used to drive the dust into the sphere is carefully metered to insure a pressure of 1 atm (0.987 bar) within the sphere at ignition time.

The data are collected and analyzed in the same fashion as for the vapor explosion apparatus. The maximum pressure and the maximum rate of pressure rise are determined, as well as the flammability limits.

Explosion characteristics. The explosion characteristics determined using the vapor and dust explosion apparatus are used in the following way.

1. The limits of flammability or explosivity are used to determine the safe concentrations for operation or the quantity of inert required to control the concentration within safe regions.
2. The maximum rate of pressure rise is indicative of the robustness of an explosion. Thus, the explosive behavior of different materials can be compared on a relative basis. It is also used to design a vent for relieving a vessel during an explosion before the pressure ruptures the vessel, or to establish the time interval for adding an explosion suppressant (water, carbon dioxide, or Halon) to stop the combustion process.

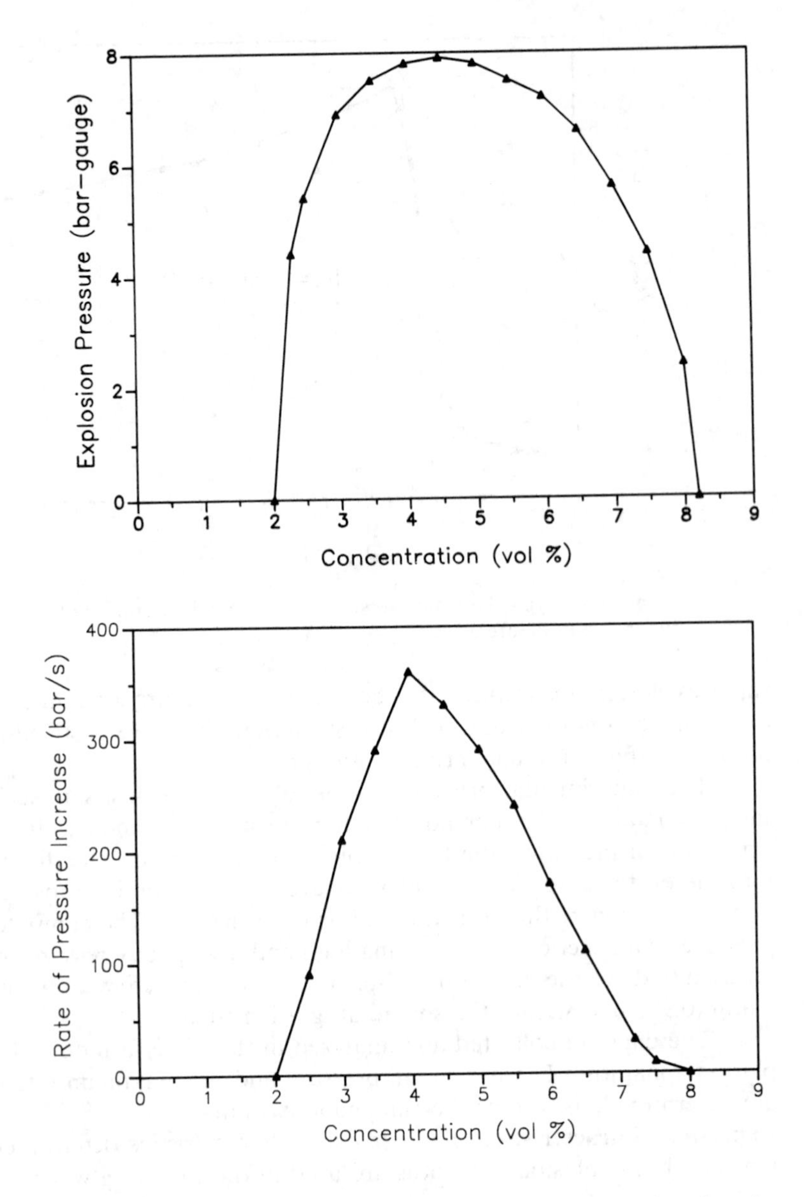

Figure 6-7 Pressure rate and maximum explosion pressure as a function of vapor concentration. The maximum pressure rate does not necessarily occur at the maximum pressure.

A plot of the log of the maximum pressure slope versus the log of the vessel volume frequently produces a straight line of slope −1/3 as shown in Figure 6-9. This relationship is called the "Cubic Law."

$$(dP/dt)_{\max} V^{1/3} = \text{constant} = K_g \tag{6-12}$$

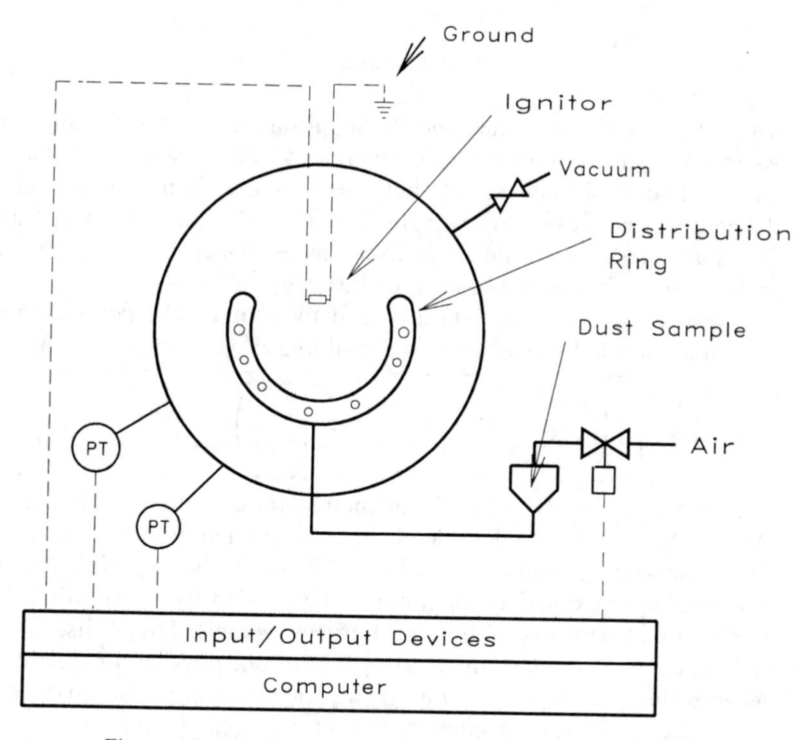

Figure 6-8 Test apparatus for acquiring dust explosion data.

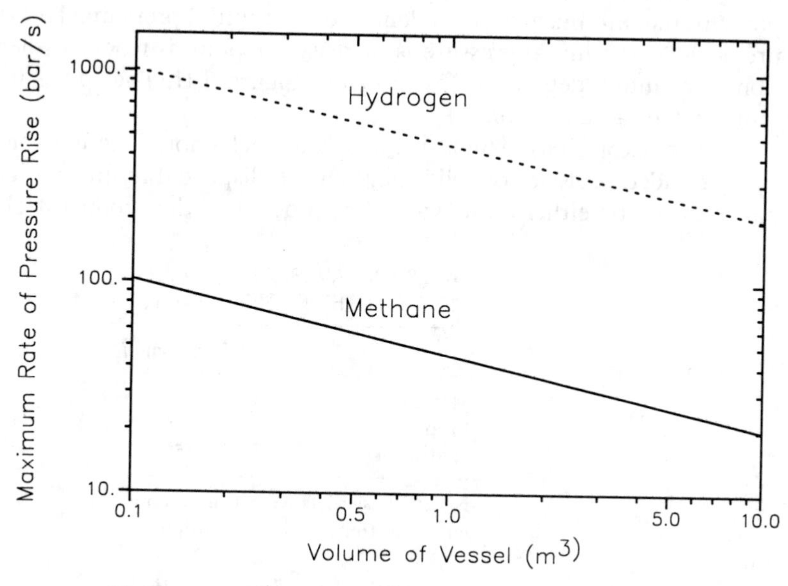

Figure 6-9 Typical explosion data exhibiting the cubic law.

$$\boxed{(dP/dt)_{\max} V^{1/3} = K_{St}} \tag{6-13}$$

where K_g and K_{st} are called the deflagration indices for gas and dust respectively. As the robustness of an explosion increases, the deflagration indices K_g and K_{st} increase. The cubic law states that the pressure front takes longer to propagate through a larger vessel. A few values for K_g and K_{st} are given in Tables 6-5 and 6-6. Dusts are further classified into four classes, depending on the value of the deflagration index. These *St classes* are shown in Table 6-6.

Equations 6-12 and 6-13 are used to estimate the consequences of an explosion in a confined space, such as a building or vessel, as follows.

$$\boxed{\left[\left(\frac{dP}{dt}\right)_{\text{MAX}} V^{1/3}\right]_{\text{IN VESSEL}} = \left[\left(\frac{dP}{dt}\right)_{\text{MAX}} V^{1/3}\right]_{\text{EXPERIMENTAL}}} \tag{6-14}$$

The subscript "IN VESSEL" is for the reactor or building. The subscript "EXPERIMENTAL" applies to data determined in the laboratory using either the vapor or dust explosion apparatus. Equation 6-14 allows the experimental results from the dust and vapor explosion apparatus to be applied to determining the explosive behavior of materials in buildings and process vessels. This is discussed in more detail in Chapter 9. The constants K_g and K_{st} are not physical properties of the material because they are dependent on (a) the composition of the mixture, (b) the mixing within the vessel, (c) the shape of the reaction vessel, and (d) the energy of the ignition source. It is, therefore, necessary to run the experiments as close as possible to the actual conditions under consideration.

Experimental studies indicate that the maximum explosion pressure is usually not affected by changes in volume and the maximum pressure and the maximum pressure rate are linearly dependent upon the initial pressure. This is shown in Figure 6-10. As the initial pressure is increased, a point is reached where the deflagration turns into a detonation, as shown in Figure 6-11. The spikes in the curves are indicative of a detonation.

Dust explosions demonstrate unique behavior. These explosions occur if finely divided particles of solid material are dispersed in air and ignited. The dust particles can be either an unwanted by-product or the product itself.

TABLE 6-5 AVERAGE K_g VALUES FOR SELECTED GASES[1]

Gas	K_g (bar m/s)
Methane	55
Propane	75
Hydrogen	550

Note: These tests were run at ambient conditions with an ignition energy of 10 J.

[1]W. Bartknecht, *Explosions* (New York: Springer-Verlag, 1981), p 10.

TABLE 6-6 AVERAGE K_{St} VALUES FOR SELECTED DUSTS[1]

Dust	P_{Max} (bar)	K_{St} (bar m/s)
PVC	6.7 - 8.5	27 - 98
Milk Powder	8.1 - 9.7	58 - 130
Polyethylene	7.4 - 8.8	54 - 131
Sugar	8.2 - 9.4	59 - 165
Resin Dust	7.8 - 8.9	108 - 174
Brown Coal	8.1 - 10.0	93 - 176
Wood Dusts	7.7 - 10.5	83 - 211
Cellulose	8.0 - 9.8	56 - 229
Pigments	6.5 - 10.7	28 - 344
Aluminum	5.4 - 12.9	16 - 750

St classes for dusts

Deflagration index, K_{St}	St class
0	St-0
1 - 200	St-1
200 - 300	St-2
> 300	St-3

[1]Bartknecht, *Explosions,* p. 35.

Explosions involving dusts are most common in the flour milling, grain storage, and coal mining industries. Accidents involving dust explosions can be quite substantial; a series of grain silo explosions in Westwego near New Orleans in 1977 killed thirty-five people.[12]

An initial dust explosion can cause secondary explosions. The primary explosion sends a shock wave through the plant, stirring up additional dust which may result in a secondary explosion. In this fashion the explosion "leapfrogs" its way through a plant. Many times the secondary explosions are more damaging than the primary.

Dust explosions are even more difficult to characterize than gaseous explosions. For a gas, the molecules are very small and of well defined size. For dust particles, the particles are of varying size and many orders of magnitude larger than molecules. Gravity also affects dust particle behavior.

For dusts, deflagrations appear to be much more common than detonations.[13] The pressure waves from dust deflagrations, however, are powerful enough to destroy structures and kill or injure people.

To be explosive, a dust mixture must have the following characteristics.

- The particles must be below a certain minimum size.
- The particle loading must be between certain limits.
- The dust loading must be reasonably uniform.

[12]Lees, *Loss Prevention in the Process Industries,* p.619.

[13]Lees, *Loss Prevention in the Process Industries,* p.619.

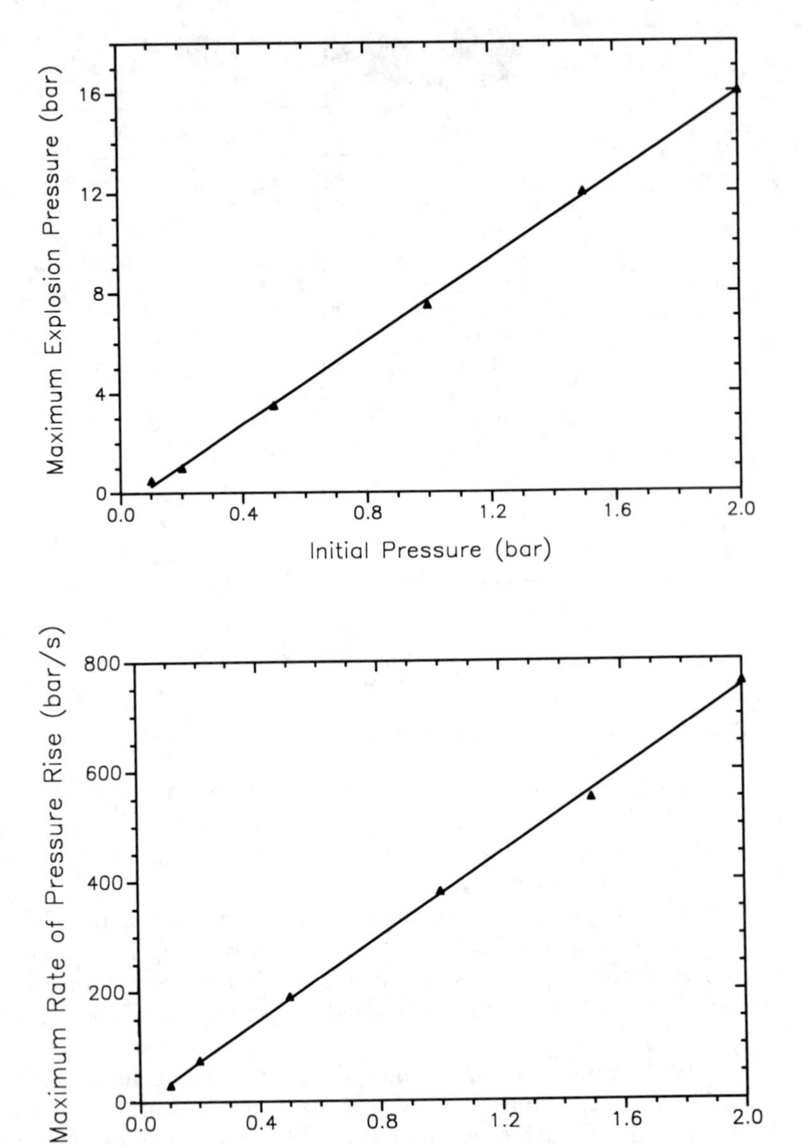

Figure 6-10 Effect of initial pressure on maximum explosion pressure and rate. (Data from Bartknecht, 1981.)

For most dusts,[14] the lower explosion limit is between 20 and 60 gm/m^3 and the upper explosion limit between 2 and 6 kg/m^3.

[14]W. Bartknecht, *Explosions* (New York: Springer-Verlag, 1981), p. 27.

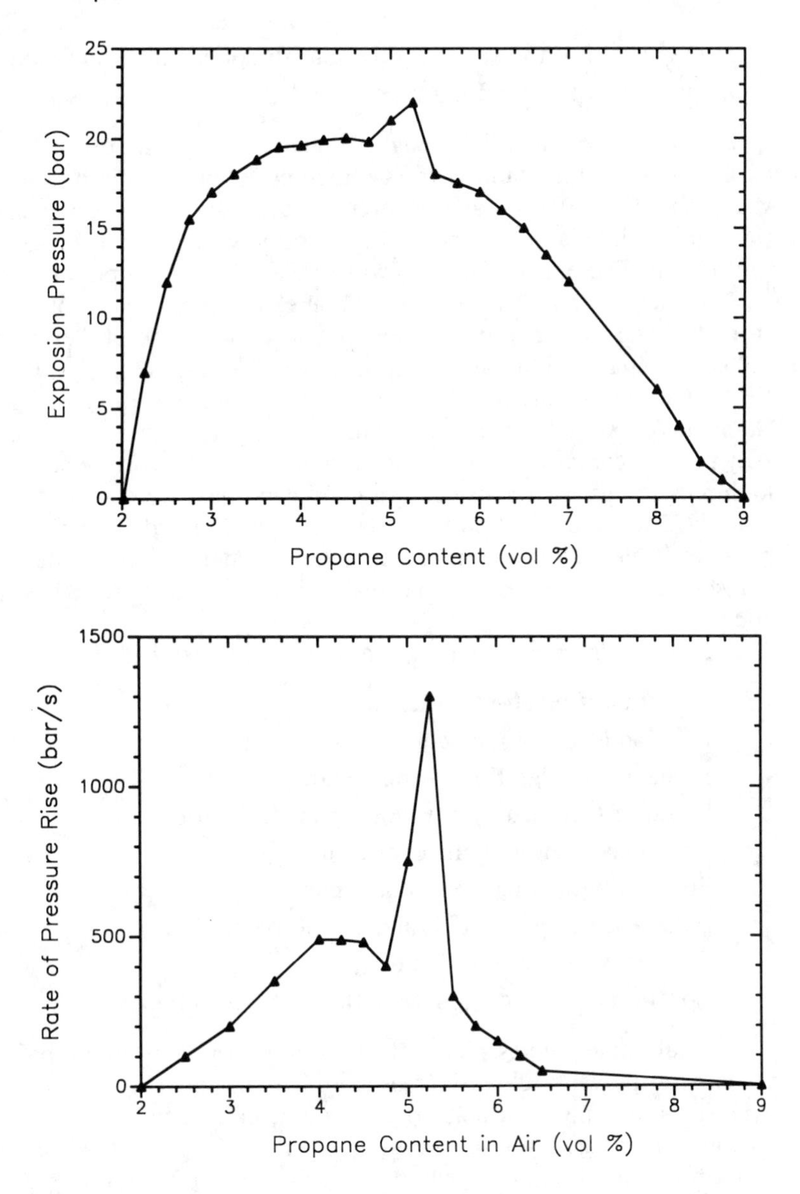

Figure 6-11 Explosion data for propane showing peaks indicative of the onset of detonation. (Data from Bartknecht, 1981.)

Vapor Cloud Explosions (VCE)

The most dangerous and destructive explosions in the chemical process industries are vapor cloud explosions (VCE). These explosions occur by a sequence of steps:

1. Sudden release of a large quantity of flammable vapor. Typically this occurs when a vessel, containing a superheated and pressurized liquid, ruptures.

2. Dispersion of the vapor throughout the plant site while mixing with air.
3. Ignition of the resulting vapor cloud.

The accident at Flixborough, England is a classic example of a vapor cloud explosion. A sudden failure of a 20-inch cyclohexane line between reactors led to vaporization of an estimated 30-tons of cyclohexane. The vapor cloud dispersed throughout the plant site and was ignited by an unknown source 45-seconds after the release. The entire plant site was leveled and 28 people were killed.

A summary of twenty-nine vapor cloud explosions[15] over the period 1974 through 1986 shows property losses for each event of between $5,000,000 to $100,000,000 and 140 fatalities (an average of almost 13 per year).

VCEs have increased in number due to an increase in inventories of flammable materials in process plants and operations at more severe conditions. Any process containing quantities of liquefied gases, volatile superheated liquid, or high pressure gases is considered a good candidate for a VCE.

VCEs are difficult to characterize, primarily due to the large number of parameters needed to describe an event. Accidents occur under uncontrolled circumstances. Data collected from real events are mostly unreliable and difficult to compare.

Some of the parameters that affect VCE behavior[16] are

Quantity of material released,
Fraction of material vaporized,
Probability of ignition of the cloud,
Distance travelled by the cloud prior to ignition,
Time delay before ignition of cloud,
Probability of explosion rather than fire,
Existence of a threshold quantity of material,
Efficiency of explosion, and
Location of ignition source with respect to release.

Qualitative studies[17] have shown that (a) the ignition probability increases as the size of the vapor cloud increases, (b) vapor cloud fires are more common than explosions, (c) the explosion efficiency is usually small; approximately 2% of the combustion energy is converted into a blast wave, and (d) turbulent mixing of vapor and air, and ignition of the cloud at a point remote from the release, increases the impact of the explosion.[18]

From a safety standpoint the best approach is to prevent the release of material. A large cloud of combustible material is very dangerous and almost impossible to control, despite any safety systems installed to prevent ignition.

[15]Richard W. Prugh, "Evaluation of Unconfined Vapor Cloud Explosion Hazards," *International Conference on Vapor Cloud Modeling* (New York: American Institute of Chemical Engineers, 1987), p. 713.

[16]Frank P. Lees, *Loss Prevention in the Process Industries,* p. 582.

[17]Lees, *Loss Prevention in the Process Industries,* p. 583.

[18]Prugh, "Evaluation of Unconfined Vapor Cloud Explosion Hazards," p. 714.

Methods which are used for preventing VCEs include keeping low inventories of volatile, flammable materials; using process conditions which minimize flashing if a vessel or pipeline is ruptured; using analyzers to detect leaks at very low concentrations; and installing automated block valves to shut systems down while the spill is in the incipient stage of development.

Boiling Liquid Expanding Vapor Explosions (BLEVE)[19]

A boiling liquid expanding vapor explosion (BLEVE, pronounced ble′-vee) is a special type of accident that can release large quantities of materials. If the materials are flammable, a VCE might result; if toxic, a large area might be subjected to toxic materials. For either situation, the energy released by the BLEVE process itself can result in considerable damage.

A BLEVE occurs when a tank containing a liquid held above its atmospheric pressure boiling point ruptures; resulting in the explosive vaporization of a large fraction of the tank contents.

BLEVEs are caused by the sudden failure of the container due to any cause. The most common type of BLEVE is caused by fire. The steps are as follows.

1. A fire develops adjacent to a tank containing a liquid.
2. The fire heats the walls of the tank.
3. The tank walls below liquid level are cooled by the liquid, increasing the liquid temperature and the pressure in the tank.
4. If the flames reach the tank walls or roof where there is only vapor and no liquid to remove the heat, the tank metal temperature rises until it loses it structural strength.
5. The tank ruptures, explosively vaporizing its contents.

If the liquid is flammable and a fire is the cause of the BLEVE, it may ignite as the tank ruptures. Often, the boiling and burning liquid behaves as a rocket fuel, propelling vessel parts for great distances. If the BLEVE is not caused by a fire, a vapor cloud might form, resulting in a VCE. The vapors might also be hazardous to personnel via skin burns or toxic effects.

When a BLEVE occurs in a vessel, only a fraction of the liquid vaporizes; the amount depends on the physical and thermodynamic conditions of the vessel contents. The fraction vaporized is estimated using the methods discussed in Section 4-6.

Blast Damage Due to Overpressure

The explosion of a dust or gas (either as a deflagration of detonation) results in a reaction front moving outwards from the ignition source preceded by a shock wave or pressure front. After the combustible material is consumed, the reaction front terminates, but the pressure wave continues its outward movement. A blast wave is composed of the pressure wave and subsequent wind. It is the blast wave that causes most of the damage.

[19]Lees, *Loss Prevention in the Process Industries,* p. 589; Bodurtha, *Industrial Explosion Prevention and Protection,* p. 99.

Blast damage is based on the determination of the peak overpressure resulting from the pressure wave impacting on a structure. In general, the damage is also a function of the rate of pressure rise and the duration of the blast wave. Good estimates of blast damage, however, are obtained using just the peak overpressure.

Damage estimates based on overpressures are given in Table 6-7. As illustrated, significant damage is expected for even small overpressures.

Experiments with explosives have demonstrated[20] that the overpressure can be estimated using an equivalent mass of TNT, denoted by m_{TNT}, and using the distance from the ground zero point of the explosion, denoted by r. The empirically derived scaling law is

$$z_e = \frac{r}{m_{TNT}^{1/3}} \tag{6-15}$$

The equivalent energy of TNT is 1120 cal/gm.

Figure 6-12 provides the correlation for overpressure (psi) versus the scaling parameter, z_e (ft/lb$^{1/3}$) of Equation 6-15. Figure 6-13 provides the correlation in SI units: the overpressure is in kPa and the scaling parameter, z_e, is in m/kg$^{1/3}$

Weather conditions also affect the overpressure. The upper and lower functions of Figure 6-12 are used for abnormal weather conditions while the center function is used for normal conditions.

TABLE 6-7 DAMAGE PRODUCED BY OVERPRESSURE[1]

Overpressure (PSIG)	Damage
0.03	Large glass windows which are already under strain broken.
0.04	Loud noise. Sonic boom glass failure.
0.15	Typical pressure for glass failure.
0.3	95% probability of no serious damage.
0.5-1	Large and small windows usually shattered.
0.7	Minor damage to house structures.
1	Partial demolition of houses, made uninhabitable
1.3	Steel frame of clad building slightly distorted.
2-3	Non-reinforced concrete or cinder walls shattered.
2.3	Lower limit of serious structural damage.
3	Steel frame building distorted and pulled from foundations.
3-4	Rupture of oil storage tanks.
5	Wooden utility poles snapped.
5-7	Nearly complete destruction of houses.
7	Loaded train wagons overturned.
9	Loaded train boxcars completely demolished.
10	Probable total destruction of buildings.
300	Limit of crater lip.

[1]V. J. Clancey, "Diagnostic Features of Explosion Damage," *Sixth Int. Mtg. of Forensic Sciences,* Edinburgh (1972).

[20]W. E. Baker, *Explosions in Air* (Austin: Univ. of Texas Press, 1973); S. Glasstone, *The Effects of Nuclear Weapons* (Washington: U.S. Atomic Energy Comm., 1962).

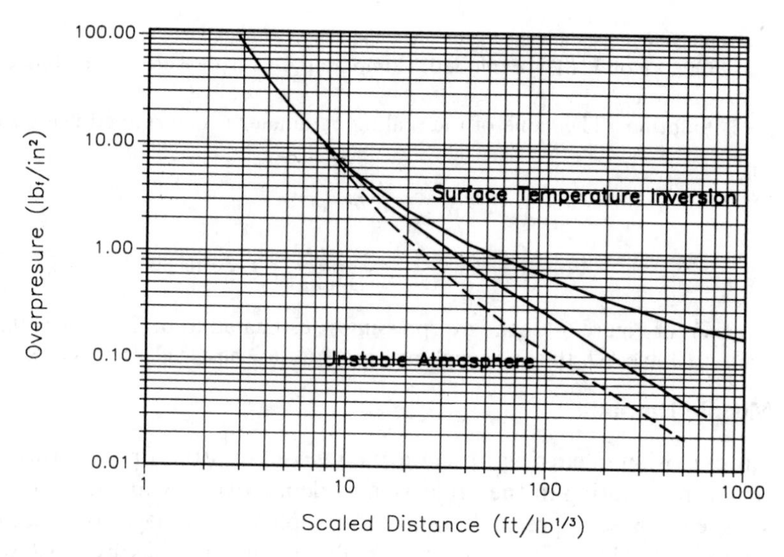

Figure 6-12 Correlation between overpressure and scaled distance, English engineering units.

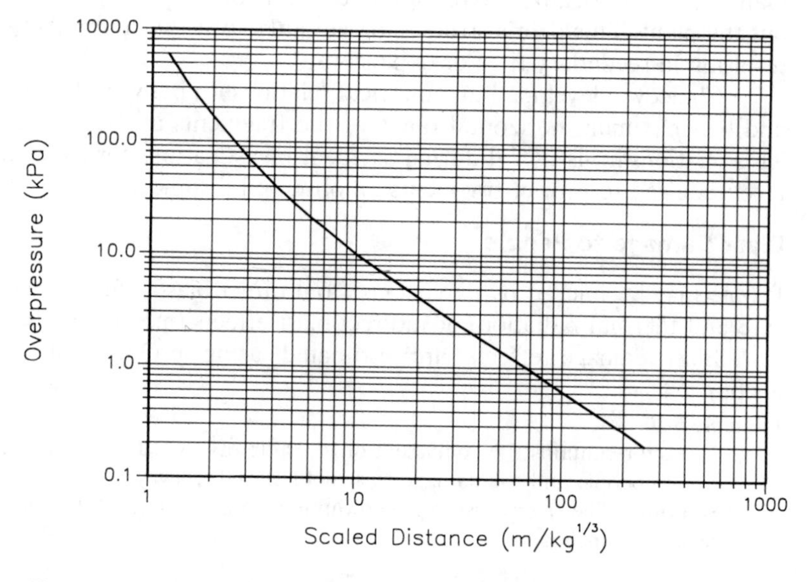

Figure 6-13 Correlation between overpressure and scaled distance, SI units.

The procedure for estimating the overpressure at any distance, r, due to the explosion of a mass of material is as follows: (1) compute the energy of the explosion using established thermodynamic procedures, (2) convert the energy to an equivalent amount of TNT, (3) use the scaling law and the correlations of Figures 6-12 and 6-13 to estimate the overpressure, and (4) use Table 6-7 to estimate the damage.

Example 6-9

One kg of TNT is exploded. Compute the overpressure at a distance of 30 m from the explosion.

Solution The value of the scaling parameter is determined using Equation 6-15.

$$z_e = \frac{r}{m_{TNT}^{1/3}}$$

$$z_e = \frac{30 \text{ m}}{(1.0 \text{ kg})^{1/3}} = 30 \text{ m kg}^{-1/3}$$

From Figure 6-12, the overpressure is estimated to be 2.4 kPa = 0.35 psi. According to Table 6-7, this is enough overpressure to shatter glass windows.

Missile Damage

An explosion occurring in a confined vessel or structure can rupture the vessel or structure resulting in the projection of debris over a wide area. This debris, or missiles, can cause appreciable injury to people and damage to structures and process equipment. Unconfined explosions also create missiles by blast wave impact and subsequent translation of structures.

Missiles are frequently a means by which an accident propagates throughout a plant facility. A localized explosion in one part of the plant projects debris throughout the plant. This debris strikes storage tanks, process equipment, and pipe lines, resulting in secondary fires or explosions.

Clancey[21] developed an empirical relationship between the mass of explosive and the maximum horizontal range of the fragments as illustrated in Figure 6-14. This relationship is useful during accident investigations for calculating the energy level required to project fragments an observed distance.

Blast Damage to People

People may be injured by explosions from direct blast effects (including overpressure and thermal radiation) or indirect blast effects (mostly missile damage).

Blast damage effects are estimated using probit analysis, discussed in Section 2-6.

Example 6-10

A reactor contains the equivalent of 10,000 lb of TNT. If it explodes, estimate the injury to people and the damage to structures 500 ft away.

Solution The overpressure is determined using Equation 6-15 and Figure 6-12. The scaled distance is

$$z_e = \frac{r}{m_{TNT}^{1/3}}$$

$$z_e = \frac{500 \text{ ft}}{(10{,}000 \text{ lb})^{1/3}}$$

$$z_e = 23.2 \text{ ft/lb}^{1/3}$$

[21] V. J. Clancey, "Diagnostic Features of Explosion Damage," *Sixth Int. Mtg. of Forensic Sciences,* Edinburgh, 1972.

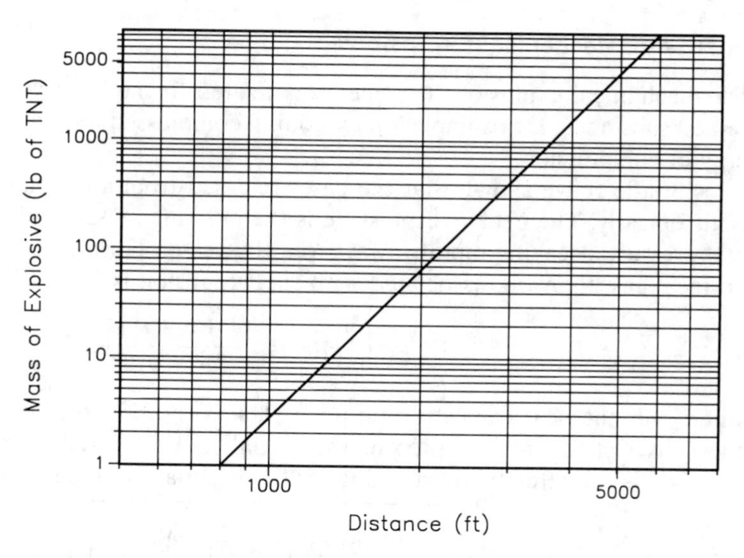

Figure 6-14 Maximum horizontal range of blast fragments. (Data from V. Clancey, "Diagnostic Features of Explosive Damage," *Sixth International Meeting of Forensic Science,* Edinburgh, 1972.)

From Figure 6-12 the overpressure is 1.8 psi. Table 6-7 indicates that houses will be severely damaged at this location.

Injury to personnel is determined using probit equations from Table 2-5. The probit equation for deaths due to lung hemorrhage is

$$Y = -77.1 + 6.91 \ln P$$

and the probit equation for eardrum rupture is

$$Y = -15.6 + 1.93 \ln P$$

where P is the overpressure in $\mathrm{N/m^2}$.

$$P = \left(\frac{1.8 \text{ psi}}{14.7 \text{ psi/atm}}\right)\left(101{,}325 \frac{\mathrm{N/m^2}}{\mathrm{atm}}\right)$$

$$P = 12{,}400 \mathrm{\ N/m^2}$$

Substituting into the probit equations,

$$Y_{\text{Deaths}} = -77.1 + 6.91 \ln(12{,}400) = -11.9$$

$$Y_{\text{Eardrums}} = -15.6 + 1.93 \ln(12{,}400) = 2.59$$

Table 2-4 converts the probit to percentages. The result shows that there are no deaths and less than 0.1 percent of the exposed people suffer eardrum ruptures. This assumes complete conversion of explosion energy.

Based on Figure 6-14, this explosion could project blast fragments a maximum distance of 6000 ft, resulting in probable injuries and damage due to blast fragments.

Energy of Mechanical Explosions

For mechanical explosions the energy is related to the energy content of the contained substance; for example, an exploding compressed air cylinder releases its energy of compression.

Studies have shown that the gases released from an exploding vessel expand isentropically. The peak overpressure is the bursting pressure of the vessel with the overpressure decaying rapidly away from the vessel. For an isentropic expansion of an ideal gas, the work associated with the expansion is

$$W_e = \int_1^2 PdV = \frac{(P_2V_2 - P_1V_1)}{(1-\gamma)} \tag{6-16}$$

where γ is the heat capacity ratio (C_p/C_v), P is the pressure, V is the volume, and the subscripts 1 and 2 represent the initial and final states, respectively. Equation 6-16 is rewritten by substituting for V_2, the final expanded volume. The result is

$$\boxed{W_e = \left(\frac{P_1 V_1}{\gamma - 1}\right)\left[1 - \left(\frac{P_2}{P_1}\right)^{(\gamma - 1)/\gamma}\right]} \tag{6-17}$$

where V_1 is the volume of gas within the container prior to rupture. Equation 6-17 is valid for compressed gases.

The energy released by compressed liquids is usually small since liquids are mostly incompressible. For this reason, pressure vessels are normally tested with liquids; very little energy is released if a liquid filled vessel ruptures.

Example 6-11

Compute the explosive energy of 1 kg-mole of air at 100 atm gauge and 20°C. Assume $\gamma = 1.4$. Determine the equivalent amount of TNT.

Solution From the ideal gas law

$$V_1 = \frac{R_g T}{P_1} = \frac{(82.06 \text{ liter atm/kg-mole °K})(20 + 273)\text{K}}{(101 \text{ atm})}$$

$$= 238 \text{ liters}$$

Substituting into Equation 6-17 to compute the expansion work

$$W_e = \left(\frac{P_1 V_1}{\gamma - 1}\right)\left[1 - \left(\frac{P_2}{P_1}\right)^{(\gamma - 1)/\gamma}\right]$$

$$W_e = \left[\frac{(101 \text{ atm})(238 \text{ l})}{(1.4 - 1)}\right]\left[1 - \left(\frac{1 \text{ atm}}{101 \text{ atm}}\right)^{0.4/1.4}\right]$$

$$= 44{,}000 \text{ liter-atm} = 1.06 \times 10^6 \text{ cal}$$

The equivalent amount of TNT due to this explosion is

$$\text{TNT equivalent} = 1.06 \times 10^6 \text{ cal/(1120 cal/gm TNT)}$$

$$= 0.95 \text{ kg TNT}$$

Tanks containing compressed gases at high pressures are relatively common in chemical plants and laboratories. Tanks of this type must be treated carefully since they contain a considerable amount of explosive energy.

Energy of Chemical Explosions

The blast wave following a chemical explosion is generated by two mechanisms: thermal heating of the reaction products and change in the number of moles by reaction.

If the explosion occurs so rapidly that the volume is unchanged, and the gas is ideal, the peak pressure after explosion is

$$P_f = \frac{P_i n_f T_f}{n_i T_i} \tag{6-18}$$

where P is the pressure, n is the number of moles, T is the temperature, and the subscripts i and f denote the initial and final states, respectively.

For the complete combustion of propane in air, the following stoichiometry applies:

$$C_3H_8 + 5\ O_2 + 18.8\ N_2 \rightarrow 3\ CO_2 + 4\ H_2O + 18.8\ N_2$$

The initial and final mole values are

$$n_i = 24.8$$

$$n_f = 25.8$$

In this case only a small pressure increase due to a change in the number of moles is expected and almost all of the blast wave energy is due to thermal effects. This is typical of many gases, particularly gases which are exploded in air.

The energy released during a reaction explosion is computed using standard thermodynamics. The released energy is equal to the work, Equation 6-16, required to expand the gases. This work is based on a constant temperature process with all the chemical energy being converted to PdV work; therefore, it is the maximum work derived for this process. In reality, there will be some increases in temperature; the actual work will be less than that determined using Equation 6-16.

Equation 6-16 is not convenient for making energy calculations. A more useful approach is to relate this process to the initial and final thermodynamic states. In this case, the maximum energy released is given by the Helmholtz free energy A,

$$dA = -SdT - PdV \tag{6-19}$$

For a constant temperature process

$$\boxed{dA = -PdV} \tag{6-20}$$

Other thermodynamic expressions which are needed for computing the energy for a chemical explosion include

$$\Delta A = (\Delta A_f^o)_P - (\Delta A_f^o)_R \tag{6-21}$$

$$\Delta A = \Delta U - T\Delta S \tag{6-22}$$

$$\Delta U = (\Delta U_f^o)_P - (\Delta U_f^o)_R \tag{6-23}$$

$$\Delta S = S_P - S_R \tag{6-24}$$

$$S = S^o - R_g \sum x_i \ln x_i \tag{6-25}$$

where the subscripts P and R denote products and reactants respectively, and

ΔA_f^o = Helmholtz free energy at the standard state,

ΔU_f^o = internal energy at the standard state,

x_i = mole fraction of species i,

S^o = entropy at the standard state, and

R_g = ideal gas constant.

The ln term in Equation 6-25 represents the entropy change of mixing.

Since most explosions occur at ambient conditions, the explosion energy is usually computed at standard conditions, 25°C and 1 atm. This is especially convenient because the standard state properties are usually available. Table 6-8 contains thermodynamic data useful for these calculations.

There are two methods for computing the energy of a chemical explosion using the above relationships: (1) If Helmholtz free energy data for reactants and products are available, use Equations 6-20 and 6-21, or (2) use Equations 6-20 and 6-22 to 6-25. In this case the internal energies and entropies for each species at the standard states must be known.

Frequently, however, thermodynamic data is not available for computing the energy of explosion. A number of approximations are useful. First, if internal energy data are not available, the Helmholtz free energy, A, cannot be computed using Equation 6-21. A good approximation is to assume that the internal energy change

TABLE 6-8 SELECTED THERMODYNAMIC PARAMETERS FOR EXPLOSION CALCULATIONS[1]

Species	State	ΔU_f^o	ΔH_f^o	ΔA_f^o	ΔG_f^o	S^o	ΔH_c^o
C(graphite)	s	0	0	0	0	1.3609	−94.052
CO	g	−26.722	−26.416	−33.104	−32.808	47.301	−67.636
CO_2	g	−94.052	−94.052	−94.260	−94.260	51.061	—
H_2	g	0	0	0	0	31.211	−68.317
H_2O	g	−57.502	−57.798	−56.339	−54.636	45.106	—
N_2	g	0	0	0	0	45.767	—
O_2	g	0	0	0	0	49.003	—
Acetylene	g		54.194		50.000	47.997	−310.615
Benzene	g		19.820		30.989	64.34	−789.080
n-Butane	g		−30.15		−4.10	74.12	−687.982
Cyclohexane	g		−29.43		7.59	71.28	−936.880
Ethane	g		−20.236		−7.860	54.85	−372.820
Ethylene	g		12.496		16.282	52.45	−337.234
Methane	g		−17.889		−12.140	44.50	−212.798
Propane	g		−24.820		−5.614	64.51	−530.605
Propylene	g		4.879		14.990	63.80	−491.987
Styrene	g		35.22		51.10	82.48	—
TNT	s	−13.0				65.0	

UNITS: ΔU_f^o, ΔH_f^o, ΔG_f^o, ΔH_c^o in kcal/mole, S^o in cal/mole°K

[1]Lees, *Loss Prevention in the Process Industries*, p. 562.

ΔH_{vap} H_2O = 10.5 Kcal/gmol

is equal to the enthalpy change. Similarly, the Gibbs free energy can be used instead of the Helmholtz free energy. Finally, unless the change in molecular weight during the reaction is large, the term $T\Delta S$ is small and can be neglected.

Example 6-12[22]

Compute the energy of explosion for TNT assuming the following reaction:

$$C_7H_5O_6N_3 \rightarrow C + 6\ CO + 2.5\ H_2 + 1.5\ N_2$$

The molecular weight of TNT is 227.

Solution The thermodynamic data is obtained from Table 6-8. ΔU of explosion is computed directly from the data using Equation 6-23,

$$\Delta U = (\Delta U_f^o)_P - (\Delta U_f^o)_R$$

$$= \{(1 \times 0) + [6(-26.722)] + (2.5 \times 0) + (1.5 \times 0)\} - [1(-13.0)]$$

$$= -147 \text{ kcal/mole}$$

$$= -648 \text{ cal/gm of TNT}$$

The entropy of the explosion is computed from Equations 6-24 and 6-25.

$$\Delta S = (S^o)_P - R_g \sum x_i \ln x_i - (S)_R$$

and the entropy of mixing terms are

Species	x_i	$x_i \ln x_i$
CO	0.6	−0.3065
H_2	0.25	−0.3466
N_2	0.15	−0.2846
		−0.9377

$$-R_g \sum x_i \ln x_i = -(1.987 \text{ cal/gm-mole°K})(-0.9377)$$

$$= 1.863 \text{ cal/gm-mole°K}$$

The change in entropy of explosion is

$$\Delta S = 1 \times 1.361 + 6 \times 47.301 + 2.5 \times 31.211 + 1.5 \times 45.767 + 10 \times 1.863 - 1 \times 65$$

$$= 385.5 \text{ cal/mole°K}$$

The energy of explosion is given by Equation 6-22.

$$\Delta A = \Delta U - T\Delta S$$

$$= -147{,}000 - (298 \times 385.5)$$

$$= -262{,}000 \text{ cal/mole}$$

$$= -1154 \text{ cal/gm of TNT}$$

The actual experimental value is 1120 cal/gm.

[22] Lees, *Loss Prevention in the Process Industries,* p. 563.

Other thermodynamic routes to the pdV expansion work are described in Lees;[23] the methods described above are the most commonly used.

SUGGESTED READING

W. BARTKNECHT, *Explosions* (New York: Springer-Verlag, 1980).

FRANK T. BODURTHA, *Industrial Explosion Prevention and Protection* (New York: McGraw-Hill Book Company, 1980).

FRANK P. LEES, *Loss Prevention in the Process Industries* (London: Butterworths, 1986), Chapter 16: Fire, Chapter 17: Explosion.

DANIEL R. STULL, "Fundamentals of Fire and Explosion," *AICHE Monograph Series,* No. 10, Vol 73 (New York: American Institute of Chemical Engineers, 1977).

PROBLEMS

6-1. Estimate the flash point of a solution of 50 mole % water and 50 mole % methanol.

6-2. Estimate the flash point of a solution of 50 mole % water and 50 mole % ethanol.

6-3. Estimate the LFL and UFL of the following mixtures:

	All in volume %			
	a	b	c	d
Hexane	0.5	0.0	1.0	0.0
Methane	1.0	0.0	1.0	0.0
Ethylene	0.5	0.5	1.0	1.0
Acetone	0.0	1.0	0.0	1.0
Ethyl Ether	0.0	0.5	0.0	1.0
Total combustibles	2.0	2.0	3.0	3.0
Air	98.0	98.0	97.0	97.0

6-4. Estimate the LFL and UFL of Problem 6-3a at 50°C, 75°C, and 100°C.

6-5. Estimate the UFL of Problem 6-3a at 1 atm, 5 atm, 10 atm, and 20 atm of pressure.

6-6. Estimate the LFL and UFL using the stoichiometric concentrations for methane, propylene, ethyl ether, and acetone. Compare these estimates to actual values.

6-7. Estimate the MOC of propane, hydrogen, and methane.

6-8. Determine the MOC of a mixture of 2% hexane, 3% propane, and 2% methane, by volume.

6-9. Determine the minimum compression ratio required to raise the temperature of air over hexane to its AIT. Assume an initial temperature of 100°C.

6-10. What will be the LFL of hexane in the presence of hexane mists with drops larger than 0.01 mm?

6-11. Why do staged hydrogen compressors need interstage coolers?

6-12. Why do hot engines sometimes continue to run after the ignition is turned off?

[23]Lees, *Loss Prevention in the Process Industries,* pp. 561-564.

6-13. A set of experiments are run on a flammable gas in a spherical vessel. The following data are obtained for two different vessel volumes. Estimate the value of K_g for this combustible gas.

$V = 1\ m^3$		$V = 20\ m^3$	
Time (sec)	P (bar)	Time (sec)	P (bar)
0.0	0.0	0.0	0.0
0.1	0.2	0.2	0.15
0.2	0.5	0.3	0.35
0.3	1.2	0.4	0.6
0.35	1.6	0.5	0.9
0.40	3.2	0.6	1.4
0.425	4.7	0.7	2.2
0.450	6.5	0.8	4.1
0.475	6.9	0.85	5.0
0.500	7.1	0.90	6.2
0.550	7.4	0.95	7.1
0.600	7.3	1.00	7.0
0.650	7.0	1.05	7.2
0.700	6.4	1.10	6.7
0.750	6.1	1.15	6.25
0.800	5.7	1.20	5.90
0.900	5.1	1.30	5.40
1.000	4.7	1.40	5.00
		1.50	5.60

6-14. Determine the energy of explosion for 1-lb of gaseous n-butane. What is the TNT equivalent?

6-15. A gas cylinder contains 50 lbs of propane. The cylinder accidently falls over and ruptures, vaporizing the entire contents of the cylinder. The cloud is ignited and an explosion occurs. Determine the overpressure from this explosion 100 feet away. What type of damage is expected?

6-16. A VCE with methane destroyed a house structure 100 feet away from the ignition source. Estimate the amount of methane released.

6-17. A large cloud of propane is released and eventually ignited producing a VCE. Estimate the quantity of propane released if the blast shattered windows three miles from the source of the ignition.

Designs to Prevent Fires and Explosions

7

A two-fold strategy is used to limit the potential damage from fires and explosions: prevent the initiation of the fire or explosion and minimize the damage after a fire or explosion has occurred. This strategy is presented in this chapter. The specific topics include

- Inerting
- Static electricity
- Controlling static electricity
- Ventilation
- Explosion proof equipment and instruments
- Sprinkler systems
- Miscellaneous design features for preventing fires and explosions

For any fire or combustion explosion to occur, three conditions must be met (as shown in the Fire Triangle of Figure 6-1). First, a combustible or explosive material must be present. Second, oxygen must be present to support the combustion reaction. Finally, a source of ignition must be available to initiate the reaction. If any of the three conditions of the fire triangle are eliminated, the triangle is broken and it is impossible for a fire or combustion explosion to result. This is the basis for the first four design methods listed above.

Damage due to fires and explosions is minimized by stopping fires or explosions as quickly as possible, and also by designing the process equipment (and control centers) to withstand their effects. This is the basis for design methods five and six listed above.

The last topic on miscellaneous design features gives a summary of additional safety design methods which are usually included in the early design phases of new projects and are often the basis for safety improvements in existing plants.

7-1 INERTING

Inerting is the process of adding an inert gas to a combustible mixture to reduce the concentration of oxygen below the minimum oxygen concentration (MOC). The inert gas is usually nitrogen or carbon dioxide, although steam is sometimes used. For many gases the MOC is approximately 10 percent and for many dusts approximately 8 percent.

Inerting begins with an initial purge of the vessel with inert gas to bring the oxygen concentration down to safe concentrations. A commonly used control point is 4 percent below the MOC, that is 6 percent oxygen if the MOC is 10 percent.

After the empty vessel has been inerted, the flammable material is charged. An inerting system is required to maintain an inert atmosphere in the vapor space above the liquid. This system, ideally, should include an automatic inert gas addition feature to control the oxygen concentration below the MOC. This control system should have an analyzer to continuously monitor the oxygen concentration in relationship to the MOC, and a controlled inert gas feed system to add inert gas when the oxygen concentration approaches the MOC. More frequently, however, the inerting system consists only of a regulator designed to maintain a fixed positive inert pressure in the vapor space; this insures that inert is always flowing out of the vessel rather than air flowing in. The analyzer system, however, results in a significant savings in inert gas usage without sacrificing safety.

Consider an inerting system designed to maintain the oxygen concentration below 10%. As oxygen leaks into the vessel and the concentration rises to 8 percent, a signal from the oxygen sensor opens the inert gas feed valve. Once again the oxygen level is adjusted to 6 percent. This closed loop control system, with high (8 percent) and low (6 percent) inerting set points, maintains the oxygen concentration at safe levels with a reasonable margin of safety.

There are several purging methods used to *initially* reduce the oxygen concentration to the low set point, as described below.

Vacuum Purging

Vacuum purging is the most common inerting procedure for vessels. This procedure is not used for large storage vessels because they are usually not designed for vacuums, and usually only withstand a pressure of a few inches of water.

Reactors, however, are often designed for full vacuum, that is -760 mm Hg gauge or 0.0 mm Hg absolute. Consequently, vacuum purging is a common procedure for reactors. The steps in a vacuum purging process include (1) draw a vacuum on the vessel until the desired vacuum is reached, (2) relieve the vacuum with an inert, such as, nitrogen or carbon dioxide to atmospheric pressure, (3) repeat steps 1 and 2 until the desired oxidant concentration is reached.

The initial oxidant concentration under vacuum (y_o) is the same as the initial concentration, and the number of moles at the initial high pressure (P_H) and initial low pressure or vacuum (P_L) are computed using an equation of state.

Assuming ideal gas behavior, the total moles at each pressure are

$$n_H = \frac{P_H V}{R_g T} \tag{7-1}$$

$$n_L = \frac{P_L V}{R_g T} \tag{7-2}$$

where n_H and n_L are the total moles under the atmospheric and vacuum states, respectively.

The number of moles of oxidant for the low pressure P_L and high pressure P_H are computed using Dalton's Law

$$(n_{oxy})_{1L} = y_0 n_L \tag{7-3}$$

$$(n_{oxy})_{1H} = y_0 n_H \tag{7-4}$$

where 1L and 1H is the first atmospheric and first vacuum states respectively.

When the vacuum is relieved with pure nitrogen, the moles of oxidant are the same as in the vacuum state, and the moles of nitrogen increase. The new (lower) oxidant concentration is

$$y_1 = \frac{(n_{oxy})_{1L}}{n_H} \tag{7-5}$$

where y_1 is the oxygen concentration after the first purge with nitrogen. Substituting Equation 7-3 into Equation 7-5,

$$y_1 = \frac{(n_{oxy})_{1L}}{n_H} = y_o\left(\frac{n_L}{n_H}\right)$$

If the vacuum and inert relief process is repeated, the concentration after the second purge is

$$y_2 = \frac{(n_{oxy})_{2L}}{n_H} = y_1\frac{n_L}{n_H} = y_0\left(\frac{n_L}{n_H}\right)^2$$

This process is repeated as often as required to decrease the oxidant concentration to a desired level. The concentration after j purge cycles, vacuum and relief, is given by the following general equation

$$y_j = y_0\left(\frac{n_L}{n_H}\right)^j = y_0\left(\frac{P_L}{P_H}\right)^j \tag{7-6}$$

This equation assumes that the pressure limits, P_H and P_L are identical for each cycle.

The total moles of nitrogen added for each cycle is constant. For j cycles the total nitrogen is given by

$$\Delta n_{N_2} = j\,(P_H - P_L)\frac{V}{R_g T} \tag{7-7}$$

Example 7-1

Use a vacuum purging technique to reduce the oxygen concentration within a 1000 gallon vessel to 1 ppm. Determine the number of purges required and the total nitrogen used. The temperature is 75°F and the vessel is originally charged with air under ambient conditions. A vacuum pump is used which reaches 20 mm Hg absolute, and the vacuum is subsequently relieved with pure nitrogen until the pressure returns to 1 atm absolute.

Solution The concentration of oxygen at the initial and final states is

$$y_0 = 0.21 \text{ lb-moles } O_2/\text{total moles}$$

$$y_f = 1 \text{ ppm} = 1 \times 10^{-6} \text{ lb-moles } O_2/\text{total moles}$$

The required number of cycles is computed using Equation (7-6)

$$y_j = y_0 \left(\frac{P_L}{P_h}\right)^j$$

$$\ln\left(\frac{y_j}{y_0}\right) = j \ln\left(\frac{P_L}{P_H}\right)$$

$$j = \frac{\ln(10^{-6}/0.21)}{\ln(20 \text{ mm Hg}/760 \text{ mm Hg})} = 3.37$$

Number of purges $= j = 3.37$

Four purge cycles are required to reduce the oxygen concentration to 1 ppm.

The total nitrogen used is determined from Equation 7-7. The final pressure, P_L is

$$P_L = \left(\frac{20 \text{ mm Hg}}{760 \text{ mm Hg}}\right)(14.7 \text{ psia}) = 0.387 \text{ psia}$$

$$\Delta n_{N_2} = j\,(P_H - P_L)\frac{V}{R_g T}$$

$$= 4\,(14.7 - 0.387) \text{ psia} \frac{(1000 \text{ gal})(1 \text{ ft}^3/7.48 \text{ gal})}{(10.73 \text{ psia ft}^3/\text{lb-mole°R})(75 + 460)\text{°R}}$$

$$= 1.33 \text{ lb-mole} = 37.2 \text{ lb of nitrogen}$$

Pressure Purging

Vessels may be pressure purged by adding inert gas under pressure. After this added gas is diffused throughout the vessel, it is vented to the atmosphere, usually down to atmospheric pressure. More than one pressure cycle may be necessary to reduce the oxidant content to the desired concentration.

The relationship used for this purging process is identical to Equation 7-6, where n_L is now the total moles at atmospheric pressure(low pressure) and n_H is the total moles under pressure (high pressure). In this case, however, the initial concentration of oxidant in the vessel (y_o) is computed after the vessel is pressurized (the first pressurized state). The moles for this pressurized state is n_H and the moles for the atmospheric case is n_L.

Equation 7-6 can also be used for a purging process including combined pressure and vacuum purging. In this case the moles under pressure is n_H and the moles under vacuum is n_L.

One practical advantage of pressure purging versus vacuum purging is the potential for cycle time reductions. The pressurization process is much more rapid compared to the relatively slow process of developing a vacuum. Also, the capacity of vacuum systems decreases significantly as the absolute vacuum is decreased. Pressure purging, however, uses more inert gas. Therefore the best purging process is selected based on cost and performance.

Example 7-2

Use a pressure purging technique to reduce the oxygen concentration in the same vessel discussed in Example 7-1. Determine the number of purges required to reduce the oxygen concentration to 1 ppm using pure nitrogen at a pressure of 80 psig, and at a temperature of 75°F. Also determine the total nitrogen required. Compare the quantities of nitrogen required for the two purging processes.

Solution Equation 7-6 is used to determine the number of cycles required. The initial mole fraction of oxygen, y_o, is now the concentration of oxygen at the end of the first pressurization cycle. The composition at the high pressure condition is determined using the following equation,

$$y_o = (0.21)(P_o/P_H)$$

where P_o is the starting pressure (here atmospheric).
Substituting the numbers provided,

$$y_o = (0.21)[14.7 \text{ psia}/(80 + 14.7) \text{ psia}] = 0.0326$$

The final oxygen concentration (y_f) is specified to be 1 ppm or 10^{-6} lb-mole oxygen/total lb-moles. The number of cycles required is computed using Equation 7-6

$$y_j = y_0 \left(\frac{P_L}{P_H}\right)^j$$

$$j = \frac{\ln(10^{-6}/0.0326)}{\ln[14.7 \text{ psia}/(80 + 14.7) \text{ psia}]}$$

The number of purge cycles is $j = 5.6$. Thus, six pressure purges are required, compared to four for the vacuum purge process. The quantity of nitrogen used for this inerting operation is determined using Equation 7-7.

$$\Delta n_{N_2} = j(P_H - P_L)\frac{V}{R_g T}$$

$$= 6(94.7 - 14.7) \text{ psia} \frac{133.7 \text{ ft}^3}{(10.73 \text{ psia ft}^3/\text{lb-mole}^\circ\text{R})(535^\circ\text{R})}$$

$$= 11.1 \text{ lb moles} = 311 \text{ lb of nitrogen.}$$

Pressure purging requires six purges and a total of 311 lb of nitrogen, compared to four purges and a total of 37.2 lb of nitrogen for vacuum purging. This result illustrates the need for a cost performance comparison to determine if the time saved in pressure purging justifies the added cost for nitrogen.

Sweep-Through Purging

The sweep-through purging process adds purge gas into a vessel at one opening, and withdraws the mixed gas from the vessel to the atmosphere (or scrubber) from another opening. This purging process is commonly used when the vessel or equipment is not rated for pressure or vacuum; the purge gas is added and withdrawn at atmospheric pressure.

Purging results are defined by assuming perfect mixing within the vessel, constant temperature, and constant pressure. Under these conditions the mass or volumetric flow rate for the exit stream is equal to the inlet stream. The material balance around the vessel is

$$V\frac{dC}{dt} = C_0Q_v - CQ_v \tag{7-8}$$

where

V is the vessel volume,
C is the concentration of oxidant within the vessel (mass or volumetric units),
C_0 is the inlet oxidant concentration (mass or volumetric units),
Q_v is the volumetric flow rate, and
t is time.

The mass or volumetric flow rate of oxidant into the vessel is C_0Q_v and the flow rate of oxidant exiting is CQ_v. Equation 7-8 is rearranged and integrated.

$$Q_v\int_0^t dt = V\int_{C_1}^{C_2}\frac{dC}{(C_0 - C)} \tag{7-9}$$

The volumetric quantity of inert required to reduce the oxidant concentration from C_1 to C_2 is Q_vt, and is determined from Equation 7-9.

$$\boxed{Q_vt = V\ln\left(\frac{C_1 - C_0}{C_2 - C_0}\right)} \tag{7-10}$$

For many systems $C_o = 0$.

Example 7-3

A storage vessel contains 100% air by volume and must be inerted with nitrogen until the oxygen concentration is below 1.25% by volume. The vessel volume is 1000 cubic ft. How much nitrogen must be added, assuming the nitrogen contains 0.01% oxygen?

Solution The volume of nitrogen required, Q_vt, is determined using Equation 7-10.

$$Q_vt = V\ln\left(\frac{C_1 - C_0}{C_2 - C_0}\right)$$

$$Q_vt = (1000\ \text{ft}^3)\ln\left(\frac{21.0 - 0.01}{1.25 - 0.01}\right)$$

$$Q_vt = 2830\ \text{ft}^3$$

This is the quantity of contaminated nitrogen added (containing 0.01% oxygen). The quantity of pure nitrogen required to reduce the oxygen concentration to 1.25% is

$$Q_v t = (1000 \text{ ft}^3) \ln\left(\frac{21.0}{1.25}\right) = 2821 \text{ ft}^3.$$

Siphon Purging

As illustrated by Example 7-3, the sweep-through process requires large quantities of nitrogen. This could be expensive when purging large storage vessels. Siphon purging is used to minimize this type of purging expense.

The siphon purging process starts by filling the vessel with liquid—water or any liquid compatible with the product. The purge gas is subsequently added to the vapor space of the vessel as the liquid is drained from the vessel. The volume of purge gas is equal to the volume of the vessel and the rate of purging is equivalent to the volumetric rate of liquid discharge.

When using the siphon purging process, it may be desirable to first fill the vessel with liquid and then use the sweep-through purge process to remove oxygen from the residual head space. Using this method, the oxygen concentration is decreased to very low concentrations with only a small added expense for the additional sweep-through purging.

7-2 STATIC ELECTRICITY

A common ignition source within chemical plants is sparks due to static charge buildup and sudden discharge. It is perhaps the most elusive of ignition sources. Despite considerable efforts, serious explosions and fires due to static ignition continue to plague the chemical process industry.

The best design methods for preventing this type of ignition source are developed by understanding the fundamentals relevant to static charge and using these fundamentals to design specific features within a plant to prevent the accumulation of static charge, or to recognize situations where the build-up of static is inevitable and unavoidable. For unavoidable static buildup, design features are added to continuously and reliably inert the atmosphere around the regions where static sparks are likely.

Fundamentals of Static Charge

Static charge buildup is a result of physically separating a poor conductor from a good conductor or another poor conductor. When different materials touch each other, the electrons move across the interface from one surface to the other. Upon separation, more of the electrons remain on one surface than the other; one material becomes positively charged and the other negatively charged.

If both of the materials are good conductors, the charge buildup as a result of separation is small because the electrons are able to scurry between the surfaces. If, however, one or both of the materials are insulators or poor conductors, electrons are not as mobile and are trapped on one of the surfaces, and the magnitude of the charge is much greater.

Household examples which result in a buildup of a static charge are walking across a rug, placing different materials in a tumble dryer, removing a sweater, and combing hair. The clinging fabrics and sometimes audible sparks are the result of the build-up of a static charge.

Common industrial examples are pumping a nonconductive liquid through a pipe, mixing immiscible liquids, pneumatically conveying solids, and leaking steam impacting an ungrounded conductor. The static charges in these examples accumulate to develop large voltages. Subsequent grounding produces large and energetic sparks.

For industrial operations where flammable vapors may be present, any charge accumulation exceeding 350 volts and 0.1 mJ is considered dangerous. Static charges of this magnitude are easy to generate; the static buildup due to walking across a carpet averages about 20 mJ and exceeds several thousand volts.

Basic electrostatic relationships are used to understand and investigate the situations described above. These relationships may include field strengths produced by static charges, electrostatic potential, capacitance, relaxation times, currents and potentials in flow systems and many more. A more detailed explanation is found in a review article by F. G. Eichel.[1]

Streaming Current

A streaming current, I_s, is the flow of electricity produced by transfering electons from one surface to another by a flowing fluid or solid.

When a fluid flows through a pipe there is an uneven distribution of electrons at the interface of the fluid and the pipe. A charged layer of liquid at the wall develops a voltage gradient; this layer is called a double layer. The double layer is defined as the distance from the wall ($x = 0$) to a distance away from the wall where the velocity gradient is zero ($du/dx = 0$).

The streaming current is obtained by integrating the product of this voltage gradient and the fluid velocity over the thickness of the double layer (see Eichel[2] for this derivation). The resulting streaming current is calculated from

$$I_s = \frac{f\rho\bar{u}^2}{2} \cdot \frac{\varepsilon_r \varepsilon_o \zeta}{\mu} \tag{7-11}$$

where

I_s is the streaming current (amps),
f is the Fanning friction factor (unitless),
ρ is the fluid density (mass/volume),
$\bar{u}$ is the average velocity in the pipe (length/time),
ε_r is the relative dielectric constant (unitless),
ε_o is the permittivity constant (charge^2/[force length^2]),
ζ is the zeta potential (volts), and
μ is the fluid viscosity.

[1] F. G. Eichel, "Electrostatics," *Chemical Enginering,* March 13, 1967, pp. 153-167.
[2] Eichel, "Electrostatics," p. 153.

For liquids under laminar flow conditions and with practical engineering units, Equation 7-11 is converted to

$$I_s = \left[\frac{4.24 \times 10^{-12}\ \text{amp}}{(\text{ft/s})\ \text{volt}}\right] \cdot f\ Re\ \bar{u}\ \varepsilon_r\ \zeta \tag{7-12}$$

where

f is the Fanning friction factor; see Equations 4-24 to 4-29 (unitless),
Re is the Reynolds number (unitless),
$\bar{u}$ is the fluid velocity (ft/s),
ε_r is the relative dielectric constant (unitless),
ζ is the zeta potential (volts), and
I_s is the streaming current (amps).

Typical values for dielectric constants and zeta potentials are given in Tables 7-1 and 7-2.

The unitless Reynolds number is defined as

$$Re = \frac{d\bar{u}\rho}{\mu} \tag{7-13}$$

TABLE 7-1 PROPERTIES FOR ELECTROSTATIC CALCULATIONS[1]

Liquids	Specific conductivity[2] (mho/cm)	Dielectric constant
Benzene	7.6×10^{-8} to $< 1 \times 10^{-18}$	2.3
Toluene	$< 1 \times 10^{-14}$	2.4
Xylene	$< 1 \times 10^{-15}$	2.4
Heptane	$< 1 \times 10^{-18}$	2.0
Hexane	$< 1 \times 10^{-18}$	1.9
Methanol	4.4×10^{-7}	33.7
Ethanol	1.5×10^{-7}	25.7
Isopropanol	3.5×10^{-6}	25.0
Water	5.5×10^{-6}	80.4
Other materials and air		
Air		1.0
Cellulose	1.0×10^{-9}	3.9 to 7.5
Pyrex	1.0×10^{-14}	4.8
Paraffin	10^{-16} to 0.2×10^{-18}	1.9 to 2.3
Rubber	0.33×10^{-13}	3.0
Slate	1.0×10^{-8}	6.0 to 7.5
Teflon	0.5×10^{-13}	2.0
Wood	10^{-10} to 10^{-13}	3.0

[1] J. H. Perry, *Chemical Engineers' Handbook,* 3rd Edition, (New York: McGraw Hill, 1950), p. 1734.

[2] Resistance = 1/conductivity = 1/(mho/cm) = ohm cm

TABLE 7-2 ACCEPTED ELECTROSTATIC VALUES FOR CALCULATIONS[1]

Voltage to produce spark between needle points 1/2 in apart	14,000 volts
Voltage to produce spark between plates 0.01 mm apart	350 volts
Maximum charge density prior to corona discharge	2.65×10^{-9} coulomb/cm^2
Minimum ignition energies, millijoule	
Vapors in air	0.1
Mists in air	1.0
Dusts in air	10.0
Approximate capacitances, *C*, in micro-microfarads	
Humans	100 to 400
Automobiles	500
Tank Truck (2000 gallon)	1000
Tank (12 ft diameter with insulation)	100,000
Contact zeta potentials	0.01 to 0.1 volt

[1]F. G. Eichel, "Electrostatics," *Chemical Engineering,* March 13, 1967, p. 163.

where the pipe diameter d, velocity $\bar{u}$, density ρ, and visocity μ have consistant engineering units.

The streaming current for turbulent flow is given by

$$I_s = \left[\frac{5.89 \times 10^{-14} \text{ amp}}{(\text{ft/s}) \text{ volt}}\right] \cdot \frac{d\, \varepsilon_r\, \zeta\, \bar{u}}{\delta} \tag{7-14}$$

where

I_s, ε_r, ζ, and $\bar{u}$ are defined above,
d is the pipe diameter (inches), and
δ is the double-layer thickness (inches).

The double layer thickness, δ, is computed from

$$\delta = \sqrt{D_m \tau} \tag{7-15}$$

where

D_m is the molecular diffusivity (in^2/s) and
τ is the relaxation time (seconds).

The relaxation time is the time required for a charge to dissipate by leakage. It is determined using

$$\tau = \frac{\varepsilon_r \varepsilon_0}{\gamma_c} \tag{7-16}$$

where

ε_r is the relative dielectric constant (unitless),

ε_0 is the permittivity constant

$$\left[8.85 \times 10^{-12} \frac{\text{coulomb}^2}{\text{N m}^2} = 8.85 \times 10^{-14} \frac{\text{s}}{\text{ohm cm}}\right],$$

γ_c is the specific conductivity (mho/cm).

Specific conductivities are listed in Table 7-1, and molecular diffusivities are given in Table 7-3.

Charges also accumulate when solids are transported. The buildup results from the separation of solid particle surfaces. Since solid geometries are almost always ill defined, electrostatic calculations for solids are handled emperically. The charge build-up characteristics are determined using generally accepted guidelines (see Tables 7-4 and 7-5).

TABLE 7-3 EXPERIMENTAL DIFFUSIVITIES OF LIQUIDS[1]

A	B	T(°C)	Mole fraction(A)	Diffusivity (cm^2/sec)
Chlorobenzene	Bromobenzene	10	0.0332	1.007×10^{-5}
			0.5122	1.007×10^{-5}
			0.9652	1.291×10^{-5}
		40	0.332	1.584×10^{-5}
			0.512	1.806×10^{-5}
			0.965	1.996×10^{-5}
Methanol	Water	25	0.05	1.13×10^{-5}
			0.50	0.9×10^{-5}
			0.95	2.20×10^{-5}
Water	n-Butanol	30	0.131	1.24×10^{-5}
			0.358	0.56×10^{-5}
			0.524	0.267×10^{-5}

[1]P. A. Johnson and A. L. Babb, "Liquid Diffusion in Non-Electrolytes," *Chem. Revs.*, 56, 1956, pp. 387-453; and R. Byron Bird, Warren E. Stewart, and Edwin N. Lightfoot, *Transport Phenomenon* (New York: John Wiley and Sons, 1960), p.504.

TABLE 7-4 STATIC CHARGE DENSITIES FOR VARIOUS OPERATIONS[1]

	Coulomb/cm^2
Sliding Contact	$<0.212 \times 10^{-9}$
Rolling Contact	$<0.212 \times 10^{-9}$
Dispersion of Dusts	0.0265 to 0.265×10^{-9}
Pneumatic Transport of Solids	$<1.59 \times 10^{-9}$
Sheets Pressed Together	$<1.59 \times 10^{-9}$
Close Machining	$<2.65 \times 10^{-9}$

[1]F. G. Eichel, "Electrostatics," *Chemical Engineering*, March 13, 1967, p.163.

TABLE 7-5 CHARGE BUILDUP FOR VARIOUS OPERATIONS[1]

Process	Charge (coulomb/kg)
Sieving	10^{-9} to 10^{-11}
Pouring	10^{-7} to 10^{-9}
Grinding	10^{-6} to 10^{-7}
Micronizing	10^{-4} to 10^{-7}
Sliding down on incline	10^{-5} to 10^{-7}
Pneumatic transport of solids	10^{-5} to 10^{-7}

[1]R. A. Mancini, "The Use (and Misuse) of Bonding for Control of Static Ignition Hazards," *Plant/Operations Progress,* Vol. 7, No. 1 (Jan. 1988), p.24.

Electrostatic Voltage Drops

Figure 7-1 illustrates a tank with a feed line. Fluid flows through the feed line and drops into the tank. The streaming current builds-up a charge and voltage in the feed line to the vessel and in the vessel itself. The voltage from the electrical ground in the metal line to the end of the glass pipe is calculated using

$$\boxed{V = I_s R} \tag{7-17}$$

The resistance, R, in ohms, is computed using the conductivity of the fluid, γ_C, in mho/cm, the length of the conductor, L, in cm, and the area, A, of the conductor in cm^2.

$$\boxed{R = \frac{L}{\gamma_C A}} \tag{7-18}$$

This relationship shows that as the area of the conductor increases the resistance decreases, and if the conductor length increases the resistance increases.

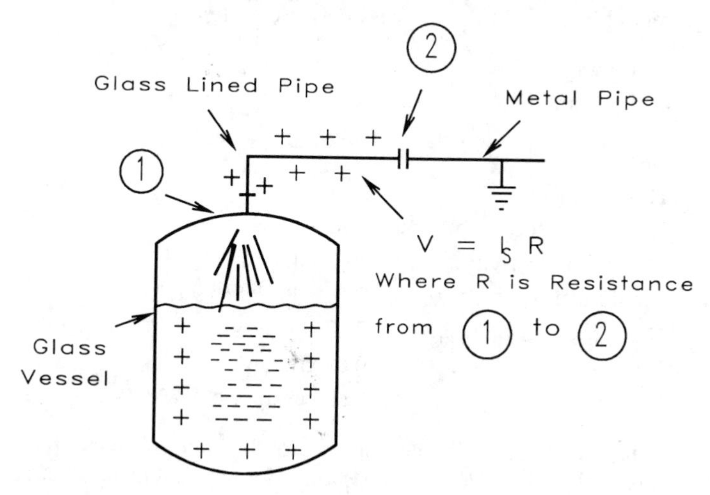

Figure 7-1 Electrical charge accumulation in a feed line due to fluid flow.

Energy of Charged Capacitors

The amount of work required to increase the charge on a capacitor from Q to $Q + dQ$ is $dJ = VdQ$, where V is the potential difference and the charge is Q. Since $V = Q/C$

$$J = \frac{Q^2}{2C} \tag{7-19}$$

$$J = \frac{CV^2}{2} \tag{7-20}$$

$$J = \frac{QV}{2} \tag{7-21}$$

The units used in the above equations are usually C in farads, V in volts, and Q in coulombs.

Capacitances of various materials used in the chemical industry are given in Table 7-6.

Charges can accumulate as a result of a streaming current $dQ/dt = I_s$. Assuming a constant streaming current,

$$Q = I_s t \tag{7-22}$$

where I_s is in amps, and t is in seconds. Equation 7-22 assumes the system starts with no accumulation of charge, only one constant source of charge I_s, and no current or charge loss terms (see the section on Balance of Charges for more complex systems).

Electrostatic properties for determining charge build-up are given in Table 7-2.

Example 7-4

Determine the voltage developed between a charging nozzle and a grounded tank, as shown in Figure 7-2. Also compute the energy stored in the nozzle. Explain the potential hazards in this process for a flowrate of

a. 1 gpm
b. 150 gpm

DATA

Hose length:	20 ft
Hose diameter:	2 in
Liquid conductivity:	10^{-8} mho/cm
Liquid diffusivity:	2.2×10^{-5} cm^2 sec^{-1}
Dielectric constant, ε_r:	25.7
Density:	0.88 gm/cm^3
Viscosity:	0.60 centipoise

Solution **a.** Since the hose and nozzle are not grounded, the voltage generated at the nozzle tip is $V = IR$. The resistance is computed using Equation 7-18 for the conducting fluid with a resistance length equivalent to the hose length (from the ground near

TABLE 7-6 CAPACITANCE OF VARIOUS OBJECTS[1]

Object	Capacitance (farad)
Small scoop, beercan, tools	5×10^{-12}
Buckets, small drums	20×10^{-12}
50-100 gallon containers	100×10^{-12}
Person	200×10^{-12}
Automobile	500×10^{-12}
Tank Truck	1000×10^{-12}

[1]R. A. Mancini, "The Use (and Misuse) of Bonding for Control of Static Ignition Hazards," *Plant/Operations Progress,* Vol. 7, No. 1 (Jan. 1988), p. 24.

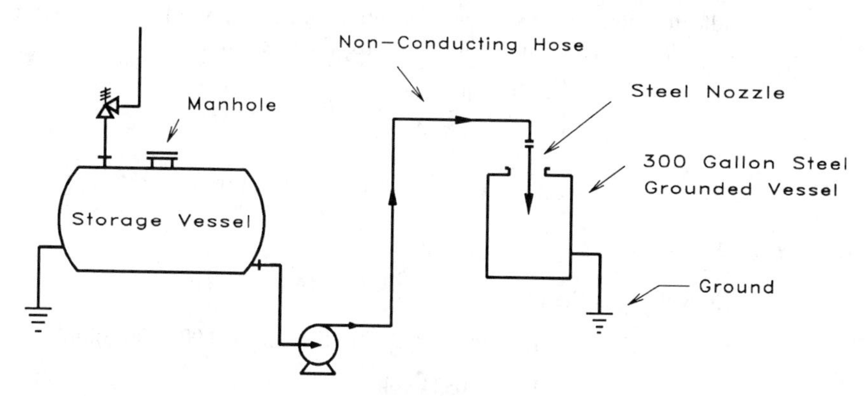

Figure 7-2

the pump to the nozzle) and a resistance area equivalent to the cross sectional area of the conducting fluid.

$$L = (20 \text{ ft})(12 \text{ in/ft})(2.54 \text{ cm/in}) = 610 \text{ cm}$$

$$A = \pi r^2 = (3.14)(1 \text{ in})^2(2.54 \text{ cm/in})^2 = 20.3 \text{ cm}^2$$

Using Equation 7-18

$$R = \left(\frac{1}{\gamma_c}\right)\left(\frac{L}{A}\right)$$

$$R = (10^8 \text{ ohm cm})\left(\frac{610 \text{ cm}}{20.3 \text{ cm}^2}\right)$$

$$R = 3.00 \times 10^9 \text{ ohm}$$

The streaming current is a function of the velocity, the Reynolds number, and the friction factor f. The average velocity in the pipe is

$$\bar{u} = \left(\frac{1 \text{ gallon/min}}{3.14 \text{ in}^2}\right)\left(\frac{\text{ft}^3}{7.48 \text{ gal}}\right)\left(\frac{144 \text{ in}^2}{\text{ft}^2}\right)\left(\frac{1 \text{ min}}{60 \text{ s}}\right)$$

$$\bar{u} = 0.102 \text{ ft/s}$$

The Reynolds number, Equation 7-13, is modified for use with commonly used units.

$$Re = \left[7750 \frac{\text{centipoise}}{\text{in (ft/s) (gm/cm}^3)}\right]\left(\frac{d\bar{u}\rho}{\mu}\right) \tag{7-23}$$

Substituting

$$Re = 7750 \text{ (2 in) (0.102 ft/s) (0.88 gm/cm}^3\text{)/0.60 cp}$$

$$= 2319$$

The friction factor is computed using Equation 4-24

$$f = \frac{16}{Re} = \frac{16}{2319}$$

$$= 6.9 \times 10^{-3}$$

The streaming current is computed using Equation 7-12 since the Reynolds number indicates laminar flow. A contact zeta potential, ζ, of 0.1 volt is selected from the data of Table 7-2. This value will maximize the streaming current.

$$I_s = \left[\frac{4.24 \times 10^{-12} \text{ amp}}{\text{(ft/s) volt}}\right] f \, Re \, \bar{u} \, \varepsilon_r \, \zeta$$

$$I_s = 4.24 \times 10^{-12}(6.9 \times 10^{-3}) (2319) \left(0.102 \frac{\text{ft}}{\text{s}}\right) (25.7) (0.1 \text{ volt})$$

$$I_s = 1.78 \times 10^{-11} \text{ amp}$$

The voltage is therefore

$$V = IR = (1.78 \times 10^{-11} \text{ amp}) (3.00 \times 10^9 \text{ ohm})$$

$$V = 0.0534 \text{ volt}$$

The accumulated charge is computed using Equation 7-22,

$$Q = I_s t$$

with the time equal to the filling time of the vessel

$$t = \text{(300 gal/1 gpm) (60 s/min)} = 18{,}000 \text{ s}$$

Substituting to compute the charge,

$$Q = (1.78 \times 10^{-11} \text{ amp}) (18{,}000 \text{ s})$$

$$Q = 3.2 \times 10^{-7} \text{ coulomb}$$

The capacitance is computed using $C = Q/V$

$$C = \frac{3.2 \times 10^{-7} \text{ coulomb}}{0.0534 \text{ volt}}$$

$$C = 6.00 \times 10^{-6} \text{ farads}$$

and the accumulated energy is determined using Equation 7-19.

$$J = \frac{Q^2}{2C} = \frac{(3.2 \times 10^{-7} \text{ coulomb})^2}{2(6.00 \times 10^{-6} \text{ farad})}$$

$$J = 8.53 \times 10^{-9} \text{ Joule}$$

Although the practice illustrated in this problem is not recommended (inlet nozzle not grounded), the voltage (0.0534 volt) is too low to initiate sparking, and the energy (8.53×10^{-9} Joule) is far below the normal minimum ignition energy (MIE) required to ignite a flammable gas.

b. This case is identical to case a except the flow rate is higher, 150 gpm versus 1 gpm in case a. Now,

$$\bar{u} = (0.102 \text{ ft/s})\left(\frac{150 \text{ gpm}}{1 \text{ gpm}}\right)$$

$$\bar{u} = 15.3 \text{ ft/s}$$

The resistance is the same as for case a (3.00×10^9 ohm).

The Reynolds number is 150/1 times the case a Reynolds number (Re = 348,000).

Since the flow regime is turbulent, Equation 7-14 is used to compute I_s, and the double layer thickness is computed using Equation 7-15.

$$\delta = \sqrt{D_m \tau}$$

The diffusivity and dielectric constant are given.

$$D_m = 2.2 \times 10^{-5} \text{ cm}^2 \text{ s}^{-1}$$

$$\varepsilon_r = 25.7$$

The relaxation time is estimated using Equation 7-16.

$$\tau = \frac{\varepsilon_r \varepsilon_o}{\gamma_c}$$

$$\tau = \frac{(25.7)(8.85 \times 10^{-14} \text{ mho s/cm})}{10^{-8} \text{ mho/cm}}$$

$$\tau = 22.7 \times 10^{-5} \text{ s}$$

The double layer thickness is now computed

$$\delta = \sqrt{(2.2 \times 10^{-5} \text{ cm}^2 \text{ s}^{-1})(22.7 \times 10^{-5} \text{ s})}$$

$$\delta = 7.07 \times 10^{-5} \text{ cm} = 2.78 \times 10^{-5} \text{ in}$$

The streaming current is calculated using Equation 7-14.

$$I_s = \left[5.89 \times 10^{-14} \frac{\text{amp}}{(\text{ft/s}) \text{ volt}}\right]\left(\frac{d\, \varepsilon_r\, \zeta\, \bar{u}}{\delta}\right)$$

$$I_s = 5.89 \times 10^{-14}\left[\frac{(2 \text{ in})(25.7)(0.1 \text{ volt})\left(15.3 \dfrac{\text{ft}}{\text{s}}\right)}{2.78 \times 10^{-5} \text{ in}}\right]$$

$$I_s = 1.66 \times 10^{-7} \text{ amp}$$

Since the current and resistances are now known, the accumulated voltage is calculated:

$$V = I_s R = (1.66 \times 10^{-7} \text{ amp})(3.00 \times 10^9 \text{ ohm}) = 498 \text{ volts}$$

The accumulated current is $Q = I_s t$, where t is

$$t = \frac{(300 \text{ gal})(60 \text{ s/min})}{150 \text{ gal/min}} = 120 \text{ s}$$

$$Q = (1.66 \times 10^{-7} \text{ amp})(120 \text{ s}) = 1.99 \times 10^{-5} \text{ coulomb}$$

Therefore the energy accumulated, Equation 7-21, during this filling process is

$$J = \frac{QV}{2} = \frac{(1.99 \times 10^{-5} \text{ coulomb})(498 \text{ volt})}{2} = 4.9 \text{ mJ}$$

In this case, the increased velocity raised the voltage (498 volt) and the accumulated energy (4.9 mJ) to values which significantly exceed the minimum values required for sparks and ignition (350 volt and 0.1 mJ respectively). If the charged nozzle approaches the grounded vessel, a spark will discharge with an energy exceeding most MIEs. If the surrounding vapors are flammable, a fire and/or explosion will be initiated.

Capacitance of a Body

The buildup of a charge on one surface relative to another produces a capacitor. In the chemical industry, the properties of the developed capacitor are estimated by assuming parallel flat plate or spherical geometries. For example, the capacitance of a tank or a person is estimated by assuming spherical geometries, and the capacitance of a person's shoe sole, or a noncorrosive tank lining is estimated assuming parallel flat plates. Several examples are shown in Figure 7-3.

The capacitance, C, of a body is Q/V. For a sphere with radius r the voltage developed when a charge Q is accumulated is derived in elementary physics

$$V = \frac{1}{4\pi\varepsilon_0}\frac{Q}{\varepsilon_r r} \tag{7-24}$$

Therefore, since $C = Q/V$,

$$\boxed{C = 4\pi\varepsilon_r\varepsilon_0 r} \tag{7-25}$$

where

ε_r = relative dielectric constant (unitless),

ε_0 = permittivity, $8.85 \times 10^{-12} \dfrac{\text{coulomb}^2}{\text{N m}^2} = 2.7 \times 10^{-12} \dfrac{\text{coulomb}}{\text{volt ft}}$,

r = sphere radius, and

C = capacitance.

For two parallel plates,

$$V = \frac{QL}{\varepsilon_r\varepsilon_0 A} \tag{7-26}$$

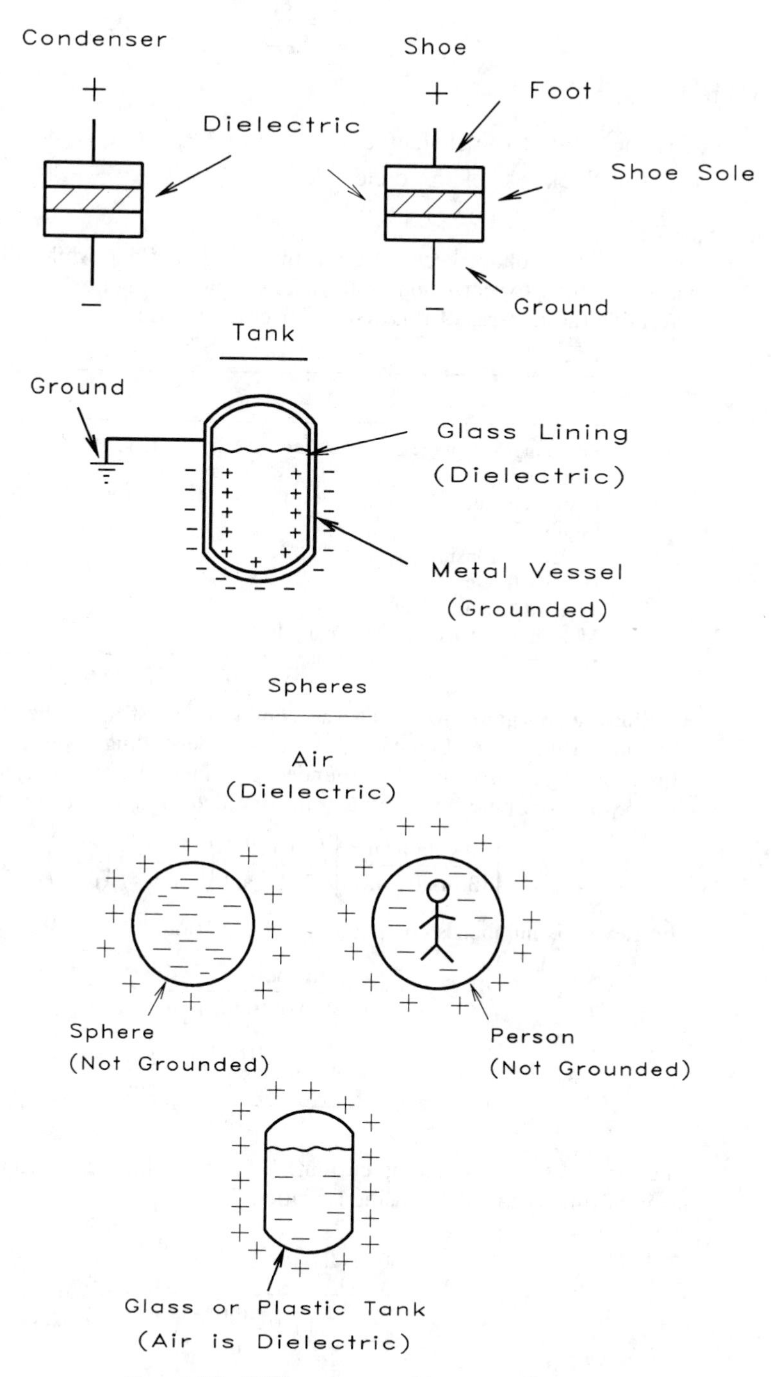

Figure 7-3 Different types of industrial capacitors.

Therefore,

$$C = \frac{\varepsilon_r \varepsilon_0 A}{L} \tag{7-27}$$

where

A is the area of the surface and
L is the thickness of the dielectric.

Example 7-5

Determine the voltage generated and the energy buildup while filling a vessel which has a glass liner for corrosion protection as shown in Figure 7-4. Using the data below determine the hazards of this system for cases a and b.

Case	a	b
Pipe length, storage to vessel (ft)	100	100
Flow rate (gallons/min)	40	40
Liquid conductivity, γ_c,(mho/cm)	10^{-10}	10^{-6}
Liquid	Organic A	Organic B
Dielectric constant, ε_r	30	30
Density (gm/cm^3)	0.87	0.87
Viscosity (centipoise)	0.7	0.7
Molecular diffusivity, D_m, (cm^2/s)	10^{-5}	10^{-5}

Solution a. (Organic A) Under the conditions illustrated, the streaming current is computed using either Equation 7-12 or 7-14, depending upon the Reynolds number. This equation is also used for grounded pipelines; the transported fluid builds up a charge while the pipe remains neutral. First determine the average velocity

$$\bar{u} = \left[\frac{40 \text{ gallon/min}}{(3.14)(0.5 \text{ in})^2}\right]\left(\frac{1 \text{ ft}^3}{7.48 \text{ gal}}\right)\left(\frac{144 \text{ in}^2}{\text{ft}^2}\right)\left(\frac{\text{min}}{60 \text{ s}}\right) = 16.3 \frac{\text{ft}}{\text{s}}$$

The Reynolds number is computed using Equation 7-23.

$$Re = 7750 \frac{\text{centipoise}}{\text{in (ft/s)(gm/cm}^3)} \frac{d\bar{u}\rho}{\mu}$$

$$Re = 7750 \frac{(1 \text{ in})\left(16.3 \frac{\text{ft}}{\text{s}}\right)\left(0.87 \frac{\text{gm}}{\text{cm}^3}\right)}{0.7 \text{ cp}} = 157{,}000$$

The turbulent flow streaming current, I_s, is determined using Equation 7-14. The relaxation time is calculated using Equation 7-16.

$$\tau = \frac{\varepsilon_r \varepsilon_0}{\gamma_c} = \frac{(30.0)\left(8.85 \times 10^{-14} \frac{\text{mho s}}{\text{cm}}\right)}{10^{-10} \frac{\text{mho}}{\text{cm}}} = 2.65 \times 10^{-2} \text{ s}$$

The double layer thickness is now determined using Equation 7-15.

$$\delta = \sqrt{D_m \tau} = \sqrt{(1 \times 10^{-5}\ \text{cm}^2/\text{s})(2.65 \times 10^{-2}\ \text{s})} = 5.15 \times 10^{-4}\ \text{cm} = 2.03 \times 10^{-4}\ \text{in}$$

The streaming current is computed using Equation 7-14.

$$I_s = \left[5.89 \times 10^{-14} \frac{\text{amp}}{(\text{ft/s})\ \text{volt}}\right]\left(\frac{d\varepsilon_r \zeta \bar{u}}{\delta}\right)$$

$$I_s = \left[5.89 \times 10^{-14} \frac{\text{amp}}{(\text{ft/s})\ \text{volt}}\right]\frac{(1\ \text{in})(30.0)(0.1\ \text{volt})(16.3\ \text{ft/s})}{2.03 \times 10^{-4}\ \text{in}}$$

$$I_s = 1.42 \times 10^{-8}\ \text{amp}$$

The total leakage resistance is the sum of the leakage resistance through the fluid and the leakage resistance through the glass liner. These resistances are in series. The leakage resistance of the tank contents, from the center of the tank through the fluid is, from Equation 7-18,

$$R = \frac{L}{\gamma_c A} = \left(\frac{1}{\gamma_c}\right)\left(\frac{L}{A}\right)$$

where

L is the distance from the center of the tank to the tank wall and
A is the surface area of the tank,

Substituting,

$$R = (10^{10}\ \text{ohm cm})\frac{(2.77\ \text{ft})}{2(3.14)(2.77\ \text{ft})(5.54\ \text{ft})}\left(\frac{1\ \text{ft}}{12\ \text{in}}\right)\left(\frac{1\ \text{in}}{2.54\ \text{cm}}\right)$$

$$R = 9.43 \times 10^6\ \text{ohm}$$

The resistance of the glass liner is determined using Equation 7-18. For Pyrex, $1/\gamma_c = 1 \times 10^{14}$ ohm cm,

$$R = \frac{L}{\gamma_c A} = \frac{(1 \times 10^{14}\ \text{ohm cm})(0.5\ \text{in})}{2(3.14)(2.77\ \text{ft})(5.54\ \text{ft})}\left(\frac{1\ \text{ft}}{12\ \text{in}}\right)^2\left(\frac{1\ \text{in}}{2.54\ \text{cm}}\right)$$

$$R = 1.42 \times 10^9\ \text{ohm}$$

The net resistance is the sum of the two resistances, or 1.42×10^9 ohms. The resistance is almost entirely due to the pyrex liner. The accumulated voltage is

$$V = I_s R = (1.42 \times 10^{-8}\ \text{amp})(1.42 \times 10^9\ \text{ohms}) = 20.2\ \text{volts}$$

The capacitance of the tank is estimated assuming a spherical geometry surrounded by air ($\varepsilon_r = 1$). Equation 7-25 is used.

$$C = 4\pi\varepsilon_r\varepsilon_0 r$$

$$C = 4(3.14)(1)\left(2.7 \times 10^{-12} \frac{\text{coulomb}}{\text{volt ft}}\right)(2.77\ \text{ft})$$

$$C = 9.39 \times 10^{-11}\ \text{farad}$$

The accumulated energy, J, is computed using Equation 7-20.

$$J = \frac{CV^2}{2} = \frac{(9.39 \times 10^{-11} \text{ farad})(20.2 \text{ volts})^2}{2}$$

$$J = 1.92 \times 10^{-5} \text{ mJ}$$

In summary, the voltage (20.2 volts) and stored energy (1.92×10^{-5} mJ) are considerably below the minimum requirements for ignition, 350 V and 0.1 mJ.

b. (Organic B) All the conditions are the same except the physical properties for organic B are used: $\bar{u} = 16.3$ ft/s and the flow is still turbulent.

Then,

$$\tau = \frac{\varepsilon_r \varepsilon_0}{\gamma_c} = \frac{(30.0)\left(8.85 \times 10^{-14} \dfrac{\text{mho s}}{\text{cm}}\right)}{10^{-6} \text{ mho/cm}} = 2.65 \times 10^{-6} \text{ s}$$

$$\delta = \sqrt{D_m \tau} = \sqrt{(1 \times 10^{-5} \text{ cm}^2/\text{s})(2.65 \times 10^{-6} \text{ s})} = 5.15 \times 10^{-6} \text{ cm} = 2.03 \times 10^{-6} \text{ in}$$

$$I_s = (1.42 \times 10^{-8} \text{ amp})\left(\frac{2.03 \times 10^{-4} \text{ in}}{2.03 \times 10^{-6} \text{ in}}\right) = 1.42 \times 10^{-6} \text{ amp}$$

The leakage resistance of the glass liner is 1.42×10^9 ohm.

$$V = I_s R = (1.42 \times 10^{-6} \text{ amp})(1.42 \times 10^9 \text{ ohm}) = 2016 \text{ volts}$$

The capacitance of the tank is the same as part a, 9.39×10^{-11} farad. The stored energy is

$$J = \frac{CV^2}{2} = \frac{(9.39 \times 10^{-11} \text{ farads})(2016 \text{ volts})^2}{2} = 0.191 \text{ mJ}$$

In summary, for part b, the voltage and stored energy are large enough to cause ignition. The hazard of this particular system (see Figure 7-4) is the result of a spark which will be generated when the organic liquid approaches the grounded thermocouple. If the atmosphere above the liquid is explosive, the ignition will cre-

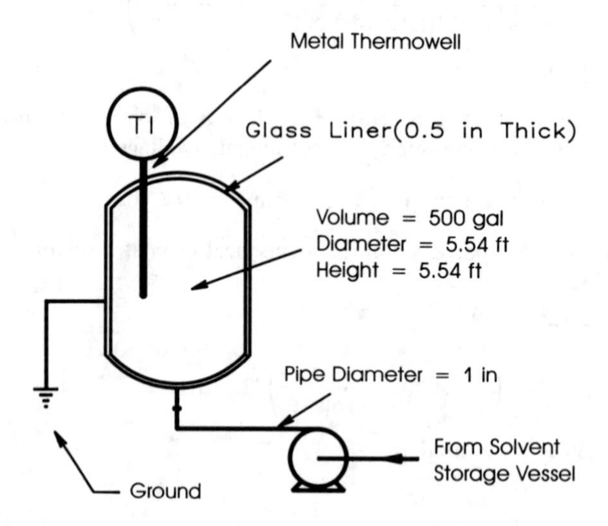

Figure 7-4 Charge build-up in insulated vessel (lined).

ate a deflagration within the vessel. If the vessel is not designed to withstand the pressure a catastrophy will result.

In this example the ignition hazard increases as the fluid conductivity increases, primarily because the streaming current, I_s, increases and the leakage resistance remains the same. The controlling resistance is the glass wall. If the glass wall were absent, the more conductive fluid would have a lower leakage resistance to the grounded wall, and less charge would accumulate. Also, if the fluid were pumped into the vessel from the bottom, leakage current could flow out through the inlet line. In this case the more conductive fluid would have less leakage resistance.

In cases similar to Examples 7-4 and 7-5, it is common practice to add a grounded nobel metal plug on the liquid side of the liner. As illustrated in these examples, however, resistance through nonconductive liquids is large enough to minimize the effect of this single ground. Even if the entire surface is grounded, high voltages and energies are generated when the liquids are pumped at a high rate and when the liquids have low conductivities.

Example 7-6

Estimate the capacitance of a person (6 ft 2 in tall) standing on a dry wooden floor.

Solution This person's capacitance is estimated assuming the person's shape is spherical and the "sphere" is surrounded by air (ε_r is 1.0 for air). Using Equation 7-25 for a sphere,

$$C = 4\pi\varepsilon_r\varepsilon_0 r$$

$$C = 4(3.14)(1.0)\left(2.7 \times 10^{-12} \frac{\text{coulomb}}{\text{volt ft}}\right)\left(\frac{6.17 \text{ ft}}{2}\right)$$

$$C = 1.05 \times 10^{-10} \frac{\text{coulomb}}{\text{volt}}$$

The calculated capacitance is close to the value listed for a person in Table 7-6.

Example 7-7

Estimate the capacitance of a person standing on a conductive floor. Assume the person's shoe soles separate the person from the floor; that is, the shoe sole is the dielectric of the capacitor. Given

Shoe sole area (ft^2) = 2 shoes (0.4 ft^2 each)
Shoe sole thickness (in) = 0.2 in
Dielectric constant of shoe soles = 3.5

Solution Use Equation 7-27 which is for flat parallel plates.

$$C = \frac{\varepsilon_r\varepsilon_0 A}{L}$$

$$C = \frac{(0.8 \text{ ft}^2)(3.5)\left(2.7 \times 10^{-12} \dfrac{\text{coulomb}}{\text{volt ft}}\right)}{\left(\dfrac{0.2 \text{ in}}{12 \text{ in/ft}}\right)}$$

$$C = 4.54 \times 10^{-10} \text{ farads}$$

Example 7-8

Estimate the charge buildup, accumulated energy, and accumulated voltage as a result of a person (insulated from the floor) charging 30 pounds of a dry powder, using a scoop, into a 20 gallon insulated drum. Assume the person's capacitance is 300×10^{-12} farad.

Solution This operation is a sliding-contact type operation. From Table 7-4, this operation gives a charge density of 2.21×10^{-10} coulomb/cm^2. Since the charge is accumulated on the person, the magnitude of the charge is based on the area of a person.

$$\text{Area (front and back)} = (1.5\text{ ft})(6.2\text{ ft})(2) = 18.6\text{ ft}^2$$

$$\text{Area (sides)} = (0.5\text{ ft})(6.2\text{ ft})(2) = 6.2\text{ ft}^2$$

$$\text{Area (top and bottom)} = (1.5\text{ ft})(0.5\text{ ft})(2) = 1.5\text{ ft}^2$$

$$\text{Total area} = 26.3\text{ ft}^2 = 24{,}400\text{ cm}^2$$

Therefore the charge buildup is

$$Q = \left(2.21 \times 10^{-10}\,\frac{\text{coulomb}}{\text{cm}^2}\right)(24{,}400\text{ cm}^2) = 5.39 \times 10^{-6}\text{ coulomb}$$

The accumulated energy, using Equation 7-19 is

$$J = \frac{Q^2}{2C} = \frac{(5.39 \times 10^{-6}\text{ coulomb})^2}{2(300 \times 10^{-12}\text{ farad})} = 48.4\text{ mJ}$$

and the accumulated voltage is

$$V = \frac{Q}{C} = \frac{5.39 \times 10^{-6}\text{ coulomb}}{300 \times 10^{-12}\text{ farad}} = 18{,}000\text{ volt}$$

These results illustrate that the energy and voltage exceed the requirements for generating a spark capable of igniting a flammable gas. This spark would be discharged if the person approached a ground with a hand or with the scoop.

An equal and opposite charge is also accumulated in the powder in the insulated drum. Therefore, the charged powder is another ignition source. For example, if a grounded object of any kind is placed close to the solids, an energetic spark could be generated.

Balance of Charges

Some systems are more complex than those previously discussed; for example, a vessel may have several inlet lines and several outlet lines. An example is illustrated in Figure 7-5.

For this type of system, a charge balance is required to establish the charge, voltage, and accumulated energy as a function of time. The charge balance is developed by considering the streaming currents in, the charge carried away by flows going out, and charge loss due to relaxation. The result is

$$\boxed{\frac{dQ}{dt} = \sum^{n} (I_s)_{i,\text{in}} - \sum^{m} (I_s)_{j,\text{out}} - \frac{Q}{\tau}} \tag{7-28}$$

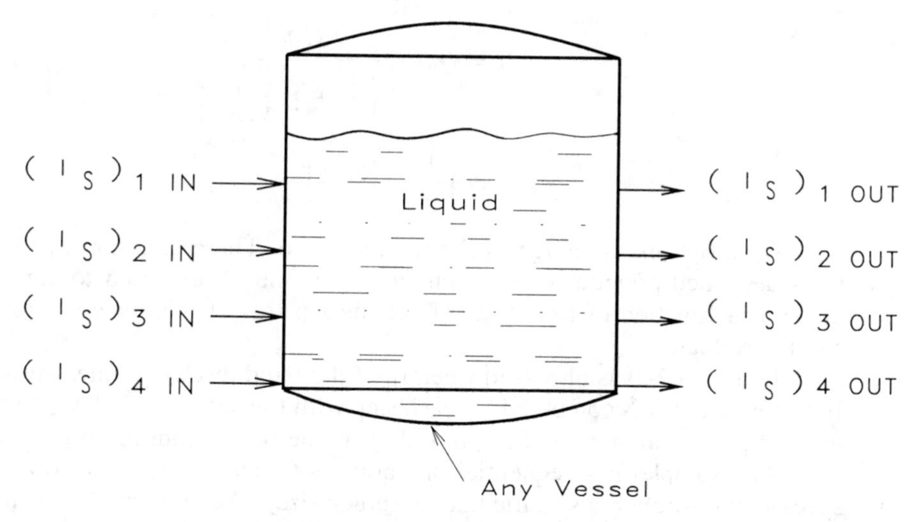

Figure 7-5 Vessel with multiple inlets and outlets.

The term $(I_s)_{i,\,\mathrm{in}}$ is the streaming current entering the tank through a specific inlet line, i, from a set of n lines; $(I_s)_{j,\,\mathrm{out}}$ is the current leaving with one specific outlet line, j, from a set of m lines; Q/τ is the charge loss due to relaxation; and τ is the relaxation time.

$(I_s)_{j,\,\mathrm{out}}$ is a function of the charge accumulated in the tank and the rate of discharge F from the specific outlet nozzle j:

$$\boxed{(I_s)_{j,\,\mathrm{out}} = \frac{F_j}{V_c} Q} \tag{7-29}$$

where

V_c is the container or tank volume and
Q is the total charge in the tank.

Substituting Equation 7-29 into 7-28 gives

$$\frac{dQ}{dt} = \sum (I_s)_{i,\,\mathrm{in}} - \sum \frac{F_j}{V_c} Q - \frac{Q}{\tau} \tag{7-30}$$

If the flows, streaming currents, and relaxation times are constant, Equation 7-30 is a linear differential equation solvable using standard techniques. The result is

$$\boxed{Q = A + B\, e^{-Ct}} \tag{7-31}$$

where

$$A = \frac{\sum (I_s)_{i,\,\mathrm{in}}}{\left(\dfrac{1}{\tau} + \sum \dfrac{F_n}{V_c}\right)}$$

$$B = Q_0 - \frac{\sum (I_s)_{i,\text{in}}}{\left(\frac{1}{\tau} + \sum \frac{F_n}{V_c}\right)}$$

$$C = \left(\frac{1}{\tau} + \sum \frac{F_n}{V_c}\right)$$

Q_0 is the initial charge in the tank at $t = 0$. These equations, plus the equations described previously (equations for I_s, V, and J), are used to compute Q, V, and J as a function of time. Therefore, the hazards of relatively complex systems can be evaluated.

Equation 7-31 is also used when the filling and discharge rates are sequential. In this case the Q is computed for each step with the specified $\Sigma (I_s)_{i,\text{in}}$ and $\Sigma (F_n/V_c)$ for that particular step, and the initial Q_0 is the result from the previous step.

An example of a sequential operation is (1) charge benzene to a vessel at a specific rate through a specific line of known size, (2) charge methanol and toluene through two different lines at different rates, (3) hold the batch for a specified time, and (4) discharge the batch through a different line at a specified rate. If the line sizes, rates, and materials of construction are known, the potential hazard of each step of the operation can be estimated.

Example 7-9

A large vessel (50,000 gal) is being filled with toluene. Compute Q, V, and J during the filling operation when the vessel is half full (25,000 gal) and where

$$F = 100 \text{ gpm},$$

$$I_s = 1.5 \times 10^{-7} \text{ amp},$$

$$\text{Liquid conductivity} = 10^{-14} \text{ mho cm}^{-1}, \text{ and}$$

$$\text{Dielectric constant} = 2.4.$$

Solution Since there is only one inlet line and no outlet lines, Equation 7-28 reduces to

$$\frac{dQ}{dt} = I_s - \frac{Q}{\tau}$$

Therefore,

$$Q = I_s\tau + (Q_0 - I_s \tau)\, e^{-t/\tau}$$

Since the vessel is initially empty, $Q_0 = 0$. The relaxation time is computed using Equation 7-16.

$$\tau = \frac{\varepsilon_r \varepsilon_0}{\gamma_c} = \frac{(2.4)\left(8.85 \times 10^{-14} \frac{\text{s}}{\text{ohm cm}}\right)}{(10^{-14} \text{ mho cm}^{-1})} = 21.2 \text{ s}$$

The charge buildup as a function of time is

$$Q(t) = I_s\tau(1 - e^{-t/\tau}) = \left(1.5 \times 10^{-7} \frac{\text{coulomb}}{\text{s}}\right)(21.2 \text{ s})(1 - e^{-t/21.2})$$

When the vessel contains 25,000 gal the elapsed time is 15,000 sec. Therefore,

$$Q(15{,}000 \text{ sec}) = 3.19 \times 10^{-6} \text{ coulombs}$$

The capacitance of this vessel is estimated assuming a spherical geometry surrounded by air.

$$V_t = \frac{4}{3}\pi r^3$$

$$r = \left(\frac{3V_t}{4\pi}\right)^{1/3}$$

$$r = \left(\frac{3}{4\pi}\frac{25{,}000 \text{ gal}}{7.48 \text{ gal ft}^{-3}}\right)^{1/3} = 9.27 \text{ ft}$$

Using Equation 7-25 and assuming a dielectric of 1 for air,

$$C = 4\pi\varepsilon_r\varepsilon_0 r = 4(3.14)(1.0)\left(2.7 \times 10^{-12}\frac{\text{coulomb}}{\text{volt ft}}\right)(9.27 \text{ ft})$$

$$C = 3.14 \times 10^{-10} \text{ farads}$$

The energy stored in this vessel (25,000 gal of toluene) is computed using Equation 7-19.

$$J = \frac{Q^2}{2C} = \frac{(3.19 \times 10^{-6} \text{ coulomb})^2}{2(3.14 \times 10^{-10} \text{ farad})} = 16.2 \text{ mJ}$$

The accumulated voltage is estimated using Equation 7-21.

$$V = \frac{2J}{Q} = \frac{2(1.62 \times 10^{-2} \text{ Joule})}{3.19 \times 10^{-6} \text{ coulomb}} = 10{,}100 \text{ volts}$$

The minimum conditions for an ignition are 0.10 mJ and 350 volts; therefore, the operating conditions for this vessel are extremely hazardous.

Example 7-10

Figure 7-6 shows an in-line trap for removing water from a process stream. Compute

a. Q, V, and J when the vessel fluid just reaches the overflow line (start with an empty vessel).

b. Q, V, and J under equilibrium conditions ($t = \infty$).

c. The time required to reduce the accumulated charge to 1/2 the equilibrium charge if the flows are stopped after equilibrium conditions are reached.

d. The charge removed with the discharge under equilibrium conditions.

Given:

Volume of vessel	5 gal
Flow rate	100 gpm toluene
Streaming current, I_s	1.5×10^{-7} amp (high value due to filter in line)
Liquid conductivity	10^{-14} mho/cm
Dielectric constant	2.4
Initial vessel charge	2×10^{-7} coulomb

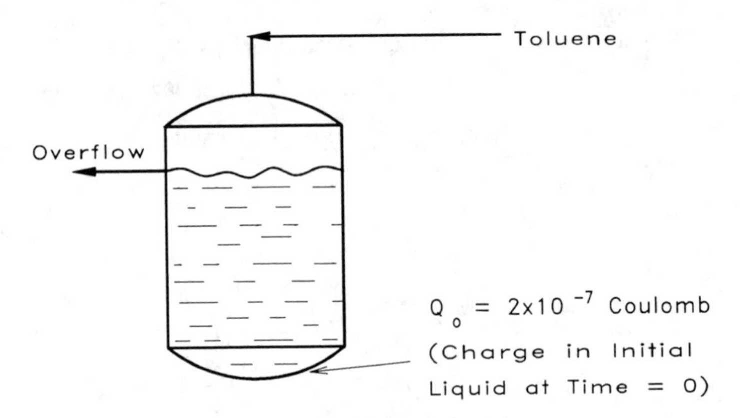

Figure 7-6 Charge buildup with complex vessel system.

Solution **a.** The residence and relaxation times of this vessel are

$$\text{Residence time} = \left(\frac{5 \text{ gal}}{100 \text{ gpm}}\right)\left(\frac{60 \text{ s}}{\text{min}}\right) = 3.00 \text{ s}$$

The relaxation time is determined using Equation 7-16.

$$\tau = \frac{\varepsilon_r \varepsilon_0}{\gamma_c} = \frac{(2.4)\left(8.85 \times 10^{-14} \dfrac{\text{s}}{\text{ohm cm}}\right)}{10^{-14} \dfrac{\text{mho}}{\text{cm}}} = 21.2 \text{ s}$$

During the filling operation, before the liquid level reaches the discharge line, Equations 7-30 and 7-31 reduce to

$$\frac{dQ}{dt} = I_s - \frac{Q}{\tau}$$

$$Q(t) = I_s\tau + (Q_0 - I_s \tau)\, e^{-t/\tau}$$

$$Q(t) = 1.5 \times 10^{-7} \frac{\text{coulomb}}{\text{s}} (21.2 \text{ s})$$

$$+ [2 \times 10^{-7}\text{coulomb} - 1.5 \times 10^{-7} \text{ amp } (21.24 \text{ s})]\, e^{-t/21.2}$$

$$Q(t) = 3.18 \times 10^{-6} - 2.98 \times 10^{-6} e^{-t/21.2}$$

with $Q(t)$ in coulombs and t in seconds. At three seconds,

$$Q\ (t = 3 \text{ s}) = 5.93 \times 10^{-7} \text{ coulombs}$$

This is the charge buildup just prior to reaching the overflow line.

The vessel capacitance is calculated assuming a spherical geometry with the surrounding air serving as the dielectric. Since 5 gallons = 0.668 ft^3, the radius of this sphere is

$$r = \left[\frac{3(0.668 \text{ ft}^3)}{4\pi}\right]^{1/3} = 0.542 \text{ ft}$$

The capacitance is estimated using Equation 7-25.

$$C = 4\pi\varepsilon_r\varepsilon_0 r = 4\pi(1.0)\left(2.7 \times 10^{-12}\ \frac{\text{coulomb}}{\text{volt ft}}\right)(0.542\ \text{ft})$$

$$= 1.84 \times 10^{-11}\ \text{farads}$$

The energy accumulated in this vessel is estimated using Equation 7-19.

$$J = \frac{Q^2}{2C} = \frac{(5.93 \times 10^{-7}\ \text{coulomb})^2}{2(1.84 \times 10^{-11}\ \text{farads})} = 9.55\ \text{mJ}$$

The accumulated voltage is computed with Equation (7-21).

$$V = \frac{2J}{Q} = \frac{2(9.55 \times 10^{-3}\ \text{Joule})}{5.93 \times 10^{-7}\ \text{coulomb}} = 32{,}200\ \text{volts}$$

The accumulated energy (9.55 mJ) and the voltage (32,200 volts) greatly exceed the quantities required for ignition of flammables. This system is operating under very hazardous conditions.

b. This vessel will gradually level off to steady state equilibrium conditions when the operating time significantly exceeds the relaxation time; therefore, the exponential term of Equation 7-31 is 0. Equation 7-31 for this case reduces to

$$Q\ (t = \infty) = \frac{I_s}{\left(\frac{1}{\tau} + \frac{F}{V_c}\right)} = \frac{(1.5 \times 10^{-7}\ \text{amps})}{\left(\frac{1}{21.2} + \frac{1}{3}\right)\text{s}^{-1}} = 3.94 \times 10^{-7}\ \text{coulomb}$$

From part a, the capacitance is $C = 1.84 \times 10^{-11}$ farads.
The energy and voltage are determined using Equations 7-19 and 7-21, respectively.

$$J = \frac{Q^2}{2C} = \frac{(3.94 \times 10^{-7}\ \text{coulombs})^2}{2(1.84 \times 10^{-11}\ \text{farads})} = 4.22\ \text{mJ}$$

$$V = \frac{2J}{Q} = \frac{2(4.22 \times 10^{-3}\ \text{Joules})}{3.94 \times 10^{-7}\ \text{coulombs}} = 21{,}400\ \text{volts}$$

Although there is an additional loss of charge with the overflowing liquid, the system is still operating under hazardous conditions.

c. After the inlet flow is stopped, $(I_s)_{in}$ and $(I_s)_{out}$ are zero, and Equation 7-31 reduces to

$$Q = Q_0 e^{-t/\tau}$$

For $Q/Q_0 = 0.5$, from the problem definition,

$$0.5 = e^{-t/\tau}$$

$$t = (21.2\ \text{s})\ \ln 2 = 14.7\ \text{s}$$

Therefore, it only takes about 15 seconds to reduce the accumulated charge to one-half of its original charge.

d. Under equilibrium conditions, Equation 7-30 is set to zero

$$\frac{dQ}{dt} = I_s - \left(\frac{1}{\tau} + \frac{F}{V_c}\right)Q = 0$$

and from part b, $Q(t = \infty) = 3.94 \times 10^{-7}$ coulomb, and

$$\text{Charge loss via relaxation} = \frac{Q}{\tau} = 1.86 \times 10^{-8} \frac{\text{coulomb}}{\text{s}}$$

$$\text{Charge loss via the overflow} = \frac{F}{V_c} Q = 1.31 \times 10^{-7} \frac{\text{coulomb}}{\text{s}}$$

For this example the charge loss due to flow out of a system is greater than the loss due to relaxation.

Sparks due to static charge and discharge continue to cause major fires and explosions within the chemical industry. The examples and fundamentals developed in these sections were designed to emphasize the importance of this subject. Hopefully this emphasis on the fundamentals will make the subject less elusive and destructive.

7-3 CONTROLLING STATIC ELECTRICITY

Charge buildup, resulting sparks, and the ignition of flammables is an inevitable event if control methods are not appropriately used. In practice, however, design engineers recognize this problem and install special features to prevent sparks by eliminating the buildup and accumulation of static charge and prevent ignition by inerting the surroundings.

Inerting, Section 7-1, is the most effective and reliable method for preventing ignition. It is always used when working with flammable liquids which are 5°C (or less) below the flash point (closed cup). Methods for preventing charge buildup are described in the following paragraphs.

Relaxation

When pumping fluids into a vessel through a pipe on top of the vessel, the separation process produces a streaming current I_s which is the basis for charge buildup. It is possible to substantially reduce this electrostatic hazard by adding an enlarged section of pipe just prior to entering the tank. This "hold" provides time for charge reduction by relaxation. The residence time in this relaxation section of pipe should be about twice the relaxation time determined from Equation 7-16.

In actual practice,[3] it was found that a hold time equal to or greater than one-half the calculated relaxation time is sufficient to eliminate charge buildup. The "twice the relaxation time" rule, therefore, provides a safety factor of 4.

Bonding and Grounding

The voltage difference between two conductive materials is reduced to zero by bonding the two materials, that is, bonding one end of a conducting wire to one of the materials and the other end to the second material.

[3]F. G. Eichel, "Electrostatics," *Chemical Engineering,* March 13, 1967, p. 153.

When comparing sets of bonded materials, the sets may have different voltages. The voltage difference between sets is reduced to zero by bonding each set to ground, that is, *grounding.*

Bonding and grounding reduces the voltage of an entire system to ground level or zero voltage. This also eliminates the charge buildup between various parts of a system, eliminating the potential for static sparks. Examples of grounding and bonding are illustrated in Figures 7-7 and 7-8.

Glass and plastic lined vessels are grounded using tantalum inserts or metal probes as illustrated in Figure 7-9. This technique, however, is not very effective when handling liquid with low conductivity. In this case, the fill line should extend to the bottom of the vessel (see Figure 7-10), to help eliminate the charge generation (and accumulation) due to separation during the filling operation. Also, the inlet velocities should be low enough to minimize the charge generation via streaming current I_s.

Dip Pipes

An extended line, sometimes called a dip leg or dip pipe, reduces the electrical charge that accumulates when liquid is allowed to free fall. When using dip pipes, however, care must be taken to prevent siphoning back when the inlet flow is

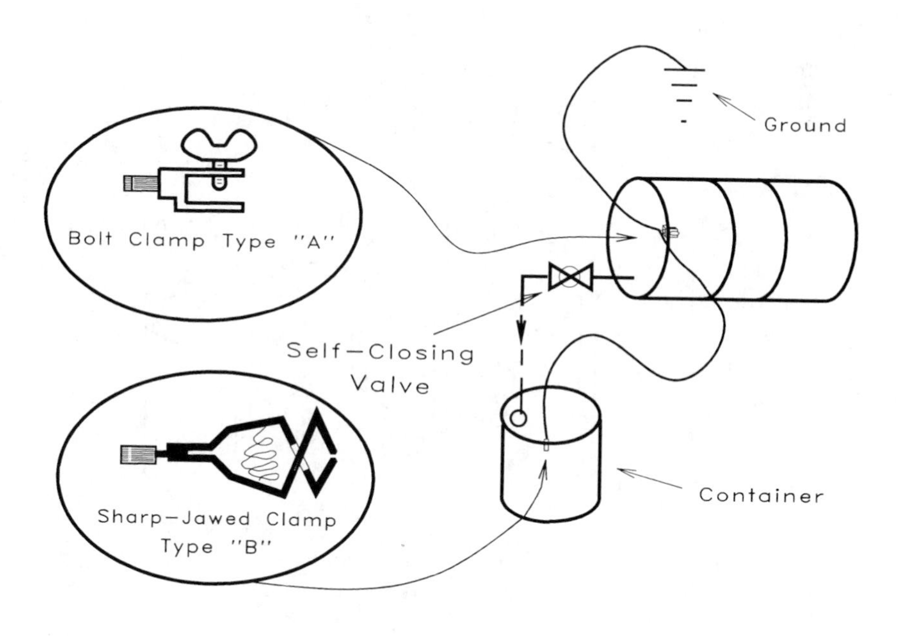

Figure 7-7 Bonding and grounding procedures for tanks and vessels. (Adapted from F.G. Eichel, "Electrostatics," *Chemical Engineering,* March 13, 1967, p. 153.)

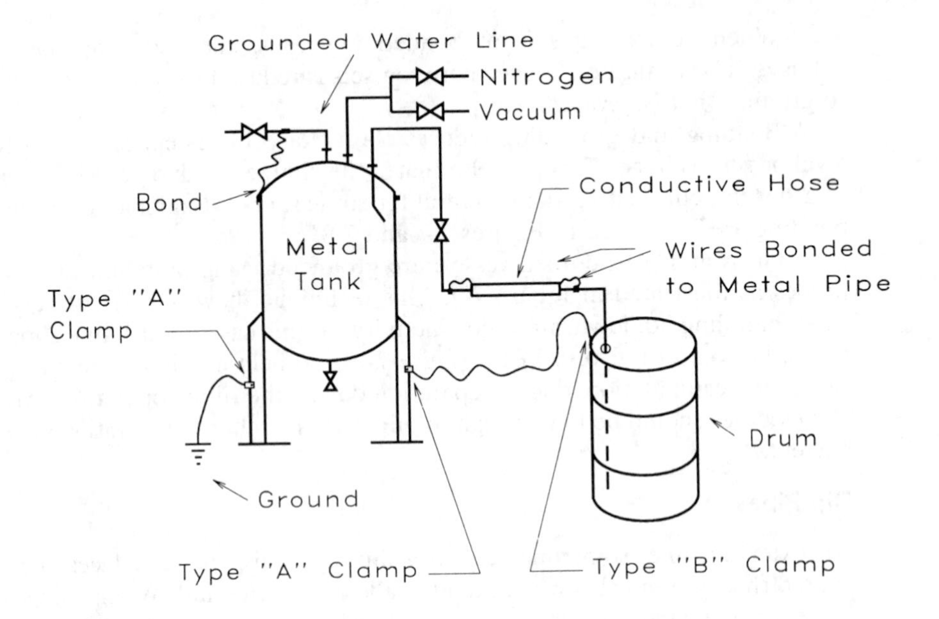

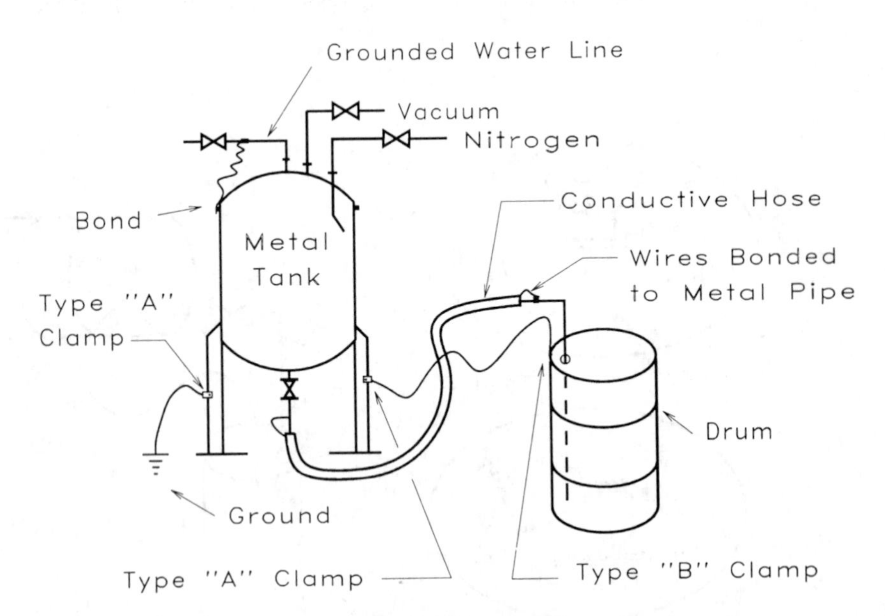

Figure 7-7 (continued)

stopped. A commonly used method is to place a hole in the dip pipe near the top of the vessel. Another technique is to use an angle iron instead of a pipe and let the liquid flow down the angle iron (see Figure 7-10). These methods are also used when filling drums.

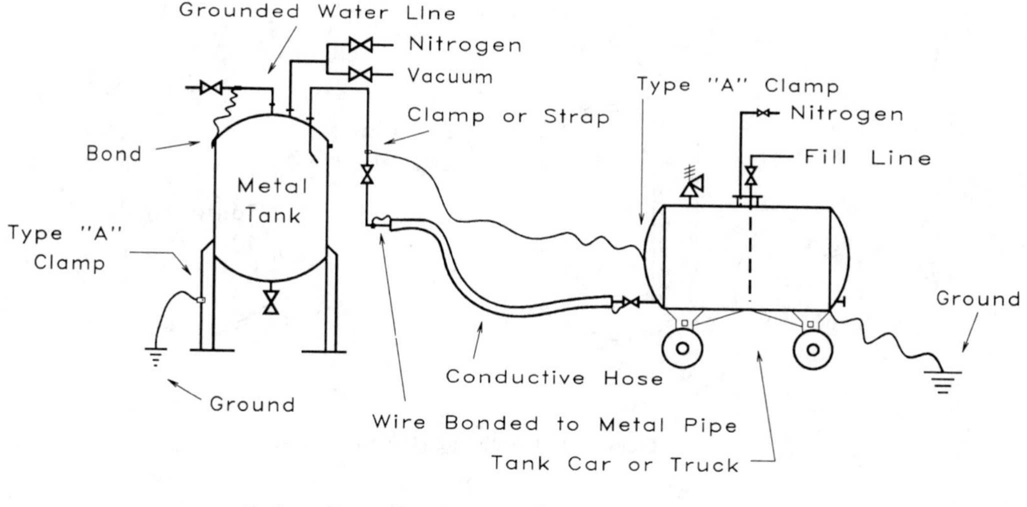

Figure 7-7 (continued)

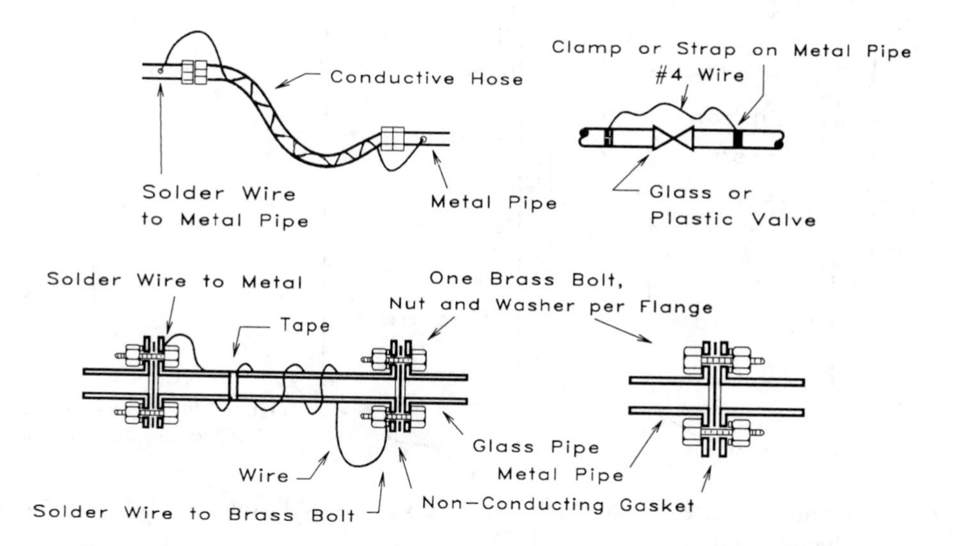

Figure 7-8 Bonding procedures for valves, pipes and flanges. (Adapted from F. G. Eichel, "Electrostatics," *Chemical Engineering,* March 13, 1967, p. 153.)

Increasing Conductivity with Additives

The conductivity of nonconducting organics can sometimes be increased using additives called antistatic additives. Examples of antistats include water or polar solvents, such as alcohols. Water is only effective when it is soluble in the offending

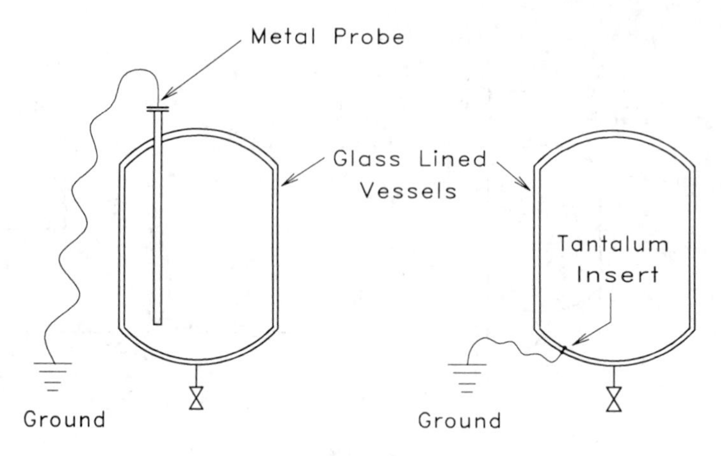

Figure 7-9 Grounding glass-lined vessels.

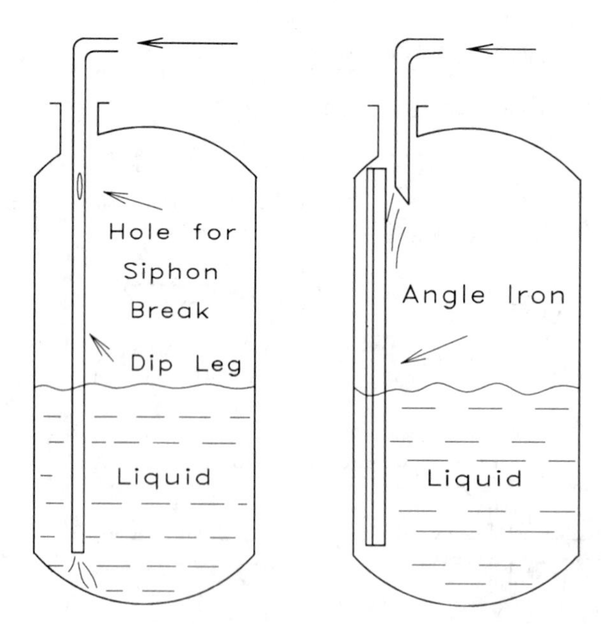

Figure 7-10 Dip legs to prevent free fall and accumulation of static charge.

liquid, because an insoluble phase gives an additional source of separation and charge buildup.

Handling Solids without Flammable Vapors

Charging solids with a nongrounded and conductive chute can result in a buildup of a charge on the chute. This charge can accumulate and finally produce a spark which may ignite a dispersed and flammable dust.

Solids are transferred safely by bonding and grounding all conductive parts and/or using nonconductive parts (drum and chute). See illustrations shown in Figure 7-11.

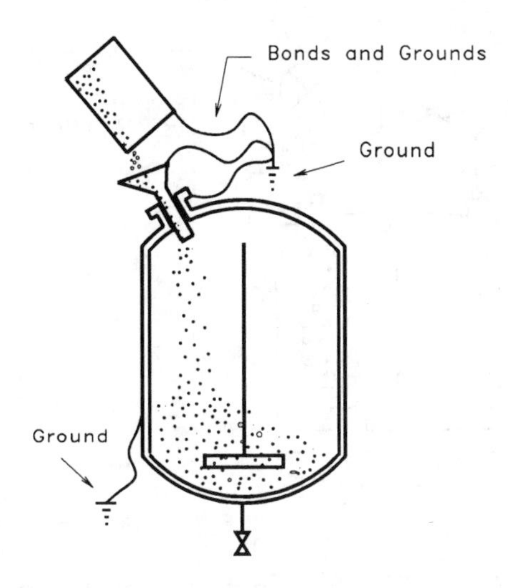

Figure 7-11 Handling solids with no flammable vapors present. (Adapted from Expert Commission for Safety in the Swiss Chemical Industry, "Static Electricity: Rules for Plant Safety," *Plant/Operations Progress,* January 1988, p 19.)

Handling Solids with Flammable Vapors

A safe design for this operation includes closed handling of the solids and liquids under an inert atmosphere, see Figure 7-12.

For solvent-free solids, the use of nonconductive containers is permitted. For solids containing flammable solvents, only conductive and grounded containers are recommended.[4]

7-4 EXPLOSION PROOF EQUIPMENT AND INSTRUMENTS

All electrical devices are inherent ignition sources. Special design features are required to prevent the ignition of flammable vapors and dusts. The fire and explosion hazard is directly proportional to the number and type of electrically powered devices in a process area.

Most safety practices for electrical installations are based on the National Electric Code (NEC)[5]. Although states, municipalities, and insurance companies may have their own installation requirements, they are usually based on the NEC.

Process areas are divided into two major types of environments: XP and non-XP. XP, for eXplosion Proof, means flammable materials (particularly vapors) might be present at certain times. Non-XP means that flammable materials are not

[4]Expert Commission for Safety in Swiss Chemical Industry, "Static Electricity: Rules for Plant Safety," *Plant/Operations Progress,* Vol. 7, No. 1, January 1988. p.1.

[5]Peter J. Schram, ed., *The National Electrical Code Handbook* (Quincy, MA: National Fire Protection Association, 1986).

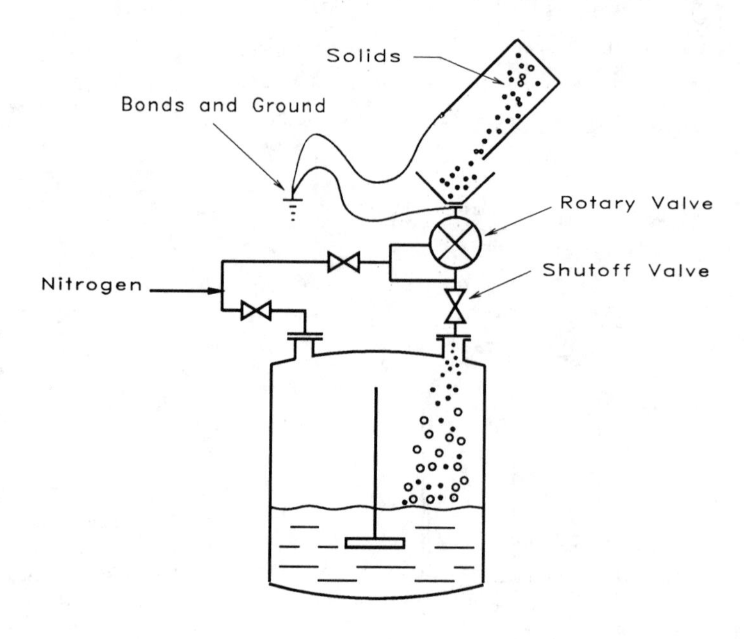

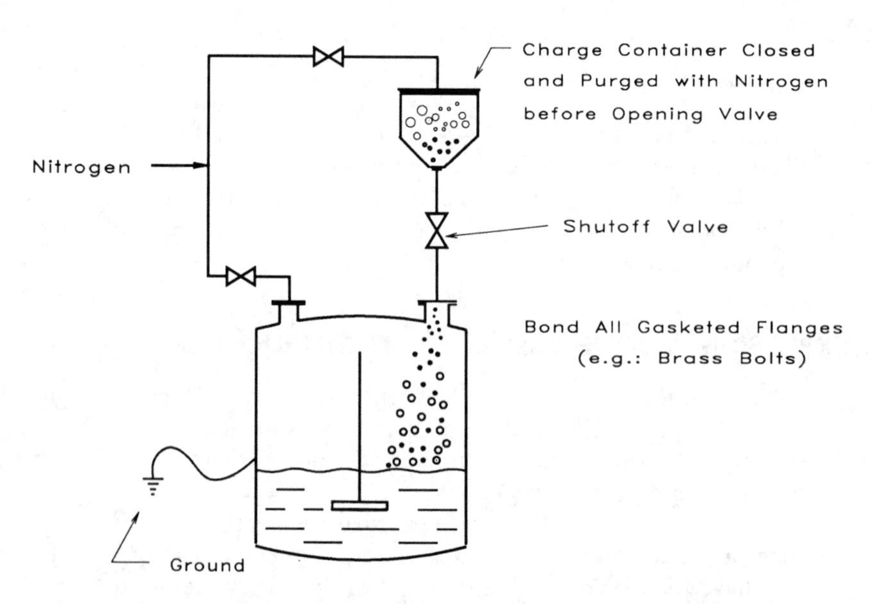

Figure 7-12 Handling solids with flammable vapors present. (Expert Commission for Safety in the Swiss Chemical Industry, "Static Electricity: Rules for Plant Safety," *Plant/Operations Progress,* January, 1988, p. 19. Reprinted by permission of the American Institute of Chemical Engineers, New York.)

present even under abnormal conditions. For non-XP designated areas, open flames, heated elements, and other sources of ignition may be present.

Explosion-Proof Housings

In an XP area, the electrical equipment and some instrumentation must have special explosion proof housings. The housings are not designed to prevent flammable vapors and gases from entering but are designed to withstand an internal explosion and prevent the combustion from spreading beyond the inside of the enclosure. A motor starter, for example, is enclosed in a heavy cast walled box with the strength needed to withstand explosive pressures.

The explosion proof design includes the use of conduit with special sealed connections around all junction boxes.

Area and Material Classification

The design of electrical equipment and instrumentation is based on the nature of the process hazards or specific process classifications. The classification method is defined in the National Electrical Code; it is a function of the nature and degree of the process hazards within a particular area. The rating method includes Classes I, II, and III, Groups A-G, and Divisions 1 or 2.

The classes are related to the nature of the flammable material.

Class I: Locations where flammable gases or vapors are present.

Class II: Same for combustible dusts.

Class III: Hazard locations where combustible fibers or dusts are present but not likely to be in suspension.

The groups designate the presence of specific chemical types. Chemicals which are grouped have equivalent hazards

Group A: acetylene

Group B: hydrogen
ethylene

Group C: carbon monoxide
hydrogen sulfide

Group D: butane
ethane
ethyl alcohol

Group E: aluminum dust

Group F: carbon black

Group G: flour

Division designations are categorized in relationship to the probability of the material being within the flammable or explosive regions.

Division 1: Probability of ignition is high; that is, flammable concentrations are normally present.

Division 2: Hazardous only under abnormal conditions. Flammables are normally contained in closed containers or systems.

Design of XP Area

When designing an XP area, all pieces of electrical equipment and instrumentation are specified for the class, group, and division as discussed previously. All pieces of equipment and instrumentation within an area must be appropriately specified and installed. The overall classification is only as good as the piece of equipment in an area with the lowest classification.

7-5 VENTILATION

Proper ventilation is another method used to prevent fires and explosions. The purpose of ventilation is to dilute the explosive vapors with air to prevent explosion and to confine the hazardous flammable mixtures.

Open Air Plants

Open air plants are recommended because the average wind velocities are high enough to safely dilute volitile chemical leaks which may exist within a plant. While safety precautions are always practiced to minimize leaks, accidental releases from pump seals and other process points may occur.

Example 7-11

A plant handling substantial quantities of flammable toluene is located 1000 feet from a residential area. There is some concern that a sizable leak of flammable vapors will form a flammable cloud with subsequent ignition in the residential area. Determine the minimum mass flow rate of toluene leakage required to produce a vapor cloud in the residential area with a concentration equal to the LFL. Assume a 5 mph wind and *D* atmospheric stability.

Solution Assume a continuous leak at ground level. The plume concentration directly downwind along the cloud centerline is given by Equation 5-48.

$$\langle C \rangle = \frac{Q_{\mathrm{m}}}{\pi \sigma_{\mathrm{y}} \sigma_{\mathrm{z}} u}$$

Solving for Q_{m}, the mass flow rate from the leak,

$$Q_{\mathrm{m}} = \langle C \rangle \pi \sigma_{\mathrm{y}} \sigma_{\mathrm{z}} u$$

The LFL for toluene is 1.2% in air (from Table 6-1). Converting the units,

$$\left(0.012 \frac{\mathrm{m^3\ toluene}}{\mathrm{m^3\ air}}\right)\left(\frac{1\ \mathrm{gm\text{-}mole\ toluene}}{22.4 \times 10^{-3}\ \mathrm{m^3\ toluene}}\right)\left(\frac{92\ \mathrm{gm\ toluene}}{1\ \mathrm{gm\text{-}mole\ toluene}}\right) = 49.3\ \mathrm{gm/m^3}$$

The wind speed is 5 mph = 2.23 m/s.

The distance downwind is 1000 feet = 304 meters.

From Figure 5-10, σ_y = 22 meters.

From Figure 5-11, σ_z = 12 meters.

Substituting,

$$Q_m = (49.3 \text{ gm/m}^3)(3.14)(22 \text{ m})(12 \text{ m})(2.23 \text{ m/s})$$
$$= 9.11 \times 10^4 \text{ gm/s}$$
$$= 201 \text{ lb/s}$$

Any leak with a flow rate greater than this is capable of producing a flammable cloud in the residential area. Of course, the toxic effects of this cloud must also be considered. The LEL of 49.3 gm/m^3 is much above the TLV of 0.375 gm/m^3.

Plants Inside Buildings

Frequently, processes cannot be constructed outside. In this case, local and dilution ventilation systems are required. These ventilation systems were discussed in detail in Chapter 3, Section 3-4.

Local ventilation is the most effective method for controlling flammable gas releases. Dilution ventilation, however, is also used because the potential points of release are usually numerous and it may be mechanically or economically impossible to cover every potential release point with only local ventilation.

There are empirically determined design criteria for designing ventilation systems for flammables inside storage and process areas. These design criteria are given in Table 7-7.

The effectiveness of a ventilation system is determined using material balance equations described in Chapter 3 under **Estimating worker exposures to toxic vapors,** p. 61, and as illustrated in the following example.

Example 7-12

Determine the concentration of toluene over a diked area (100 ft^2) which contains toluene as a result of a spill. Assume that the process area (2500 ft^2) is designed to handle Class I flammables, and the liquid and air temperature is 65°F. The vapor pressure of toluene at 65°F is 20 mm Hg. The LEL is 1.27% by volume.

TABLE 7-7 VENTILATION DATA FOR HANDLING FLAMMABLES[1]

I. Ventilation for inside storage areas
Rate: Complete room air change at least six times per hour.
Conditions: (a) Control switch and lights must be outside the storage area.
(b) Vent control and lights must be operated by same switch.
(c) A pilot light must be installed next to switch if Class I flammables are dispensed in area.[2]
II. Ventilation for processing area inside building
Rate: 1 ft^3/min/ft^2 of solid floor area.
Conditions: (a) Make-up air must not short circuit ventilation system.
(b) Ventilation must limit flammable concentrations not more than five feet from equipment exposing Class I liquids to air.

[1] *Dust and Fume Control,* OSHA Bulletin No. 1910.106.

[2] Class I: Flash-point(closed cup) below 37.8°C (100°F).
Class II: Flash-point from 37.8°C to 60°C (100° to 140°F).
Class III: Flash-point above 60°C (140°F).

Solution The source models for spills are described in Chapter 3, Equations 3-14 and 3-18. The concentration of volatiles in a ventilated area due to the evaporation from a pool is given by Equation 3-14.

$$C_{\text{ppm}} = \frac{KAP^{\text{sat}}}{kQ_v P} \times 10^6$$

where

K is the mass transfer coeeficient determined using Equation 3-18,
A is the area of the pool,
P^{sat} is the saturation vapor pressure of the liquid
k is the nonideal mixing factor,
Q_v is the volumetric ventilation rate, and
P is the pressure.

The ventilation capacity for this process area is based on the design criteria of 1 $\text{ft}^3/\text{min}/\text{ft}^2$ (Table 7-7); therefore,

$$Q_v = \left(\frac{1 \text{ ft}^3}{\text{min ft}^2}\right)(2500 \text{ ft}^2) = 2500 \frac{\text{ft}^3}{\text{min}}$$

Also,

$$M = 92$$

$$P^{\text{sat}} = 20 \text{ mm Hg}$$

$$A = 100 \text{ ft}^2$$

The mass transfer coefficient is computed using Equation 3-18 with M_0, and K_0 for water; that is, 18 and 0.83 cm/s respectively.

$$K = K_0 \left(\frac{M_0}{M}\right)^{1/3} = 0.83 \left(\frac{18}{92}\right)^{1/3} = 0.482 \text{ cm/s} = 0.948 \text{ ft/min}$$

The nonideal mixing factor, k, ranges between 0.1 and 0.5. Since no information is given about the ventilation, k will be used as a parameter. Substituting into Equation 3-14,

$$kC_{\text{ppm}} = \frac{KAP^{\text{sat}} \times 10^6}{Q_v P}$$

$$kC_{\text{ppm}} = \frac{(0.948 \text{ ft/min})(100 \text{ ft}^2)(20 / 760) \text{ atm} \times 10^6}{(2500 \text{ ft}^3/\text{min})(1 \text{ atm})} = 998 \text{ ppm}$$

The concentration range is estimated to be,

For $k = 0.5$, $C_{\text{ppm}} = 1996$ ppm $= 0.1996\%$ by volume.

For $k = 0.1$, $C_{\text{ppm}} = 9980$ ppm $= 0.998\%$ by volume.

These concentrations are considerally below the LFL of 1.27% by volume, which illustrates that the specified ventilation rate for Class I liquids is satisfactory for handling relatively large spills of flammable materials. The concentrations will, however, exceed the TLV for this substance.

7-6 SPRINKLER SYSTEMS

Sprinkler systems are an effective way to contain fires. The system consists of an array of sprinkler heads connected to a water supply. The heads are mounted in a high location (usually near ceilings) and disperse a fine spray of water over an area when activated. The heads are activated by a variety of methods. A common approach activates the heads individually by the melting of a fusible link holding a plug in the head assembly. Once activated they cannot be turned off unless the main water supply is stopped. This approach is called a closed head area system. These systems are used for storage areas, laboratories, control rooms and small pilot areas. Another approach activates the entire sprinkler array from a common control point. The control point is connected to an array of heat and/or smoke detectors that start the sprinklers when an abnormal condition is detected. If a fire is detected, the entire sprinkler array within an area is activated, possibly in areas not even affected by the fire. This approach is called an open head area system. This system is used for plant process areas and larger pilot plants.

Sprinkler systems may cause considerable water damage when activated, depending upon the contents of the building or process structure. Statistically, the amount of water damage is never as great as the damage from fires in areas that should have had sprinklers.

Sprinkler systems require maintenance to insure they remain in service and have an adequate and uninterrupted water supply.

There are various fire classes which require different sprinkler designs. The detailed descriptions of these classes and sprinkler specifications are in the National Fire Code.[6] An average chemical plant is classified as an Ordinary Hazard (Group 3) area. Various sprinkler specifications for this type area are given in Table 7-8.

Sometimes vessels need special water protection to keep the vessel walls cool during fires. High surface temperatures may result in metal failure at pressures far below the vessel's maximum allowable working pressure (MAWP) with potentially disastrous consequences. In hydrocarbon spill fires, unprotected vessels (no insulation or water spray) may fail within minutes.

A water spray protection system around vessels is recommended to prevent this type of failure. These water spray protection systems, commonly called deluge systems, are designed to keep the vessel cool, flush away potentially hazardous spills, and help to knock down gas clouds.[7] The deluge systems may alternatively provide enough time to transfer the material out of the storage tank into another (and safe) area.

These vessel deluge systems are usually designed as open head systems, which are activated when a fire is detected and/or a flammable gas mixture is detected. The deluge system is usually opened when the flammable gas concentration is a fraction of the LFL (approximately 25%) or when a fire is detected through heat. Table 7-8 provides design specifications for these deluge systems.

[6]*National Fire Codes,* Vol 1 (Quincy, MA: National Fire Protection Association, 1986).

[7]D. C. Kirby and J. L. De Roo, "Water Spray Protection for a Chemical Processing Unit: One Company's View," *Plant/Operations Progress,* Vol. 3, No. 4, October 1984.

TABLE 7-8 FIRE PROTECTION FOR CHEMICAL PLANTS[1]

Closed Head Area Systems for small storage areas, laboratories, control rooms, and small pilot plants.

(a) 0.25 gpm/ft^2 of floor area over an area of 3000 ft^2 for normal hydrocarbons such as, hexane, ethanol, toluene, etc.
(b) 0.35 gpm/ft^2 of floor area over an area of 3000 ft^2 for reactive hydrocarbons, such as styrene, butadiene, ethylene oxide, etc.
(c) For areas greater than 3000 ft^2 the system is designed for the most hydraulically remote 3000 ft^2 of the system. For example, if a warehouse area is 10,000 ft^2, the total water requirement is usually based on the most distant 3000 ft^2 area.

Open Head Area Systems for process areas including larger pilot plants.

(a) 0.25 gpm/ft^2 of floor area for normal hydrocarbons such as, hexane, ethanol, toluene, etc.
(b) 0.35 gpm/ft^2 of floor area for reactive hydrocarbons, such as styrene, butadiene, ethylene oxide, etc.
(c) An area covered is based on a particular hazard or potential spill which could include several vessels.

Deluge Water Spray Systems for vessels, heat exchangers, etc. These systems are similiar to open head area systems.

(a) Same as open head area system above, except area is based on surface area of the vessels covered.
(b) Maximum spacing around perimeter of vessel is 8 feet.
(c) Maximum distance from vessel surface is 2 feet.

Nominal Discharge Capacities of approved sprinklers having a nominal 1/2 inch orifice.

Gpm:	18	25	34	50	58
Psi:	10	20	35	75	100

Fire Monitors (usually fixed)

(a) Rate is 500 to 2000 gpm.
(b) Area coverage is 150 ft radius.

Spacings between Nozzles are based on vendors' specifications.

Piping sizes are based on nozzle specifications, nozzle layout, and conventional hydraulic calculations.

[1]*National Fire Codes,* Vol. 1, (Quincy, MA: National Fire Protection Association, 1986); *NFPA Fire Protection Handbook,* 16th ed. (Boston: National Fire Protection Association, 1986); and *SFPE Handbook of Fire Protection Engineering,* 1st ed. (Quincy, MA: National Fire Protection Association / Society of Fire Protection Engineering, 1988).

Monitors are fixed water hydrants with an attached discharge gun. They are also installed in process areas and storage tank areas. Fire hydrants and monitors are spaced 150 to 250 ft apart around process units, located so that all areas of the plant can be covered by two streams. The monitor is usually located 50 ft from the equipment being protected.[8] The specifications for monitors are also in Table 7-8.

Example 7-13

Determine the sprinkler requirements for a chemical process area within a building with an area of 100 ft by 30 ft which handles reactive solvents. Determine the number of sprinkler spray nozzles and pump specifications. Assume 1/2 inch orifice sprinklers with 35 psig at each nozzle, a 10 psig frictional loss within the system and a 15 ft elevation of the sprinkler system above the pump.

[8]Orville M. Slye, "Loss Control Association," Paper presented at AIChE Symposium, New Orleans, Louisiana, March 6-10, 1988.

TABLE 7-9 MISCELLANEOUS DESIGNS FOR PREVENTING FIRES AND EXPLOSIONS[1]

Feature	Explanation
Maintenance Programs	The best way to prevent fires and explosions is to stop the release of flammables. Preventative maintenance programs are designed to up-grade system before failures occur.
Fireproofing	Insulate vessels, pipes, and structures to minimize damage due to fires. Add deluge systems and design to withstand some damage due to fires and explosions; e.g. use multiple deluge systems with separate shut-offs.
Control Rooms	Design control rooms to withstand explosions.
Water Supplies	Provide supply for maximum demand. Consider many deluge systems running simultaneously. Diesel-engine pumps are recommended.
Control Valves for Deluge	Place shut-offs well away from process areas.
Manual Fire Protection	Install hydrants, monitors, and deluge systems. Add good drainage.
Separate Units	Separate(space) plants on a site, and units within plants. Provide access from two sides.
Utilities	Design steam, water, electricity, and air supplies to be available during emergencies. Place substations away from process areas.
Personnel Areas	Locate personnel areas away from hazardous process and storage areas.
Group Units	Group units in rows. Design for safe operation and maintenance. Create islands of risk by concentrating hazardous process units in one area. Space units so "hot work" can be performed on one group while another is operating.
Isolation Valves	Install isolation valves for safe shutdowns. Install in safe and accessible locations at edge of unit or group.
Railroads and Flares	Process equipment should be separated from flares and railroads.
Compressors	Place gas compressors downwind and separated from fired heaters.
Dikes	Locate flammable storage vessels at periphery of unit. Dike vessels to contain and carry away spills.
Block Valves	Automated block valves should be placed to stop and/or control flows during emergencies. Ability to transfer hazardous materials from one area to another should be considered.
On-line Analyzers	Add appropriate on-line analyzers to (a) monitor the status of the process, (b) detect problems at the incipient stage, and (c) take appropriate action to minimize effects of problems while still in initial phase of development.
Fail Safe Designs	All controls need to be designed to fail safely. Add safeguards for automated and safe shut-downs during emergencies.

[1]John A. Davenport, "Prevent Vapor Cloud Explosions," *Hydrocarbon Processing,* March 1977, pp. 205-214, and Orville M. Slye, "Loss Prevention Fundamentals for Process Industry," *AIChE Loss Prevention Symposium,* New Orleans, LA, March 6-10, 1988.

Solution Data for designing this system is found in Table 7-8.

$$\text{Total water requirement} = (0.35 \text{ gpm/ft}^2)(100 \text{ ft})(30 \text{ ft})$$
$$= 1050 \text{ gpm}$$

$$\text{Number of sprinkler nozzles} = \frac{(1050 \text{ gpm})}{(34 \text{ gpm/nozzle})} = 30.9$$

which is rounded up to the next even number for layout convenience, or 32.

The pressure required at the pump is the sum of the minimum pressure at the nozzle (specified as 35 psi), the pressure loss due to friction (10 psi), and the pressure due to the pipe elevation over the pump (15 ft water or 6.5 psi). Therefore, the total pressure is 51.5 psi which is rounded up to 52 psi. The pump power is now determined.

$$\frac{\text{ft-lb}_f}{\text{sec}} = \left(\frac{52 \text{ lb}_f}{\text{in}^2}\right)\left(\frac{144 \text{ in}^2}{\text{ft}^2}\right)\left(\frac{1050 \text{ gal}}{\text{min}}\right)\left(\frac{\text{min}}{60 \text{ s}}\right)\left(\frac{\text{ft}^3}{7.48 \text{ gal}}\right) = 17{,}520$$

$$\text{The horsepower} = (17{,}520 \text{ ft-lb}_f/\text{s})\left(\frac{\text{HP}}{550 \dfrac{\text{ft-lb}_f}{\text{s}}}\right) = 31.8 \text{ HP}$$

Therefore, this sprinkler requires a pump with a capacity of 1050 gpm and a 31.8 HP motor, assuming an efficiency of 100%.

Actually, fire pumps are usually designed with discharge pressures of 100 to 125 psig so that the hose and monitor nozzle streams will have an effective reach. In addition, the size of the monitor is governed by requirements in the fire codes.[9]

7-7 MISCELLANEOUS DESIGNS FOR PREVENTING FIRES AND EXPLOSIONS

The successful prevention of fires and explosions in chemical plants requires a combination of many design techniques including those mentioned previously and many more. A complete description of these techniques is far beyond the scope of this text. A partial list, shown in Table 7-9, is given to illustrate that safety technology is relatively complex (the appropriate application requires significant knowledge and experience) and to serve as a check list for engineers to help them include the critical features for preventing fires and explosions.

SUGGESTED READING

R. Beach, "Preventing Static Electricity Fires" *Chemical Engineering,* Dec. 21, 1964, pp. 73-78; Jan. 4, 1965, pp. 63-73; Feb. 2, 1965, pp. 85-88.

Fire Protection Handbook, 14th Edition (Boston: National Fire Protection Association, 1976), Chapter 5: Control of Electrostatic Ignition Sources.

H. Deichelmann, "The Electrostatic Charge of Glass-Lined Vessels and Piping," *Pfaudler PWAG Report* 326e.

[9]*National Fire Codes,* Vol. 1 (Quincy, MA: National Fire Protection Association, 1986).

J. S. DORSEY, "Static Sparks: How to Exorcise the 'Go Devils,'" *Chemical Engineering,* Sept, 13, 1976, pp. 203-205.

S. K. GALLYM, "Elements of Static Electricity," *Gas,* March 1949, pp. 12-46.

H. HAASE, *Electrostatic Hazards* (New York: Verlag Chemie-Weinheim, 1977).

T. M. KIRBY, "Overcoming Static Electricity Problems in Lined Vessels," *Chemical Engineering,* Dec. 27, 1971.

T. A. KLETZ, *What Went Wrong* (Houston: Gulf Publishing Company, 1985).

A. KLINKENBERG and J. L. VAN DER MINE, *Electrostatics in the Petroleum Industry* (New York: Elsevier Press, 1958).

L. B. LOEB, "The Basic Mechanisms of Static Electrification," *Science,* Dec. 7, 1945, pp. 573-576.

"Loss Prevention," *Chemical Engineering Progress,* Vol 11, 1977.

S. S. MAC KEOWN and V. WOUK, "Electrical Charges Produced by Flowing Gasoline," *Industrial Engineering Chemistry,* June 1942, pp. 659-664.

Recommended Practice on Static Electricity, NFPA #77-1972, (Boston: National Fire Protection Association, 1972).

D. I. SALETAN, "Static Electricity Hazards," *Chemical Engineering,* June 1, 1959, pp. 99-102; June 29, 1959, pp. 101-106.

F. B. SILSBEE, *Static Electricity,* Circular C-438 (Washington, DC: National Bureau of Standards, 1942).

Static Electricity, Bulletin No. 256 (Washington, DC: U.S. Dept. of Labor, 1963).

PROBLEMS

7-1. Develop a list of steps needed to convert a common kitchen into an XP area.

7-2. What bonding and grounding procedures must be followed to transfer a drum of flammable solvent into a storage tank?

7-3. Ethylene oxide is a flammable liquid having a normal boiling temperature below room temperature. Describe a system and a procedure for transferring ethylene oxide from a tank car through a pumping system to a storage tank. Include both inerting and purging as well as bonding and grounding procedures.

7-4. Flammable liquid is being pumped out of a drum into a bucket using a hand pump. Describe an appropriate grounding and bonding procedure.

7-5. Using the sweep-through purging method, inert a 100 gallon vessel containing 100% air until the oxygen concentration is 1%. What volume of nitrogen is required? Assume nitrogen with no oxygen and a temperature of 77°F.

7-6. A 150 cubic ft tank containing air is to be inerted to 1% oxygen concentration. Pure nitrogen is available for the job. Since the tank's maximum allowable working pressure is 150 psia, it is possible to use either the sweep-through or a pressurization technique. For the pressurization technique, multiple pressurization cycles might be necessary, with the tank being returned to atmospheric pressure at the end of each cycle. The temperature is 80°F.

a. Determine the volume of nitrogen required for each technique.

b. For the pressurization technique, determine the number of cycles required if the pressure purge includes increasing the pressure to 140 psia with nitrogen and then venting to 0 psig.

7-7. Use a vacuum purging technique to purge oxygen out of a 150 cubic ft tank containing air. Reduce the oxygen concentration to 1% using pure nitrogen as the inert. The temperature is 80°F. Assume the vacuum purge goes from atmospheric pressure to 20 mm Hg. absolute. Determine the number of purge cycles required, and the total moles of nitrogen used.

7-8. Repeat Problem 7-7 using a combined vacuum and pressure purge. Use a vacuum of 20 mm Hg. absolute and a pressure of 200 psig.

7-9. Use the sweep-through purging technique to reduce the concentration of toluene from an initial 20% to 1% in a room with a volume of 25,000 cubic ft. Assume the room is purged with air at a rate of 6 room volumes per hour. How long will it take to complete this purge process?

7-10. Use the system described in Figure 7-2 to determine the voltage developed between the charging nozzle and the grounded tank, and the energy stored in the nozzle. Explain the potential hazard for cases a and b below.

Case	a	b
Hose length (ft)	20	20
Hose diameter (in)	2	2
Flow rate (gpm)	25	25
Liquid conductivity (mho/cm)	10^{-8}	10^{-18}
Dielectric constant	2.4	19
Density (gm/cm^3)	0.8	0.8
Viscosity (centipoise)	0.6	0.6
Diffusivity, D_m (cm^2/s)	2.2×10^{-5}	2.2×10^{-5}

7-11. Use the systems described in Problem 7-10 to determine the hose diameter required to eliminate the potential hazard due to static buildup.

7-12. Repeat Example 7-2 with a 40,000 gallon storage vessel. Assume the vessel height is equal to the diameter.

7-13. Review Problem 7-10, Part b. What is the most effective way to reduce the hazard of this situation. Select either (i) increasing the inlet diameter, or (ii) decreasing the flow rate.

7-14. Estimate the charge buildup, accumulated energy, and accumulated voltage as a result of pneumatically conveying a dry powder through a 50 ft long teflon duct which is 6 inch in diameter. The powder is collected in an insulated vessel. Repeat the calculation for a transport rate of 50 and 100 pounds per minute for transport times of 1 and 5 hours. Discuss ways to improve the safety of this situation.

7-15. Compute the accumulated charge, voltage, and energy for a 100,000 gallon vessel being filled with a fluid at a rate of 200 gpm and having a streaming current of 2×10^{-6} amp. Make the calculation for a fluid having a conductivity of 10^{-18} mho/cm and a dielectric constant of 2.0. Repeat the calculation for (a) a half full vessel, (b) a full vessel, and (c) a full vessel with an overflow line.

7-16. For Problem 7-15, Part (c), if the inlet flow is stopped, compute the accumulated charge, voltage, and energy after 5 hours, and 20 hours. Discuss the consequences of these results.

7-17. Some large storage vessels have a floating head, a flat cover which floats on the liquid surface. As the liquid volume increases and decreases, the floating head rises and falls within the cylindrical shell of the vessel. What are the reasons for this design?

7-18. What electrical classification would be specified for motors which are needed in an area normally having flammable concentrations of hydrogen and propane?

7-19. What electrical classification would be specified for an area which has Classes I and II, Groups A and E, and Divisions 1 and 2 motors?

7-20. Determine the recommended distance between a process area with toluene and an area with an open flame. Toluene leaks as large as 200 gpm have been recorded. Assume an average wind speed of 5 mph and a stability class of D.

7-21. Determine the recommended ventilation rate for an inside process area (30,000 ft^3) which will handle Class I liquids and gases.

7-22. For the process area described in Problem 7-21, determine the concentration of propane in the area as a function of time if at $t = 0$ a 3/4 inch propane line breaks (the propane main header is at 100 psig). The temperature is 80°F. See Chapter 4 for the appropriate source model and Chapter 3 for material balance models.

7-23. Using results of Problem 7-22, describe what safety features should be added to this process area.

7-24. Determine the fire water requirements (gpm, number of sprinkler heads and pump horsepower) to protect an inside storage area of 10,000 square ft which stores nonreactive solvents. Assume the sprinkler nozzles have a 1/2 inch orifice and the nozzle pressure is 75 psig.

7-25. Repeat Problem 7-24 assuming the nozzle pressure is 100 psig.

7-26. Determine the water requirement (gpm) and number of nozzles for a deluge system required to protect a 10,000 gallon storage tank which has a diameter of 15 ft. Use 1/2 inch nozzles with a nozzle pressure of 35 psig, and assume the vessel contains a reactive solvent.

7-27. Determine the sprinkler requirements for a chemical process area 150 ft by 150 ft. Determine the number of sprinkler heads and the pump specifications for this system (HP and gpm). Assume the friction loss from the last sprinkler head to the pump is 50 psi, and the nozzles (1/2 inch orifice) are at 75 psig. Also assume this process area contains a large pilot plant with nonreactive solvents.

7-28. Acetone (C_3H_6O) is to be stored in a cylindrical process vessel with a diameter of 5 feet and a height of 8 feet. The vessel must be inerted with pure nitrogen prior to storage of the acetone. A limited supply of pure nitrogen is available at 80 psig and 80°F. A vacuum is available at 30 mm Hg absolute pressure.

a. Determine the target oxygen concentration for the inerting procedure.

b. Decide whether a pressure or vacuum purge, or a combination of both, is the best procedure.

c. Determine the number of cycles required for your selected procedure.

d. Determine the total amount of nitrogen used. The final pressure in the tank after the inerting procedure is atmospheric. The ambient temperature is 80°F.

Introduction to Reliefs

8

Despite many safety precautions within chemical plants, equipment failures or operator errors may cause increases in process pressures beyond safe levels. If pressures rise too high, they may exceed the maximum strength of pipelines and vessels. This can result in rupturing of process equipment, causing major releases of toxic or flammable chemicals.

The first line of defense against this type of accident is to prevent the accident in the first place. This usually means better process control. A major effort is always directed towards controlling the process within safe operating regions. Dangerous high pressure excursions must be prevented or minimized.

The second line of defense against excessive pressures is to install relief systems to relieve liquids or gases before excessive pressures are developed. The relief system is composed of the relief device and the associated downstream process equipment to safely handle the material ejected.

The methodology used for the safe installation of pressure relief devices is illustrated in Figure 8-1. The first step in the procedure is to specify where relief devices must be installed. Definitive guidelines are available. Second, the appropriate relief device type must be selected. The type depends mostly on the nature of the material relieved and the relief characteristics required. Third, scenarios are developed describing the various ways in which a relief can occur. The motivation is to determine the material mass flow rate through the relief and the physical state of the material (liquid, vapor, or two-phase). Next, data are collected on the relief process, including physical property on the ejected material, and the relief is sized. Finally, the worst case scenario is selected and the final relief design is achieved.

Every step in this methodology is critical in the development of a safe design; an error in any step of this procedure can result in catastrophic failures.

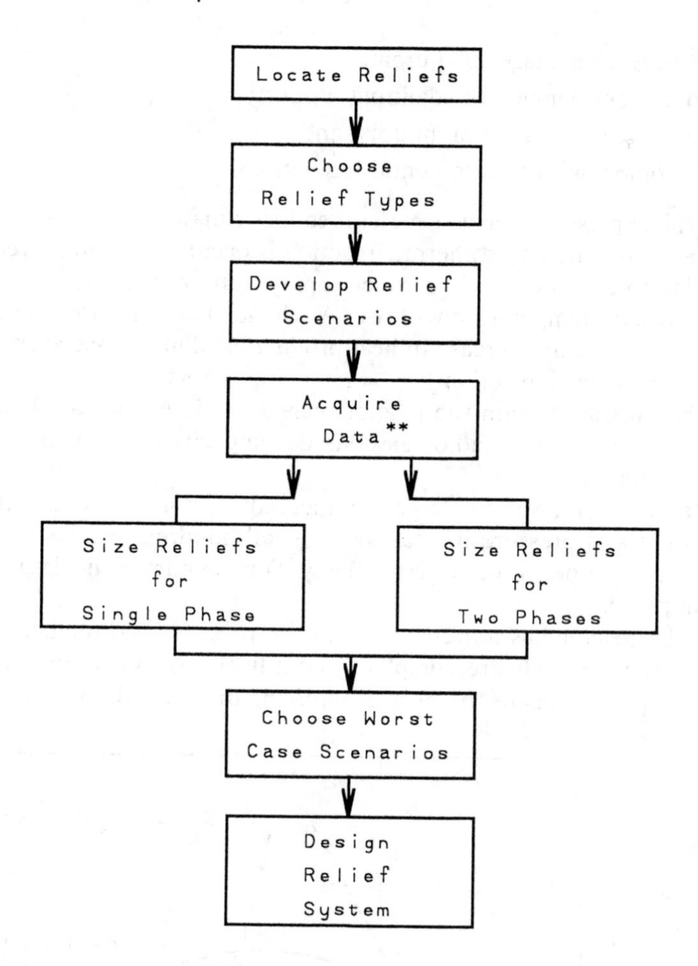

Figure 8-1 Relief methodology.

This chapter introduces relief fundamentals and the steps in the relief design procedure. Relief sizing methods are covered in Chapter 9.

8-1 RELIEF CONCEPTS

Pressure relief systems are required for the following reasons:[1]

- To protect personnel from the dangers of overpressurizing equipment,
- To minimize chemical losses during pressure upsets,

[1]Marx Isaacs, "Pressure Relief Systems," *Chemical Engineering,* Feb. 22, 1971, pp. 113-124.

- To prevent damage to equipment,
- To prevent damage to adjoining property,
- To reduce insurance premiums, and
- To comply with governmental regulations.

Typical pressure versus time curves for "runaway reactions" are illustrated in Figure 8-2. Assume an exothermic reaction is occurring within a reactor. If cooling is lost due to a loss of cooling water supply, failure of a valve, or other scenario, then the reactor temperature will rise. As the temperature rises the reaction rate increases, leading to an increase in heat production. This self-accelerating mechanism results in a runaway reaction.

The pressure within the reactor increases due to increased vapor pressure of the liquid components, and/or gaseous decomposition products as a result of the high temperature.

Reaction runaway for large commercial reactors can occur in minutes, with temperature and pressure increases of several hundred degrees per minute or several hundred psi per minute, respectively. For the curves in Figure 8-2 the cooling is lost at $t = 0$.

If the reactor has no relief system, the pressure and temperature continue to rise until the reactants are completely consumed, as shown on curve C. After the reactants are consumed, the heat generation stops and the reactor cools; the pres-

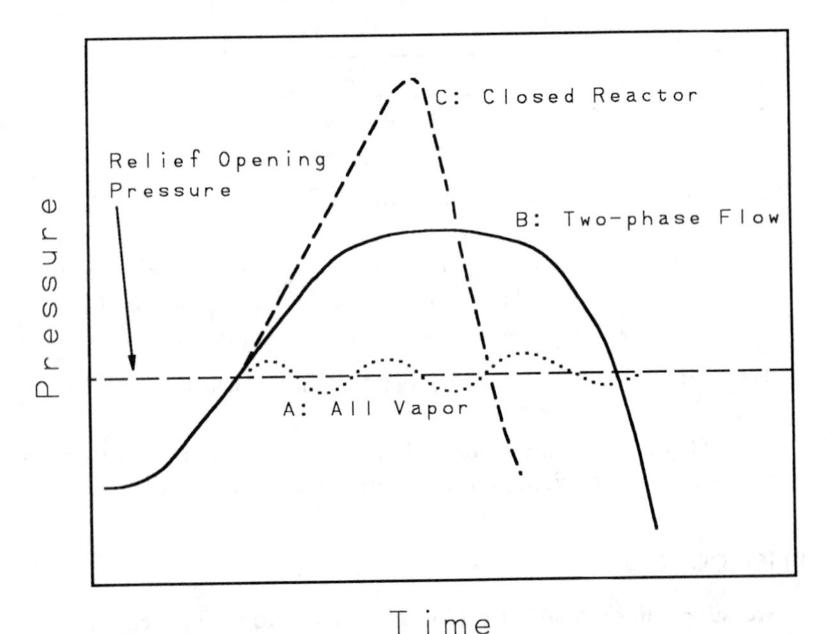

Figure 8-2 Pressure versus time for runaway reaction: (A) relieving vapor, (B) relieving froth, (C) closed reaction vessel. (Adapted from Dieter Gartner, Hartmut Giesbrecht, and Wolfgang Leuckel, "Thermodynamics and Fluid-dynamic Effects in Emergency Blow-Downs of Chemical Reactors," *Ger. Chemical Engineering,* Vol. 2 (1979), pp. 122-130.)

sure subsequently drops. Curve C assumes the reactor is capable of withstanding the full pressure of the runaway.

If the reactor has a relief device, the pressure response is dependent upon the relief device characteristics and the properties of the fluid discharged through the relief. This is illustrated by curve A for vapor relief only and curve B for a two-phase froth (vapor and liquid). The pressure will increase inside the reactor until the relief device activates at the pressure indicated.

When froth is discharged (curve B of Figure 8-2) the pressure continues to rise as the relief valve opens. The incremental pressure increase over the initial relief pressure is called overpressure.

Curve A is for vapor or gas discharged through the relief valve. The pressure drops immediately when the relief device opens because only a small amount of vapor discharge is required to decrease the pressure. The pressure drops until the relief valve closes; this pressure difference is called the blowdown .

Since the relief character of two-phase vapor-liquid material is markedly different from vapor relief, the nature of the relieved material must be known in order to design a proper relief.

8-2 DEFINITIONS[3]

Definitions which are commonly used within the chemical industry to describe reliefs are given below.

Set Pressure: The pressure at which the relief device begins to activate.

Maximum allowable working pressure (MAWP): The maximum gauge pressure permissible at the top of a vessel for a designated temperature. This is sometimes called the design pressure.

Operating pressure: The gauge pressure during normal service, usually 10% below the MAWP.

Accumulation: The pressure increase over the maximum allowable working pressure of a vessel during the relief process. It is expressed as a percentage of the maximum allowable working pressure.

Overpressure: The pressure increase in the vessel over the set pressure during the relieving process. Overpressure is equivalent to the accumulation when the set pressure is at the MAWP. It is expressed as a percentage of the set pressure.

Back pressure: The pressure at the outlet of the relief device during the relief process due to pressure in the discharge system.

Blow-down: The pressure difference between the relief set pressure and the relief reseating pressure. It is expressed as a percentage of the set pressure.

[3]API RP 521, *Recommended Practice for the Design and Installation of Pressure-Relieving Systems in Refineries,* 2nd ed. (Washington, D.C.: American Petroleum Institute, 1982), pp. 1-3.

Maximum allowable accumulated pressure: The sum of the maximum allowable working pressure plus the allowable accumulation.

Relief system: The network of components around a relief device, including the pipe to the relief, the relief device, discharge pipelines, knock-out drum, scrubber, flare, or other types of equipment which assist in the safe relief process.

The relationship between these terms is illustrated in Figure 8-3.

8-3 LOCATION OF RELIEFS[4]

The procedure for specifying the location of reliefs requires the review of every unit operation in the process and every process operating step. The engineer must anticipate the potential problems which may result in increased pressures. Pressure relief devices are installed at every point identified as potentially hazardous, that is, at points where upset conditions create pressures which may exceed the maximum allowable working pressure.

The type of questions asked in this review process are

What happens with loss of cooling, heating, or agitation?

What happens if the process is contaminated, or has a mischarge of a catalyst or monomer?

What happens if the operator makes an error?

What is the consequence of closing valves (block valves) on vessels or in lines which are filled with liquids and exposed to heat or refrigeration?

What happens if a line fails, for example, a failure of a high pressure gas line into a low pressure vessel?

What happens if the unit operation is engulfed in a fire?

What conditions cause runaway reactions, and how are relief systems designed to handle the discharge as a result of runaway reactions?

Some guidelines for locating reliefs are summarized in Table 8-1.

TABLE 8-1 GUIDELINES FOR SPECIFYING RELIEF POSITIONS[1]

All vessels other than steam generators need reliefs, including reactors, storage tanks, towers, and drums.
Blocked-in sections of cool liquid-filled lines which are exposed to heat (like the sun) or refrigeration need reliefs.
Positive displacement pumps, compressors, and turbines need reliefs on the discharge side.
Storage vessels need pressure and vacuum reliefs to protect against pumping in or out of a blocked-in vessel, or against the generation of a vacuum by condensation.
Vessel steam jackets are often rated for low pressure steam. Reliefs are installed in jackets to prevent excessive steam pressures due to operator error or regulator failure.

[1]Marx Isaacs, "Pressure-Relief Systems," *Chemical Engineering,* Feb. 22, 1971, pp. 113-124.

[4]Robert Kern, "Pressure-Relief Valves for Process Plants," *Chemical Engineering,* Feb. 28, 1977, pp.187-194.

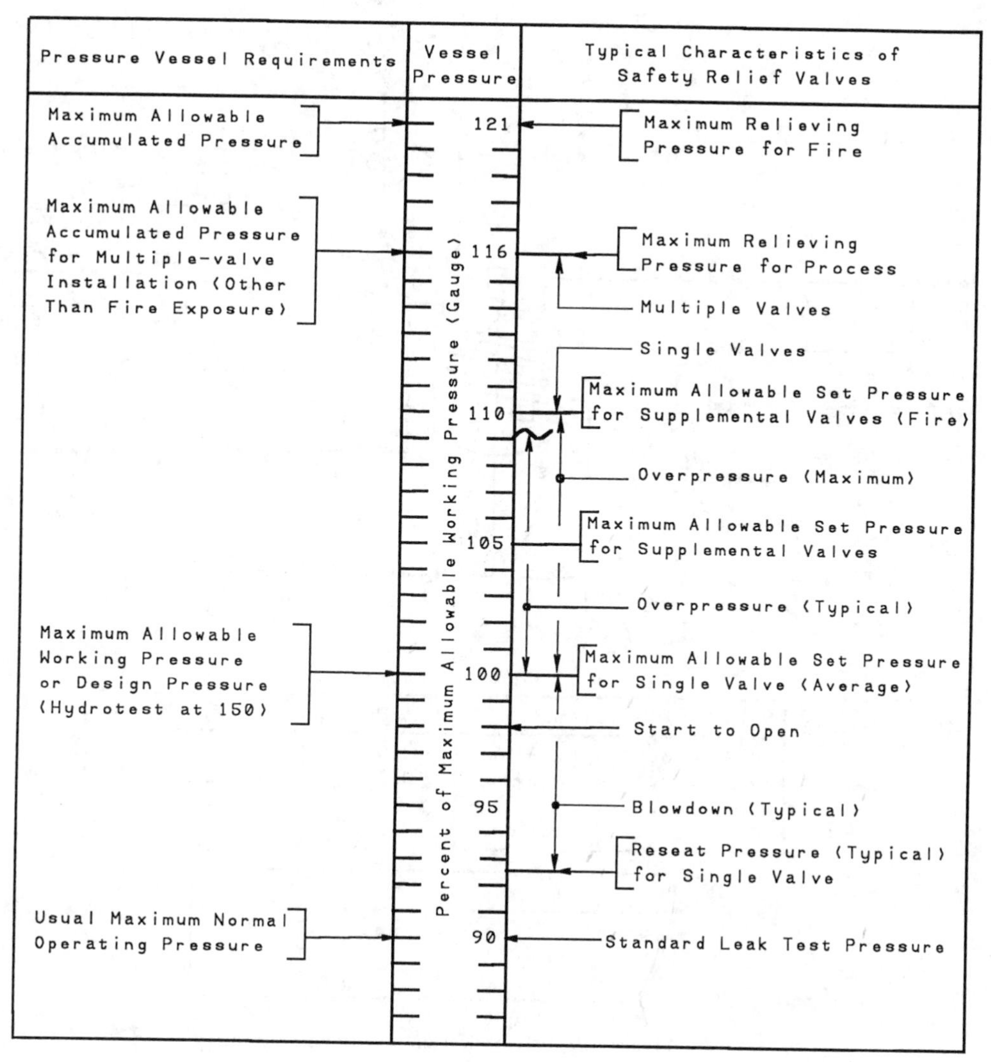

Notes:

1. The operating pressure may be any lower pressure required.
2. The set pressure and all other values related to it may be moved downward if the operating pressure permits.
3. This figure conforms with the requirements of the ASME Boiler and Pressure Vessel Code, Section VIII, "Pressure Vessels," Division 1.

Figure 8-3 Guidelines for relief pressures. (Adapted from API RP 521, *Guide for Pressure-Relieving and Depressurizing Systems,* 2nd ed. Washington, DC: American Petroleum Institute, 1982, p. 23.)

Example 8-1

Specify the location of reliefs in the simple polymerization reactor system illustrated in Figure 8-4. The major steps in this polymerization process include (1) pumping 100 pounds of initiator into the reactor R-1, (2) heating to the reaction temperature of 240°F, (3) adding monomer for a period of 3 hours, and (4) stripping the residual

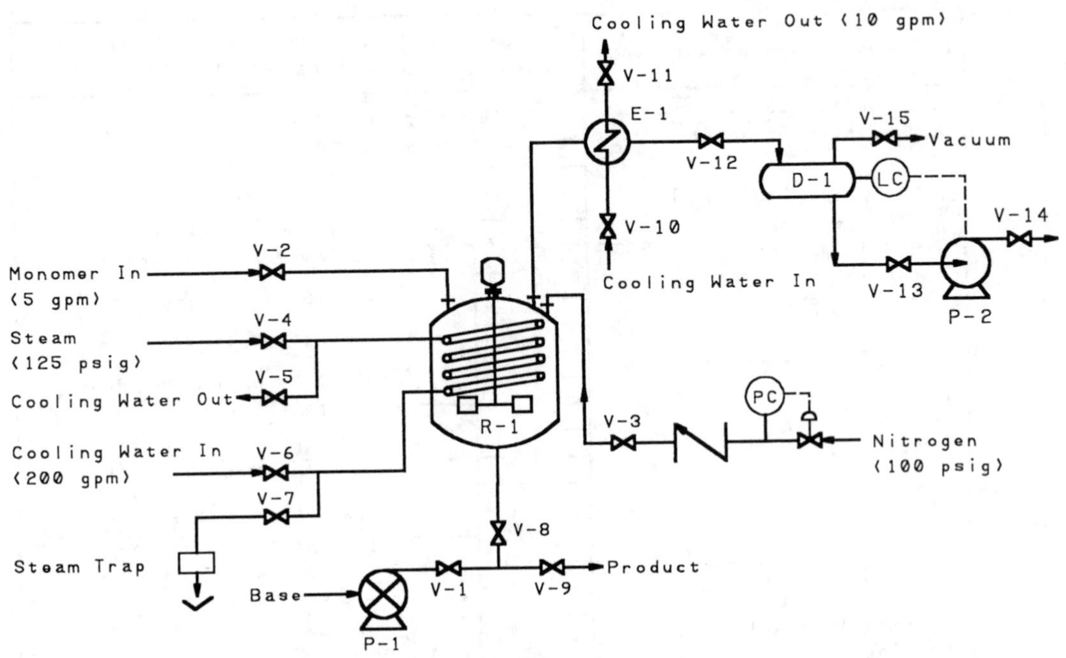

Specifications

Name	Description	Max. psig	Gpm at 50 psig
D-1	100 Gal Drum	50	—
R-1	1000 Gal Reactor	50	—
P-1	Gear Pump	100	100
P-2	Centrifugal Pump	50	20

Piping:	Size
Steam and Water Lines	2 inch
Nitrogen	1 inch
Vapor Lines	0.5 inch

Figure 8-4 Polymerization reactor without safety reliefs.

monomer via vacuum using valve V-15. Since the reaction is exothermic, cooling during monomer addition with cooling water is necessary.

Solution The review methodology for specifying the location of reliefs follows. Refer to Figures 8-4 and 8-5 for relief locations.

a. Reactor (R-1): A relief is installed on this reactor because, in general, every process vessel needs a relief. This relief is labeled PSV-1 for pressure safety valve number 1.

b. Positive displacement pump (P-1): Positive displacement pumps are overloaded, overheated, and damaged if they are dead-headed without a pressure relieving device (PSV-2). This type of relief discharge is usually recycled back to the feed vessel.

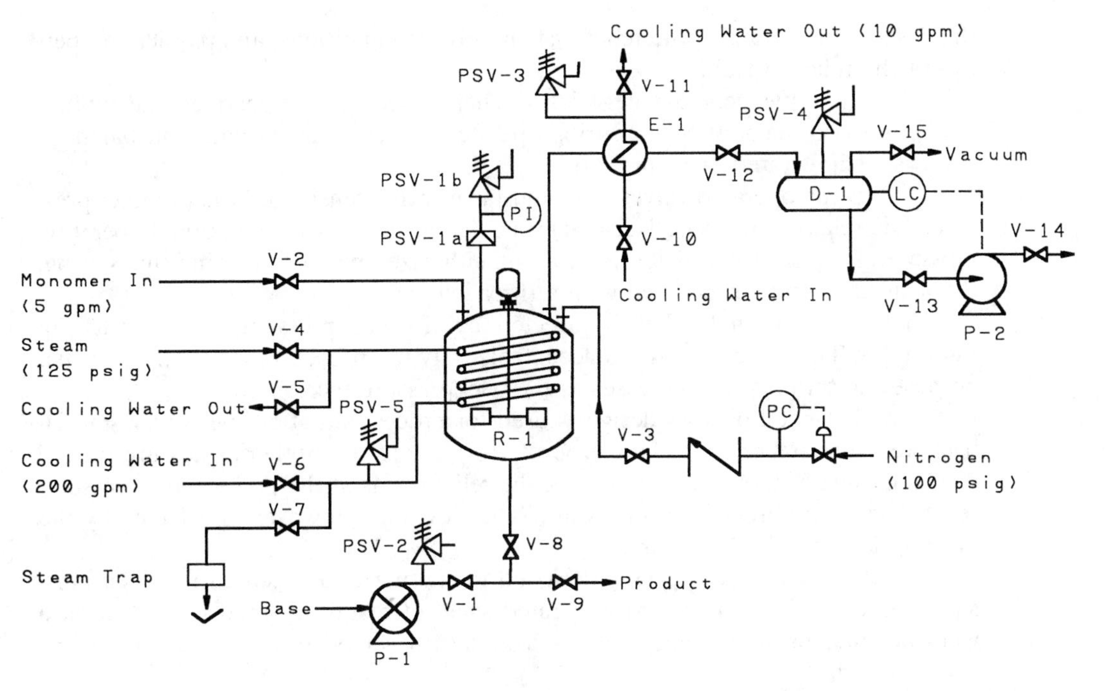

Figure 8-5 Polymerization reactor with safety reliefs.

c. Heat exchanger (E-1): Heat exchanger tubes can rupture from excessive pressures when water is blocked in (V-10 and V-11 are closed) and the exchanger is heated (by steam, for example). This hazard is eliminated by adding PSV-3.

d. Drum (D-1): Again, all process vessels need relief valves, PSV-4.

e. Reactor coil: This reactor coil can be pressure ruptured when water is blocked in (V-4, V-5, V-6, and V-7 are closed) and the coil is heated with steam or even the sun. Add PSV-5 to this coil.

This completes the specification of the relief locations for this relatively simple process. The reason for the two relief devices, PSV-1A and PSV-1B, will be described in the next section of this chapter.

The above example illustrates the engineering rationale for installing relief valves at various locations within a chemical plant. After the relief locations are specified, then the type of relief is chosen depending on the specific application.

8-4 RELIEF TYPES

Specific types of relief devices are chosen for specific applications such as for liquids, gases, liquids and gases, solids, and corrosive materials; they may be vented to the atmosphere or vented to containment systems (scrubber, flare, condenser, incinerator, and the like). In engineering terms, the type of relief device is specified on the basis of the details of the relief system, process conditions, and physical properties of the relieved fluid.

There are two general categories of relief devices (spring operated and rupture discs) and two major types of spring operated valves (conventional and balanced bellows) as illustrated in Figure 8-6.

On spring operated valves, the adjustable spring tension offsets the inlet pressure. The relief set pressure is usually specified at 10% above the normal operating pressure. To avoid the possibility of an unauthorized person changing this setting, the adjustable screw is covered with a threaded cap.

The conventional relief is affected by the backpressure as illustrated in Figure 8-6. This type of valve is acceptable only when backpressures are minimal because the set pressure increases as the backpressure increases.

The balanced bellows design is used for process situations where substantial backpressures are present. This type of valve keeps atmospheric pressure on the discharge side of the relief. Therefore, the relief opens at the preset relief pressure, regardless of the process backpressure. The flow rate, however, is affected by the magnitude of the backpressure.

Rupture discs are specially designed to rupture at a specified relief set pressure. They usually consist of a calibrated sheet of metal designed to rupture at a well-specified pressure. They are used alone, in series, or in parallel to spring

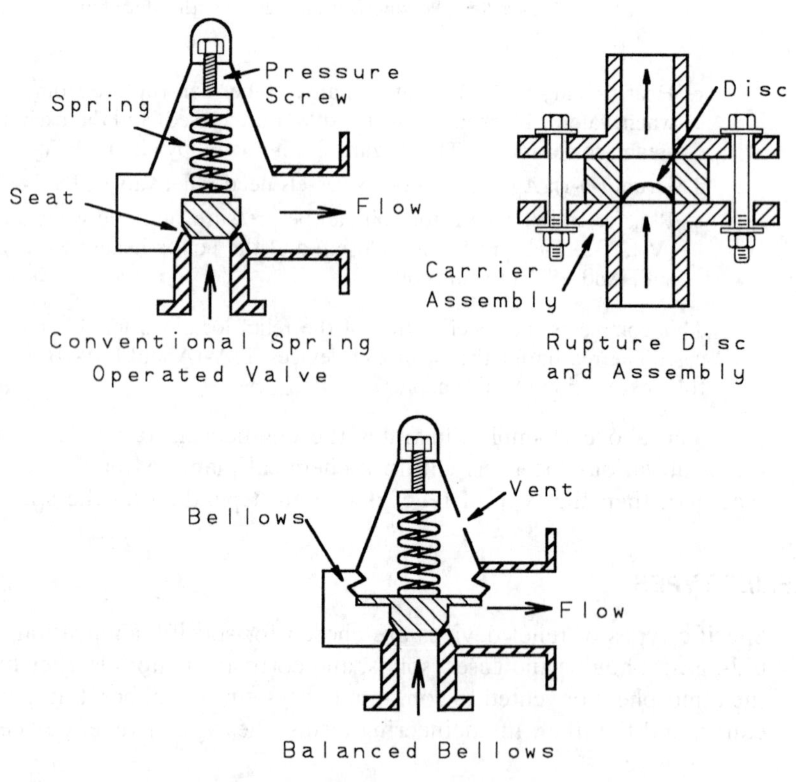

Figure 8-6 Major types of relief devices.

loaded relief devices. They can be made from a variety of materials, including exotic corrosion resistant materials.

Rupture discs are used alone when it is desired to keep the relief line open after the disc has ruptured. An example of this requirement is the relief of exploding gases or exploding dusts.

Rupture discs are frequently installed in series to a spring loaded relief (a) to protect an expensive spring loaded device from a corrosive environment, (b) to give absolute isolation when handling extremely toxic chemicals (spring loaded reliefs may weep), (c) to give absolute isolation when handling flammable gases, (d) to protect the relatively complex parts of a spring loaded device from reactive monomers which could cause plugging, and (e) to relieve slurries which may plug spring loaded devices.

When rupture discs are used before a spring loaded relief, a pressure gauge is installed between the two devices. This "tell-tale gauge" is an indicator which shows when the disc ruptures. The failure can be the result of a pressure excursion, or the result of a pin hole due to corrosion. In either case this telltale indicates that the disc needs to be replaced.

There are three subcategory types of spring loaded pressure reliefs.

1. The **relief valve** is primarily for liquid service. The relief valve (liquid only) begins to open at the set pressure. This valve reaches full capacity when the pressure reaches 25% overpressure. The valve closes as the pressure returns to the set pressure.
2. The **safety valve** is for steam, gas, and vapor service. Safety valves pop open when the pressure exceeds the set pressure. This is accomplished by using a discharge nozzle that directs high velocity material towards the valve seat. After blowdown of the excess pressure, the valve reseats at approximately 4% below the set pressure; the valve has a 4% blowdown.
3. The **safety relief valve** is used for liquid and vapor service. Safety relief valves function as relief valves for liquids and as safety valves for vapors.

Example 8-2

Specify the types of relief devices needed for the polymerization reactor in Example 8-1 (see Figure 8-5).

Solution Each relief is reviewed in relationship to the relief system and the properties of the relieved fluids:

a. PSV-1a is a rupture disc to protect PSV-1b from the reactive monomers (plugging via polymerization).

b. PSV-1b is a safety relief valve because a runaway reaction will give two phase flow, both liquid and vapor.

c. PSV-2 is a relief valve because this relief is in a liquid service line. A conventional valve is satisfactory.

d. PSV-3 is a relief valve because it is for liquid only. A conventional relief device is satisfactory in this service.

e. PSV-4 is a safety relief valve since liquid or vapor service is possible. Since this vent will go to a scrubber with possibly large backpressures, a balanced bellows is specified.

f. PSV-5 is a relief valve for liquid service only. This relief provides protection for the following scenario: the liquid is blocked in by closing all valves; the heat of reaction increases the temperature of the surrounding reactor fluid; and pressures are increased inside the coil due to thermal expansion.

After specifying the location and type of all relief devices, the relief scenarios are developed.

8-5 RELIEF SCENARIOS

A relief scenario is a description of one specific relief event. Usually each relief has more then one relief event, and the worst case scenario is the scenario or event which requires the largest relief vent area. Examples of relief events are

1. a pump is dead headed; the pump relief is sized to handle the full pump capacity at its rated pressure,
2. the same pump relief is in a line with a nitrogen regulator; the relief is sized to handle the nitrogen if the regulator fails, and
3. the same pump is connected to a heat exchanger with live steam; the relief is sized to handle steam injected into the exchanger under uncontrolled conditions, for example, a steam regulator failure.

This is a list of scenarios for one specific relief. The relief vent area is subsequently computed for each event (scenario) and the worst case scenario is the event requiring the largest relief vent area. The worst cases are a subset of the overall developed scenarios for each relief.

For each specific relief all possible scenarios are identified and catalogued. This step of the relief methodology is extremely important: The identification of the actual worst case scenario frequently has a more significant effect on the relief size compared to the accuracy of relief sizing calculations.

The scenarios developed for the reactor system described in Figure 8-5 are summarized in Table 8-2. The worst case scenarios are identified later via the computed maximum relief area for each scenario and relief (see Chapter 9). In

TABLE 8-2 RELIEF SCENARIOS FOR EXAMPLE 8-2 (SEE FIGURE 8-5)

Relief identifications	Scenarios
PSV-1a & PSV-1b	(a) Vessel full of liquid and pump (P-1) is accidentally actuated. (b) Cooling coil is broken and water enters at 200 gpm and 50 psig. (c) Nitrogen regulator fails giving critical flow through 1 inch line. (d) Loss of cooling during reaction (runaway).
PSV-2	V-1 is accidentally closed; system needs relief for 100 gpm at 50 psig.
PSV-3	Confined water line is heated with 125 psig steam.
PSV-4	(a) Nitrogen regulator fails giving critical flow through 0.5 inch line. (b) Note: the other R-1 scenarios will be relieved via PSV-1.
PSV-5	Water blocked inside coil, and heat of reaction causes thermal expansion.

Table 8-2, only three reliefs have multiple scenarios which require the comparative calculations to establish the worst cases. The other three reliefs have only single scenarios; therefore, they are the worst case scenarios.

8-6 DATA FOR SIZING RELIEFS

Physical property data and sometimes reaction rate characteristics are required for making relief sizing calculations. Data estimated using engineering assumptions is almost always acceptable when designing unit operations because the only result will be poorer yields or poorer quality. In the relief design, however, these types of assumptions are not acceptable because an error may result in catastrophic and hazardous failures.

When designing reliefs for gas or dust explosions, special deflagration data for the scenario conditions is required. This data is acquired with the apparatus already described in Section 6-12.

A runaway reaction is another scenario which requires special data.

It is known that runaway reactions nearly always result in two-phase flow reliefs.[5] The two phases discharge through the relief system similar to a champagne and carbon dioxide mixture exiting a freshly opened bottle. If the champagne is heated prior to opening, the entire contents of the bottle may be "relieved." This result has also been verified for runaway reactions in the chemical industry.

Two-phase flow calculations are relatively complex, especially when conditions change rapidly as in a runaway reaction scenario. As a result of this complexity, special methods were developed for acquiring the relevant data[6] and for making the relief vent sizing calculations[7] (see Chapter 9).

The data required for making two-phase flow relief calculations are determined with a specially designed calorimeter called a Vent Sizing Package (VSP) illustrated in Figure 8-7. The VSP is essentially an adiabatic calorimeter. A small amount of the material to be tested (30 to 80 mg) is loaded into a thin-walled reactor vessel. A series of controlled heaters increases the sample temperature to the runaway conditions. During the runaway the VSP device tracks the pressure inside the can and maintains a similiar pressure in the main containment vessel; this prevents the thin-walled sample container from rupturing.

The data acquired with this calorimeter is shown in Figures 8-8 and 8-9. Results of particular importance for relief sizing calculations include the temperature rate $(dT/dt)_s$ at the set pressure and the temperature increase, ΔT, corresponding to the overpressure, ΔP. Since the calorimeter starts with known weights and known compositions, the heat of reaction can also be determined from this T versus t data (assuming the heat capacities of the monomers and products are known).

[5]Harold G. Fisher, "DIERS Research Program on Emergency Relief Systems," *Chem. Engr. Prog.,* Aug. 1985, pp. 33-36.

[6]J. C. Leung, H. K. Fauske, and H. G. Fisher, "Thermal Runaway Reactions in a Low Thermal Inertia Apparatus," *Thermochimica Acta.,* No. 104 (1986), pp. 13-29.

[7]J. C. Leung, "Simplified Vent Sizing Equations for Emergency Relief Requirements in Reactors and Storage Vessels," *AIChE Journal,* Oct. 1986, pp. 1622-34.

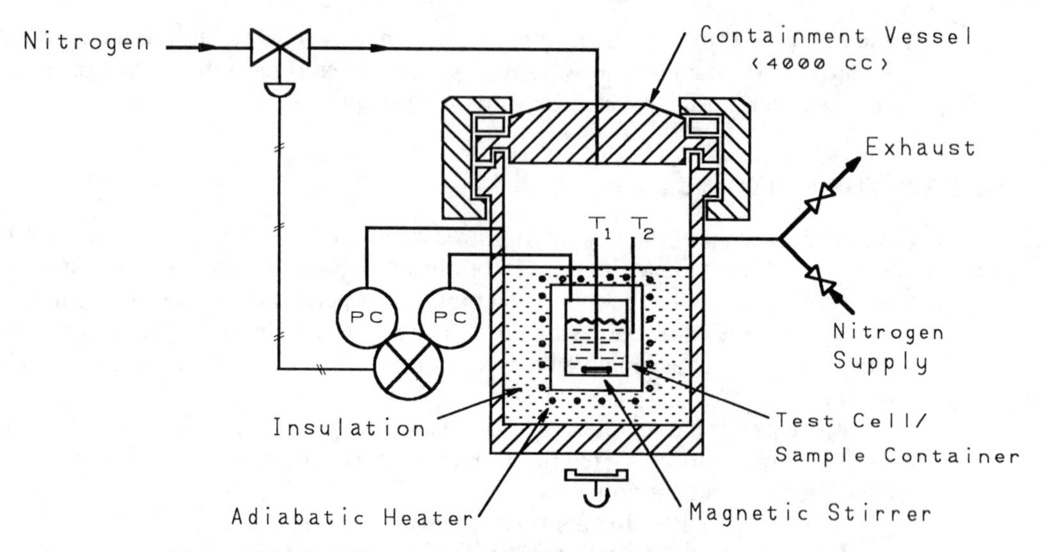

Figure 8-7 Vent sizing package (VSP) for acquiring runaway reaction data.

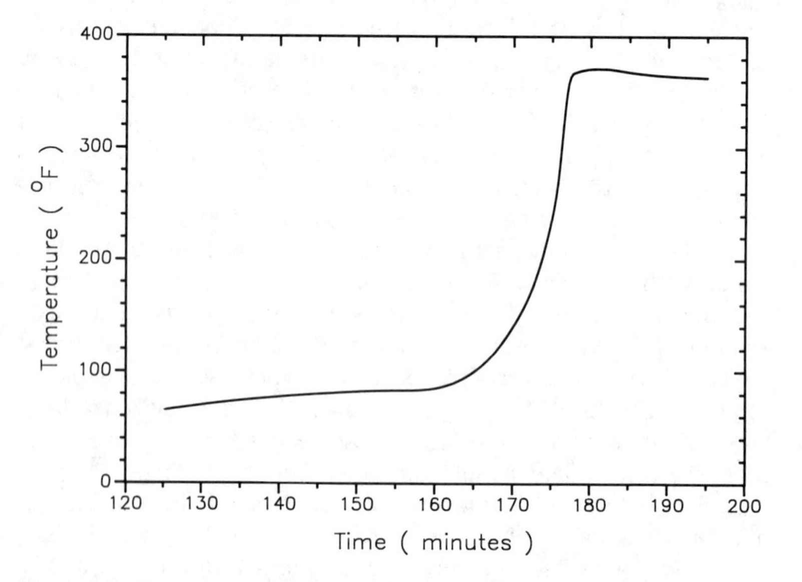

Figure 8-8 Runaway reaction temperature data acquired using the vent sizing package (VSP).

8-7 RELIEF SYSTEMS

After the relief type is chosen and the relief size is computed, the engineer takes the responsibility for completing the design of the relief system, including

a. Deciding how to install the relief in the system and

b. Deciding how to dispose of the exiting liquids and vapors.

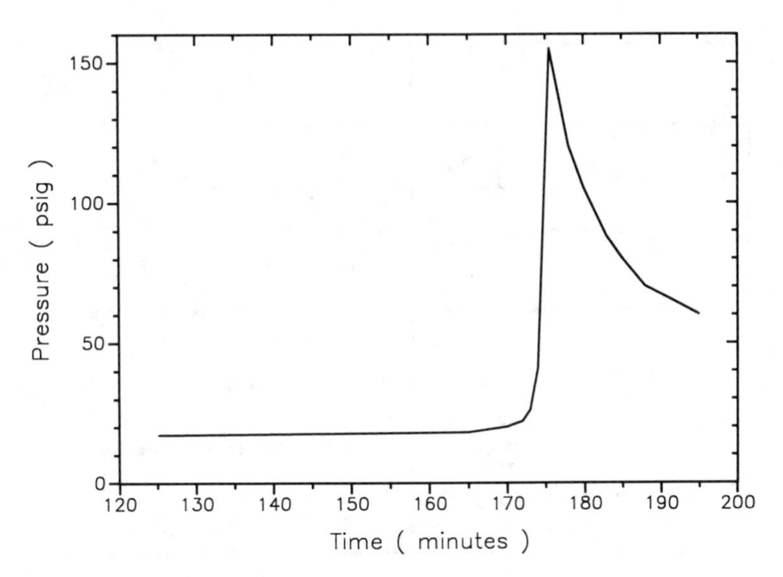

Figure 8-9 Runaway reaction pressure data acquired using the vent sizing package (VSP).

Pressure relieving systems are unique when compared to the other systems within a chemical plant; hopefully they will never need to operate, but when they do, they must do so flawlessly. Other systems, such as extraction and distillation systems usually evolve to their optimum performance and reliability. This evolution requires creativity, practical knowledge, hard work, time, and the cooperative efforts of the plant, design, and process engineers. This same effort and creativity is essential when developing relief systems, however, in this case the relief system development must be optimally designed and demonstrated within a research environment prior to the plant start-up.

In order to develop the necessary optimum and reliable relief systems, it is essential to understand this technology. The objective of this section is to give students and design engineers the details necessary for understanding relief systems.

Relief Installation Practices

Regardless of how carefully the relief is sized, specified, and tested, a poor installation can result in completely unsatisfactory relief performance. Some installation guidelines are illustrated in Figure 8-10. During field construction, sometimes expediency or construction convenience leads to modifications and deviations from acceptable practice. The engineer must take the responsibility for adhering to standard practices, especially when installing relief systems.

Relief Design Considerations

A designer of relief systems must be familiar with governmental codes, industrial standards, and insurance requirements. This is particularly important because local government standards may vary. Codes of particular interest are published by the

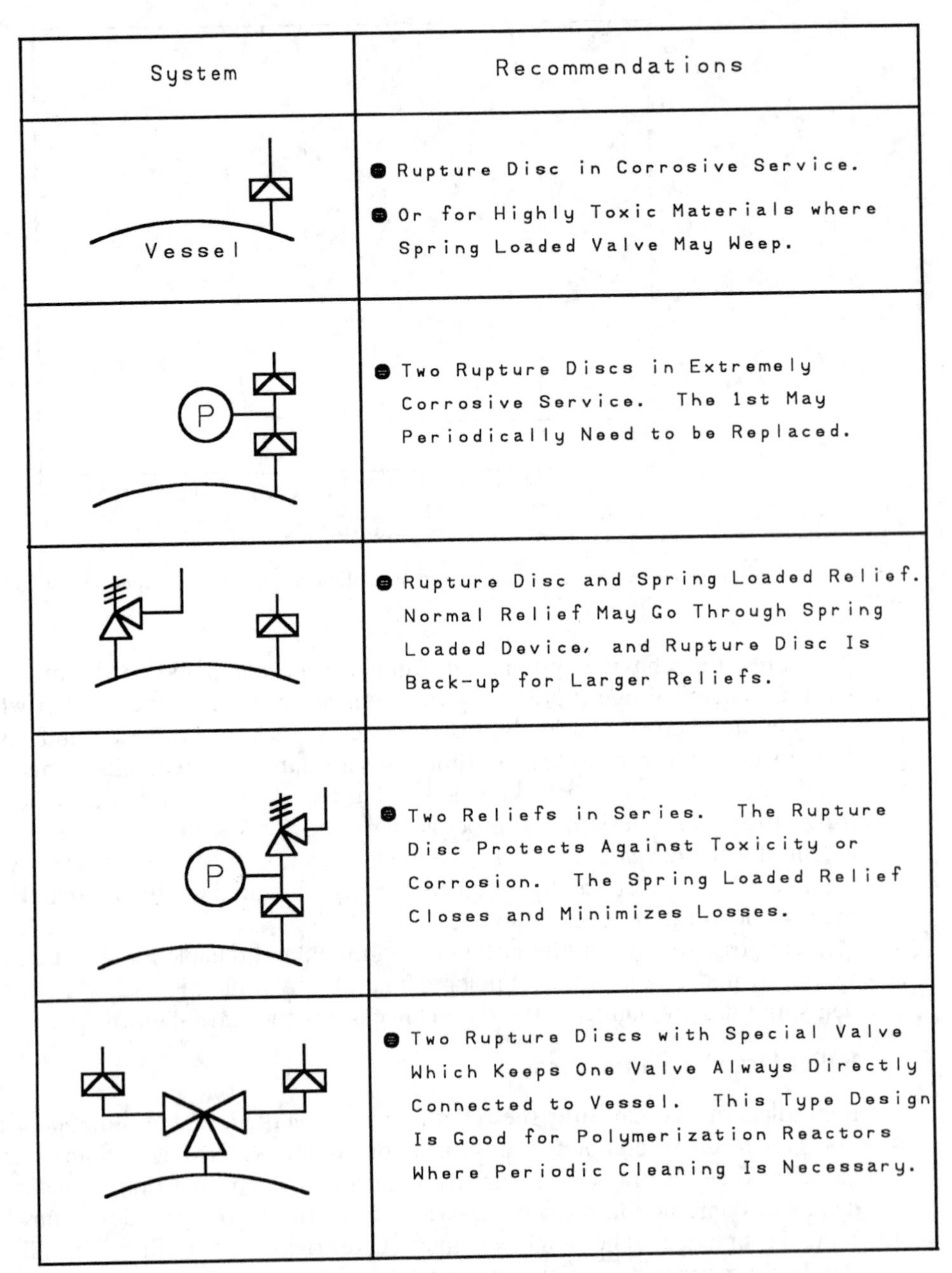

System	Recommendations
Vessel	• Rupture Disc in Corrosive Service. • Or for Highly Toxic Materials where Spring Loaded Valve May Weep.
P	• Two Rupture Discs in Extremely Corrosive Service. The 1st May Periodically Need to be Replaced.
	• Rupture Disc and Spring Loaded Relief. Normal Relief May Go Through Spring Loaded Device, and Rupture Disc Is Back-up for Larger Reliefs.
P	• Two Reliefs in Series. The Rupture Disc Protects Against Toxicity or Corrosion. The Spring Loaded Relief Closes and Minimizes Losses.
	• Two Rupture Discs with Special Valve Which Keeps One Valve Always Directly Connected to Vessel. This Type Design Is Good for Polymerization Reactors Where Periodic Cleaning Is Necessary.

(a)

Figure 8-10 Relief installation practices. (Adapted from Eric Jennett, "Components of Pressure-Relieving Systems," *Chemical Engineering,* August 19, 1963, pp. 151-158.)

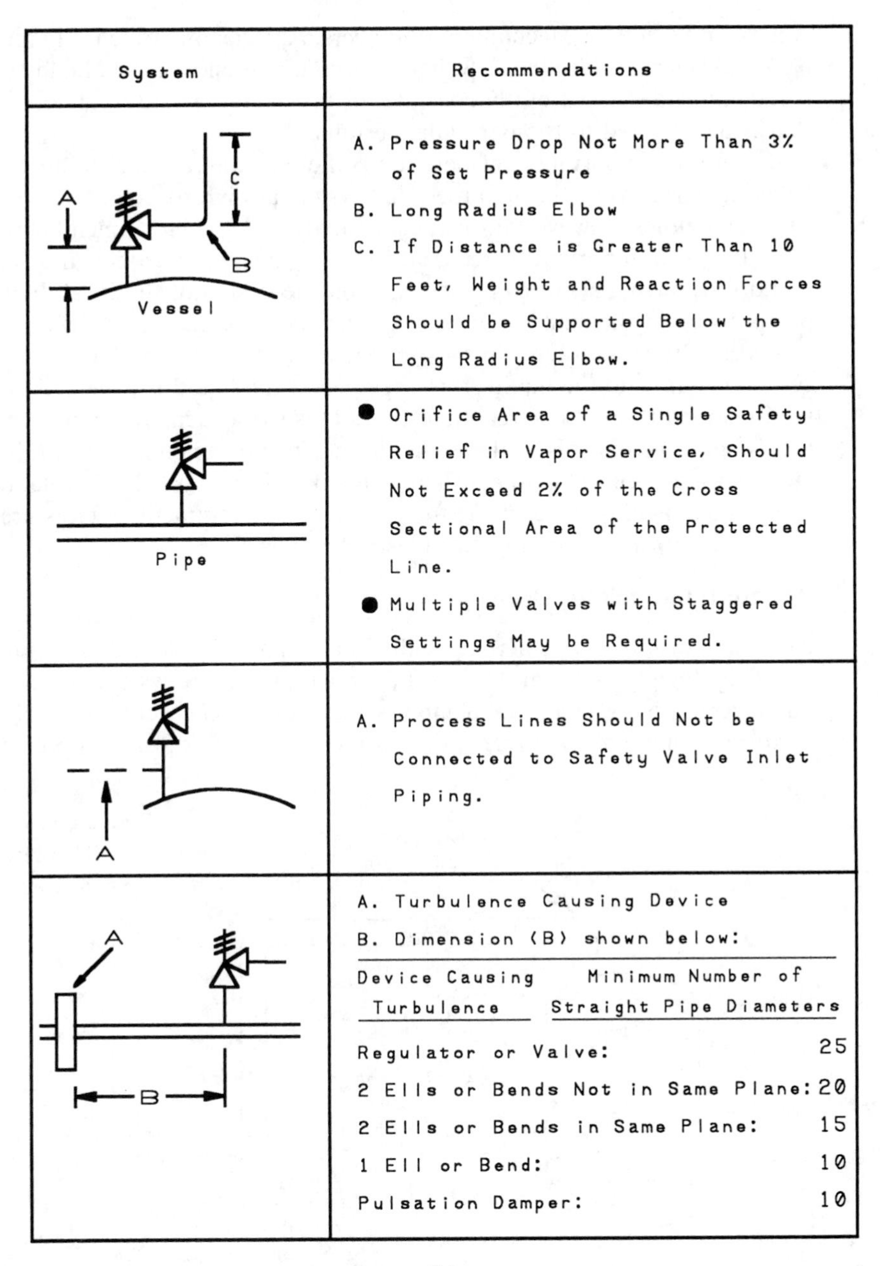

System	Recommendations
A C B Vessel	A. Pressure Drop Not More Than 3% of Set Pressure B. Long Radius Elbow C. If Distance is Greater Than 10 Feet, Weight and Reaction Forces Should be Supported Below the Long Radius Elbow.
Pipe	● Orifice Area of a Single Safety Relief in Vapor Service, Should Not Exceed 2% of the Cross Sectional Area of the Protected Line. ● Multiple Valves with Staggered Settings May be Required.
A	A. Process Lines Should Not be Connected to Safety Valve Inlet Piping.
A B	A. Turbulence Causing Device B. Dimension (B) shown below:

Device Causing Turbulence	Minimum Number of Straight Pipe Diameters
Regulator or Valve:	25
2 Ells or Bends Not in Same Plane:	20
2 Ells or Bends in Same Plane:	15
1 Ell or Bend:	10
Pulsation Damper:	10

(b)

Figure 8-10 (continued)

American Society of Mechanical Engineers, American Petroleum Institute, and the National Board of Fire Underwriters. Specific references have already been sited. It is recommended to carefully consider all codes and, where feasible, select the one that is most suited to the particular installation.

Another important consideration is the reaction forces generated when the relieved materials flow through the relief system at high speed. The API RP 520[8] has some guidelines, however, normal stress analysis is the recommended methodology.

It is also important to recognize that company philosophy and the regulatory authorities have a significant influence on the design of the final disposal system, primarily from the standpoint of pollution. For this reason, reliefs are now rarely vented to the atmosphere. In most cases, a relief is first discharged to a knockout system to separate the liquid from vapor; here the liquid is collected and the vapor is discharged to another treatment unit. This subsequent vapor treatment unit depends upon the hazards of the vapor; it may include a condenser, scrubber, incinerator, flare, or a combination of them. This type system is called a total containment system; one is illustrated in Figure 8-11. Total containment systems are commonly used and they are becoming an industrial standard.

Horizontal Knockout Drum

Knockout drums are sometimes called catchtanks or blowdown drums. As illustrated in Figure 8-11, this horizontal knock-out drum system serves as a vapor-liquid separator as well as a holdup vessel for the disengaged liquid. The two-phase mixture usually enters at one end, and the vapor leaves at the opposite end. Inlets

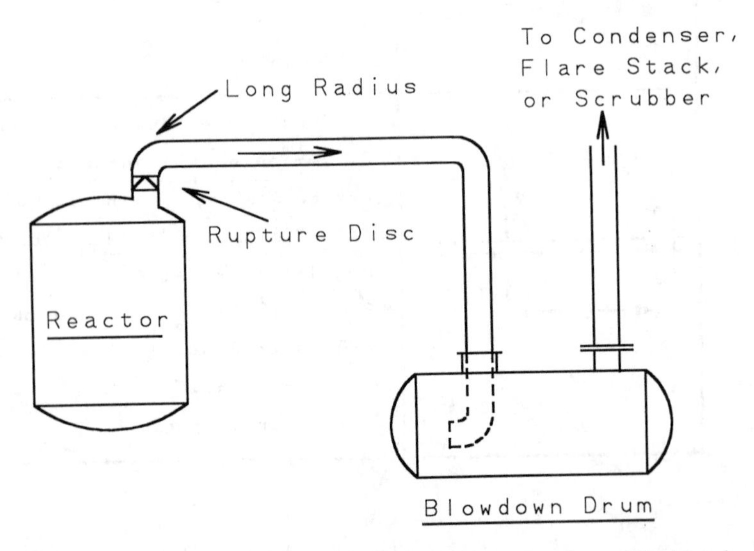

Figure 8-11 Relief containment system with blowdown drum. The blowdown drum separates the vapor from the liquid.

[8]API RP 520, *Recommended Practice for the Design and Installation of Pressure-Relieving Systems in Refineries,* 4th ed. (Washington, D.C.: American Petroleum Institute, 1976).

may be provided at each end, with a vapor exit in the center to minimize vapor velocities. When space within a plant is limited, a tangential knock-out drum is used as shown in Figure 8-12.

The design method for sizing this type of system was published by Grossel[9] and in API 521[10]. The method is based on the maximum allowable velocity for minimizing liquid entrainment. The dropout velocity of a particle in a stream is

$$u_d = 1.15 \sqrt{\frac{g d_p (\rho_L - \rho_V)}{\rho_V C}} \tag{8-1}$$

where

u_d is the dropout velocity,
g is the acceleration due to gravity,
d_p is the particle diameter,
ρ_L is the liquid density,
ρ_V is the vapor density, and
C is the drag coefficient given by Figure 8-13.

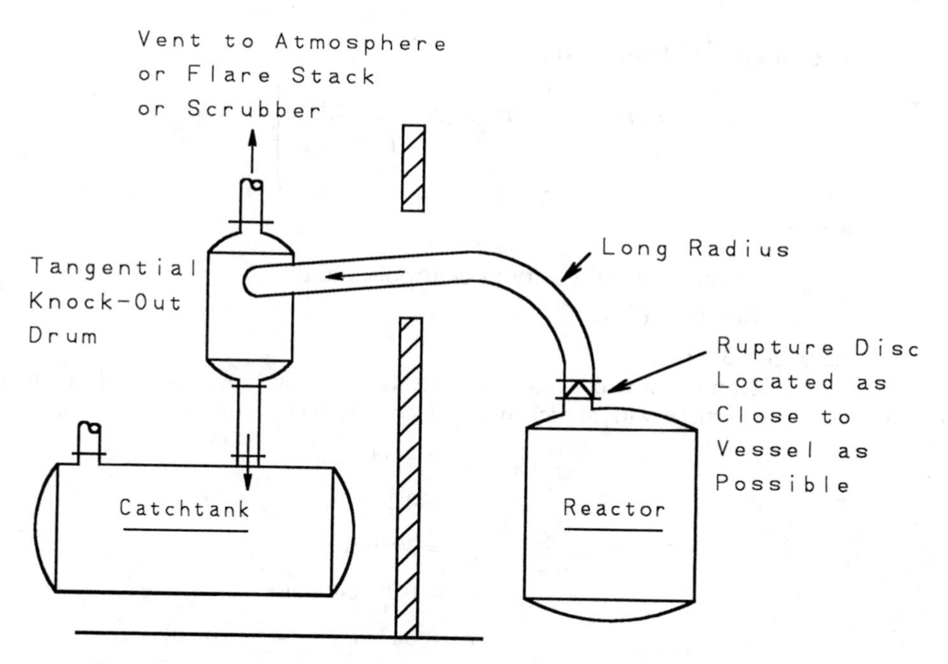

Figure 8-12 Tangential inlet knock-out drum with separate liquid catchtank.

[9] S. S. Grossel, "Design and Sizing of Knockout Drums/Catchtanks for Reactor Emergency Relief Systems," *Plant/Operations Progress,* July 1986.

[10] API RP 521, *Guide for Pressure-Relieving and Depressurizing Systems,* 2nd ed. (Washington, D.C.: American Petroleum Institute, 1982), pp.51-52.

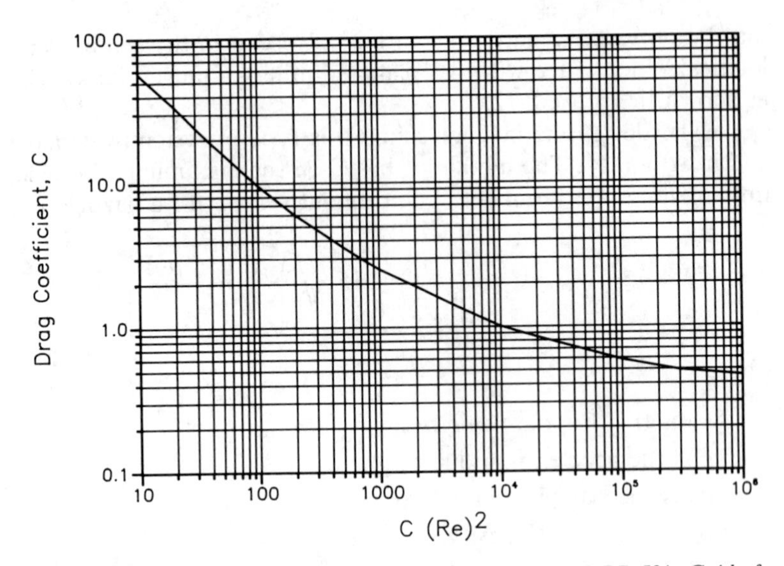

Figure 8-13 Drag coefficient correlation. (Data from API RP 521, *Guide for Pressure-Relieving and Depressurizing Systems,* 2nd ed. Washington, DC: American Petroleum Institute, 1982.)

The abscissa of Figure 8-13 is

$$C(Re)^2 = \left[0.95 \times 10^8 \frac{\text{centipoise}^2}{\left(\frac{\text{lb}}{\text{ft}^3}\right)^2 \text{ft}^3}\right] \frac{\rho_V d_p^3(\rho_L - \rho_V)}{\mu_V^2} \tag{8-2}$$

where

μ_v is the vapor viscosity in centipoise, and
$C(Re)^2$ is unitless.

Example 8-3

Determine the maximum vapor velocity in a horizontal knockout drum to dropout liquid particles with particle diameters of 300 micron, where

$$\text{Vapor rate} = 170 \text{ lb/hr}$$
$$\rho_V = 0.20 \text{ lb/ft}^3$$
$$\rho_L = 30 \text{ lb/ft}^3$$
$$\mu_V = 0.01 \text{ centipoise}$$
$$d_p = 300 \text{ micron} = 9.84 \times 10^{-4} \text{ ft}$$

Solution To determine the dropout velocity, the drag coefficient is first determined using Figure 8-13. The graph abscissa is computed using Equation 8-2.

$$C(Re)^2 = \left[0.95 \times 10^8 \frac{\text{centipoise}^2}{\left(\frac{\text{lb}}{\text{ft}^3}\right)^2 \text{ft}^3}\right] \frac{\rho_V d_p^3(\rho_L - \rho_V)}{\mu_V^2}$$

$$C(Re)^2 = \left[0.95 \times 10^8 \frac{cp^2}{\left(\frac{\mathrm{lb_m}}{\mathrm{ft}^3}\right)^2 \mathrm{ft}^3} \right]$$

$$\times \frac{(0.2\ \mathrm{lb_m/ft^3})(9.84 \times 10^{-4}\ \mathrm{ft})^3(30 - 0.2)\ \mathrm{lb_m/ft^3}}{(0.01\ \mathrm{centipoise})^2}$$

$$C(Re)^2 = 5394$$

Using Figure 8-13, $C = 1.3$.

The dropout velocity is determined using Equation 8-1.

$$u_d = 1.15 \sqrt{\frac{g d_p(\rho_L - \rho_V)}{\rho_V C}}$$

$$u_d = 1.15 \sqrt{\frac{(32.2\ \mathrm{ft/s^2})(9.84 \times 10^{-4}\ \mathrm{ft})(30 - 0.2)\ \mathrm{lb/ft^3}}{(0.2\ \mathrm{lb/ft^3})(1.3)}} = 2.19\ \frac{\mathrm{ft}}{\mathrm{s}}$$

The required vapor space area, perpendicular to the vapor path, is subsequently computed using the velocity and the volumetric flow rate of the vapor. The entire vessel design is determined as a function of this vapor area plus the liquid hold volume, and the general geometric configuration of the vessel.

Flares[11]

Flares are sometimes used after knock-out drums. The objective of a flare is to burn the combustible or toxic gas to produce combustion products which are neither toxic nor combustible. The diameter of the flare must be suitable to maintain a stable flame and prevent a blowout (when vapor velocities are greater than 20% of the sonic velocity).

The height of a flare is fixed on the basis of the heat generated and the resulting potential damage to equipment and humans. The usual design criteria is that the heat intensity at the base of the stack is not to exceed 1500 Btu/hr/sq ft. The effects of thermal radiation are shown below:

Heat intensity (Btu/hr/sq ft)	Effect
2000	Blisters in 20 seconds
5300	Blisters in 5 seconds
3000-4000	Vegetation and wood are ignited
350	Solar radiation

Using the fundamentals of radiation, the heat intensity, q, at a specific point is a function of the heat generated, Q_f, by the flame, the emissivity ε, and the distance, R, from the flame,

$$q = \frac{\varepsilon Q_f}{4\pi R^2} \tag{8-3}$$

[11]Soen H. Tan, "Flare System Design Simplified," *Hydrocarbon Processing*, Jan. 1967.

Assuming a flame height of 120 d_f, an emissivity $\varepsilon = 0.048\sqrt{M}$, and a heating value of 20,000 Btu/lb, Equation (8-3) is algebraically modified to give the flare height H_f as a function of the flare diameter d_f, the desired heat intensity q_f, at a point X_f from the base of the flare, for a burning fuel with a molecular weight M, and a vapor rate Q_m.

$$H_f = -60d_f + 0.5\sqrt{(120\ d_f)^2 - \left(\frac{4\pi q_f X_f^{\ 2} - 960\ Q_m\sqrt{M}}{\pi q_f}\right)} \tag{8-4}$$

where

H_f is the flare height (ft),
d_f is the diameter of flare stack (ft),
q_f is the heat intensity (Btu/hr/ft^2),
X_f is the distance from the stack at grade (ft),
Q_m is the vapor rate (lb/hr), and
M is the molecular weight of the vapor.

Example 8-4

Determine the stack height required to give a heat intensity of 1500 Btu/hr/sq ft at a distance of 410 ft from the base of the flare. The flare diameter is 4 ft, the flare load is 970,000 lb/hr, and the molecular weight of the vapor is 44.

Solution The flare height is computed using Equation 8-4. The units are consistent with those required.

$$H_f = -60d_f + 0.5\sqrt{(120\ d_f)^2 - \left(\frac{4\pi q_f X_f^{\ 2} - 960\ Q_m\sqrt{M}}{\pi q_f}\right)}$$

$$H_f = -(60)(4) + 0.5\sqrt{[(120)(4)]^2 - \left[\frac{(4)(3.14)(1500)(410)^2 - (960)(970{,}000)\sqrt{44}}{(3.14)(1{,}500)}\right]}$$

$$H_f = 381 \text{ ft}$$

Scrubbers

The fluid from reliefs, sometimes two phase flow, must first go to a knockout system where the liquids and vapors are separated. Liquids are subsequently collected and the vapors may or may not be vented. If the vapors are nontoxic and nonflammable, they may be vented unless some regulation prohibits this type of discharge.

If the vapors are toxic, a flare (described previously) or a scrubber system may be required. Scrubber systems may be packed columns, plate columns, or venturi type. Details of scrubber designs are covered by Treybal.[12]

Condensers

A simple condenser is another possible alternative for treating exiting vapors. This alternative is particularly attractive if the vapors have a relatively high boiling point

[12] R. E. Treybal, *Mass Transfer Operations,* 3rd ed. (New York: McGraw-Hill, 1958).

and if the recovered condensate is valuable. This alternative should always be evaluated because it is simple, usually less expensive, and it minimizes the volume of material which may need additional post treatment. The design of condenser systems is covered by Kern.[13]

SUGGESTED READING

General Articles on Relief Valves and Systems

FLOYD E. ANDERSON, "Pressure Relieving Devices," *Safe and Efficient Plant Operations and Maintenance,* Richard Greene, ed. (New York: McGraw-Hill, 1980), p. 207.

MARX ISAACS, "Pressure-Relief Systems," *Chemical Engineering,* February 22, 1971, p. 113.

ROBERT KERN, "Pressure-Relief Valves for Process Plants," *Chemical Engineering,* February 28, 1977, p. 187.

PROBLEMS

8-1. Can gate valves be placed between a vessel relief and its vessel?

8-2. Describe the process of creating a vacuum in a storage vessel as a result of condensation. Develop an example to illustrate the potential magnitude of the vacuum.

8-3. In the future, it is anticipated that insurance rates will be set as a function of the safety of a plant. Illustrate the kinds of plant statistics which you would cite to reduce your insurance costs.

8-4. Give four examples of situations requiring a combination of spring operated reliefs in series with rupture discs.

8-5. PSV-2 of Figure 8-5 is a relief to protect the positive displacement pump P-1. If the fluid being handled is extremely volatile and flammable, what design modifications would you make to this relief system?

8-6. The first defense against runaway reactions is better process control. Using the system illustrated in Figure 8-5, what control features (safe guards) would you add to this reactor system?

8-7. If a scrubber is installed after PSV-1b, and it has a pressure drop of 30 psig, how would this affect the size (qualitatively) of this relief system?

8-8. Referring to Problem 8-7, qualitatively describe the algorithm you would use to compute the relief size for this system.

8-9. Review Figure P8-9 and determine the locations for relief devices.

8-10. Review Figure P8-10 and determine the locations for relief devices.

8-11. Review Figure P8-9 and Problem 8-9 to determine what types of relief devices should be used at each location.

8-12. Review Figure P8-10 and Problem 8-10 to determine what types of relief devices should be used at each location.

[13] D. Q. Kern, *Process Heat Transfer* (New York: McGraw-Hill, 1950).

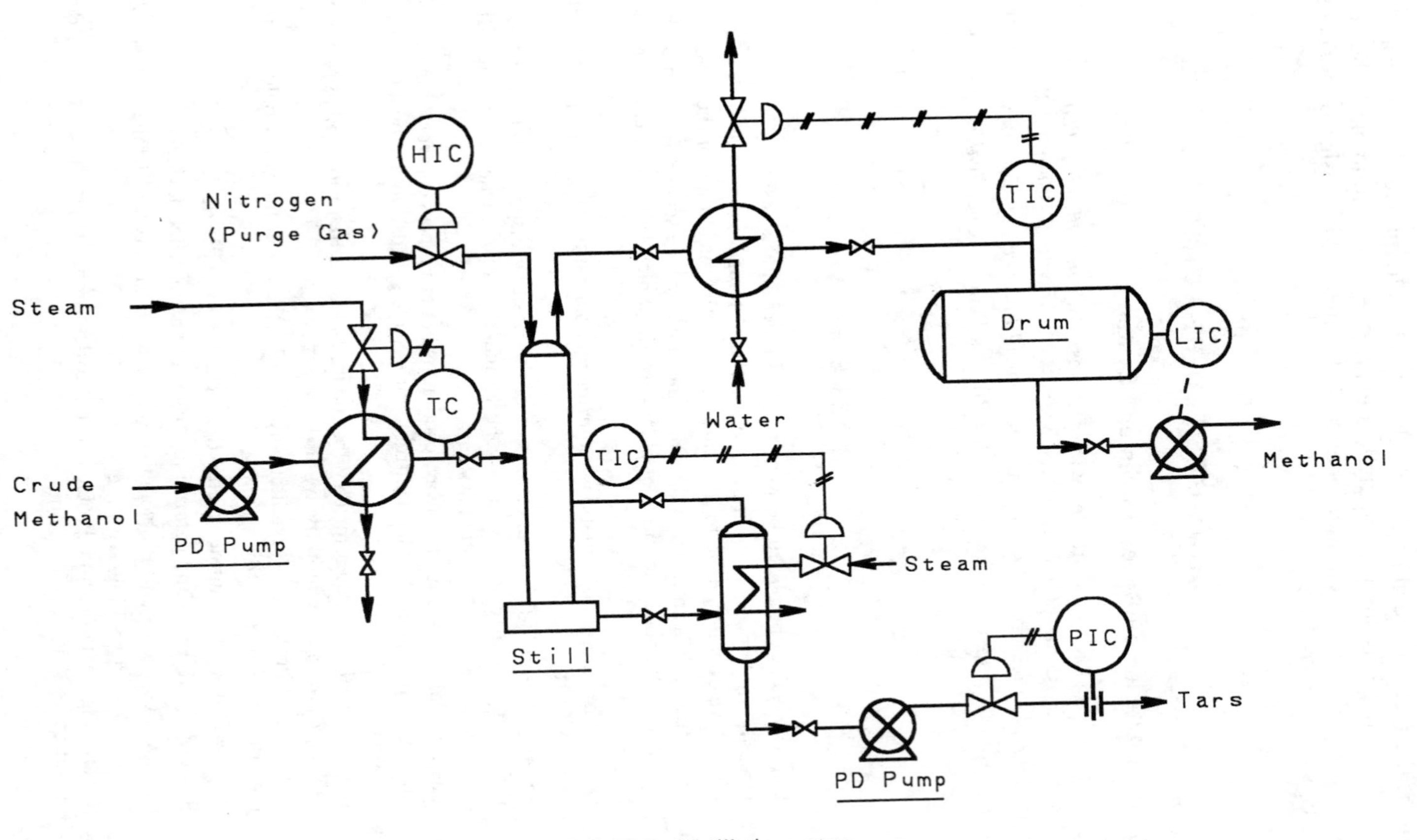

Figure P8-9 Distillation system.

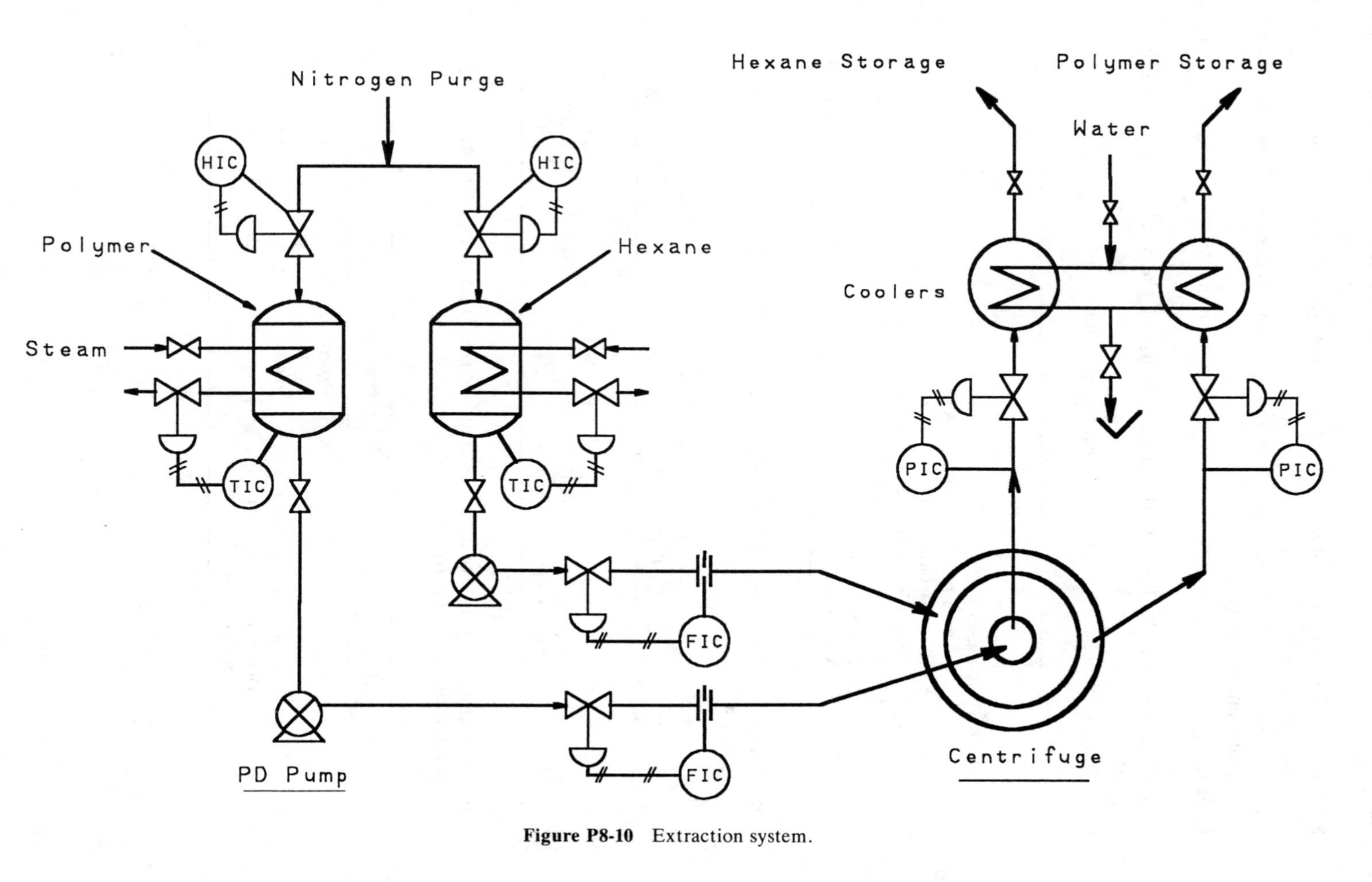

Figure P8-10 Extraction system.

8-13. Review Figure P8-9, Problems 8-9 and 8-11, and make recommendations for total containment systems.

8-14. Review Figure P8-10, Problems 8-10 and 8-12, and make recommendations for total containment systems.

8-15. Using results of Problems 8-9 and 8-11, determine the relief scenarios for each relief device.

8-16. Using results of Problems 8-10 and 8-12, determine the relief scenarios for each relief device.

8-17. Develop sketches of reactor vent systems assuming

Part	a	b	c	d
Reactor relief is vapor only	X			X
Reactor relief is two phase flow		X	X	
Reactor contents are corrosive		X		X
Reactor contents are plugging type		X		
Relieved vapors are toxic	X			X
Relieved vapors are high boilers	X	X		
Vapors are low boilers			X	X

8-18. Determine the vapor velocity inside a horizontal knockout drum for the following systems:

Part	a	b	c
ρ_V (lb/ft^3)	0.03	0.04	0.05
ρ_L (lb/ft^3)	64.0	64.5	50.0
Vapor visc.(centipoise)	0.01	0.02	0.01
Particle diameter(micron)	300	400	350

8-19. Determine the height of a flare assuming various maximum heat intensities at ground level at the specified distances from the flare.

Part	a	b	c
Vapor flow (lb/hr)	60,000	70,000	80,000
Molecular weight	30	60	80
Heat intensity (Btu/hr/ft^2)	2,000	3,000	4,000
Distance from base (ft)	5	10	50
Stack diameter (ft)	2	3	5

Relief Sizing

9

Relief sizing calculations are performed to determine the vent area of the relief device.

The relief sizing calculation procedure involves, first, using an appropriate source model to determine the rate of material release through the relief device (see Chapter 4) and, second, using an appropriate equation based on fundamental hydrodynamic principles to determine the relief device vent area.

The relief vent area calculation is dependent upon

- the type of flow, liquid, vapor, or two-phase, and
- the type of relief device, spring or rupture disc.

Chapter 8 showed that for liquids and two-phase relief, the relieving process begins at the relief set pressure with the pressure normally continuing to rise past the set pressure (see curve B on Figure 8-2). These overpressures frequently exceed the set pressure by 25% or more. A relief device designed to maintain the pressure at the set pressure could require an excessively large vent area. As shown in Figure 9-1, the relief vent area is reduced substantially as the overpressure increases. This is one example which illustrates this typical result. The optimal vent area for a particular relief depends on the specific application. The overpressure specification is part of the relief design. Normally, relief devices are specified for overpressures from 10 to 25%, depending on the requirements of the equipment protected and the type of material relieved.

Spring relief devices require 25 to 30% of maximum flow capacity to maintain the valve seat in the open position. Lower flows will result in "chattering," caused by rapid opening and closing of the valve disc. This can lead to destruction of the

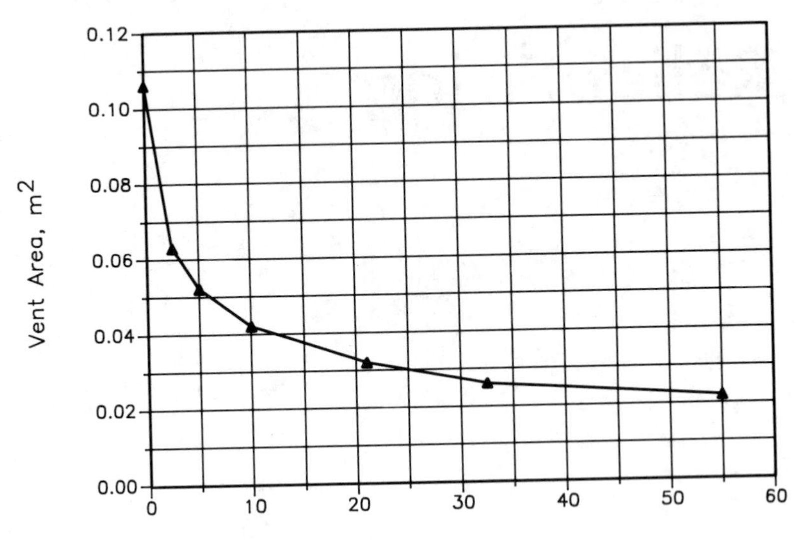

Figure 9-1 Required vent area as a function of overpressure for two-phase flow. The vent area is decreased appreciably as the overpressure increases. (Data from J. C. Leung, "Simplified Vent Sizing Equations for Emergency Relief Requirements in Reactors and Storage Vessels," *AICHE Journal*, Vol. 32, No. 10, 1986.)

relief device and a dangerous situation. A relief device with an area that is too large for the required flow may chatter. For this reason, reliefs must be designed with the proper vent area, neither too small or too large.

Experimental data at the actual relief conditions is recommended for sizing relief vents for runaway reaction scenarios. As always, manufacturers' technical specifications are used for selection, purchase and installation.

This chapter will present methods for calculating the relief device vent areas for the following configurations:

- conventional spring operated reliefs in liquid or vapor-gas service
- rupture discs in liquid or vapor-gas service
- two-phase flow during runaway reactor relief
- reliefs for dust and vapor explosions
- reliefs for fires external to process vessels
- reliefs for thermal expansion of process fluids

9-1 CONVENTIONAL SPRING OPERATED RELIEFS IN LIQUID SERVICE

Flow through spring type reliefs is approximated as flow through an orifice. An equation representing this flow is derived from the mechanical energy balance,

Equation 4-1. The result is similiar to Equation 4-6, except that the pressure is represented by a pressure difference across the spring relief,

$$\bar{u} = C_o \sqrt{\frac{2g_c \Delta P}{\rho}} \tag{9-1}$$

where

$\bar{u}$ is the liquid velocity through the spring relief,
C_o is the discharge coefficient,
ΔP is the pressure drop across the relief, and
ρ is the liquid density.

The volumetric flow, Q_v, of liquid is the product of the velocity times the area, or $\bar{u}A$. Substituting Equation 9-1 and solving for the vent area of the relief, A,

$$A = \frac{Q_v}{C_o \sqrt{2g_c}} \sqrt{\frac{\rho}{\Delta P}} \tag{9-2}$$

A working equation with fixed units is derived from Equation 9-2 by (1) replacing the density ρ by the specific gravity, (ρ/ρ_{ref}), and (2) making the appropriate substitutions for the unit conversions. The result is

$$\boxed{A = \left[\frac{\text{in}^2 \, (\text{psi})^{1/2}}{38.0 \text{ gpm}}\right] \frac{Q_v}{C_0} \sqrt{\frac{(\rho/\rho_{ref})}{\Delta P}}} \tag{9-3}$$

where

A is the computed relief area (in^2),
Q_v is the volumetric flow through the relief (gallons per minute),
C_0 is the discharge coefficient (unitless),
(ρ/ρ_{ref}) is the specific gravity of the liquid (unitless), and
ΔP is the pressure drop across the spring relief (lb_f/in^2).

In reality, flow through a spring type relief is different from flow through an orifice. As the pressure increases, the relief spring is compressed, increasing the discharge area and increasing the flow. A true orifice has a fixed area. Equation 9-3 does not consider the viscosity of the fluid. Many process fluids have high viscosities. The relief vent area must increase as the fluid viscosity increases. Equation 9-3 does not consider the special case of a balanced bellows type relief.

Equation 9-3 has been modified by the American Petroleum Institute to include corrections for the above situations. The result[1] is

$$\boxed{A = \left[\frac{\text{in}^2 \, (\text{psi})^{1/2}}{38.0 \text{ gpm}}\right] \frac{Q_v}{C_0 K_v K_p K_b} \sqrt{\frac{(\rho/\rho_{ref})}{1.25P_s - P_b}}} \tag{9-4}$$

[1] *API RP 520, Recommended Practice for the Design and Installation of Pressure Relieving Systems in Refineries* (Washington, DC: American Petroleum Institute, 1976).

where

A is the computed relief area (in^2),
Q_v is the volumetric flow through the relief (gallons per minute),
C_0 is the discharge coefficient (unitless),
K_v is the viscosity correction (unitless),
K_p is the overpressure correction (unitless),
K_b is the backpressure correction (unitless),
(ρ/ρ_{ref}) is the specific gravity of the liquid (unitless),
P_s is the gauge set pressure (lb_f/in^2), and
P_b is the gauge back pressure (lb_f/in^2).

Note that the ΔP term in Equation 9-3 has been replaced by a term involving the difference between the set pressure and the back pressure. Equation 9-4 appears to assume a maximum pressure equal to 1.25 times the set pressure. Discharge at other maximum pressures is accounted for in the overpressure correction term, K_b.

C_0 is the discharge coefficient. Specific guidelines for the selection of an appropriate value are given in Chapter 4, Section 4-1. If this value is uncertain, a conservative value of 0.61 is used to maximize the relief vent area.

The viscosity correction, K_v, corrects for the additional frictional losses due to flow of high viscosity material through the valve. This correction is given in Figure 9-2. The required relief vent area becomes larger as the viscosity of the liquid increases (lower Reynolds numbers). Since the Reynolds number is required to determine the viscosity correction and the vent area is required to calculate the Reynolds number, the procedure is iterative. For most reliefs the Reynolds number is greater than 5000 and the correction is near unity. This assumption is frequently used as an initial estimate to begin the calculations.

The overpressure correction, K_p, includes the effect of discharge pressures greater than the set pressure. This correction is given in Figure 9-3. The overpressure correction, K_p, is a function of the overpressure specified for the design. As the overpressure specified becomes smaller, the correction value decreases, resulting in a larger relief area. Designs incorporating less than 10% overpressure are not recommended.

The backpressure correction, K_b, is used only for balanced bellows type spring reliefs, and is given in Figure 9-4. This correction compensates for the absence of backpressure on the back of the relief vent disc.

Example 9-1

A positive displacement pump pumps water at 200 gpm, at a pressure of 200 psig. Since a dead-headed pump can be easily damaged, compute the area required to relieve the pump assuming a backpressure of 20 psig and (a) a 10% overpressure and (b) a 25% overpressure.

Solution **a.** The set pressure is 200 psig. The backpressure is specified as 20 psig and the overpressure is 10% of the set pressure, or 20 psig.

The discharge coefficient, C_0, is not specified. However, for a conservative estimate a value of 0.61 is used.

The quantity of material relieved is the total flow of water; so $Q_v = 200$ gpm.

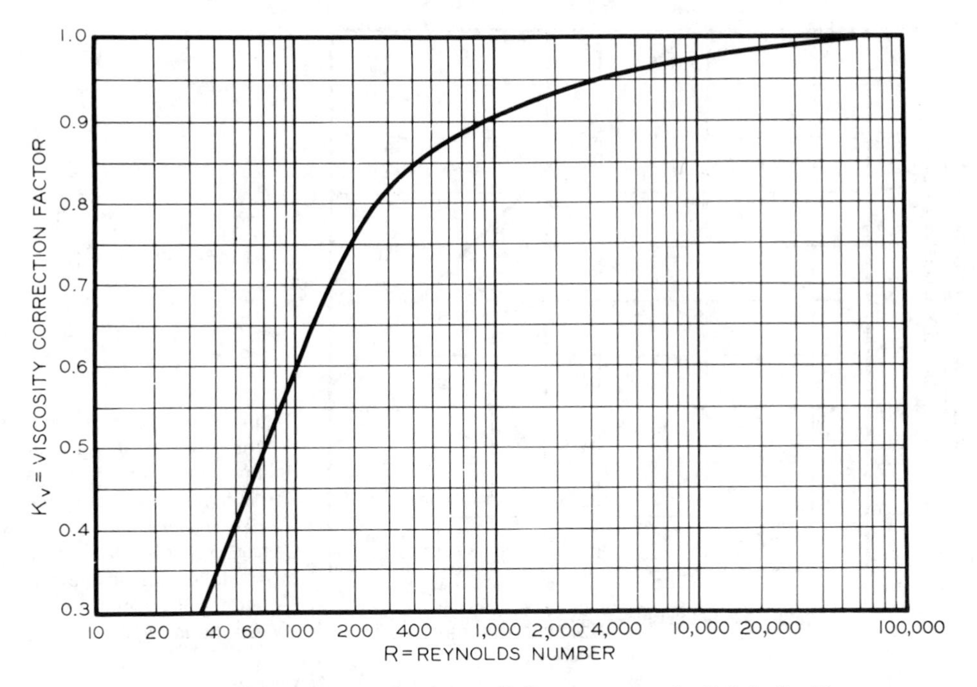

Figure 9-2 Viscosity correction factor, K_v for conventional reliefs in liquid service. (API RP 520, *Recommended Practice for the Design and Installation of Pressure-Relieving Systems in Refineries,* 4th ed., p. 45, 1976. Used by permission of the American Petroleum Institute, Washington, DC.)

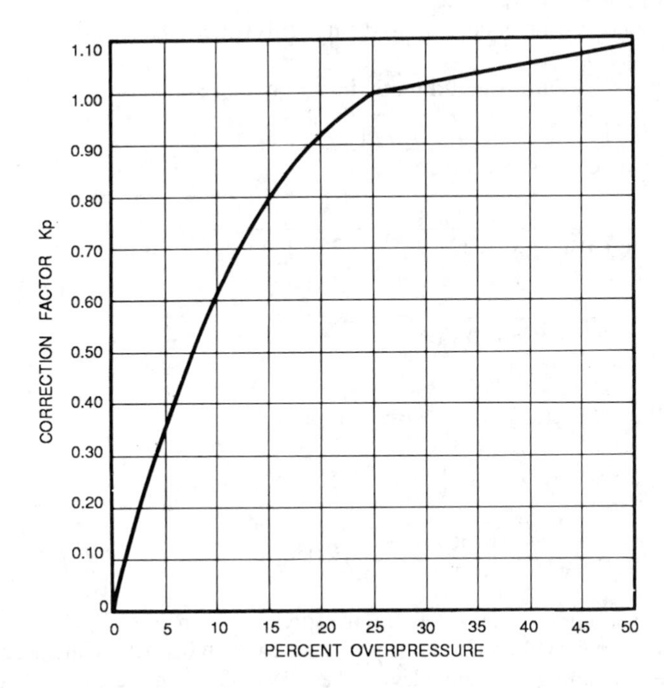

Figure 9-3 Overpressure correction, K_p for spring operated reliefs in liquid service. (API RP 520, *Recommended Practice for the Design and Installation of Pressure-Relieving Systems in Refineries,* 4th ed., p. 44, 1976. Used by permission of the American Petroleum Institute, Washington, DC.)

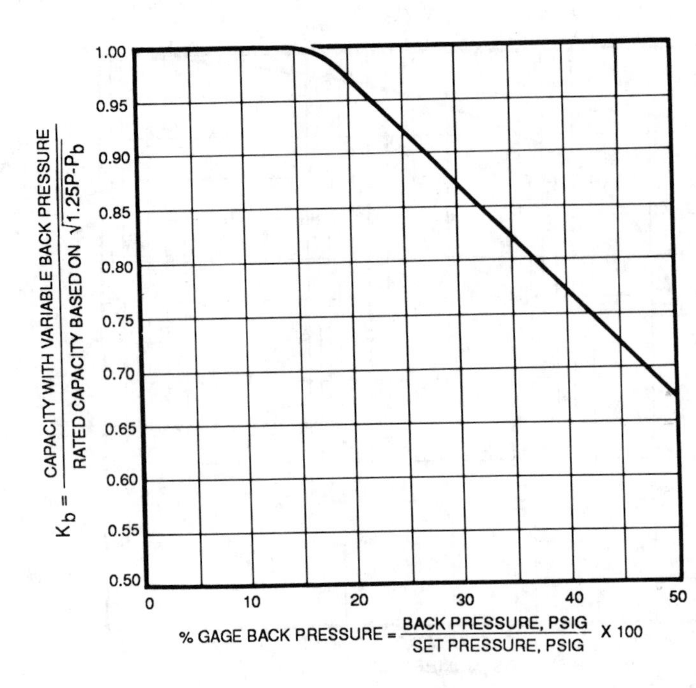

Figure 9-4 Backpressure correction, K_b for 25% overpressure on balanced bellows reliefs in liquid service. (API RP 520, *Recommended Practice for the Design and Installation of Pressure-Relieving Systems in Refineries,* 4th ed., p. 44, 1976. Used by permission of the American Petroleum Institute, Washington, DC.)

The Reynolds number through the relief device is not known. However, at 200 gpm the Reynolds number is assumed to be greater than 5000. Thus, the viscosity correction is, from Figure 9-2, $K_v = 1.0$.

The overpressure correction, K_p, is given in Figure 9-3. Since the percent overpressure is 10%, from Figure 9-3, $K_p = 0.6$

The backpressure correction is not required since this is not a balanced bellows spring relief. Thus, $K_b = 1.0$.

The above numbers are substituted directly into Equation 9-4.

$$A = \left[\frac{\text{in}^2\ (\text{psi})^{1/2}}{38.0\ \text{gpm}}\right] \frac{Q_v}{C_0 K_v K_p K_b} \sqrt{\frac{(\rho/\rho_{\text{ref}})}{1.25P_s - P_b}}$$

$$A = \left[\frac{\text{in}^2\ (\text{psi})^{1/2}}{38.0\ \text{gpm}}\right] \frac{200\ \text{gpm}}{(0.61)(1.0)(0.6)(1.0)} \sqrt{\frac{1.0}{(1.25)(200\ \text{psig}) - 20\ \text{psig}}}$$

$$A = 0.948\ \text{in}^2$$

$$d = \sqrt{\frac{4A}{\pi}} = \sqrt{\frac{(4)(0.948\ \text{in}^2)}{(3.14)}} = 1.10\ \text{in}$$

b. For an overpressure of 25%, $K_p = 1.0$ (Figure 9-3), and

$$A = (0.948\ \text{in}^2)\left(\frac{0.6}{1.0}\right) = 0.569\ \text{in}^2$$

$$d = \sqrt{\frac{(4)(0.569\ \text{in}^2)}{(3.14)}} = 0.851\ \text{in}$$

As expected, the relief vent area decreases as the overpressure increases.

Manufacturers do not provide relief devices to the nearest 0.01 inch. Thus, a selection must be made depending on relief device sizes available commercially. The next largest available size is normally selected. For all relief devices, the manufacturers' technical specifications must be checked before selection and installation.

9-2 CONVENTIONAL SPRING OPERATED RELIEFS IN VAPOR OR GAS SERVICE

For most vapor discharges through spring reliefs the flow is critical. However, the downstream pressure must be checked to insure it is less than the choked pressure computed using Equation 4-39. Thus, for an ideal gas, Equation 4-40 is valid,

$$(Q_{\mathrm{m}})_{\mathrm{choked}} = C_{\mathrm{o}}AP\sqrt{\frac{\gamma g_{\mathrm{c}}M}{R_{\mathrm{g}}T}\left(\frac{2}{\gamma+1}\right)^{(\gamma+1)/(\gamma-1)}} \tag{4-40}$$

where

$(Q_{\mathrm{m}})_{\mathrm{choked}}$ is the discharge mass flow,
C_{o} is the discharge coefficient,
A is the area of the discharge,
P is the absolute upstream pressure,
γ is the heat capacity ratio for the gas,
g_{c} is the gravitational constant,
M is the molecular weight of the gas,
R_{g} is the ideal gas constant, and
T is the absolute temperature of the discharge.

Equation 4-40 is solved for the area of the relief vent given a specified mass flow rate, Q_{m}.

$$A = \frac{Q_{\mathrm{m}}}{C_{\mathrm{o}}P}\sqrt{\frac{T}{M}\bigg/\left[\frac{\gamma g_{\mathrm{c}}}{R_{\mathrm{g}}}\left(\frac{2}{\gamma+1}\right)^{(\gamma+1)/(\gamma-1)}\right]} \tag{9-5}$$

Equation 9-5 is simplified by defining a function χ.

$$\chi = \sqrt{\frac{\gamma g_{\mathrm{c}}}{R_{\mathrm{g}}}\left(\frac{2}{\gamma+1}\right)^{(\gamma+1)/(\gamma-1)}} \tag{9-6}$$

Then the required relief vent area for an ideal gas is computed using a simplified form of Equation 9-5.

$$\boxed{A = \frac{Q_{\mathrm{m}}}{C_{\mathrm{o}}\chi P}\sqrt{\frac{T}{M}}} \tag{9-7}$$

For nonideal gases and real vents, Equation 9-7 is modified by (1) including the compressibility factor, z, to represent a nonideal gas and (2) including a backpressure correction, K_{b}. The result is the following.

$$\boxed{A = \frac{Q_m}{C_0\chi K_{\mathrm{b}}P}\sqrt{\frac{Tz}{M}}} \tag{9-8}$$

where

A is the area of the relief vent,
Q_m is the discharge flow,
C_0 is the effective discharge coefficient, usually 0.975 (unitless),
K_b is the backpressure correction (unitless),
P is the maximum absolute discharge pressure,
T is the absolute temperature,
z is the compressibility factor (unitless), and
M is the average molecular weight of the discharge material.

The constant χ is represented by Equation 9-6. It is conveniently calculated using the following fixed unit expression.

$$\chi = 519.5\sqrt{\gamma\left(\frac{2}{\gamma + 1}\right)^{(\gamma+1)/(\gamma-1)}} \tag{9-9}$$

If Equation 9-9 is used, Equation 9-8 must have the following fixed units: Q_m in lb_m/hour, P in psia, T in °R and M in lb_m/lb-mole. The area computed will be in in^2.

K_b is the backpressure correction and is dependent on the type of relief used. Values are given in Figure 9-5 for conventional spring reliefs and Figure 9-6 for balanced bellows reliefs, respectively.

The pressure used in Equation 9-8 is the maximum absolute relieving pressure. It is given, for the fixed unit case, by,

$$P = P_{max} + 14.7 \tag{9-10}$$

where P_{max} is the maximum gauge pressure in psig. For vapor reliefs, the following guidelines are recommended.[2]

$$\begin{aligned} P_{max} &= 1.1\ P_s \text{ for unfired pressure vessels} \\ P_{max} &= 1.2\ P_s \text{ for vessels exposed to fire} \\ P_{max} &= 1.33\ P_s \text{ for piping} \end{aligned} \tag{9-11}$$

For vapor flows that are not choked by sonic flow, the area is determined using Equation 4-38. The downstream pressure, P, is now required and the discharge coefficient, C_o, must be estimated.

Example 9-2

A nitrogen regulator fails and allows nitrogen to enter a reactor through a 6-inch diameter line. The source of the nitrogen is at 70°F and 150 psig. The reactor relief is set at 50 psig. Determine the diameter of a balanced bellows spring type vapor relief required to protect the reactor from this incident. Assume a relief backpressure of 20 psig.

[2] *ASME Boiler and Pressure Vessel Code* (New York: American Society of Mechanical Engineers, 1965).

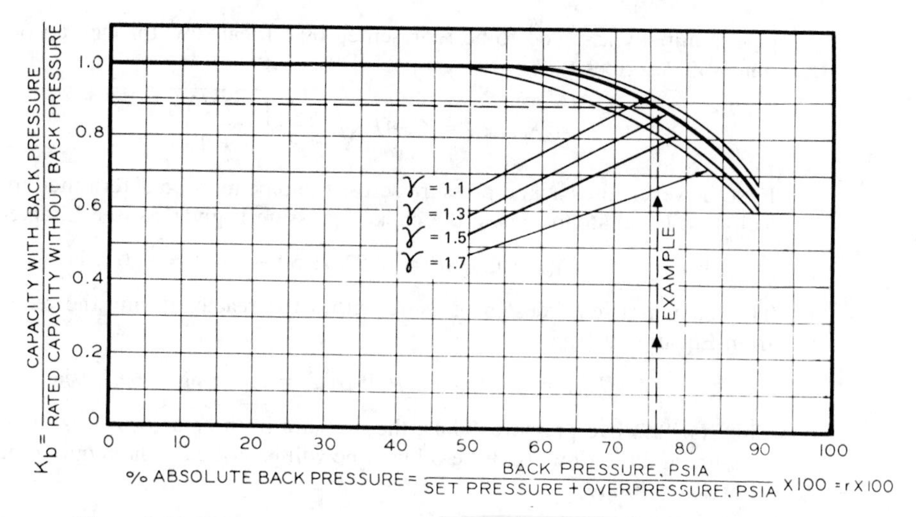

Figure 9-5 Backpressure correction, K_b for conventional spring type reliefs in vapor or gas service. (API RP 520, *Recommended Practice for the Design and Installation of Pressure-Relieving Systems in Refineries,* 4th ed., p. 40, 1976. Used by permission of the American Petroleum Institute, Washington, DC.)

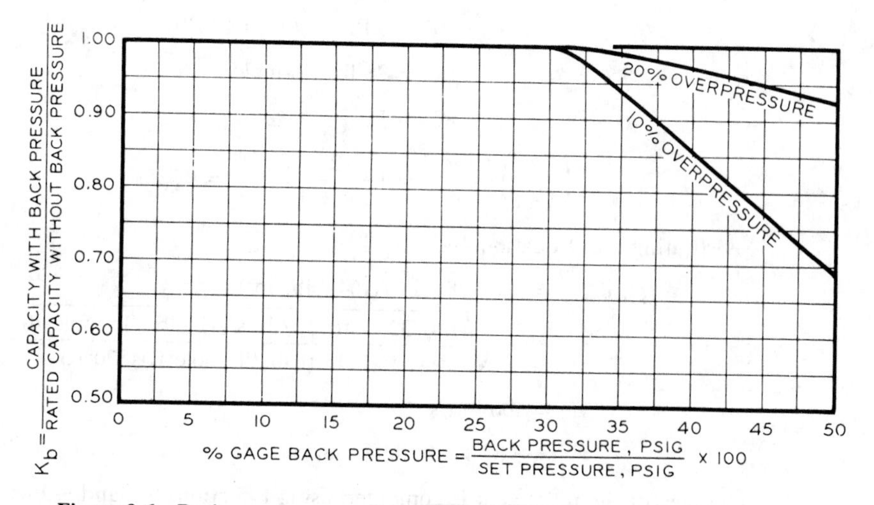

Figure 9-6 Backpressure correction, K_b for balanced bellows reliefs in vapor or gas service. (API RP 520, *Recommended Practice for the Design and Installation of Pressure-Relieving Systems in Refineries,* 4th ed., p. 41, 1976. Used by permission of the American Petroleum Institute, Washington, DC.)

Solution The nitrogen source is at 150 psig. If the regulator fails the nitrogen will flood the reactor, increasing the pressure to a point where the vessel will fail. A relief vent must be installed to vent the nitrogen as fast as it is supplied through the 6-inch line. Since no other information on the piping system is provided, the flow from the pipe is initially assumed to be represented by critical flow through an orifice. Equation 4-40 describes this.

$$(Q_m)_{choked} = C_o A P \sqrt{\frac{\gamma g_c M}{R_g T}\left(\frac{2}{\gamma + 1}\right)^{(\gamma+1)/(\gamma-1)}}$$

First, however, the choked pressure across the pipe must be determined to insure critical flow. For diatomic gases, the choked pressure is given as (see Chapter 4).

$$P_{choked} = 0.528P = (0.528)(150 + 14.7) = 87.0 \text{ psia}$$

The maximum relief design pressure within the reactor during the relief venting is, from Equation 9-11,

$$P_{max} = 1.1\ P_s = (1.1)(50 \text{ psig}) = 55.0 \text{ psig} = 69.7 \text{ psia}$$

This is a 10% overpressure. Thus, the pressure in the reactor is less than the choked pressure and the flow from the 6-inch line will be critical. The required quantities for Equation 4-40 are

$$A = \frac{\pi d^2}{4} = \frac{(3.14)(6\text{-inch})^2}{4} = 28.3 \text{ in}^2$$

$$P = 150 + 14.7 = 164.7 \text{ psia}$$

$$\gamma = 1.40 \text{ for diatomic gases}$$

$$T = 70°\text{F} + 460 = 530°\text{R}$$

$$M = 28 \text{ lb}_m/\text{lb-mole}$$

$$C_o = 1.0$$

$$\left(\frac{2}{\gamma+1}\right)^{(\gamma+1)/(\gamma-1)} = \left(\frac{2}{1.4 + 1}\right)^{(2.4/0.4)} = 0.335$$

Substituting into Equation 4-40,

$$(Q_m)_{choked} = (1.0)(28.3 \text{ in}^2)(164.7 \text{ lb}_f/\text{in}^2) \times \sqrt{\frac{(1.4)(32.17 \text{ ft lb}_m/\text{lb}_f \text{ s}^2)(28 \text{ lb}_m/\text{lb-mole})(0.335)}{(1545 \text{ ft lb}_f/\text{lb-mole}°\text{R})(530°\text{R})}}$$

$$= 106 \text{ lb}_m/\text{s}$$

$$= 3.82 \times 10^5 \text{ lb}_m/\text{hour}$$

The area of the relief vent is computed using Equations 9-8 and 9-9 with a backpressure correction, K_b, determined from Figure 9-6. The backpressure is 20 psig. Thus,

$$\left(\frac{\text{Backpressure, psig}}{\text{Set Pressure, psig}}\right) \times 100 = \left(\frac{20 \text{ psig}}{50 \text{ psig}}\right) \times 100 = 40\%$$

From Figure 9-6, $K_b = 0.86$ for an overpressure of 10%. The effective discharge coefficient is assumed to be 0.975. The gas compressibility factor, z, is approximately 1 at these pressures. The pressure, P, is the maximum absolute pressure. Thus, $P = 69.7$ psia. The constant χ is computed from Equation 9-9,

$$\chi = 519.5\sqrt{\gamma\left(\frac{2}{\gamma + 1}\right)^{(\gamma+1)/(\gamma-1)}} = 519.5\sqrt{(1.4)(0.335)} = 356$$

The required vent area is computed using Equation (9-8)

$$A = \frac{Q_m}{C_0 \chi K_b P}\sqrt{\frac{Tz}{M}}$$

$$= \frac{3.82 \times 10^5 \text{ lb}_m/\text{hour}}{(0.975)(356)(0.86)(69.7 \text{ psia})}\sqrt{\frac{(530°\text{R})(1.0)}{(28 \text{ lb}_m/\text{lb-mole})}}$$

$$A = 79.9 \text{ in}^2$$

The required vent diameter is,

$$d = \sqrt{\frac{4A}{\pi}} = \sqrt{\frac{(4)(79.9 \text{ in}^2)}{(3.14)}} = 10.1 \text{ in}$$

Manufacturers provide relief devices only at convenient sizes. The next largest diameter closest to the one required is selected. This would likely be 10-1/8 inches (10.125 inches).

9-3 RUPTURE DISC RELIEFS IN LIQUID SERVICE

For liquid reliefs through rupture discs without significant lengths of downstream piping, the flow is represented by Equation 9-2 or Equation 9-3 for flow through a sharp-edged orifice. No corrections are suggested.

Equations 9-2 and 9-3 apply to rupture discs discharging directly to the atmosphere. For rupture discs discharging into a relief system (which might include knock-out drums, scrubbers or flares), the rupture disc is considered a flow restriction and the flow through the entire pipe system must be considered. The calculation is performed identically to regular pipe flow (see Chapter 4). The calculation to determine the rupture disc area is iterative for this case. Isaacs[3] recommends assuming the rupture disc equivalent to 50 pipe diameters in the calculation.

9-4 RUPTURE DISC RELIEFS IN VAPOR OR GAS SERVICE

Flow of vapor through rupture disks is described using an orifice equation similiar to Equation 9-8, but without the additional correction factors. The result is,

$$\boxed{A = \frac{Q_m}{\chi P}\sqrt{\frac{Tz}{M}}} \tag{9-12}$$

Equation 9-12 assumes a discharge coefficient, C_o, of 1.0.

If appreciable backpressures exist from downstream relief systems, a procedure similiar to the procedure used for liquid reliefs through rupture disks is required. The procedure is iterative.

Example 9-3

Determine the diameter of a rupture disk required to relieve the pump of Example 9-1, Part a.

[3]Marx Isaacs, "Pressure Relief Systems," *Chemical Engineering,* February 22, 1971, p. 113.

Solution The pressure drop across the rupture disk is

$$\Delta P = P_{\text{max}} - P_{\text{b}} = 220 \text{ psig} - 20 \text{ psig} = 200 \text{ psig}$$

The specific gravity of the water, (ρ/ρ_{ref}), is 1.0.
A conservative discharge coefficient of 0.61 is assumed.
Substituting into Equation 9-3,

$$A = \left[\frac{\text{in}^2 \,(\text{psi})^{1/2}}{38.0 \text{ gpm}}\right] \frac{Q_{\text{v}}}{C_0} \sqrt{\frac{(\rho/\rho_{\text{ref}})}{\Delta P}}$$

$$= \left[\frac{\text{in}^2 \,(\text{psi})^{1/2}}{38.0 \text{ gpm}}\right] \frac{200 \text{ gpm}}{0.61} \sqrt{\frac{1.0}{200 \text{ psia}}} = 0.610 \text{ in}^2$$

The relief vent diameter is

$$d = \sqrt{\frac{4A}{\pi}} = \sqrt{\frac{(4)\,(0.610 \text{ in}^2)}{(3.14)}} = 0.881 \text{ in}$$

This compares to a spring relief vent area of 1.10 in.

Example 9-4

Compute the rupture disk vent diameter required to relieve the process of Example 9-2.

Solution The solution is provided by Equation 9-8. The solution is identical to Example 9-2, with the exception of a deletion of the correction factor K_{b}. The area is, therefore,

$$A = (79.9 \text{ in}^2)\,(0.86) = 68.7 \text{ in}^2$$

The rupture disc diameter is

$$d = \sqrt{\frac{4A}{\pi}} = \sqrt{\frac{(4)\,(68.7 \text{ in}^2)}{(3.14)}} = 9.35 \text{ in}$$

This compares to a spring relief diameter of 9.32 in.

9-5 TWO-PHASE FLOW DURING RUNAWAY REACTION RELIEF

When a runaway reaction occurs within a reactor vessel, two-phase flow should be expected during the relief process. The Vent Sizing Package (VSP) laboratory apparatus described in Chapter 8 provides the much needed temperature and pressure rise data for relief area sizing.

Figure 9-7 shows the most common type of reactor system called a tempered reactor. It is called tempered because the reactor contains a volatile liquid which vaporizes or flashes during the relieving process. This vaporization removes energy via the heat of vaporization and *tempers* the rate of temperature rise due to the exothermic reaction.

The runaway reactor is treated as entirely adiabatic. The energy terms include (1) energy accumulation due to the sensible heat of the reactor fluid as a result of its increased temperature due to overpressure, and (2) the energy removal due to the vaporization of liquid in the reactor, and subsequent discharge through the relief vent.

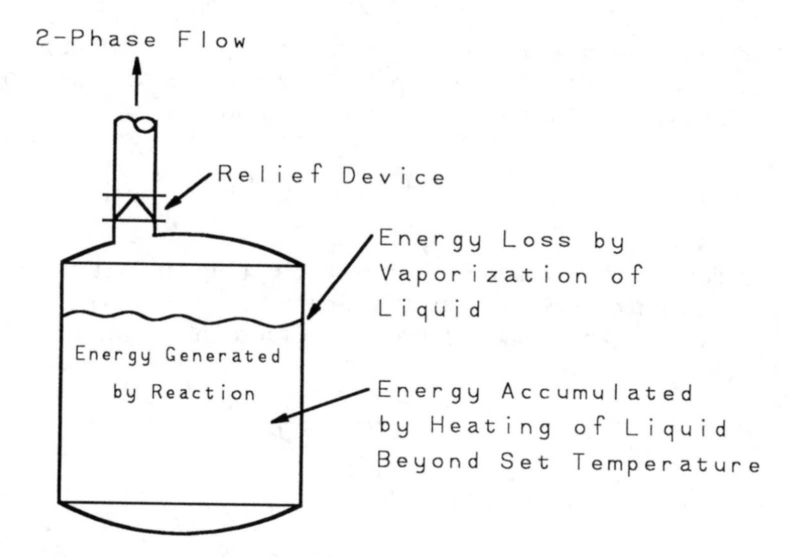

Figure 9-7 A tempered reaction system showing the important energy terms. The heat losses through the reactor walls are assumed negligible.

The first step in the relief sizing calculation for two-phase vents is to determine the mass flux through the relief. This is computed using Equation 4-86, representing choked, two-phase flow through a hole,

$$Q_m = \frac{\Delta H_V A}{v_{fg}} \sqrt{\frac{g_c}{C_P T_s}} \tag{4-86}$$

where, for this case,

Q_m is the mass flow through the relief,
ΔH_V is the heat of vaporization of the fluid,
A is the area of the hole,
v_{fg} is the change of specific volume of the flashing liquid,
C_P is the heat capacity of the fluid, and
T_s is the absolute saturation temperature of the fluid at the set pressure.

The mass flux, G_T, is given by

$$G_T = \frac{Q_m}{A} = \frac{\Delta H_V}{v_{fg}} \sqrt{\frac{g_c}{C_P T_S}} \tag{9-13}$$

Equation 9-13 applies to two-phase relief through a hole. For two-phase flow through pipes, an overall dimensionless discharge coefficient, ψ, is applied. Equation 9-13 is the so-called equilibrium rate model (ERM) for low quality choked flow.[4] Leung[5] has shown that Equation 9-13 must be multiplied by a factor 0.9 to

[4] H. K. Fauske, "Flashing Flows or: Some Practical Guidelines for Emergency Releases," *Plant Operations Progress* Vol. 4, No. 3, July, 1985.

[5] J. C. Leung, "Simplified Vent Sizing Equations for Emergency Relief Requirements in Reactors and Storage Vessels," *AICHE Journal,* Vol. 32, No. 10, p. 1622, 1986.

bring the value in line with the classic homogeneous equilibrium model (HEM). The result should be generally applicable for homogeneous venting of a reactor (low quality, not restricted just to liquid inlet condition).

$$G_T = \frac{Q_m}{A} = 0.9\,\psi \frac{\Delta H_V}{v_{fg}} \sqrt{\frac{g_c}{C_P T_S}} \tag{9-14}$$

Values for ψ are provided in Figure 9-8. For a pipe of length zero, $\psi = 1$. As the pipe length increases, the value of ψ decreases.

A somewhat more convenient expression is derived by rearranging Equation 4-84 to yield

$$\frac{\Delta H_V}{v_{fg}} = T_S \frac{dP}{dT} \tag{9-15}$$

and substituting into Equation 9-14,

$$G_T = 0.9\,\psi \frac{dP}{dT} \sqrt{\frac{g_c T_S}{C_P}} \tag{9-16}$$

The exact derivative is approximated by a finite difference derivative to yield

$$G_T \cong 0.9\,\psi \frac{\Delta P}{\Delta T} \sqrt{\frac{g_c T_S}{C_P}} \tag{9-17}$$

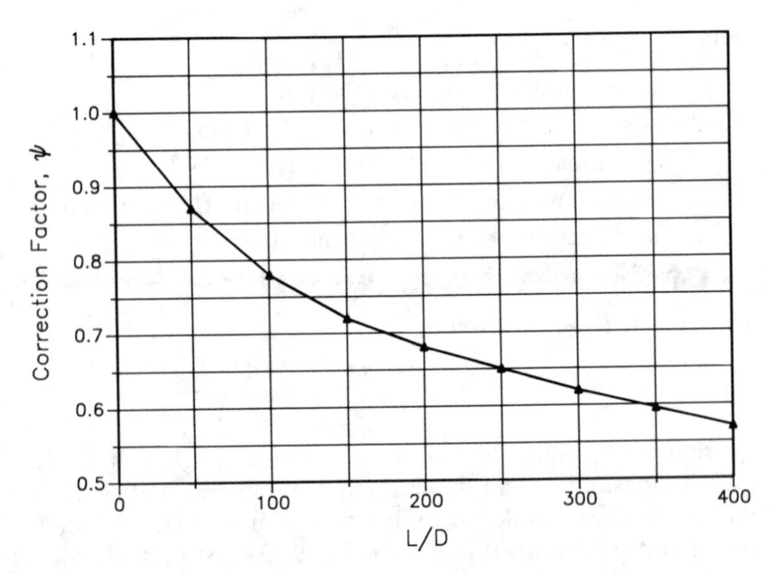

Figure 9-8 Correction factor, ψ, correcting for two-phase flashing flow through pipes. (Data from J. C. Leung and M. A. Grolmes, "The Discharge of Two-Phase Flashing Flow in a Horizontal Duct," *AICHE Journal*, 33(3), pp. 524-527.)

where

ΔP is the overpressure, and

ΔT is temperature rise corresponding to the overpressure.

The required vent area is computed by solving a particular form of the dynamic energy balance. Details are provided elsewhere.[6] The result is

$$A = \frac{m_o q}{G_T \left[\sqrt{\frac{V}{m_o} \frac{\Delta H_V}{v_{fg}}} + \sqrt{C_v \Delta T} \right]^2} \tag{9-18}$$

An alternate form is derived by applying Equation 4-84.

$$A = \frac{m_o q}{G_T \left[\sqrt{\frac{V}{m_o} T_S \frac{dP}{dT}} + \sqrt{C_V \Delta T} \right]^2} \tag{9-19}$$

For Equations 9-18 and 9-19, the following additional nomenclature is defined:

m_o is the total mass contained within the reactor vessel prior to relief,

q is the exothermic heat release rate per unit mass,

V is the volume of the vessel, and

C_V is the liquid heat capacity at constant volume.

For both Equations 9-18 and 9-19 the relief area is based on the total heat added to the system (numerator) and the heat removed or absorbed (denominator). The first term in the denominator corresponds to the net heat removed by the liquid and vapor leaving the system; the second term corresponds to the heat absorbed as a result of increasing the temperature of the liquid due to the overpressure.

The heat input, q, due to an exothermic reaction is determined using fundamental kinetic information or from the DIERS Vent Sizing Package (see Chapter 8). For data obtained using the VSP, the equation

$$q = \frac{1}{2} C_V \left[\left(\frac{dT}{dt} \right)_S + \left(\frac{dT}{dt} \right)_m \right] \tag{9-20}$$

is applied, where the derivative, denoted by the subscript S, corresponds to the heating rate at the set pressure and the derivative, denoted by the subscript m, corresponds to the temperature rise at the maximum turnaround pressure. Both derivatives are determined experimentally using the VSP.

The above equations assume

1. Uniform froth or homogeneous vessel venting occurs.
2. The mass flux, G_T, varies little during the relief.

[6]J. C. Leung, "Simplified Vent Sizing," *AICHE Journal*.

3. The reaction energy per unit mass, q, is treated as a constant.
4. Constant physical properties C_V, ΔH_V and v_{fg}.
5. The system is a tempered reactor system. This applies to the majority of reaction systems.

Units are a particular problem when using the above two-phase equations. The best procedure is to convert all energy units to their mechanical equivalents before solving for the relief area, particularly when English engineering units are used.

Example 9-5

Leung[7] reports on the data of Huff[8] involving a 3500 gallon reactor with styrene monomer undergoing adiabatic polymerization after being heated inadvertantly to 70°C. The maximum allowable working pressure (MAWP) of the vessel is 5 bar. Given the data below, determine the relief vent diameter required. Assume a set pressure of 4.5 bar and a maximum pressure of 5.4 bar absolute.

DATA

Volume (V):	3,500 gal = 13.16 m^3
Reaction mass (m_o):	9,500 kg
Set temperature (T_s):	209.4°C = 482.5 K
Data from VSP:	
Maximum temperature (T_m):	219.5°C = 492.7 K

$(dT/dt)_s$ = 29.6°C/min = 0.493 K/s
$(dT/dt)_m$ = 39.7°C/min = 0.662 K/s

PHYSICAL PROPERTY DATA

	4.5 bar set	5.4 bar peak
v_f, m^3/kg:	0.001388	0.001414
v_g, m^3/kg:	0.08553	0.07278
C_p, kJ/kg K:	2.470	2.514
ΔH_v, kJ/kg:	310.6	302.3

Solution The heating rate, q, is determined using Equation 9-20.

$$q = \frac{1}{2} C_V \left[\left(\frac{dT}{dt}\right)_s + \left(\frac{dT}{dt}\right)_m \right] \tag{9-20}$$

Assuming $C_v = C_P$,

$$q = \frac{1}{2}(2.470 \text{ kJ/kg K})(0.493 + 0.662)(\text{K/s})$$
$$= 1.426 \text{ kJ/kg s}$$

[7]Leung, "Simplified Vent Sizing," *AICHE Journal.*

[8]J. E. Huff, "Emergency Venting Requirements," *Plant/Operations Progress,* Vol. 1, No. 4, p. 211, 1982.

The mass flux is given by Equation 9-14. Assuming L/D = 0, $\psi = 1.0$.

$$G_T = 0.9\,\psi \frac{\Delta H_V}{v_{fg}} \sqrt{\frac{g_c}{C_P T_S}}$$

$$= (0.9)(1.0)\frac{(310{,}600 \text{ J/kg})[1 \text{ (N m)/J}]}{(0.08553 - 0.001388) \text{ m}^3\text{/kg}}$$

$$\times \sqrt{\frac{[1 \text{ (kg m/s}^2\text{)/N}]}{(2{,}470 \text{ J/kg K})(482.5 \text{ K})[1 \text{ (N m)/J}]}}$$

$$= 3{,}043 \text{ kg/m}^2\text{s}$$

The relief vent area is determined from Equation 9-18. The change in temperature, ΔT is $T_m - T_s = 492.7 - 482.5 = 10.2$ K

$$A = \frac{m_o q}{G_T\left[\sqrt{\frac{V}{m_o}\frac{\Delta H_V}{v_{fg}}} + \sqrt{C_v \Delta T}\right]^2}$$

$$= \frac{(9{,}500 \text{ kg})(1{,}426 \text{ J/kg s})[1 \text{ (N m)/J}]}{(3{,}043 \text{ kg/m}^2 \text{ s})}$$

$$\times \left[\sqrt{\left(\frac{13.16 \text{ m}^3}{9{,}500 \text{ kg}}\right)\left(\frac{(310{,}600 \text{ J/kg})[1 \text{ (N m)/J}]}{(0.08414 \text{ m}^3\text{/kg})}\right)}\right.$$

$$\left. + \sqrt{(2{,}470 \text{ J/kg K})(10.2 \text{ K})[1 \text{ (N m)/J}]}\right]^{-2}$$

$$A = 0.084 \text{ m}^2$$

The required relief diameter is

$$d = \sqrt{\frac{4A}{\pi}} = \sqrt{\frac{(4)(0.084 \text{ m}^2)}{3.14}} = 0.327 \text{ m}$$

Suppose that all vapor relief was assumed. The size of a vapor phase rupture disk required is determined by assuming that all of the heat energy is absorbed by the vaporization of the liquid. At the set temperature, the heat release rate, q, is

$$q = C_V\left(\frac{dT}{dt}\right)_s = (2.470 \text{ kJ/kg K})(0.493 \text{ K/s}) = 1.218 \text{ kJ/kg s}$$

The vapor mass flow through the relief is then

$$Q_m = \frac{q\, m_o}{\Delta H_v}$$

$$= \frac{(1{,}218 \text{ J/kg s})(9{,}500 \text{ kg})}{(310{,}600 \text{ J/kg})}$$

$$= 37.2 \text{ kg/s}$$

Equation 9-5 provides the required relief area. The molecular weight of styrene is 104. Assume $\gamma = 1.32$ and $C_o = 1.0$. Then,

$$A = \frac{Q_m}{C_o P}\sqrt{\frac{R_g T}{\gamma g_c M}\left(\frac{2}{\gamma + 1}\right)^{(\gamma+1)/(1-\gamma)}}$$

$$A = \frac{(37.2\ \text{kg/s})}{(1.0)(4.5\ \text{bar})(100{,}000\ \text{Pa/bar})[1\ (\text{N/m}^2)/\text{Pa}]}$$

$$\times \sqrt{\frac{(8{,}314\ \text{Pa m}^3/\text{kg-mol K})(482.5\ \text{K})[1\ (\text{N/m}^2)/\text{Pa})}{(1.32)[1\ (\text{kg m/s}^2)/\text{N}](104\ \text{kg/kg-mol})}}$$

$$\times \sqrt{\left(\frac{2}{2.32}\right)^{2.32/(-0.32)}}$$

$$A = 0.0242\ \text{m}^2$$

This requires a relief device with a diameter of 0.176 m, a significantly smaller diameter than for two-phase flow. Thus, if the relief were sized assuming all vapor relief the result would be physically incorrect and the reactor would be severely tested during this runaway event.

Simplified Nomograph Method

Fauske[9] has developed a simplified, chart driven approach to the two-phase reactor relief problem. He suggests the following equation for determining the relief area,

$$A = \frac{V\rho}{G_T\ \Delta t_v} \tag{9-21}$$

where

- A is the relief vent area,
- V is the reactor volume,
- ρ is the density of the reactants,
- G_T is the mass flux through the relief, and
- Δt_v is the venting time.

Equation 9-21 was developed by Boyle[10] by defining the required vent area as that size which would empty the reactor before the pressure could rise above some allowable overpressure for a given vessel.

The mass flux, G_T is given by Equations 9-14 or 9-17 and the venting time is given approximately by

$$\Delta t_v \cong \frac{\Delta T\ C_P}{q_s} \tag{9-22}$$

[9]Hans K. Fauske, "A Quick Approach to Reactor Vent Sizing," *Plant/Operations Progress,* Vol. 3, No. 3, 1984, and "Generalized Vent Sizing Monogram for Runaway Chemical Reactions," *Plant/Operations Progress,* Vol. 3, No. 4, 1984.

[10]W. J. Boyle, Jr., "Sizing Relief Area for Polymerization Reactors," *Chemical Engineering Progress,* Vol. 63, No. 8 (August 1967), p 61.

where

ΔT is the temperature rise corresponding to the overpressure ΔP,
T is the temperature,
C_P is the heat capacity, and
q_s is the energy release rate per unit mass at the set pressure of the relief system.

Combining Equations 9-21, 9-13, and 9-22 yields

$$A = V\rho(g_c T_s C_P)^{-1/2}\frac{q_s}{\Delta P} \tag{9-23}$$

Equation 9-23 provides a conservative estimate of the vent area required. By considering the case of 20% absolute overpressure, assuming a typical liquid heat capacity of 2510 J/kg K for most organics, and assuming a saturated water relationship, the following equation is obtained.[11]

$$A = (\text{m}^2/1000\ \text{kg}) = \frac{0.00208\left(\dfrac{dT}{dt}\right)(^0\text{C/min})}{P_s\ (\text{bar})} \tag{9-24}$$

A simple nomograph of the results can be plotted and is shown in Figure 9-9. The required vent area is determined simply from the heating rate, the set pressure and the mass of reactants.

The Fauske nomograph is useful for performing quick estimates and checking the results of the more rigorous computation.

Recent studies[12] suggest that the nomograph data of Figure 9-9 applies for a discharge coefficient of $\psi = 0.5$ representing a discharge (L/D) of 400. Use of the nomograph at other discharge pipe lengths and different ψ requires a suitable correction, as shown in the following example.

Example 9-6

Estimate the relief vent area using the Fauske nomograph approach for the reaction system of Example 9-5.

Solution The heating rate at the set temperature is specified as 29.6°C/min. The set pressure is 4.5 bar absolute.

$$P_s = (4.5\ \text{bar})(0.9869\ \text{bar/atm})(14.7\ \text{psia/atm}) = 65.3\ \text{psia}$$

From Figure 9-9, the vent area required per 1000 kg of reactant is about 1.03×10^{-2} m². Thus, the total relief area is,

$$A = (1.03 \times 10^{-2}\ \text{m}^2/1000\ \text{kg})(9500\ \text{kg})$$
$$= 0.098\ \text{m}^2$$

[11] J. C. Leung and H. K. Fauske, "Runaway System Characterization and Vent Sizing Based on DIERS Methodology," *Plant/Operations Progress,* Vol. 6, No. 2, April, 1987.

[12] Personal communication with H. G. Fisher and J. C. Leung in January, 1989.

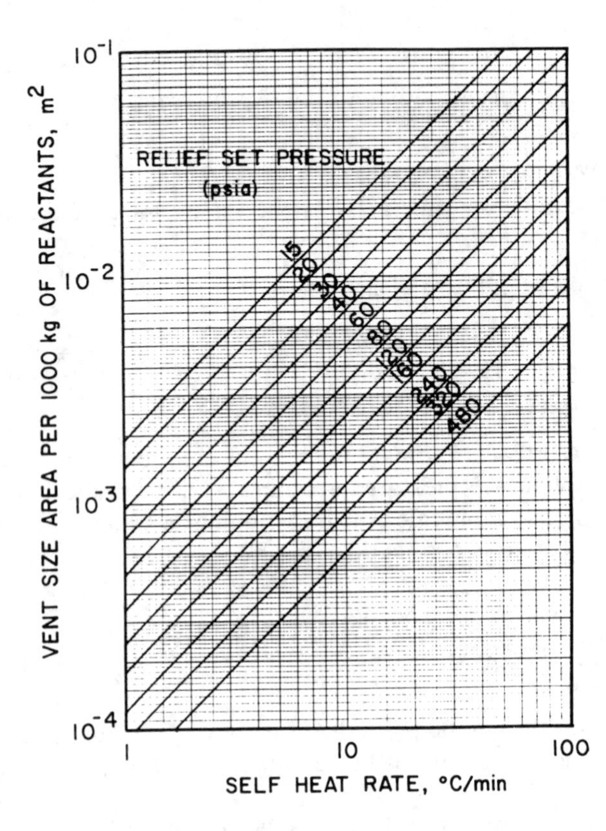

Figure 9-9 Nomograph for sizing two-phase reactor reliefs. (H. K. Fauske, "Generalized Vent Sizing Monogram for Runaway Chemical Reactions," *Plant/Operations Progress,* Vol. 3, No. 4, 1984. Used by permission of the American Institute of Chemical Engineers.)

Figure 9-9 is applicable for $\psi = 0.5$. For $\psi = 1.0$ the area is adjusted linearly,

$$A = (0.098 \text{ m}^2)\left(\frac{0.5}{1.0}\right)$$

$$= 0.049 \text{ m}^2$$

This assumes a 20% absolute overpressure. The result can be adjusted for other overpressures by multiplying the area by a ratio of 20/(new absolute percent overpressure).

This result compares to a more rigorously computed area of 0.084 m^2.

Two-phase flow through reliefs is much more complex than the introduction provided here. Furthermore, the technology is still undergoing substantial development. The equations presented here are not universally applicable; however, they do represent the most accepted methodology available today.

9-6 DEFLAGRATION VENTING FOR DUST AND VAPOR EXPLOSIONS

Loss prevention means preventing the existence of hazards. However, for some situations hazards are unavoidable. For example, during the milling process to make flour from wheat, substantial quantities of flammable dust is produced. An uncontrolled dust explosion in a warehouse, storage bin, or processing unit can eject high velocity structural debris over a considerable area, propagating the accident and resulting in increased injuries. Deflagration venting reduces the impact of dust and vapor cloud explosions by controlling the release of the explosion energy. The energy of the explosion is directed away from plant personnel and equipment.

Deflagration venting in buildings and process vessels is usually achieved by using blowout panels, as shown in Figure 9-10. The blowout panel is designed to have less strength than the walls of the structure. Thus, during an explosive event, the blowout panels are preferentially detached and the explosive energy is vented. Damage to the remaining structure and equipment is minimized. For particularly explosive dusts or vapors, it is not unusual for the walls (and perhaps roof) of the entire structure to be constructed of blowout panels.

The actual construction details of blowout panels are beyond the scope of the text. A detached blowout panel moving at high velocity can cause considerable damage. Therefore, a mechanisim must be provided to retain the panel during the deflagration process. Furthermore, thermal insulation of panels is also required. Construction details are available in manufacturers' literature.

Blowout panels are designed to provide the proper relief area, depending on a number of design factors. This includes the explosive behavior of the dust or vapor,

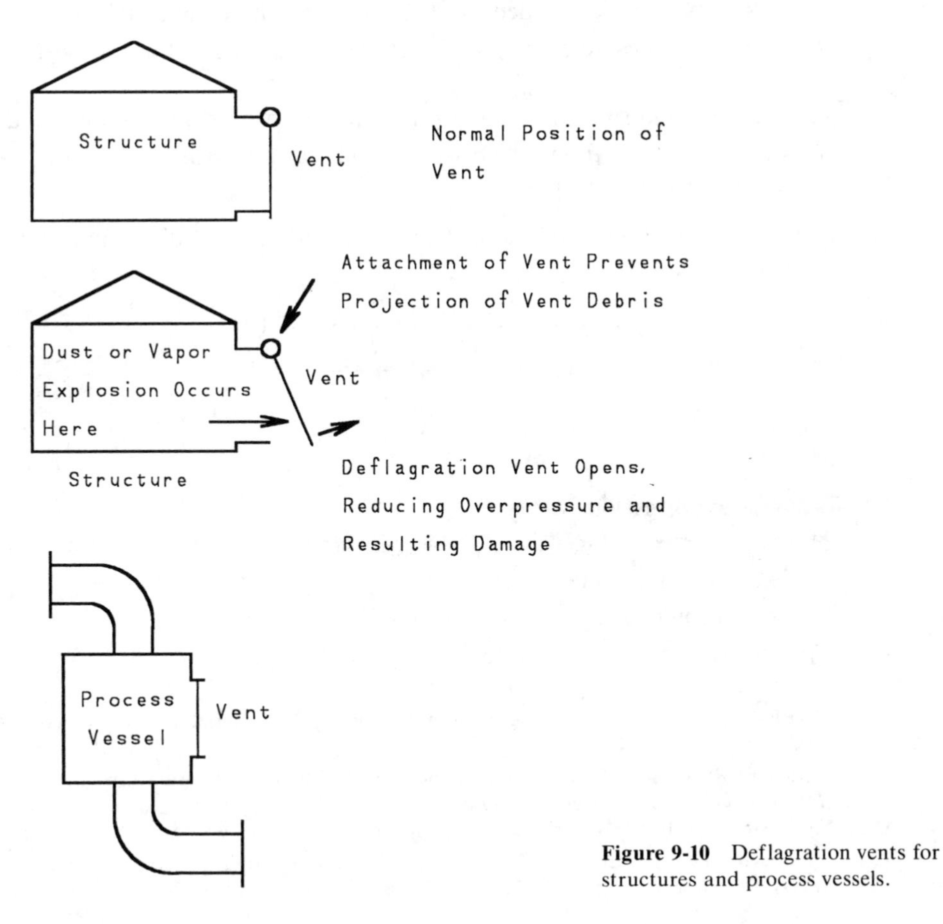

Figure 9-10 Deflagration vents for structures and process vessels.

the maximum overpressure allowable in the structure, and the volume of the structure. Design standards are available.[13]

Deflagration design is segregated into two categories: low and high pressure structures. Low pressure structures include structures with sheet metal sides and other low strength building materials. These structures are capable of withstanding not more than 1.5 psig (0.1 bar gauge). High pressure structures include steel process vessels, concrete buildings, and so forth, that are capable of withstanding pressures greater than 1.5 psig (0.1 bar gauge).

Vents for Low Pressure Structures

For low pressure structures, original design techniques were based on the Runes (pronounced Roo-ness) equation,[14]

$$A = \frac{C^*_{\text{vent}} L_1 L_2}{\sqrt{P}} \tag{9-25}$$

where

- A is the required vent area,
- C^*_{vent} is a constant, dependent on the nature of the combustible,
- L_1 is the smallest dimension of the rectangular building structure to be vented,
- L_2 is the second smallest dimension of the enclosure to be vented, and
- P is the maximum internal pressure which can be withstood by the weakest member of the enclosure.

Swift and Epstein[15] present a more detailed equation, including many important combustion features.

$$A = \frac{\dfrac{A_s}{C_o} \dfrac{\lambda S_u \rho_u}{G} \left[\left(\dfrac{P_{\max}}{P_o} \right)^{1/\gamma} - 1 \right]}{\sqrt{\dfrac{P_f}{P_o} - 1}} \tag{9-26}$$

where

- A is the required vent area,
- A_s is the inside surface area of the enclosure,
- C_o is the discharge coefficient,
- λ is the turbulent augmentation factor,
- S_u is the laminar burning velocity,

[13]NFPA 68: *Guide for Venting of Deflagrations* (Quincy, MA: National Fire Protection Association, 1988).

[14]Richard R. Schwab, "Recent Developments in Deflagration Venting Design," *Proceedings of the International Symposium on Preventing Major Chemical Accidents,* John L. Woodward, ed., (New York: American Institute of Chemical Engineers, 1987), p. 3.101

[15]Ian Swift and Mike Epstein, "Performance of Low Pressure Explosion Vents," *Plant/Operations Progress,* Vol. 6, No. 2, April, 1987.

ρ_u is the density of the unburned gas,
G is the mass flux,
P_{max} is the maximum unvented explosion pressure,
P_o is the initial pressure,
P_f is the final peak pressure during the vent, and
γ is the heat capacity ratio.

Many of the variables in Equation 9-26 can be estimated or assumed. These variables are regrouped to result in the following form,

$$A = \frac{C_{vent} A_s}{\sqrt{P}} \tag{9-27}$$

where P is the maximum internal overpressure which can be withstood by the weakest structural element. Equation 9-27 is remarkably similiar to the Runes equation, Equation 9-25.

Values for the constant, C_{vent} are given in Table 9-1.

Vents for High Pressure Structures

For high pressure structures, Equation 6-14 is used as a basis for vent design.

$$K_G \text{ or } K_{St} = \left(\frac{dP}{dt}\right)_{max} V^{1/3} \tag{9-28}$$

where

K_G is the deflagration index for gases and vapors,
K_{St} is the deflagration index for dusts,
$(dP/dt)_{max}$ is the maximum pressure rise, and
V is the volume of the vessel.

Equation 9-28, along with experimental data, was used to prepare the nomographs shown in Figures 9-11 through 9-19. In the nomographs, P_{red} is the maximum pres-

TABLE 9-1 COMBUSTIBLE CHARACTERISTIC CONSTANT FOR THE SWIFT-EPSTEIN EQUATION. NFPA 68: *VENTING OF DEFLAGRATIONS* (QUINCY, MA: NATIONAL FIRE PROTECTION ASSOCIATION, 1988).

Combustible	C_{vent} ($\sqrt{psi}$)	C_{vent} ($\sqrt{kPa}$)
Anhydrous ammonia	0.05	0.13
Methane	0.14	0.37
Aliphatic gases (excluding methane) or gases with a fundamental burning velocity less than 1.3 times that of propane:	0.17	0.45
St-1 dusts	0.10	0.26
St-2 dusts	0.12	0.30
St-3 dusts	0.20	0.51

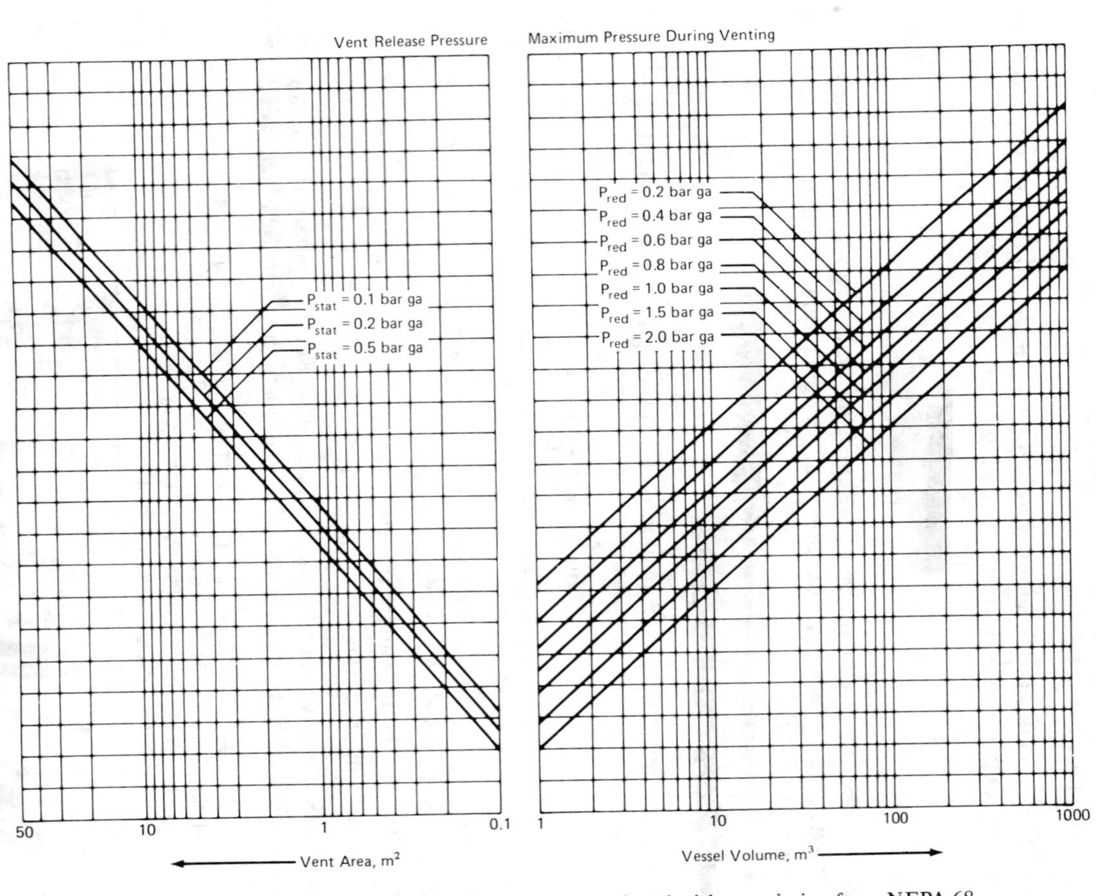

Figure 9-11 Venting nomograph for methane. (Reprinted with permission from NFPA 68-1988, *Guide for Venting of Deflagrations,* Copyright (c), 1988, National Fire Protection Association, Quincy, Mass. 02269. This reprinted material is not the complete and official position of the NFPA on the referenced subject which is represented only by the standard in its entirety.)

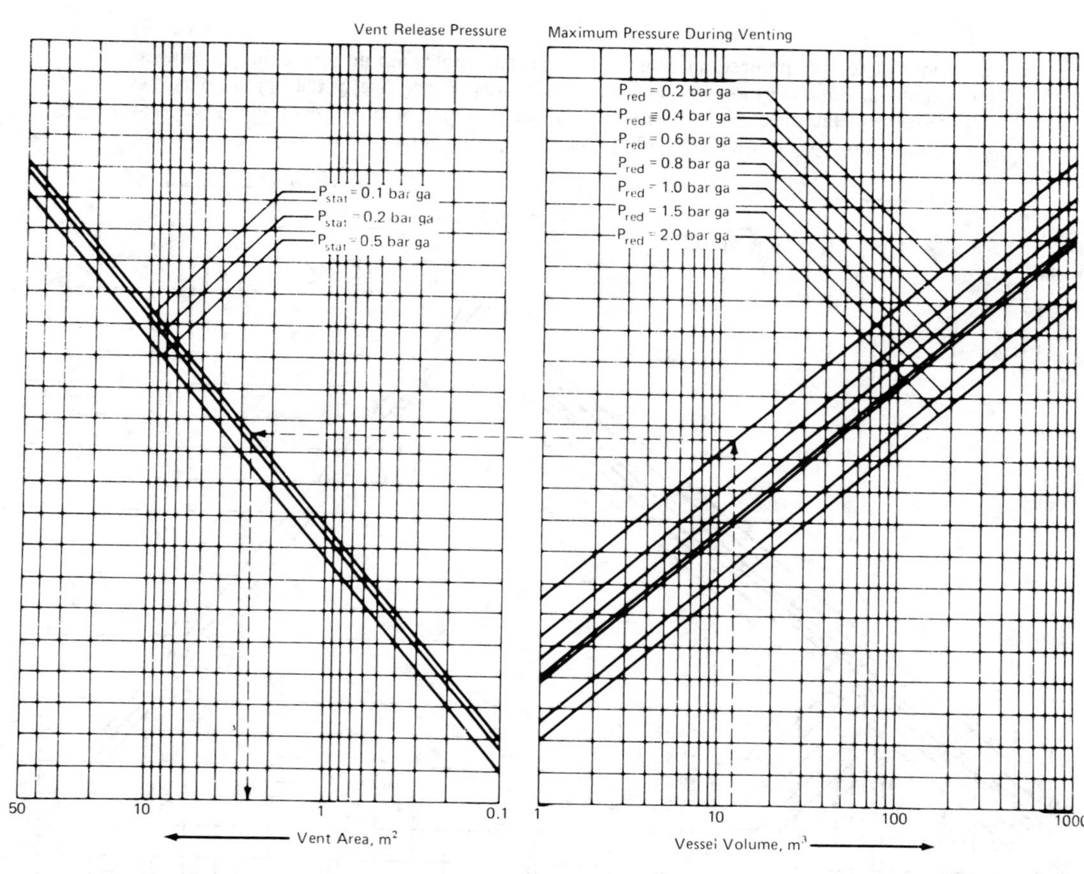

Figure 9-12 Venting nomograph for propane. (Reprinted with permission from NFPA 68-1988, *Guide for Venting of Deflagrations,* Copyright (c), 1988, National Fire Protection Association, Quincy, Mass. 02269. This reprinted material is not the complete and official position of the NFPA on the referenced subject which is represented only by the standard in its entirety.)

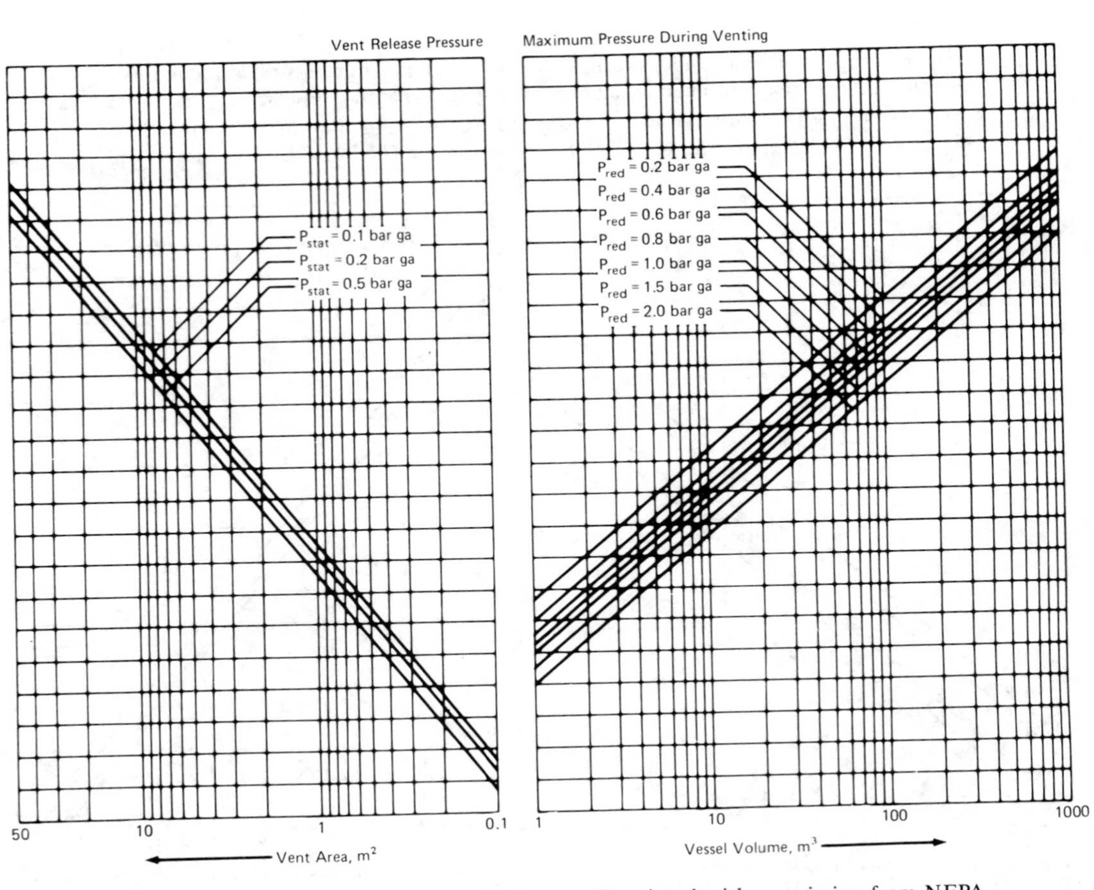

Figure 9-13 Venting nomograph for hydrogen. (Reprinted with permission from NFPA 68-1988, *Guide for Venting of Deflagrations,* Copyright (c), 1988, National Fire Protection Association, Quincy, Mass. 02269. This reprinted material is not the complete and official position of the NFPA on the referenced subject which is represented only by the standard in its entirety.)

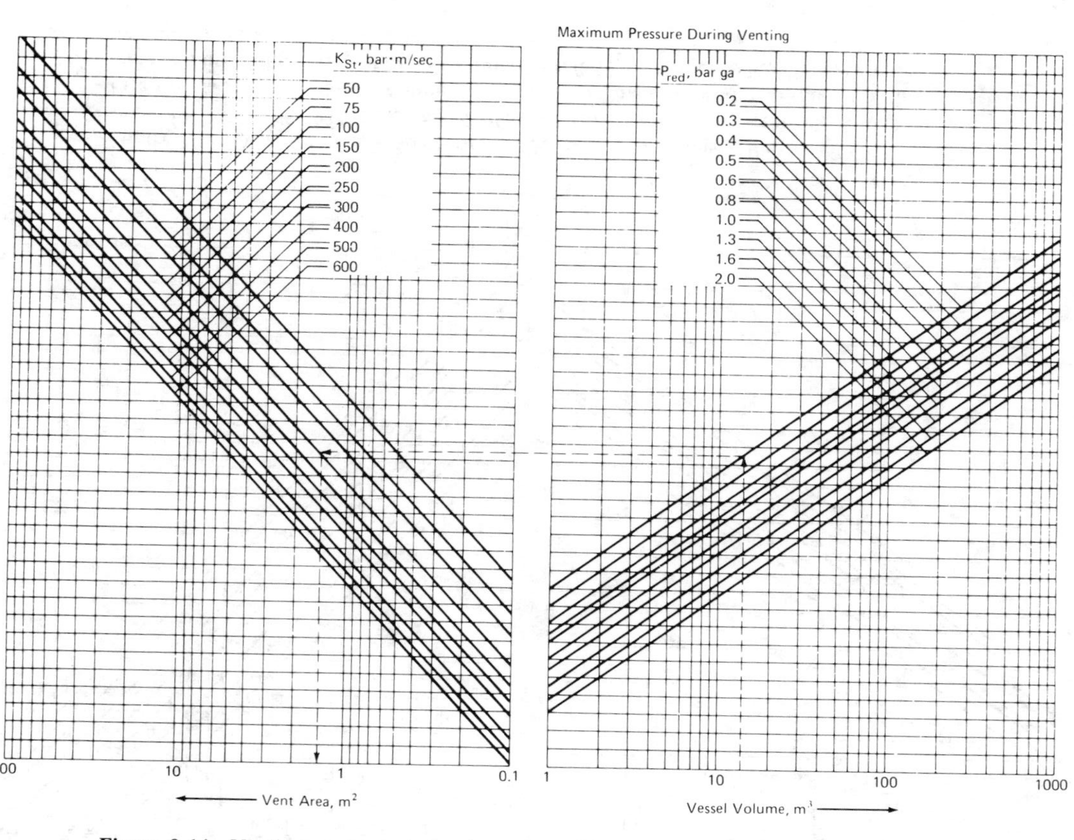

Figure 9-14 Venting nomograph for dusts, P_{stat} = 0.1 bar gauge. (Reprinted with permission from NFPA 68-1988, *Guide for Venting of Deflagrations,* Copyright (c), 1988, National Fire Protection Association, Quincy, Mass. 02269. This reprinted material is not the complete and official position of the NFPA on the referenced subject which is represented only by the standard in its entirety.)

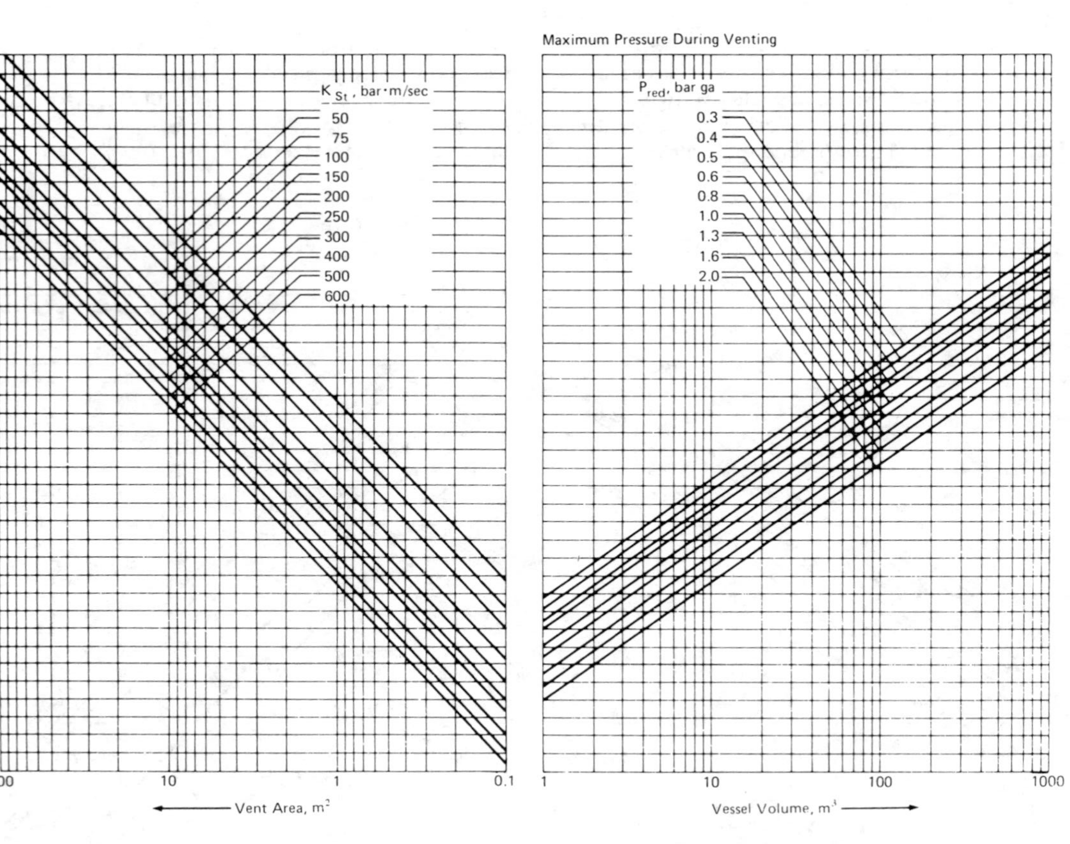

Figure 9-15 Venting nomograph for dusts, P_{stat} = 0.2 bar gauge. (Reprinted with permission from NFPA 68-1988, *Guide for Venting of Deflagrations,* Copyright (c), 1988, National Fire Protection Association, Quincy, Mass. 02269. This reprinted material is not the complete and official position of the NFPA on the referenced subject which is represented only by the standard in its entirety.)

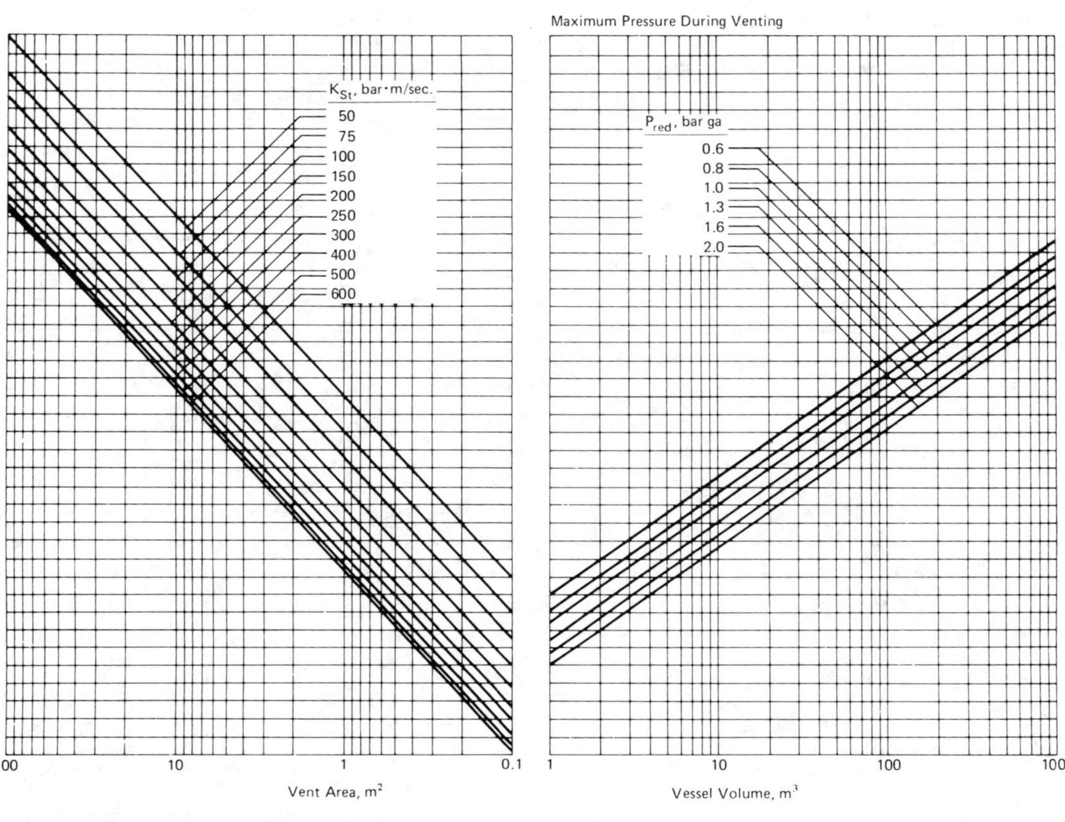

Figure 9-16 Venting nomograph for dusts, P_{stat} = 0.5 bar gauge. (Reprinted with permission from NFPA 68-1988, *Guide for Venting of Deflagrations,* Copyright (c), 1988, National Fire Protection Association, Quincy, Mass. 02269. This reprinted material is not the complete and official position of the NFPA on the referenced subject which is represented only by the standard in its entirety.)

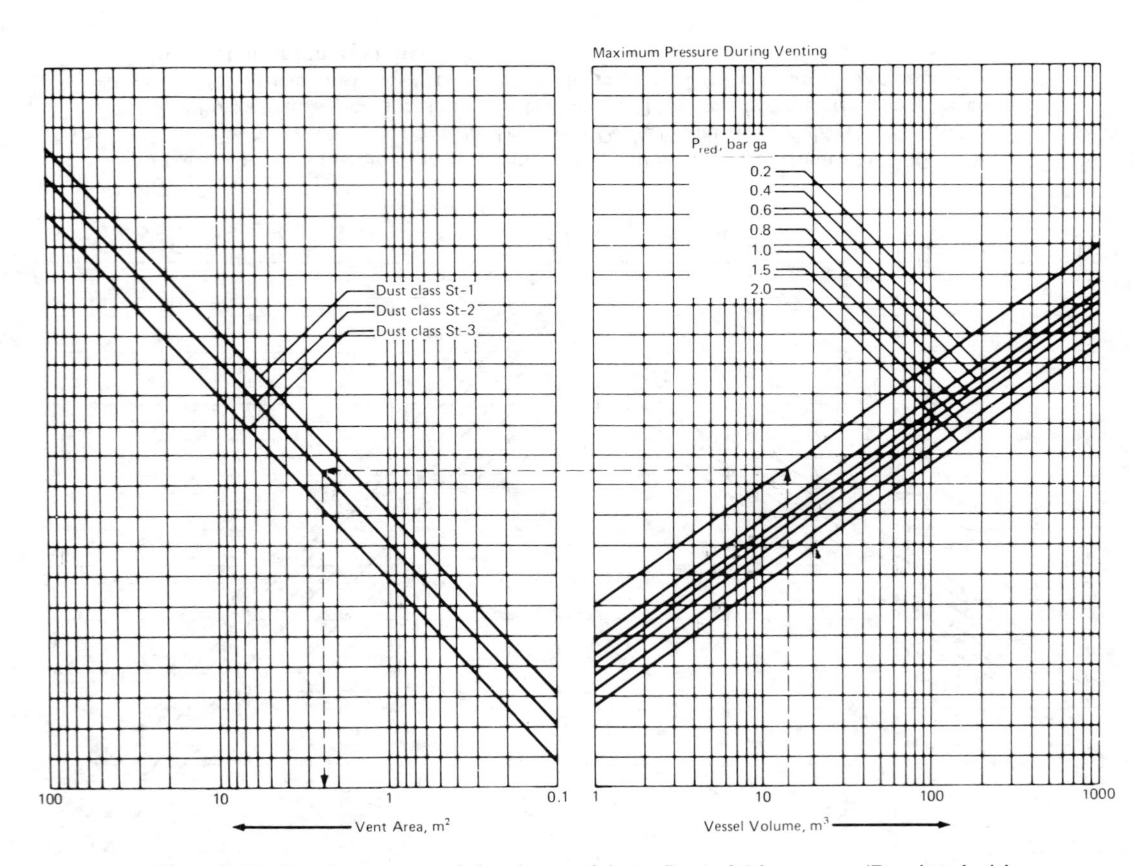

Figure 9-17 Venting nomograph for classes of dusts, $P_{stat} = 0.1$ bar gauge. (Reprinted with permission from NFPA 68-1988, *Guide for Venting of Deflagrations,* Copyright (c), 1988, National Fire Protection Association, Quincy, Mass. 02269. This reprinted material is not the complete and official position of the NFPA on the referenced subject which is represented only by the standard in its entirety.)

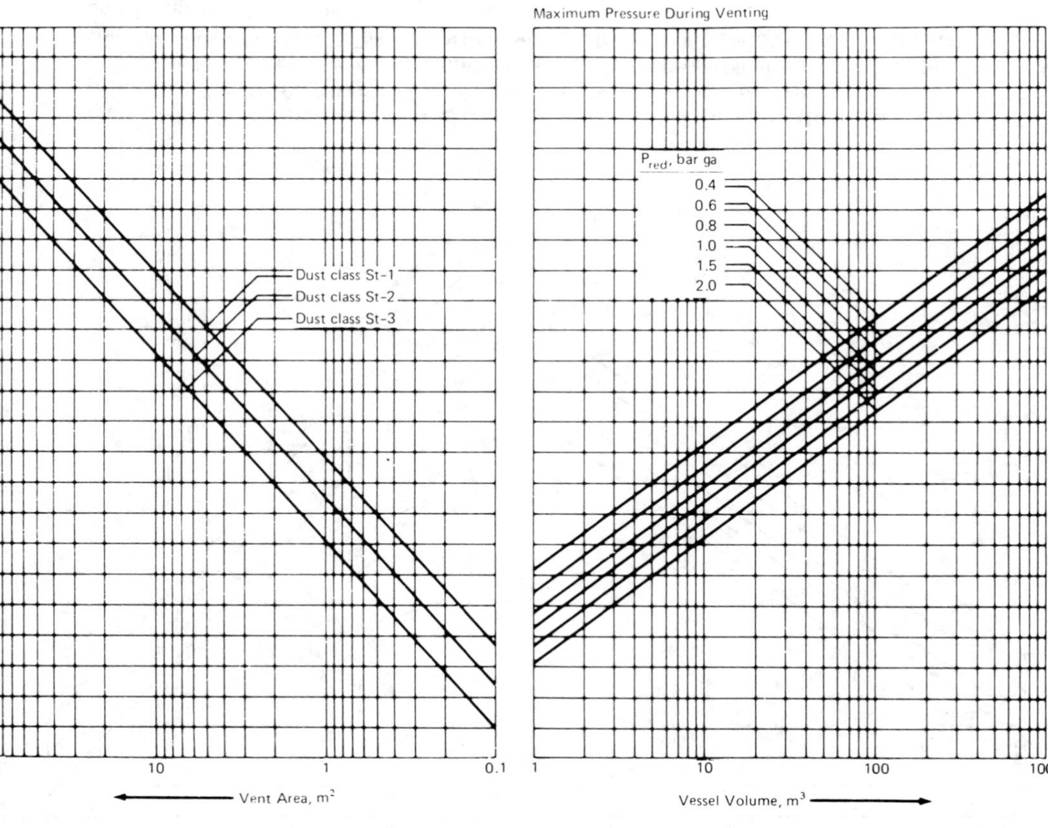

Figure 9-18 Venting nomograph for classes of dusts, P_{stat} = 0.2 bar gauge. (Reprinted with permission from NFPA 68-1988, *Guide for Venting of Deflagrations,* Copyright (c), 1988, National Fire Protection Association, Quincy, Mass. 02269. This reprinted material is not the complete and official position of the NFPA on the referenced subject which is represented only by the standard in its entirety.)

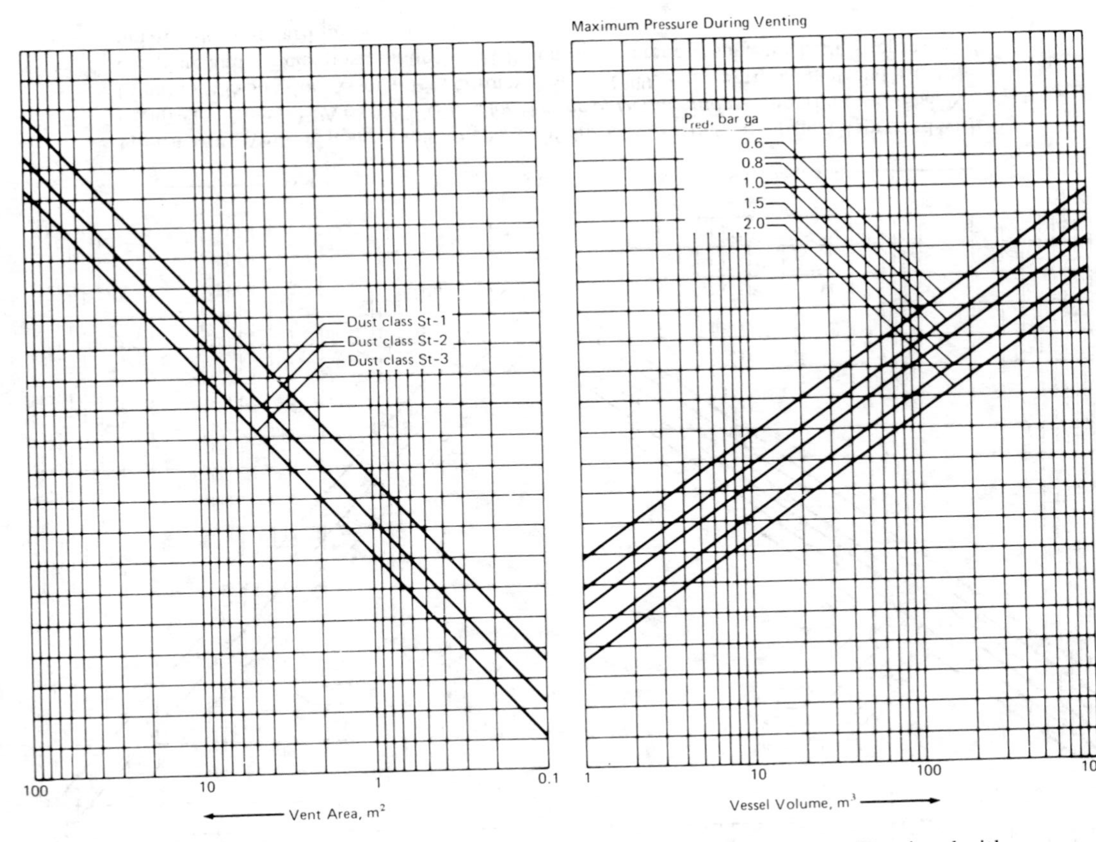

Figure 9-19 Venting nomograph for classes of dusts, $P_{stat} = 0.5$ bar gauge. (Reprinted with permission from NFPA 68-1988, *Guide for Venting of Deflagrations,* Copyright (c), 1988, National Fire Protection Association, Quincy, Mass. 02269. This reprinted material is not the complete and official position of the NFPA on the referenced subject which is represented only by the standard in its entirety.)

sure developed during the venting and P_{stat} is the pressure at which the vent releases. These nomographs are used by (1) locating the vessel volume on the right plot, (2) drawing a line from the volume vertically up to the correct parameter value, (3) drawing a line horizontally over to the correct parameter value on the left plot, and (4) drawing a line vertically down to the correct vent area. Plots are presented for various gases and dusts and in various parameter forms. The St classes for dusts are given in Table 6-6.

This procedure is used identically for dusts and gas vents using the appropriate figures.

Example 9-7

A spray dryer with a volume of 20 m^3 is used to dry a dusty powder with an St class of 1. Determine the proper size for a relief vent if the vent relief pressure is 0.1 bar gauge and the maximum allowable pressure in the vessel is 0.2 bar gauge.

Solution The solution is found directly from Figure 9-17 for a P_{stat} of 0.1 bar gauge. A vertical line is drawn from a volume of 20 m^3 on the right hand plot up to the P_{red} line for 0.2 bar gauge. A horizontal line is drawn over to the St-1 dust class and then down to the required relief area. The result for this case is 2.2 m^2.

9-7 VENTING FOR FIRES EXTERNAL TO PROCESS VESSELS

Fires external to process vessels can result in heating and boiling of process liquids, as shown in Figure 9-20. Venting is required to prevent explosion of these vessels.

Two-phase flow during these reliefs is possible, but not likely. For runaway reactor reliefs the energy is generated by reaction throughout the entire reactor liquid contents. For heating due to external fire, the heating occurs only at the surface of the vessel. Thus, liquid boiling will only occur next to the wall and the resulting two-phase foam or froth at the liquid surface will not have a substantial thickness. Two-phase flow during fire relief is therefore preventable by providing a suitable vapor space above the liquid within the vessel.

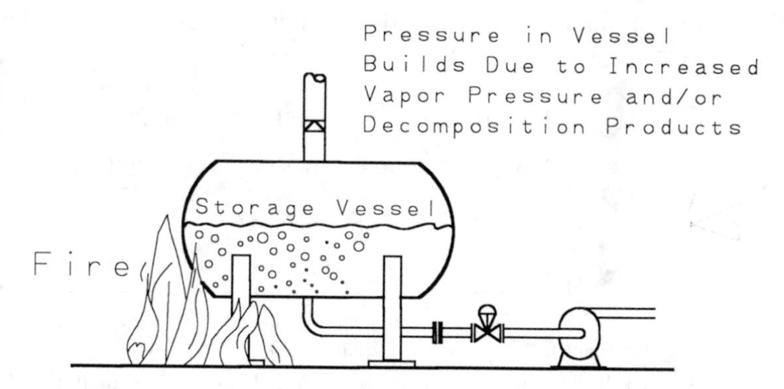

Figure 9-20 Heating of a process vessel due to external fire. Venting is required to prevent vessel rupture. For most fires, only a fraction of the external vessel is exposed to fire.

Two-phase fire relief equations are available for conservative design. Leung[16] presents an equation for the maximum temperature based on an energy balance around the heated vessel. This assumes a constant heat input rate, Q,

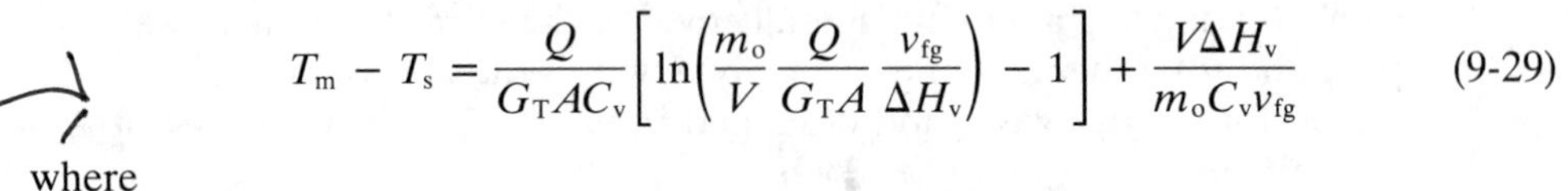

$$T_m - T_s = \frac{Q}{G_T A C_v}\left[\ln\left(\frac{m_o}{V}\frac{Q}{G_T A}\frac{v_{fg}}{\Delta H_v}\right) - 1\right] + \frac{V \Delta H_v}{m_o C_v v_{fg}} \qquad (9\text{-}29)$$

where

T_m is the maximum temperature in the vessel,
T_s is the set temperature corresponding to the set pressure,
Q is the constant heat input rate,
G_T is the mass flux through the relief,
A is the area of the relief,
C_v is the heat capacity at constant volume,
m_o is the liquid mass in the vessel,
V is the volume of the vessel,
v_{fg} is the volume difference between the vapor and liquid phases, and
ΔH_v is the heat of vaporization of the liquid.

The solution to Equation 9-29 for $G_T A$ is done by an iterative or trial and error technique. For the special case of no overpressure, $T_m = T_s$ and Equation 9-29 reduces to

$$\boxed{A = \frac{Q m_o v_{fg}}{G_T V \Delta H_v}} \qquad (9\text{-}30)$$

Various relationships have been recommended for computing the heat added to a vessel which is engulfed in fire. For regulated materials, the OSHA 1910.106 criterion[17] is mandatory. Other standards are available.[18] Crozier,[19] after analysis of the various standards, recommends the following equations for determining the total heat input, Q.

$$\begin{aligned} &\text{For } 20 < A < 200: && Q = 20{,}000\,A \\ &\text{For } 200 < A < 1{,}000: && Q = 199{,}300\,A^{0.566} \\ &\text{For } 1{,}000 < A < 2{,}800: && Q = 936{,}400\,A^{0.338} \\ &\text{For } A > 2{,}800: && Q = 21{,}000\,A^{0.82} \end{aligned} \qquad (9\text{-}31)$$

[16] Leung, "Simplified Vent Sizing Equations"

[17] OSHA 1910.106: *Flammable and Combustible Liquids* (Washington, DC: US Dept. of Labor, 1981).

[18] API Standard 2000, *Venting Atmospheric and Low-Pressure Storage Tanks (Nonrefrigerated and Refrigerated)*, 2nd ed. (Washington, DC: American Petroleum Institute, 1973), and NFPA 30: *Flammable and Combustible Liquids Code* (Quincy, MA: National Fire Protection Association, 1981).

[19] R. A. Crozier, "Sizing Relief Valves for Fire Emergencies," *Chemical Engineering*, October 28, 1985.

where

A is the area absorbing heat, in ft^2, for the following geometries:
For spheres: 55% of total exposed area.
For horizontal tanks: 75% of total exposed area.
For vertical tanks: 100% of total exposed area for first 30 ft.

Q is the total heat input to the vessel, in BTU/hour.

The mass flux, G_T is determined using Equation 9-14 or 9-17.

Example 9-8

Leung[20] reports on the computation of the required relief area for a spherical propane vessel exposed to fire. The vessel has a volume of 100 m^3, containing 50,700 kg of propane. A set pressure of 4.5 bar absolute is required. This corresponds to a set temperature, based on the saturation pressure, of 271.5 K. At these conditions the following physical property data are reported:

$$C_p = C_v = 2.41 \times 10^3 \text{ J/kg K}$$

$$\Delta H_v = 3.74 \times 10^5 \text{ J/kg}$$

$$v_{fg} = 0.1015 \text{ m}^3/\text{kg}$$

The molecular weight of propane is 44.

Solution The problem is solved assuming no overpressure during the relief. The relief vent area calculated is larger than the actual area required for a real relief device with overpressure.

The diameter of the sphere is

$$d = \left[\frac{6A}{\pi}\right]^{1/3} = \left[\frac{(6)(100 \text{ m}^3)}{(3.14)}\right]^{1/3} = 5.76 \text{ m}$$

The surface area of the sphere is

$$\pi d^2 = (3.14)(5.76 \text{ m})^2 = 104.2 \text{ m}^2 = 1{,}121 \text{ ft}^2$$

The area exposed to heat is given by the geometry factors provided with Equation 9-31.

$$A = (0.55)(1121 \text{ ft}^2) = 616 \text{ ft}^2$$

The total heat input is found using Equation 9-31.

$$Q = 199{,}300\, A^{0.566} = (199{,}300)(616 \text{ ft}^2)^{0.566} = 7.56 \times 10^6 \text{ BTU/hour}$$

$$= 2100 \text{ BTU/s} = 2.22 \times 10^6 \text{ J/s}$$

From Equation 9-14, assuming $\psi = 1.0$,

$$G_T = \frac{Q_m}{A} = 0.9\, \psi \frac{\Delta H_V}{v_{fg}} \sqrt{\frac{g_c}{C_P T_S}}$$

$$= (0.9)(1.0)\left(\frac{3.74 \times 10^5 \text{ J/kg}}{0.1015 \text{ m}^3/\text{kg}}\right)\left(\frac{1 \text{ N m}}{\text{J}}\right)$$

$$\times \sqrt{\frac{1 \text{ (kg m/s}^2)/\text{N}}{(2.41 \times 10^3 \text{ J/kg K})(271.5 \text{ K})(1 \text{ N m/J})}}$$

$$G_T = 4.10 \times 10^3 \text{ kg/m}^2 \text{ s}$$

[20] Leung, "Simplified Vent Sizing Equations"

The required vent area is determined from Equation 9-30.

$$A = \frac{Qm_o v_{fg}}{G_T V \Delta H_v}$$

$$= \frac{(2.22 \times 10^6 \text{ J/s})(50{,}700 \text{ kg})(0.1015 \text{ m}^3\text{/kg})}{(4.10 \times 10^3 \text{ kg/m}^2 \text{ s})(100 \text{ m}^3)(3.74 \times 10^5 \text{ J/kg})} \quad (9\text{-}30)$$

$$= 0.0745 \text{ m}^2$$

The required diameter is

$$d = \sqrt{\frac{4A}{\pi}} = \sqrt{\frac{(4)(0.0745 \text{ m}^2)}{(3.14)}}$$

$$= 0.308 \text{ m} = 12.1 \text{ inches}$$

An alternate way to look at the problem might be to ask the question "What initial fill fraction should be specified in the tank to avoid two-phase flow during a fire exposure incident?" No tested correlations are presently available to compute the height of a foam layer above the boiling liquid.

For fire reliefs with single phase vapor flow the equations provided in Sections 9-2 and 9-4 are used to determine the size of the relief.

9-8 RELIEFS FOR THERMAL EXPANSION OF PROCESS FLUIDS

Liquids contained within process vessels and piping will normally expand when heated. The expansion will damage pipes and vessels if the pipe or vessel is filled completely with fluid and the liquid is blocked in.

A typical situation is thermal expansion of water in cooling coils in a reactor, shown in Figure 9-21. If the coils are filled with water and accidentally blocked in, the water will expand when heated by the reactor contents, leading to damage to the cooling coils.

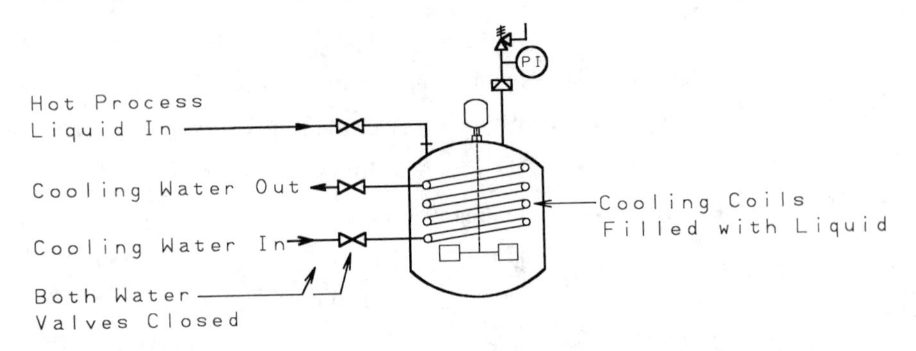

Figure 9-21 Damage to cooling coils as a result of external heating of blocked-in cooling fluid.

Relief vents are installed in these systems to prevent damage due to liquid expansion. Although this may appear to be a minor problem, damage to heat exchange systems can result in (1) contamination of product or intermediate, (2) subsequent corrosion problems, (3) substantial plant outages, and (4) large repair expenses. Failure in heat exchange equipment is also difficult to identify and repairs are time consuming.

A thermal expansion coefficient for liquids, β, is defined as

$$\beta = \frac{1}{V}\left(\frac{dV}{dT}\right) \tag{9-32}$$

where V is the volume of the fluid and T is the temperature.

Table 9-2 lists thermal expansion coefficients for a number of substances. Water behaves in an unusual fashion. The thermal expansion coefficient decreases with increasing temperature up until about 4°C, after which the thermal expansion coefficient increases with temperature. Coefficients for water are readily determined from the steam tables.

The volumetric expansion rate, Q_v, through the relief due to thermal expansion is

$$Q_v = \frac{dV}{dt} = \frac{dV}{dT}\frac{dT}{dt} \tag{9-33}$$

By applying the definition of the thermal expansion coefficient, given by Equation 9-32,

$$Q_v = \beta V \frac{dT}{dt} \tag{9-34}$$

For a pipe or process vessel heated externally by a hot fluid, an energy balance on the fluid is given by,

$$mC_P \frac{dT}{dt} = UA(T - T_a) \tag{9-35}$$

TABLE 9-2 THERMAL EXPANSION COEFFICIENTS FOR A VARIETY OF LIQUIDS. G. SHORTLEY AND D. WILLIAMS, *ELEMENTS OF PHYSICS,* 4TH ED. (PRENTICE HALL, ENGLEWOOD CLIFFS, 1965), P. 302.

Liquid	Density at 20°C (kg/m^3)	Thermal expansion coefficient ($°C^{-1}$)
Alcohol, ethyl	791	112×10^{-5}
Alcohol, methyl	792	120×10^{-5}
Benzene	877	124×10^{-5}
Carbon tetrachloride	1,595	124×10^{-5}
Ether, ethyl	714	166×10^{-5}
Glycerin	1,261	51×10^{-5}
Mercury	13,546	18.2×10^{-5}
Turpentine	873	97×10^{-5}

where

T is the temperature of the fluid,
C_P is the heat capacity of the liquid,
UA is an overall heat transfer coefficient, and
T_a is the ambient temperature.

It follows that,

$$\frac{dT}{dt} = \frac{UA}{mC_P}(T - T_a) \tag{9-36}$$

Substituting into Equation 9-34,

$$Q_v = \frac{\beta V}{mC_P} UA(T - T_a) \tag{9-37}$$

and, invoking the definition of the liquid density, ρ

$$\boxed{Q_v = \frac{\beta}{\rho C_P} UA(T - T_a)} \tag{9-38}$$

Equation 9-38 describes the fluid expansion only at the beginning of heat transfer, when it is initially exposed to the external temperature, T_a. The heat transfer will increase the temperature of the liquid, changing the value of T. However, it is apparent that Equation 9-38 provides the maximum thermal expansion rate, sufficient for sizing a relief device.

The volumetric expansion rate, Q_v, is subsequently used in an appropriate equation to determine the relief vent size.

Example 9-9

The cooling coil in a reactor has a surface area of 10,000 ft^2. Under the most severe of conditions, the coils can contain water at 32°F and be exposed to superheated steam at 400°F. Given a heat transfer coefficient of 50 BTU/hr-ft^2-°F, estimate the volumetric expansion rate of the water in the cooling coils, in gallons/minute.

Solution The expansion coefficient, β, for water at 32°F should be used. This is estimated using liquid volumetric data from the steam tables over a short range of temperatures around 32°F. However, the steam tables do not provide liquid water specific volume data below 32°F. A value between 32°F and some appropriate higher temperature will suffice. From the steam tables,

Temperature (°F)	Specific volume (ft^3/lb$_m$)
32	0.01602
50	0.01603

The expansion coefficient is computed using Equation 9-32.

$$\beta = \frac{1}{v}\frac{dv}{dT} = \frac{1}{0.016025 \text{ ft}^3/\text{lb}_m}\left(\frac{0.01602 - 0.01603}{32 - 50}\right)\left(\frac{\text{ft}^3/\text{lb}_m}{°\text{F}}\right)$$

$$\beta = 3.47 \times 10^{-5}\ °\text{F}^{-1}$$

The volumetric expansion rate is given by Equation 9-38.

$$Q_v = \frac{\beta}{\rho C_P} UA(T - T_a)$$

$$= \frac{(3.47 \times 10^{-5}/°\text{F})(50 \text{ BTU/hr ft}^2 \text{ °F})(10{,}000 \text{ ft}^2)(400 - 32)°\text{F}}{(62.4 \text{ lb}_m/\text{ft}^3)(1 \text{ BTU/lb}_m°\text{F})}$$

$$= 102 \text{ ft}^3/\text{hour} = 12.7 \text{ gallons/minute}$$

The relief device vent area must be designed to accommodate this volumetric flow.

SUGGESTED READING

Deflagration Vents

W. Bartknecht, "Pressure Venting of Dust Explosions in Large Vessels," *Plant/Operations Progress,* Vol. 5, No. 4, October, 1986, p. 196.

Frank T. Bodurtha, *Industrial Explosion Prevention and Protection* (New York: McGraw-Hill Book Company, 1980).

Ian Swift and Mike Epstein, "Performance of Low Pressure Explosion Vents," *Plant/Operations Progress,* Vol. 6, No. 2, April, 1987.

Relief Codes

API RP 520, *Recommended Practice for the Design and Installation of Pressure Relieving Systems in Refineries* (Washington, DC: American Petroleum Institute, 1963).

API RP 521, *Guide for Pressure-Relieving and Depressurizing Systems* (Washington, DC: American Petroleum Institute, 1982).

API Standard 2000, *Venting Atmospheric and Low-Pressure Storage Tanks (Nonrefrigerated and Refrigerated),* 2nd ed. (Washington, DC: American Petroleum Institute, 1973).

ASME Boiler and Pressure Vessel Code (New York: American Society of Mechanical Engineers, 1965).

NFPA 68: *Guide for Venting of Deflagrations* (Quincy, MA: National Fire Protection Association, 1988).

Two-phase Flow

H. K. Fauske, "Generalized Vent Sizing Monogram for Runaway Chemical Reactions," *Plant/Operations Progress,* Vol. 3, No. 4, October, 1984.

H. K. Fauske, "Flashing Flows or Some Practical Guidelines for Emergency Releases," *Plant/Operations Progress,* July, 1985.

H. K. Fauske and J. C. Leung, "New Experimental Technique for Characterizing Runaway Chemical Reactions," *Chemical Engineering Progress,* August, 1985.

H. K. Fauske, "Emergency Relief System (ERS) Design," *Chemical Engineering Progress,* August, 1985.

K. E. First and J. E. Huff, "Design Charts for Two-Phase Flashing Flow in Emergency Pressure Relief Systems," *1988 AICHE Spring National Meeting,* 1988.

Harold G. Fisher, "DIERS Research Program on Emergency Relief Systems," *Chemical Engineering Progress,* August, 1985, p. 33.

J. C. LEUNG, "Simplified Vent Sizing Equations for Emergency Relief Requirements in Reactors and Storage Vessels," *AICHE Journal,* Vol. 32, No. 10 (October, 1986), p. 1622.

J. C. LEUNG, "A Generalized Correlation for One-Component Homogeneous Equilibrium Flashing Choked Flow," *AICHE Journal,* Vol. 32, No. 10 (October, 1986), p. 1743.

J. C. LEUNG and M. A. GROLMES, "The Discharge of Two-Phase Flashing Flows in a Horizontal Duct," *AICHE Journal,* Vol. 33, No. 3 (March, 1987), p. 524.

J. C. LEUNG and M. A. GROLMES, "A Generalized Correlation for Flashing Choked Flow of Initially Subcooled Liquid," *AICHE Journal,* Vol. 34, No. 4 (April, 1988), p. 688.

J. C. LEUNG and H. G. FISHER, "Two-Phase Flow Venting from Reactor Vessels," *Journal of Loss Prevention,* Vol. 2, No. 2 (April 1989), p. 78.

PROBLEMS

9-1. Estimate the diameter of spring type liquid reliefs for the following conditions:

Pump capacity at ΔP (gpm)	Set pressure (psig)	Over pressure (%)	Back pressure (%)	Valve type	(ρ/ρ_{ref})
a. 100	50	20	10	Conventional	1.0
b. 200	100	20	30	Bal. bellows	1.3
c. 50	50	10	40	Bal. bellows	1.2

9-2. Determine the diameter of a spring type vapor relief for the following conditions. Assume for each case that $\gamma = 1.3$, a set pressure of 100 psia and a temperature of 100°F.

Compressibility, z	Molecular weight	Mass flow (lb/hour)	Over pressure (%)	Back pressure (%)
a. 1.0	28	50	10	10
b. 1.2	28	50	30	10
c. 1.0	44	50	10	10
d. 1.2	44	50	30	10
e. 1.0	28	100	10	30
f. 1.2	28	100	30	30

9-3. Determine the required diameter for rupture disks for the following conditions. Assume a specific gravity of 1.2 for all cases.

Liquid flow (gpm)	Pressure drop (psi)
a. 1000	100
b. 100	100
c. 1000	50
d. 100	50

9-4. Determine the required diameter for rupture disks in vapor service for the following conditions. Assume nitrogen is the vent gas and the temperature is 100°F.

Gas flow (lb/hour)	Pressure (psia)
a. 100	100
b. 200	100
c. 100	50
d. 200	50

9-5. Determine the proper relief diameter for the following two-phase flow conditions. Assume in all cases that L/D = 0.0.

	(a)	(b)	(c)	(d)
Reaction mass, lb:	10,000	10,000	10,000	10,000
Volume, ft^3:	200	500	500	500
Set pressure, psia:	100	100	100	100
Set temperature, °F:	500	500	500	500
$(dT/dt)_s$, °F/s:	0.5	0.5	2.0	2.0
Max. pressure, psia:	120	120	120	140
Max. temp., °F	520	520	520	550
$(dT/dt)_m$, °F/s:	0.66	0.66	2.4	2.6
Liquid sp. vol., ft^3/lb	0.02	~~0.2~~ .02	0.02	0.02
Vapor sp. vol., ft^3/lb:	1.4	1.4	1.4	1.4
Heat cap., BTU/lb °F:	1.1	1.1	1.1	1.1
Heat of vap., BTU/lb:	130	130	130	130

9-6. How is the overpressure included in the design of two-phase reliefs?

9-7. Determine the relief vent areas for the following two-phase fire scenarios. Assume a spherical vessel in each case.

	(a)	(b)	(c)	(d)
Molecular weight:	72	72	86	86
Volume, ft^3:	5000	5000	5000	5000
Initial mass, lb:	30,000	15,000	15,000	15,000
Set pressure, psia:	100	100	100	100
Set temperature, °F:	220	220	220	220
Max. pressure, psia:	100	100	130	150
Max. temperature, °F:	220	220	240	275
Heat of vapor, BTU/lb:	130	130	150	150
v_{fg}, ft^3/lb:	1.6	1.6	1.6	1.6
C_p, BTU/lb °F	0.40	0.40	0.52	0.52

9-8. Determine the relief size for spray dryers operating at the following conditions.

Vapors:	(a)	(b)	(c)	(d)
Volume, ft^3:	1000	1000	1000	1000
Set pressure, psia:	16.7	16.7	16.7	16.7
Max. pressure, psia:	17.6	17.6	29.4	29.4
Gas:	Methane	Hydrogen	Methane	Hydrogen

Dusts:	(a)	(b)	(c)	(d)
Volume, ft^3:	1000	1000	1000	1000
St. Class:	1	3	1	3
Set pressure, psia:	16.7	16.7	16.7	16.7
Max. pressure, psia:	20.6	20.6	29.4	29.4

9-9. Determine the size of relief required to protect the following cooling coils against thermal expansion. Water is used for each case. Assume that the tubes can withstand a pressure of 1000 psig and the normal operating pressure is 200 psig. Assume a set pressure of 500 psig, an overpressure of 20% and no backpressure.

	(a)	(b)	(c)	(d)
Blocked in area, ft^2:	10,000	10,000	10,000	10,000
Max. temp., °F:	550	550	800	550
Min. temp., °F:	70	50	32	70
Heat transfer coefficient, BTU/hr·ft^2 °F:	75	75	75	125

9-10. Consider Problem 9-9, Part a. This time use alcohol as a liquid medium with a thermal expansion coefficient of 1.12×10^{-3}/°C. The heat capacity of the alcohol is 0.58 kcal/kg °C and its density is 791 kg/m^3. Determine the relief size required.

9-11. A process vessel is equipped with a 2-inch rupture disc set at 100 psig and designed for 10% overpressure. A nitrogen line must be added to the vessel to provide the capability of purging and/or pressure discharging liquids. What size line would you select if the nitrogen is available from a 500 psig source? The temperature is 80°F.

9-12. Home hot water heaters contain relief devices to provide protection in the event that the heater controls fail and the water is heated to a high pressure.

A typical water heater contains 40 gallons of water and has a heat input of 42,000 BTU/hour. If the heater is equipped with a 150 psia spring relief device, compute the area required for relief. Hint: Two-phase flow is expected. Assume no overpressure.

Also compute the relief vent size assuming all vapor relief. Assume 20% backpressure.

9-13. A cylindrical tank 4-feet in diameter and 10 feet long is completely filled with water and blocked in. Estimate the thermal expansion rate of the water if the water is at 50°F and the steel shell of the tank is suddenly heated to 100°F by the sun. Assume a heat transfer coefficient of 50 BTU/hr ft^2 °F and that only the top half of the tank is heated.

If the tank is exposed to fire, what is the required relief area? Assume no overpressure. The tank MAWP is 200 psig.

9-14. A 10-ft wide by 10-ft long by 10-ft high square shed is used to store tanks of methane. What area deflagration vent is required? Assume a maximum internal overpressure of 0.1 psig.

9-15. A spray dryer is used to dry vitamins in powder form. The dryer consists of a cylindrical section 12.0 ft high and 5 ft in diameter. Attached to the bottom of the cylindrical section is a cone section for collecting the dried powder. The cone is 5-ft long. If the deflagration index for the vitamin powder is 80 bar m/s, determine the area required for a deflagration vent. Assume the vent opens at 0.2 bar gauge and the maximum pressure is 0.5 bar gauge.

9-16. A beverage dispensing system consists of a bottle of beverage, a dispensing hose with valve, and a CO_2 system to keep the beverage pressurized. The CO_2 system includes a small bottle of high pressure, liquified CO_2, and a regulator to regulate the gas pressure delivery to the beverage.

A typical beverage system contains a 7.75 gallon beverage bottle, a 5-lb bottle of liquefied CO_2 and a regulator set at 9 psig. The regulator is connected directly to the CO_2 bottle and a 1/2 inch ID plastic hose connects the regulator to the beverage bottle. The liquefied CO_2 saturation vapor pressure is 800 psia. Assume a temperature of 80°F.

A pressure relief system must be designed to protect the beverage bottle from overpressure. The relief device will be installed in the CO_2 line where it enters the beverage container.

a. Determine the most likely scenario contributing to overpressure of the beverage container.

b. Must two-phase flow be considered?

c. Determine the vent area required, assuming a set pressure of 20 psig and an overpressure of 10%. Also assume a spring type relief.

Hazards Identification

10

Hazards are everywhere. Unfortunately, a hazard is not always identified until an accident occurs. It is essential to identify the hazards and reduce the risk well in advance of an accident.

For each process in a chemical plant the following questions must be asked:

1. What are the hazards?
2. What can go wrong and how?
3. What are the chances?
4. What are the consequences?

The first question represents hazard identification. The last three questions are associated with risk assessment, considered in detail in Chapter 11. Risk assessment includes a determination of the events that can produce an accident, the probability of those events and the consequences. The consequences could include human injury or loss of life, damage to the environment, or loss of production and capital equipment. Question number 2 is frequently called scenario identification.

The terminology used varies considerably. Hazard identification and risk assessment are sometimes combined into a general category called hazard evaluation. Risk assessment is sometimes called hazard analysis. A risk assessment procedure that determines probabilities is frequently called probabilistic risk assessment, or PRA. The use of the words hazards identification and risk assessment, as used here, provide much clearer and specific definitions to these concepts.

Figure 10-1 illustrates the normal procedure for using hazards identification and risk assessment. After a description of the process is available, the hazards are identified. The various scenarios by which an accident can occur are then deter-

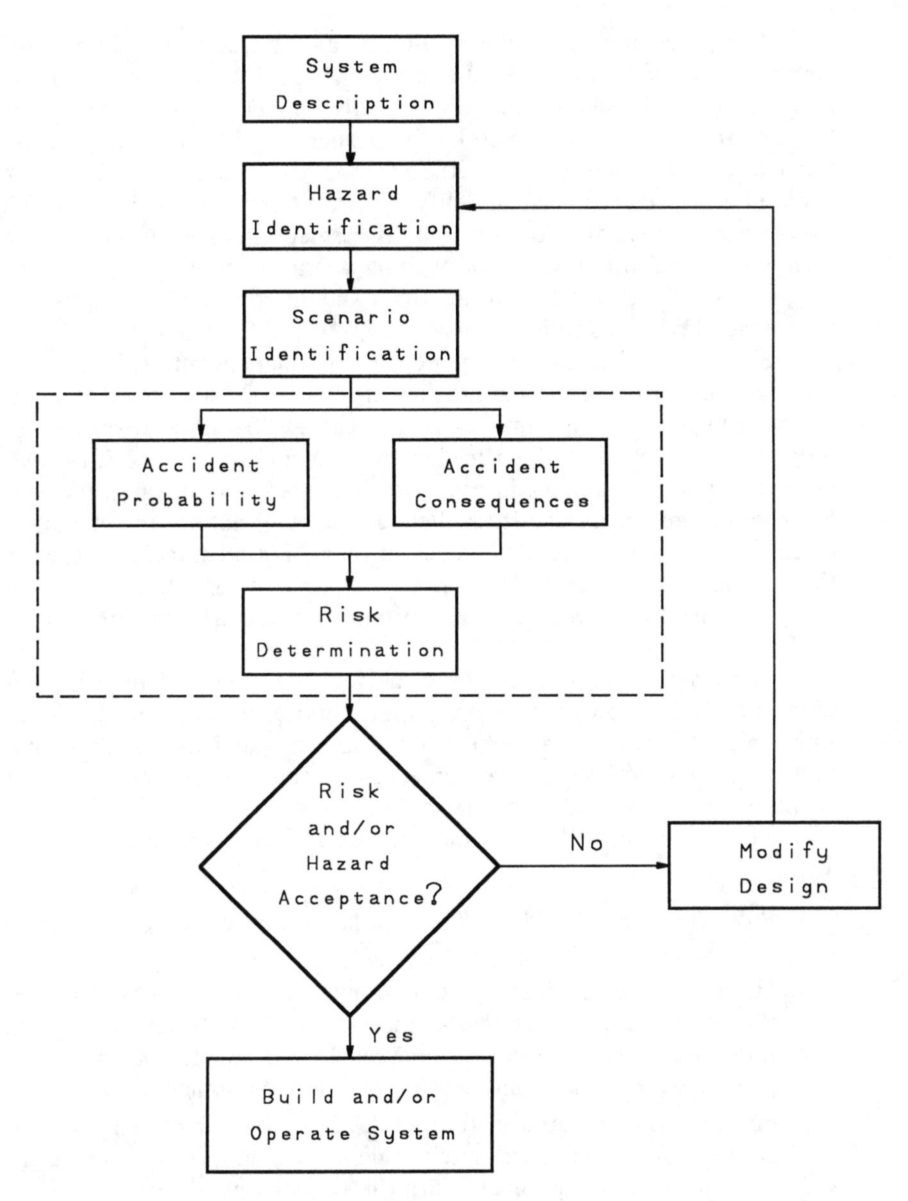

Figure 10-1 The hazards identification and risk assessment procedure. (Adapted from *Guidelines for Hazards Evaluation Procedures,* American Institute of Chemical Engineers, New York, 1985. p. 1-9.)

mined. This is followed by a concurrent study of both the probability and the consequences of an accident. This information is assembled into a final risk assessment. If the risk is acceptable, then the study is complete and the process is operated. If the risk is unacceptable, then the system must be modified and the procedure is restarted.

The procedure described by Figure 10-1 is frequently abbreviated based on circumstances. If failure rate data on the applicable equipment is not available then risk assessment procedures cannot be fully applied. For this particular situation, the area in Figure 10-1 denoted by the dashed line is removed. Also, the complete procedure is often more detailed than is practically necessary. Most plant sites (and even subunits within a plant) modify the procedure to fit their particular situation.

Hazards identification and risk assessment studies can be performed at any stage during the initial design or ongoing operation of a process. If the study is being performed with the initial design, it should be done as soon as possible. This enables modifications to be incorporated into the final design.

Hazard identification can be performed independent of risk assessment. However, the best result is obtained if they are both done together. Otherwise, hazards of low probability and minimal consequences are identified with the result that the process is "goldplated." This means that potentially unnecessary and expensive safety equipment and procedures are implemented. For instance, flying aircraft and tornadoes represent hazards to a chemical plant. What are the chances of their occurrence and what should be done about them? For most facilities the probability of these hazards is very small: No steps are required for prevention. Likewise, hazards with reasonable probability but minimal consequences are sometimes also neglected.

Many methods are available for performing hazard identification and risk assessment.[1] Only a few of the more popular approaches will be considered here. No single approach is necessarily best suited for any particular application. The selection of the best method requires experience. Most companies use these methods or adaptations to suit their particular operation.

The hazard identification methods described in this chapter include the following.

1. Process hazard checklists: This is a list of items and possible problems in the process that must be checked.
2. Hazards surveys: This can be as simple as an inventory of hazardous materials, or, it can be as detailed as the Dow Fire and Explosion Index. The Dow Index is a formal rating system, much like an income tax form, providing penalties for hazards and credits for safety equipment and procedures.
3. Hazards and Operability Studies (HAZOP): This approach allows the mind to go free in a controlled environment. Various events are suggested for a specific piece of equipment with the participants determining if and how the event could occur and if the event creates any form of risk.
4. Safety Review: A very effective but less formal type of HAZOP study. The results are highly dependent on the experience and synergism of the group reviewing the process.

[1]*Guidelines for Hazard Evaluation Procedures* (New York: American Institute of Chemical Engineers, 1985).

10-1 PROCESS HAZARDS CHECKLISTS

A process checklist is simply a list of possible problems and areas to be checked. The list reminds the reviewer or operator of the potential problem areas. A checklist can be used during the design of a process to identify design hazards, or it can be used prior to process operation.

A classical example is an automobile checklist that one might review prior to driving away on a vacation. This checklist might contain the following items.

- Check oil in engine.
- Check air pressure in tires.
- Check fluid level in radiator.
- Check air filter.
- Check fluid level in windshield washer tank.
- Check headlights and tailights.
- Check exhaust system for leaks.
- Check fluid levels in brake system.
- Check gasoline level in tank.

Checklists for chemical processes can be very detailed, involving hundreds or even thousands of items. But, as illustrated in the above example, the effort in developing and using checklists will yield significant results.

A typical process design safety checklist is shown in Figure 10-2. Note that three check-off columns are provided. The first column is used to indicate those areas that have been thoroughly investigated. The second column is used for those items that don't apply to this particular process. The last column is used to mark those areas requiring further investigation. Extensive notes on individual areas are kept separate from the checklist.

The design of the checklist depends on the intent. A checklist intended for use during the initial design of the process will be considerably different from a checklist used for a process change, or a checklist used prior to process operation.

Checklists should only be applied during the preliminary stages of hazard identification and should not be used as a replacement for a more complete hazard identification procedure. Checklists are effective in identifying hazards due to process design, plant layout, storage of chemicals, electrical systems, and so forth. Checklists are also effective for identifying improper operations or procedures, or process aberrations before a specific process is started.

10-2 HAZARD SURVEYS

A hazard survey can be as simple as an inventory of hazardous materials in a facility or as complicated as a rigorous procedure such as the Dow Fire and Explosion Index,[2] which is a most popular form of hazard survey. This is a formal system-

[2] *Dow's Fire and Explosion Index Hazard Classification Code,* 6th ed. (New York: American Institute for Chemical Engineers, 1987).

	Further study required ↓	Does not apply ↓	Completed ↓
General layout			
1. Areas properly drained?	□	□	□
2. Aisleways provided?	□	□	□
3. Fire walls, dikes and special guardrails needed?	□	□	□
4. Hazardous underground obstructions?	□	□	□
5. Hazardous overhead restrictions?	□	□	□
6. Emergency accesses and exits?	□	□	□
7. Enough headroom?	□	□	□
8. Access for emergency vehicles?	□	□	□
9. Safe storage space for raw materials and finished products?	□	□	□
10. Adequate platforms for safe maintenance operations?	□	□	□
11. Hoists and elevators properly designed and safeguarded?	□	□	□
12. Clearance for overhead power lines?	□	□	□
Buildings			
1. Adequate ladders, stairways and escapeways?	□	□	□
2. Fire doors required?	□	□	□
3. Head obstructions marked?	□	□	□
4. Ventilation adequate?	□	□	□
5. Need for ladder or stairway to roof?	□	□	□
6. Safety glass specified where necessary?	□	□	□
7. Need for fireproofed structural steel?	□	□	□
Process			
1. Consequences of exposure to adjacent operations considered?	□	□	□
2. Special fume or dust hoods required?	□	□	□
3. Unstable materials properly stored?	□	□	□
4. Process laboratory checked for runaway explosive conditions?	□	□	□
5. Provisions for protection from explosions?	□	□	□
6. Hazardous reactions possible due to mistakes or contamination?	□	□	□
7. Chemistry of processes completely understood and reviewed?	□	□	□
8. Provisions for rapid disposal of reactants in an emergency?	□	□	□
9. Failure of mechanical equipment possible cause of hazards?	□	□	□

Figure 10-2 A typical process safety checklist. A list of this type is frequently used prior to a more complete analysis. (Adapted from Henry E. Webb, "What to do When Disaster Strikes," *Safe and Efficient Plant Operation and Maintenance,* ed. Richard Greene, Mc-Graw-Hill, New York, 1980.)

	Completed ↓	Does not apply ↓	Further study required ↓
10. Hazards possible from gradual or sudden blockages in piping or equipment?	☐	☐	☐
11. Public liability risks possible from sprays, fumes, mists or noise?	☐	☐	☐
12. Provisions made for disposal of toxic materials?	☐	☐	☐
13. Hazards involved in sewering material?	☐	☐	☐
14. Material safety data sheets available for all chemical species?	☐	☐	☐
15. Hazards possible from simultaneous loss of two or more utilities?	☐	☐	☐
16. Safety factors altered by design revisions?	☐	☐	☐
17. Consequences of reasonably worst incident, or combination of incidents, reviewed?	☐	☐	☐
18. Process diagrams correct and up-to-date?	☐	☐	☐
Piping			
1. Safety showers and eye baths required?	☐	☐	☐
2. Sprinkler systems required?	☐	☐	☐
3. Provisions for thermal expansion?	☐	☐	☐
4. All overflow lines directed to safe areas?	☐	☐	☐
5. Vent lines directed safely?	☐	☐	☐
6. Piping specifications followed?	☐	☐	☐
7. Washing-down hoses needed?	☐	☐	☐
8. Check valves provided as needed?	☐	☐	☐
9. Protection and identification of fragile pipe considered?	☐	☐	☐
10. Possible deterioration of exterior of piping by chemicals?	☐	☐	☐
11. Emergency valves readily accessible?	☐	☐	☐
12. Long and large vent lines supported?	☐	☐	☐
13. Steam condensate piping safely designed?	☐	☐	☐
14. Relief valve piping designed to prevent plugging?	☐	☐	☐
15. Drains to relieve pressure on suction and discharge of all process pumps?	☐	☐	☐
16. City water lines not connected to process pipes?	☐	☐	☐
17. Flammable fluids feeding production units shut off from a safe distance in case of fire or other emergency?	☐	☐	☐
18. Personnel protective insulation provided?	☐	☐	☐
19. Hot steam lines insulated?	☐	☐	☐
Equipment			
1. Designs correct for maximum operating pressure?	☐	☐	☐
2. Corrosion allowance considered?	☐	☐	☐

Figure 10-2 (continued)

	Completed ↓	Does not apply ↓	Further study required ↓
3. Special isolation for hazardous equipment?	☐	☐	☐
4. Guards for belts, pulleys, sheaves and gears?	☐	☐	☐
5. Schedule for checking protective devices?	☐	☐	☐
6. Dikes for any storage tanks?	☐	☐	☐
7. Guard rails for storage tanks?	☐	☐	☐
8. Construction materials compatible with process chemicals?	☐	☐	☐
9. Reclaimed and replacement equipment checked structurally and for process pressures?	☐	☐	☐
10. Pipelines independently supported to relieve pumps and other equipment, as necessary?	☐	☐	☐
11. Automatic lubrication of critical machinery?	☐	☐	☐
12. Emergency standby equipment needed?	☐	☐	☐
Venting			
1. Relief valves or rupture disks required?	☐	☐	☐
2. Materials of construction corrosion resistant?	☐	☐	☐
3. Vents properly designed? (Size, direction, configuration?)	☐	☐	☐
4. Flame arrestors required on vent lines?	☐	☐	☐
5. Relief valves protected from plugging by rupture disks?	☐	☐	☐
6. Telltale pressure gauges installed between rupture disks and relief valve?	☐	☐	☐
Instrument and Electrical			
1. All controls fail safe?	☐	☐	☐
2. Dual indication of process variables necessary?	☐	☐	☐
3. All equipment properly labelled?	☐	☐	☐
4. Tubing runs protected?	☐	☐	☐
5. Safeguards provided for process control when an instrument must be taken out of service?	☐	☐	☐
6. Process safety affected by response lag?	☐	☐	☐
7. Labels for all start-stop switches?	☐	☐	☐
8. Equipment designed to permit lockout protection?	☐	☐	☐
9. Electrical failures cause unsafe conditions?	☐	☐	☐
10. Sufficient lighting for both outside and inside operations?	☐	☐	☐
11. Lights provided for all sight glasses, showers and eyebaths?	☐	☐	☐
12. Breakers adequated for circuit protection?	☐	☐	☐
13. All equipment grounded?	☐	☐	☐

Figure 10-2 (continued)

	Completed ↓	Does not apply ↓	Further study required ↓
14. Special interlocks needed for safe operation?	☐	☐	☐
15. Emergency standby power on lighting equipment required?	☐	☐	☐
16. Emergency escape lighting required during power failure?	☐	☐	☐
17. All necessary communications equipment provided?	☐	☐	☐
18. Emergency disconnect switches properly marked?	☐	☐	☐
19. Special explosion proof electrical fixtures required?	☐	☐	☐
Safety Equipment			
1. Fire extinguishers required?	☐	☐	☐
2. Special respiratory equipment required?	☐	☐	☐
3. Diking material required?	☐	☐	☐
4. Colorimetric indicator tubes required?	☐	☐	☐
5. Flammable vapor detection apparatus required?	☐	☐	☐
6. Fire extinguishing materials compatible with process materials?	☐	☐	☐
7. Special emergency procedures and alarms required?	☐	☐	☐
Raw Materials			
1. Any materials and products require special handling equipment?	☐	☐	☐
2. Any raw materials and products affected by extreme weather conditions?	☐	☐	☐
3. Any products hazardous from a toxic or fire standpoint?	☐	☐	☐
4. Proper containers being used?	☐	☐	☐
5. Containers properly labelled for toxicity, flammability, stability, etc?	☐	☐	☐
6. Consequences of bad spills considered?	☐	☐	☐
7. Special instructions needed for containers or for storage and warehousing by distributors?	☐	☐	☐
8. Does warehouse have operating instructions covering each product regarded as critical?	☐	☐	☐

Figure 10-2 (continued)

atized approach using a rating form, similar to an income tax form. The final rating number provides a relative ranking of the hazard. This approach also contains a mechanism for estimating the dollar loss in the event of an accident.

The Dow Index is designed for rating the relative hazards with the storage, handling, and processing of explosive and flammable materials. While this limits its application, similar systems have been developed for other types of hazards.[3]

The main idea of this procedure is to provide a purely systematic approach, mostly independent of judgmental factors, for determining the relative magnitude of the hazards in a chemical plant. The main forms used for the computations are shown in Figures 10-3 and 10-4.

The procedure begins with a material factor which is a function only of the type of chemical or chemicals used. This factor is adjusted for general and special process hazards. These adjustments or penalties are based on conditions such as storage above the flash or boiling point, endo- or exothermic reactions, fired heaters, and the like. Credits for various safety systems and procedures are used for estimating the consequences of the hazard, after the fire and explosion index has been determined.

The form, shown in Figure 10-3, consists of three columns of numbers. The first column is the penalty column. Penalties for various unsafe situations are placed in this column. The second column contains the penalty actually used. This allows for a reduction or increase in the penalty based on extenuating circumstances not completely covered by the form. In the event of uncertainty here, the complete penalty value from the first column is used. The final column is used for computation.

The first step in the procedure is to conceptually divide the process into separate process units. A process unit is a single pump, a reactor, or a storage tank. A large process results in hundreds of individual units. It is not practical to apply the fire and explosion index to all of these units. The usual approach is to select only the units that experience shows to have the highest likelihood of a hazard. A process safety checklist or hazards survey is frequently used to select the most hazardous units for further analysis.

The next step is to determine the material factor (MF) for use in the form shown in Figure 10-3. Table 10-1 lists material factors for a number of important compounds. This list also includes data on heat of combustion, flash and boiling point temperatures. The additional data is also used in the computation of the Dow Index. A procedure is provided in the complete Index for computing the material factor for other compounds not listed in Table 10-1 or provided in the Dow reference.

In general, the higher the value of the material factor, the more flammable and/or explosive the material. If mixtures of materials are used, the material factor is determined from the properties of the mixture. The highest value of the MF under the complete range of operating conditions is suggested. The resulting MF value for the process is written in the space provided at the top of the form in Figure 10-3.

[3]Robert A. Smith, *Chemical Exposure Index,* The Dow Chemical Company, November, 1986.

EXHIBIT A

FIRE AND EXPLOSION INDEX DOW

LOCATION	DATE

PLANT	PROCESS UNIT	EVALUATED BY	REVIEWED BY

MATERIALS AND PROCESS

MATERIALS IN PROCESS UNIT

STATE OF OPERATION	BASIC MATERIAL(S) FOR MATERIAL FACTOR
☐ START-UP ☐ SHUT-DOWN ☐ NORMAL OPERATION	

	PENALTY	PENALTY USED	
MATERIAL FACTOR (SEE TABLE I OR APPENDICES A OR B) Note requirements when unit temperature over 140 F) →			
1. GENERAL PROCESS HAZARDS	PENALTY	PENALTY USED	
BASE FACTOR →	1.00	1.00	
A. EXOTHERMIC CHEMICAL REACTIONS (FACTOR .30 to 1.25)			
B. ENDOTHERMIC PROCESSES (FACTOR .20 to .40)			
C. MATERIAL HANDLING & TRANSFER (FACTOR .25 to 1.05)			
D. ENCLOSED OR INDOOR PROCESS UNITS (FACTOR .25 to .90)			
E. ACCESS	.35		
F. DRAINAGE AND SPILL CONTROL (FACTOR .25 to .50) ______ Gals.			
GENERAL PROCESS HAZARDS FACTOR (F_1) →			
2. SPECIAL PROCESS HAZARDS			
BASE FACTOR →	1.00	1.00	
A. TOXIC MATERIAL(S) (FACTOR 0.20 to 0.80)			
B. SUB-ATMOSPHERIC PRESSURE (< 500 mm Hg)	.50		
C. OPERATION IN OR NEAR FLAMMABLE RANGE ☐ INERTED ☐ NOT INERTED			
1. TANK FARMS STORAGE FLAMMABLE LIQUIDS	.50		
2. PROCESS UPSET OR PURGE FAILURE	.30		
3. ALWAYS IN FLAMMABLE RANGE	.80		
D. DUST EXPLOSION (FACTOR .25 to 2.00) (SEE TABLE II)			
E. PRESSURE (SEE FIGURE 2) OPERATING PRESSURE ____ psig RELIEF SETTING ____ psig			
F. LOW TEMPERATURE (FACTOR .20 to .30)			
G. QUANTITY OF FLAMMABLE/UNSTABLE MATERIAL: QUANTITY ______ lbs., H_c = ______ BTU/lb			
1. LIQUIDS, GASES AND REACTIVE MATERIALS IN PROCESS (SEE FIG. 3)			
2. LIQUIDS OR GASES IN STORAGE (SEE FIG. 4)			
3. COMBUSTIBLE SOLIDS IN STORAGE, DUST IN PROCESS (SEE FIG. 5)			
H. CORROSION AND EROSION (FACTOR .10 to .75)			
I. LEAKAGE – JOINTS AND PACKING (FACTOR .10 to 1.50)			
J. USE OF FIRED HEATERS (SEE FIG. 6)			
K. HOT OIL HEAT EXCHANGE SYSTEM (FACTOR .15 to 1.15) (SEE TABLE III)			
L. ROTATING EQUIPMENT	.50		
SPECIAL PROCESS HAZARDS FACTOR (F_2) →			
UNIT HAZARD FACTOR ($F_1 \times F_2 = F_3$) →			
FIRE AND EXPLOSION INDEX ($F_3 \times MF$ = F & EI) →			

Figure 10-3 Form used in the Dow Fire and Explosion Index computation. (*Dow's Fire and Explosion Index Hazard Classification Guide,* 6th ed., 1987. Reproduced by permission of the American Institute of Chemical Engineers.)

EXHIBIT B

LOSS CONTROL CREDIT FACTORS

1. Process Control (C_1)

a) Emergency Power	.98	f) Inert Gas	.94 to .96
b) Cooling	.97 to .99	g) Operating Instructions/ Procedures	.91 to .99
c) Explosion Control	.84 to .98		
d) Emergency Shutdown	.96 to .99	h) Reactive Chemical Review	.91 to .98
e) Computer Control	.93 to .99		

C_1 Total____________ *

2. Material Isolation (C_2)

a) Remote Control Valves	.96 to .98	c) Drainage	.91 to .97
b) Dump/Blowdown	.96 to .98	d) Interlock	.98

C_2 Total____________ *

3. Fire Protection (C_3)

a) Leak Detection	.94 to .98	f) Sprinkler Systems	.74 to .97
b) Structural Steel	.95 to .98	g) Water Curtains	.97 to .98
c) Buried Tanks	.84 to .91	h) Foam	.92 to .97
d) Water Supply	.94 to .97	i) Hand Extinguishers/Monitors	.95 to .98
e) Special Systems	.91	j) Cable Protection	.94 to .98

C_3 Value____________ *

Credit Factor = $C_1 \times C_2 \times C_3$ = __________ Enter on Line D Below

UNIT ANALYSIS SUMMARY

A-1. F & EI	________	
A-2. Radius of Exposure	________ ft.	
A-3. Value of Area of Exposure		$MM________
B. Damage Factor	________	
C. Base MPPD (A-3 X B)		$MM________
D. Credit Factor	________	
E. Actual MPPD (C X D)		$MM________
F. Days Outage (MPDO)		________ days.
G. Business Interruption Loss (BI)		$MM________

* Product of all factors used.

Figure 10-4 Form used for consequences analysis. (*Dow's Fire and Explosion Index Hazard Classification Guide,* 6th ed., 1987. Reproduced by permission of the American Institute of Chemical Engineers.)

The next step is to determine the general process hazards. Penalties are applied for the following factors.

1. Exothermic reactions that might self-heat.
2. Endothermic reactions that could react due to an external heat source such as a fire.

TABLE 10-1 SELECTED DATA FOR THE DOW FIRE AND EXPLOSION INDEX.*

Compound	Material factor	Heat of combustion BTU/lb × 10^{-3}	Flash point deg. F	Boiling point deg. F
Acetone	16	12.3	−4	133
Acetylene	40	20.7	Gas	−118
Benzene	16	17.3	12	176
Bromine	1	0.0	—	—
Butane	21	19.7	Gas	31
Calcium Carbide	24	9.1	—	—
Carbon Monoxide	16	4.3	Gas	−314
Chlorine	1	0.0	—	—
Cyclohexane	16	18.7	−4	179
Cyclohexanol	4	15.0	154	322
Diesel Fuel	10	18.7	100-130	315
Ethane	21	20.4	Gas	−128
Fuel Oil #1 to #6	10	18.7	100-150	304-574
Gasoline	16	18.8	−45	100-400
Hydrogen	21	51.6	Gas	−422
Methane	21	21.5	Gas	−259
Methanol	16	8.6	52	147
Mineral Oil	4	17.0	380	680
Nitroglycerine	40	7.8		
Octane	16	20.5	56	258
Pentane	21	19.4	−40	97
Petroleum - Crude	16	21.3	20-90	—
Propylene	21	19.7	Gas	−53
Toluene	16	17.4	40	231
Vinyl Chloride	21	8.0	Gas	7
Xylene	16	17.6	81	292

*Selected from *Dow's Fire and Explosion Index Hazard Classification Guide* (New York: American Institute of Chemical Engineers, 1987).

3. Material handling and transfer, including pumping and connection of transfer lines.
4. Enclosed process units preventing dispersion of escaped vapors.
5. Limited access for emergency equipment.
6. Poor drainage of flammable materials away from the process unit.

Penalties for special process hazards are determined next.

1. Toxic materials.
2. Less than atmospheric pressure operation with a risk of outside air entering.

3. Operation in or near the flammable limits.
4. Dust explosion risks.
5. Higher than atmospheric pressure.
6. Low temperature operation with potential embrittlement of carbon steel vessels.
7. Quantity of flammable material.
8. Corrosion and erosion of process unit structures.
9. Leakage around joints and packings.
10. Use of fired heaters, providing a ready ignition source.
11. Hot oil heat exchange systems where the hot oil is above its ignition temperature.
12. Large rotating equipment, including pumps and compressors.

Detailed instructions and correlations for determining the general and special process hazards are provided in the complete Dow Fire and Explosion Index.

The general process hazard factor (F_1) and special process hazard factor (F_2) are multiplied together to produce a unit hazard factor (F_3). The fire and explosion index (F&EI) is computed by multiplying the unit hazard factor by the material factor. Table 10-2 provides the degree of hazard based on the index value.

The Dow Fire and Explosion Index (F&EI) can be used to determine the consequences of an accident. This includes the maximum probable property damage (MPPD) and the maximum probable days outage (MPDO).

The consequences analysis is completed using the worksheet form shown in Figure 10-4. The computations are completed in the Risk Analysis Summary table at the bottom of the form. The damage radius is first estimated using a correlation published in the complete Dow Index. This correlation is based on the previously determined Fire and Explosion Index (F&EI). The dollar value of the equipment within this radius is determined. Next, a damage factor (based on a correlation provided) is applied to the fraction of the equipment actually damaged by the explosion or fire. Finally, a credit factor is applied based on safety systems. The final number, in dollars, is called the Maximum Probable Property Damage (MPPD) value. This number is used to estimate the Maximum Probable Days Outage using

TABLE 10-2 DETERMINING THE DEGREE OF HAZARD FROM THE DOW FIRE AND EXPLOSION INDEX

Dow Fire and Explosion Index	Degree of hazard
1 - 60	Light
61 - 96	Moderate
97 - 127	Intermediate
128 - 158	Heavy
159 - up	Severe

a correlation. Details on the procedure are available in the complete Dow reference.

Example 10-1

A gasoline filling station contains a 10,000 gallon tank of gasoline with four gas pumps. The gasoline tank is stored underground. The tank is vented to the air. Proper vents with flame arrestors are provided to allow air to move into or out of the tank depending on filling or emptying operations.

The underground tank is fabricated of steel. Some deterioration of the tank walls by corrosion is expected.

The station has both fixed CO_2 fire extinguishing equipment and portable hand-held equipment. A remote control in the station house enables the gasoline flow from the tanks to be stopped in the event of an emergency.

Use the Dow Fire and Explosion Index to determine the relative hazard of this operation.

Solution The Dow Fire and Explosion Index calculation is shown in Figure 10-5.

The material factor (MF) is found directly in Table 10-1. The value for gasoline is 16.

Under general process hazards, only item C involving material handling and transfer is applicable. This gives a penalty of 0.50 per instructions found in the Dow Index. The storage tank is not considered an enclosed unit, although it is buried. This is more applicable to units enclosed in buildings. Furthermore, the access penalty is not applicable for the same reason. The total General Process Hazard Factor (F_1) is computed by summing all of the General Process Hazards. The resulting value is 1.50, as shown in Figure 10-5.

The special process hazards are considered next. Gasoline is toxic and is given a toxic materials penalty of 0.20 (Item A) according to instructions with the Index. The gasoline is stored at ground temperature, above its flash point of $-45°F$ (see Table 10-1). It is not above its boiling point or its autoignition point. The gasoline is stored with potentially flammable vapor above the liquid in the tank. A penalty of 0.50 is applied for this (Item C1). The tank and the pumping operations are always operated in the flammable range. A penalty of 0.80 is used for Item C3.

A penalty is also assessed for the quantity of flammable material in storage. The penalty is based on the total energy content of the material stored. This is covered under Item G2. Ten thousand gallons of gasoline with a specific gravity of 0.7 represents

$$(10{,}000 \text{ gallons})(1 \text{ ft}^3/7.48 \text{ gal})(62.4 \text{ lb}_m/\text{ft}^3)(0.7) = 58{,}400 \text{ lb}_m \text{ of gasoline}$$

The gasoline has a heat of combustion of 18,800 BTU/lb_m (Table 10-1). The total energy value for the gasoline in storage is:

$$(58{,}400 \text{ lbm})(18{,}800 \text{ BTU/lb}_m) = 1.10 \times 10^9 \text{ BTU}$$

Figure 4 in the Dow reference[4] is used to compute the penalty for this quantity of energy. Correlation B is selected because the gasoline has a flash point below 37.8°C. The result is a penalty of 0.40.

Finally, a corrosion factor is included under Item H. This is estimated at the low end, or 0.10.

The Special Hazard Factor (F_2) is the sum of all of the individual Special Hazard values. The resulting value is 3.00.

[4]*Dow Fire and Explosion Index,* p. 25.

EXHIBIT A

FIRE AND EXPLOSION INDEX DOW

LOCATION	DATE

PLANT	PROCESS UNIT	EVALUATED BY	REVIEWED BY
FILLING STATION			

MATERIALS AND PROCESS

MATERIALS IN PROCESS UNIT

10,000 GALS. OF GASOLINE

STATE OF OPERATION	BASIC MATERIAL(S) FOR MATERIAL FACTOR
[ ] START-UP [ ] SHUT-DOWN [ ] NORMAL OPERATION	

MATERIAL FACTOR (SEE TABLE I OR APPENDICES A OR B) Note requirements when unit temperature over 140 F) →			16
1. GENERAL PROCESS HAZARDS	PENALTY	PENALTY USED	
BASE FACTOR →	1.00	1.00	
A. EXOTHERMIC CHEMICAL REACTIONS (FACTOR .30 to 1.25)			
B. ENDOTHERMIC PROCESSES (FACTOR .20 to .40)			
C. MATERIAL HANDLING & TRANSFER (FACTOR .25 to 1.05)		0.50	
D. ENCLOSED OR INDOOR PROCESS UNITS (FACTOR .25 to .90)			
E. ACCESS	.35		
F. DRAINAGE AND SPILL CONTROL (FACTOR .25 to .50) ______ Gals.			
GENERAL PROCESS HAZARDS FACTOR (F_1) →		1.50	
2. SPECIAL PROCESS HAZARDS			
BASE FACTOR →	1.00	1.00	
A. TOXIC MATERIAL(S) (FACTOR 0.20 to 0.80)		0.20	
B. SUB-ATMOSPHERIC PRESSURE (< 500 mm Hg)	.50		
C. OPERATION IN OR NEAR FLAMMABLE RANGE [ ] INERTED [ ] NOT INERTED			
1. TANK FARMS STORAGE FLAMMABLE LIQUIDS	.50	0.50	
2. PROCESS UPSET OR PURGE FAILURE	.30		
3. ALWAYS IN FLAMMABLE RANGE	.80	0.80	
D. DUST EXPLOSION (FACTOR .25 to 2.00) (SEE TABLE II)			
E. PRESSURE (SEE FIGURE 2) OPERATING PRESSURE ______ psig RELIEF SETTING ______ psig			
F. LOW TEMPERATURE (FACTOR .20 to .30)			
G. QUANTITY OF FLAMMABLE/UNSTABLE MATERIAL: QUANTITY ______ lbs., H_c = ______ BTU/lb			
1. LIQUIDS, GASES AND REACTIVE MATERIALS IN PROCESS (SEE FIG. 3)			
2. LIQUIDS OR GASES IN STORAGE (SEE FIG. 4)		0.40	
3. COMBUSTIBLE SOLIDS IN STORAGE, DUST IN PROCESS (SEE FIG. 5)			
H. CORROSION AND EROSION (FACTOR .10 to .75)		0.10	
I. LEAKAGE – JOINTS AND PACKING (FACTOR .10 to 1.50)			
J. USE OF FIRED HEATERS (SEE FIG. 6)			
K. HOT OIL HEAT EXCHANGE SYSTEM (FACTOR .15 to 1.15) (SEE TABLE III)			
L. ROTATING EQUIPMENT	.50		
SPECIAL PROCESS HAZARDS FACTOR (F_2) →		3.00	
UNIT HAZARD FACTOR ($F_1 \times F_2 = F_3$) →		4.50	
FIRE AND EXPLOSION INDEX ($F_3 \times MF$ = F & EI) →			72.0

Figure 10-5 The Dow Fire and Explosion Index applied to a gasoline filling station example.

The Unit Hazard Factor is determined by multiplying the General Process Hazard (F_1) times the Special Hazard Factor (F_2). The result is a value of 4.50 for the Unit Hazard Factor (F_3).

Finally, the Fire and Explosion Index (F&EI) is computed. The Unit Hazard Factor (F_3) is multiplied by the material factor to give 72.0. Table 10-2 indicates this process is moderately hazardous.

Hazards surveys are suitable for identifying hazards associated with equipment design, layout, material storage, and so forth. They are not suitable for identifying hazards due to improper operation or upset conditions. On the other hand, this approach is fairly rigorous, requires little experience, is easy to apply, and provides a quick result.

10-3 HAZARD AND OPERABILITY STUDIES (HAZOP)

The hazard and operability study is a formal procedure used to identify hazards in a chemical process facility.[5] The procedure is very effective in identifying hazards. However, this approach provides little information on risk and consequences. As a result, this approach tends to "goldplate" a process design, resulting in a large number of hazards being identified. Many of these hazards may have low probability or consequence. However, with some experience the method is used very effectively.

The basic idea is to "let the mind go free" in a controlled fashion in order to consider all of the possible ways that process failures can occur.

The full HAZOP study is completed by a committee composed of a cross-section of experienced plant and lab people. One individual must be assigned the task of recording the results. Applied in its fullest sense, a complete HAZOP study would require a large investment in time and effort by the people involved.

The procedure for a HAZOP study is to apply a number of guide words to various parts of the process design intention. The design intention tells what the process is expected to do. These guide words are shown in Table 10-3. Frequently the application of the guide words results in a deviation that is not meaningful, is not possible, or is trivial. These cases are simply indicated as such.

The guide words AS WELL AS, PART OF, and OTHER THAN are conceptually difficult to apply. AS WELL AS means that something else happens in addition to the normal intention. This could be boiling of a liquid, transfer of some additional component, or the transfer of the fluid somewhere else than expected. PART OF means that one of the components is missing or the stream is being preferentially pumped to only a part of the process. OTHER THAN applies to situations where a material is substituted for the expected material, is transferred somewhere else or the material solidifies and cannot be transported.

An important aspect of the procedure is determining how to divide the process into small process units. Each unit should reflect a particular process intention. For instance, for an air supply the intention might be to provide a pneumatic driving force for the process control valves. The air supply would be considered a

[5] *Guidelines for Hazard Evaluation Procedures* (New York: American Institute of Chemical Engineers, 1985).

TABLE 10-3 A LIST OF GUIDE WORDS USED FOR THE HAZOP PROCEDURE

Guide words	Meanings
NO or NOT	The complete negation of the intention.
MORE LESS	Quantitative increases or decreases
AS WELL AS PART OF	A qualitative increase A qualitative decrease
REVERSE	The logical opposite of the intention
OTHER THAN	Complete substitution

process unit for this case. This aspect requires experience and familiarity with the process. The HAZOP approach is applied to each process unit and every process pipeline into or out of a unit. The investigation proceeds through the process units until the entire process has been completed.

Since the HAZOP method is very tedious, many companies use this technique only for very hazardous operations with larger units. In general, however, the method is being increasingly used.

Example 10-2

Consider the reactor system shown in Figure 10-6. The reaction is exothermic. A cooling system is provided to remove the excess energy of reaction. In the event the cooling function is lost, the temperature of the reactor would increase. This would lead to an increase in reaction rate leading to additional energy release. The result could be a runaway reaction with pressures exceeding the bursting pressure of the reactor.

The temperature within the reactor is measured and is used to control the cooling water flow rate by a valve.

Perform a HAZOP on this unit to improve the safety of the process.

Solution The guide words are applied to the cooling coil system. The design intention is cooling. The results of the investigation are shown in Table 10-4.

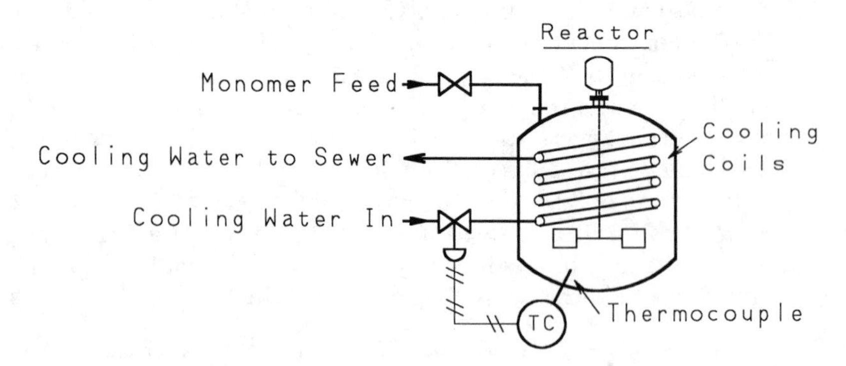

Figure 10-6 Temperature control of an exothermic reactor.

TABLE 10-4 HAZOP STUDY ON DEVIATIONS FROM COOLING FLOW. APPLIED TO REACTOR OF FIGURE 10-6

Guide word	Deviation	Possible causes	Consequences	Action
NO	No cooling	1. Control valve fails closed 2. Plugged cooling line 3. Cooling water service failure 4. Controller fails and closes valve. 5. Air pressure to drive valve fails, closing valve	1. Temperature increase in reactor 2. Possible thermal runaway	1. Install back-up control valves, or manual bypass valve. 2. Install filters to prevent debris from entering line 3. Install back-up cooling water source 4. Install back-up controller 5. Install control valve that fails open 6. Install high temperature alarm to alert operator 7. Install high temperature emergency shutdown 8. Install cooling water flow meter and low flow alarm
MORE	More cooling flow	1. Control valve fails to open 2. Controller fails and opens valve	1. Reactor cools, reactant builds-up, possible runaway on heating	1. Instruct operators on procedure
LESS	Less cooling flow	1. Control valve fails to respond 2. Partially plugged cooling line 3. Partial water source failure	1. Covered under "NO"	1. Covered under "NO"
AS WELL AS	Cooling water in reactor	Leak in cooling coils, pressure in reactor less than pressure in coils	1. Dilution of contents 2. Product ruined 3. Overfilling of reactor	1. Install high level and/or pressure alarm 2. Install proper relief 3. Check maintenance procedure and schedule

TABLE 10-4 HAZOP STUDY ON DEVIATIONS FROM COOLING FLOW. APPLIED TO REACTOR OF FIGURE 10-6 (continued)

Guide word	Deviation	Possible causes	Consequences	Action
AS WELL AS	Reactor product in coils	Leak in coils with reactor pressure greater than coil pressure	1. Product lost thru coils 2. Loss of product yield 3. Reduction in cooling function 4. Possible contamination of water	1. Check maintenance procedure and schedules 2. Install upstream check valve in cooling water source
PART OF	Partial cooling flow	Covered under "LESS COOLING FLOW"		
REVERSE	Reverse cooling flow	1. Failure of water source resulting in backward flow 2. Backflow due to backpressure	1. Improper cooling, possible runaway	1. Install check valve in cooling water line 2. Install high temperature alarm to alert operator
OTHER THAN	Another material besides cooling water	1. Water source contaminated 2. Backflow from sewer	1. Possible loss of cooling with possible runaway	1. Isolation of cooling water source 2. Install check valve to prevent reverse flow 3. Install high temperature alarm

The potential process modifications resulting from this study are the following.

- Installation of a high temperature alarm to alert the operator in the event of cooling function loss.
- Installation of a high temperature shutdown system. This system would automatically shutdown the process in the event of a high reactor temperature. The shutdown temperature would be higher than the alarm temperature to provide the operator with the opportunity to restore cooling before the reactor is shutdown.
- Installation of a check valve in the cooling line to prevent reverse flow. A check valve could be installed both before and after the reactor to prevent the reactor contents from flowing upstream and to prevent the backflow in the event of a leak in the coils.
- Periodically inspect the cooling coil to insure its integrity.

- Study of the cooling water source to consider possible contamination and interruption of supply.
- Installation of a cooling water flow meter and low flow alarm. This will provide an immediate indication of cooling loss.

In the event that the cooling water system fails (regardless of the source of the failure), the high temperature alarm and emergency shutdown system prevents a runaway. The review committee performing the HAZOP decided that the installation of a backup controller and control valve was not essential. The high temperature alarm and shutdown system prevents a runaway in this event. Similarly, a loss of coolant water source or a plugged cooling line would be detected by either the alarm or emergency shutdown system. The review committee suggested that all coolant water failures be properly reported. In the event that a particular cause occurs repeatedly then additional process modifications are warranted.

This example demonstrates that the number of process changes suggested is quite numerous although only a single process intention is considered.

The advantage to this approach is that it provides a more complete identification of the hazards, including information on how hazards can develop as a result of operating procedures and operational upsets in the process. The disadvantages are that the HAZOP approach is tedious to apply, requires considerable manpower and time, and can potentially identify all of the hazards independent of the risk.

10-4 SAFETY REVIEWS

Another method which is commonly used to identify safety problems in laboratory and process areas and to develop solutions is the safety review. There are two types of safety reviews: the informal and formal.

The *informal safety review* is used for

- Small changes to existing processes, and
- Small, bench scale or laboratory processes.

The informal safety review procedure usually involves just two or three people. It includes the individual responsible for the process and one or two others not directly associated with the process but experienced with proper safety procedures. The idea is to provide a lively dialogue where ideas can be exchanged and safety improvements developed.

The reviewers simply meet in an informal fashion to examine the process equipment and operating procedures and to offer suggestions on how the safety of the process might be improved. Significant improvements should be summarized in a memo for others to reference in the future. The improvements must be implemented before the process is operated.

Example 10-3

Consider the laboratory reactor system shown in Figure 10-7. This system is designed to react phosgene, $COCl_2$, with aniline to produce isocyanate and HCl. The reaction is shown in Figure 10-8. The isocyanate is used for the production of foams and plastics.

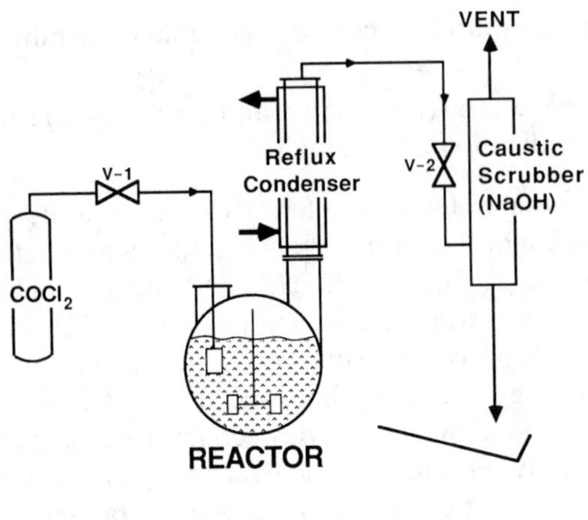

Figure 10-7 Original design of phosgene reactor prior to informal safety review.

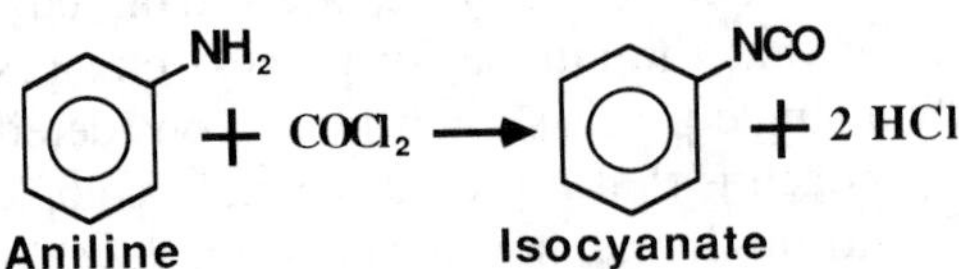

Figure 10-8 Reaction stoichiometry for phosgene reactor.

Phosgene is a colorless vapor with a boiling point of 46.8°F. Thus, it is normally stored as a liquid in a container under pressure above its normal boiling point temperature. The TLV for phosgene is 0.1 ppm and its odor threshold is 0.5 − 1 ppm, well above the TLV.

Aniline is a liquid with a boiling point of 364°F. Its TLV is 2 ppm. It is absorbed through the skin.

In the process shown in Figure 10-7, the phosgene is fed from the container through a valve into a fritted glass bubbler in the reactor. The reflux condenser condenses aniline vapors and returns them to the reactor. A caustic scrubber is used to remove the phosgene and HCl vapors from the exit vent stream. The complete process is contained in a hood.

Conduct an informal safety review on this process.

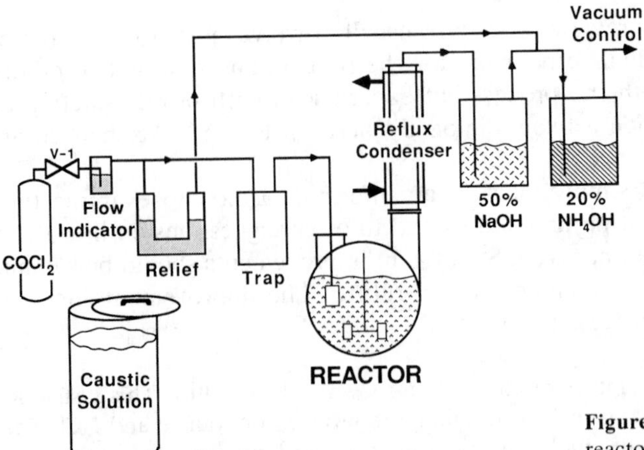

Figure 10-9 Final design of phosgene reactor after informal safety review.

Solution The safety review was completed by two individuals. The final process design is shown in Figure 10-9. The changes and additions to the process are as follows.

1. Vacuum to reduce boiling temperature.
2. Relief system with outlet to scrubber to prevent hazards due to plugged fritted glass bubbler.
3. Flow indicator provides visual indication of flow.
4. Bubblers are used instead of scrubbers since they are more effective.
5. Ammonium hydroxide bubbler is more effective for absorbing phosgene.
6. Trap catches liquid phosgene.
7. Pail of caustic: The phosgene cylinder would be dumped into this pail in the event of a cylinder or valve leak. The caustic would absorb the phosgene.

In addition, the reviewers recommended the following: (1) Hang phosgene indicator paper around the hood, room, and operating areas. This paper is normally white but turns brown when exposed to 0.1 ppm of phosgene. (2) Use a safety checklist, daily, before the process is started. (3) Post an up-to-date process sketch near the process.

The *formal safety review* is used for

- New processes,
- Substantial changes in existing processes, and
- Processes which need an updated review.

The formal safety review is a three step procedure. This includes

- Preparation of a detailed formal safety review report,
- Committee review of the report and inspection of the process, and
- Implementation of the recommendations.

The formal safety review report includes the following sections.

I. Introduction
  A. Overview or summary: Provides a brief summary of the results of the formal safety review. This is done after the formal safety review is complete.
  B. Process overview or summary: Provides a brief description of the process with an emphasis on the major hazards in the operation.
  C. Reactions and stoichiometry: Provides the chemical reaction equations and stoichiometry.
  D. Engineering data: Provides operating temperatures, pressures, and relevant physical property data for the materials used.

II. Raw Materials and Products: Refers to specific hazards and handling problems associated with the raw materials and products. Discusses procedures to minimize these hazards.

III. Equipment Setup
  A. Equipment description: Describes the configuration of the equipment. Sketches of the equipment are provided.

B. Equipment specifications: Identifies the equipment by manufacturer name and model number. Provides the physical data and design information associated with this equipment.

IV. Procedures

A. Normal operating procedures: Describes how the process is operated.

B. Safety procedures: Provides a description of the unique concerns associated with the equipment and materials and specific procedures used to minimize the risk. This includes:

1. Emergency shutdown: Describes the procedure used to shut-down the equipment if an emergency should occur. This includes major leaks, reactor runaway, loss of electricity, water, air pressure, and the like.
2. Fail safe procedures: Examines the consequences of utility failures such as loss of steam, electricity, water, air pressure, or inert padding. Describes what to do for each case so the system fails safely.
3. Major release procedures: Describes what to do in the event of a major spill of toxic or flammable material.

C. Waste disposal procedure: Describes how toxic or hazardous materials are collected, handled and disposed.

D. Clean-up procedures: Describes how to clean the process after use.

V. Safety checklist: Provides the complete safety checklist for the operator to complete prior to operation of the process. This checklist is used prior to every startup.

VI. Material safety data sheets: Provided for each hazardous material used.

Example 10-4

A toluene water wash process is shown in Figure 10-10. This process is used to clean water soluble impurities from contaminated toluene. The separation is achieved with a Podbielniak centrifuge, or Pod, due to a difference in densities. The light phase (contaminated toluene) is fed to the periphery of the centrifuge and travels to the center. The heavy phase (water) is fed to the center and travels countercurrent to the toluene to the periphery of the centrifuge. Both phases are mixed within the centrifuge and separated countercurrently. The extraction is conducted at 190°F.

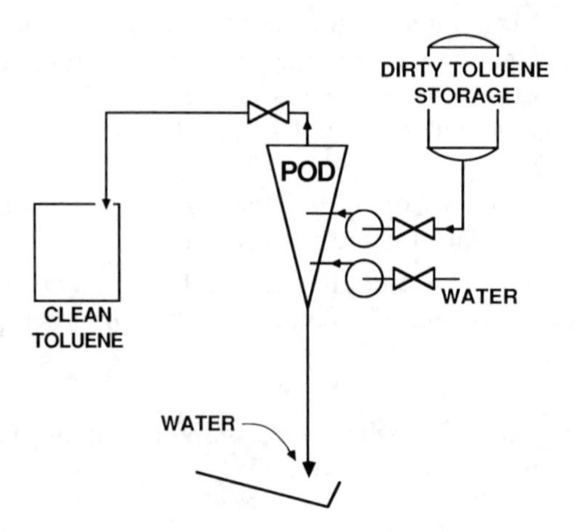

Figure 10-10 Toluene water wash process prior to formal safety review.

The contaminated toluene is fed from a storage tank into the Pod. The heavy liquid out (contaminated water) is sent to waste treatment and the light liquid out (clean toluene) is collected in a 55-gallon drum.

Perform a formal safety review on this process.

Solution The complete safety review report is provided in Appendix A. Figure 10-11 shows the modified process after the formal safety review has been completed. The significant changes or additions added as a result of the review are as follows.

1. Add gounding and bonding to all collection and storage drums, and process vessels.
2. Add inerting and purging to all drums.
3. Add elephant trunks at all drums to provide ventilation.
4. Provide dip legs in all drums to prevent the free fall of solvent resulting in the generation and accumulation of static charge.
5. Add a charge drum with grounding, bonding, inerting, and ventilation.
6. Provide a vacuum connection to the dirty toluene storage for charging.
7. Add a relief valve to the dirty toluene storage tank.
8. Add heat exchangers to all outlet streams to cool the exit solvents below their flash point. This must include temperature gauges to insure proper operation.
9. Provide a waste water collection drum to collect all waste water which might contain substantial amounts of toluene due to upset conditions.

Additional changes were made in the operating and emergency procedure. They included

1. Checking the room air periodically with colorimetric tubes to determine if any toluene vapors are present and
2. Changing the emergency procedure for spills to include (a) activating the spill alarm, (b) increasing the ventilation to high speed, and (c) throwing the sewer isolation switch to prevent solvent from entering the main sewer lines.

The formal safety review can be used almost immediately, is relatively easy to apply, and is known to provide good results. However, the committee participants

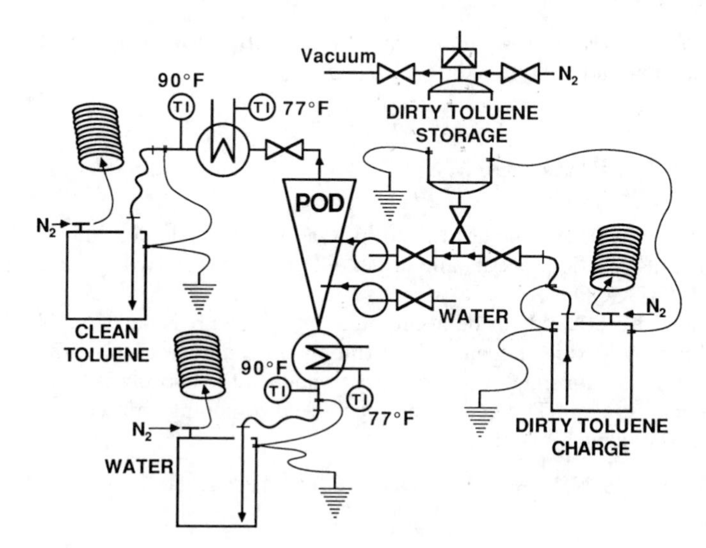

Figure 10-11 Toluene water wash process after formal safety review.

must be experienced in identifying safety problems. For less experienced committees, a more formal HAZOP may be more effective in identifying these hazards.

10-5 OTHER METHODS

Other methods that are available for identifying hazards are the following.

1. "What if" analysis: This less formal method of identifying hazards applies the words "what if" to a number of areas of investigation. For instance, the question might be "What if the flow stops?" The analysis team then decides what the potential consequences might be and how to solve any problems.
2. Human error analysis: This method is used to identify the parts and the procedures of a process that have a higher than normal probability of human error. Control panel layout is an excellent application for human error analysis since a control panel can be designed in such a fashion that human error is inevitable.
3. Failure mode, effects and criticality analysis (FMECA): This tabulates a list of equipment in the process along with all of the possible failure modes for each item. The effect of a particular failure is considered with respect to the process.

SUGGESTED READING

Dow's Fire and Explosion Index Hazard Classification Guide, 6th ed. (New York: American Institute of Chemical Engineers, 1987).

Guidelines for Hazard Evaluation Procedures (New York: American Institute of Chemical Engineers, 1985).

Trevor A. Kletz, *HAZOP and HAZAN* (Warwickshire, England: The Institution of Chemical Engineers, 1986).

Frank P. Lees, *Loss Prevention in the Process Industries* (London: Butterworths, 1986), Chapter 8: Hazard Identification and Safety Audit.

PROBLEMS

10-1. The hydrolysis of acetic anhydride is being studied in a laboratory scale continuous stirred tank reactor. In this reaction, acetic anhydride, $(CH_3CO)_2O$ reacts with water to produce acetic acid, CH_3COOH.

The concentration of acetic anhydride at any time in the CSTR is determined by titration with sodium hydroxide. Because the titration procedure requires time (relative to the hydrolysis reaction time) it is necessary to quench the hydrolysis reaction as soon as the sample is taken. The quenching is achieved by adding an excess of aniline to the sample. The quench reaction is

$$(CH_3CO)_2 + C_6H_5NH_2 \rightarrow CH_3COOH + C_6H_5NHCOCH_3$$

The quenching reaction also forms acetic acid, but in a different stoichiometric ratio than the hydrolysis reaction. Thus it is possible to determine the acetic anhydride concentration at the time the sample was taken.

The initial experimental design is shown in Figure P10-1. Water and acetic anhydride are gravity fed from reservoirs and through a set of rotameters. The water is mixed with the acetic anhydride just prior to entry into the reactor. Water is also circulated by a centrifugal pump from the temperature bath through coils in the reactor vessel. This maintains the reactor temperature at a fixed value. A temperature controller in the water bath maintains the temperature to within 1°F of desired.

Samples are withdrawn from the point shown and titrated manually in a hood.

a. Develop a safety checklist for use prior to operation of this experiment.
b. What safety equipment must be available?
c. Perform an informal safety review on the experiment. Suggest modifications to improve the safety.

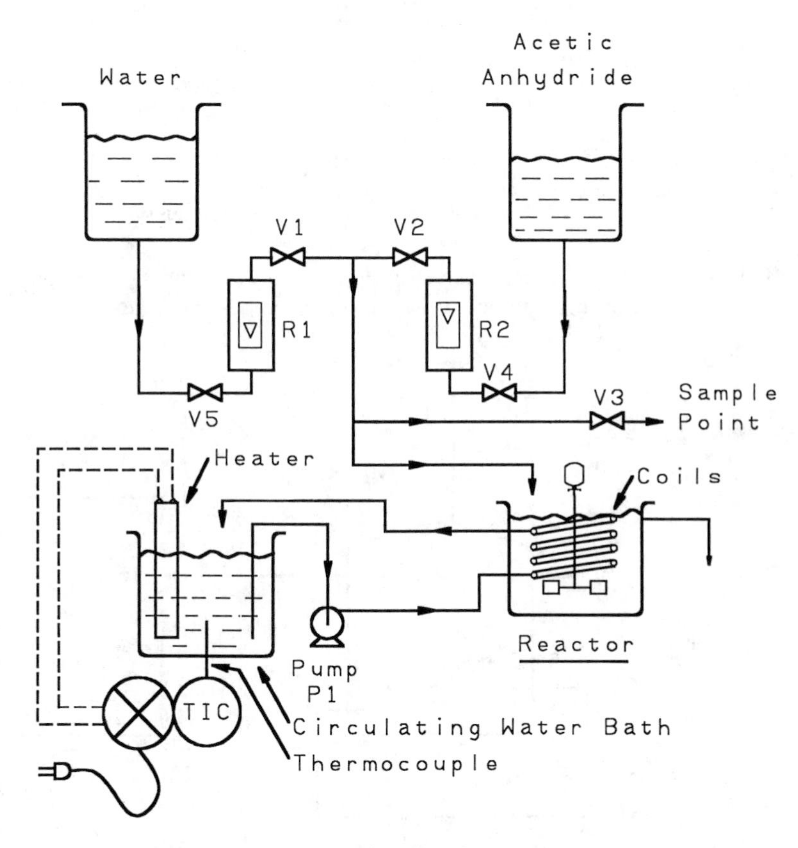

Figure P10-1 Acetic anhydride reactor system.

10-2. Perform a HAZOP study on the laboratory process of Problem 1. Consider the intention "reactant flow to reactor" for your analysis. What specific recommendations can you make to improve the safety of this experiment?

10-3. A heat exchanger is used to heat volatile solvents as shown in Figure P10-3. The temperature of the outlet stream is measured by a thermocouple and a controller/valve manipulates the amount of steam to the heat exchanger to achieve the desired setpoint temperature. Perform a HAZOP study on the intention "hot solvent from heat exchanger." Recommend possible modifications to improve the safety of this process.

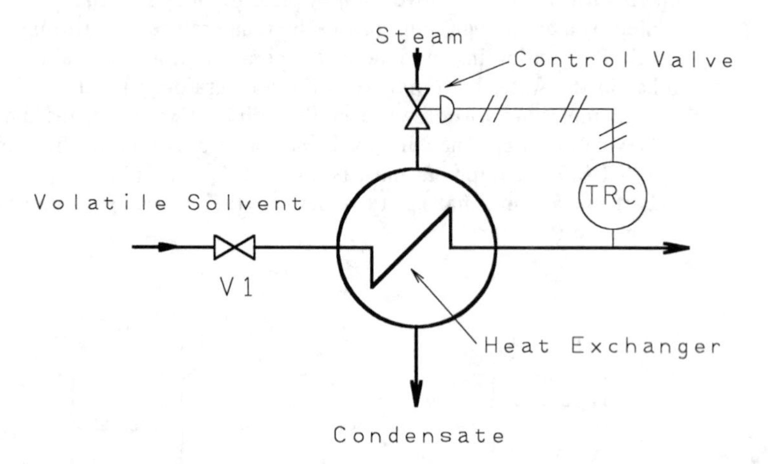

Figure P10-3 Volatile solvent heating system.

10-4. A gas-fired furnace is shown in Figure P10-4. The hot combustion gases pass through a heat exchanger to heat fresh air for space heating. The gas flow is controlled by an

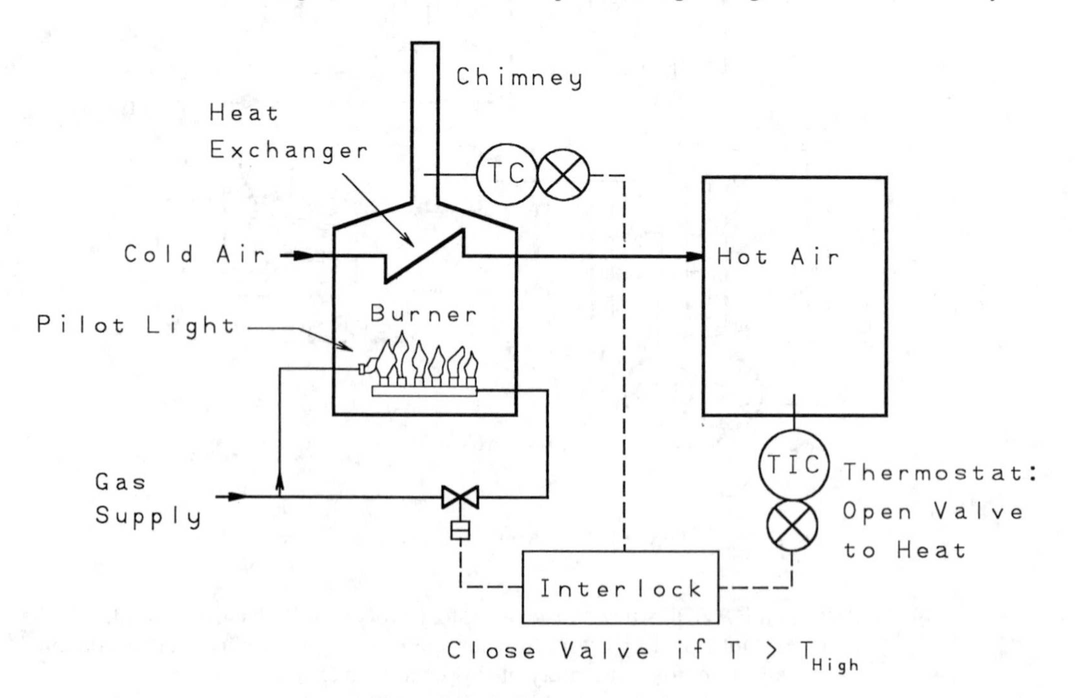

Figure P10-4 Furnace control system

electric solenoid valve connected to a thermostat. The gas is ignited by a pilot light flame. A high temperature switch shuts off all gas in the event of high temperature in the fresh air plenum.

a. Determine the various ways in which this system can fail, leading to excessive heating of the plenum and possible fire.
b. What type of valve (normally open or normally closed) is recommended for the gas supply?
c. What is the most likely failure mode?
d. A problem can also arise due to failure of the pilot light, leading to combustible gases in the furnace, heat exchanger and chimney. Suggest at least two ways to prevent this problem.

10-5. Beverage dispensers are notorious for either taking one's money or not delivering the proper beverage. Consider a beverage dispenser that delivers a paper cup, ice, and beverage (composed of syrup and water) in a sequential order. The machine also makes change.

Identify as many failure modes as possible. Use the HAZOP guidewords to identify additional possibilities.

10-6. World War II submarines used torpedo tubes with an outer and inner door. The torpedo was loaded into the tube from the torpedo room using the inner door. The inner door was then closed, the outer door opened and the torpedo launched.

One problem was insuring that the outer door was closed before the inner opened. Since no direct visible check was possible, a small pipe and valve was attached to the top of the torpedo tube in the torpedo room. Prior to opening the inner door, the valve was opened momentarily to check for the presence of pressurized water in the tube. The presence of pressurized water was a direct indication that the outer door was open.

Determine a failure mode for this system, leading to the inner door being opened when the outer door was open, resulting in flooding of the torpedo room and possible sinking of the sub.

10-7. Five process pumps are lined up in a row and numbered as shown in Figure P10-7. Can you identify the hazard? A similar layout led to a serious accident by a maintenence worker who was sprayed by hot solvent when he disconnected a pump line on the wrong pump. An accident like this might be attributed to human error but is really a hazard due to poor layout.

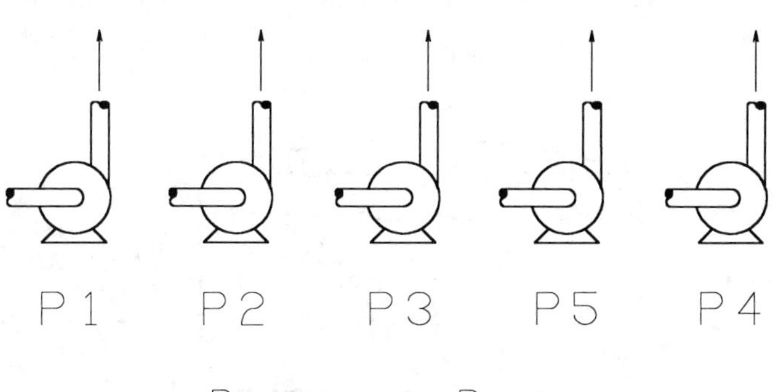

Figure P10-7 Pump layout.

10-8. A good acronym in chemical plant design is KISS—Keep It Simple, Stupid! This also applies to hazards. Complicated designs are almost always more hazardous than simple ones.

Figure P10-8 shows a sump designed to collect process fluids. The level controller and pump insure that the sump level is maintained below a maximum height. Can you suggest a much simpler system?

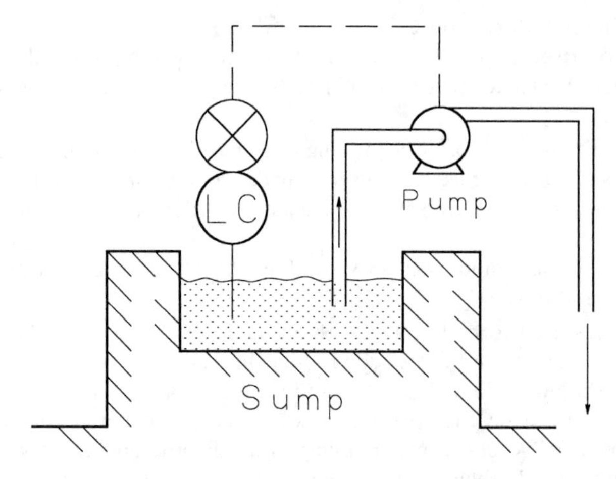

Figure P10-8 Sump level control system.

10-9. Storage tanks typically are not capable of withstanding much pressure or vacuum. Standard storage tanks are designed for a maximum of 2.5 inches of water gauge vacuum (0.1 psi) and about 6 inches of water gauge pressure (0.2 psi).

A welding operation was to occur on the roof of a storage vessel. The tank contained a flammable, volatile liquid. The roof was equipped with a vent pipe with a flame arrestor.

The foreman recognized a possible hazard due to flammable vapor escaping from the vent pipe and igniting on the sparks from the welding operation. He connected a hose to the vent at the top of the tank and ran the hose down to the ground. Since the flammable vapors were water soluble, he stuck the end of the hose in a drum full of water. During a subsequent operation which involved emptying the tank, an accident occurred. Can you explain what happened and how?

10-10. Figure P10-10 shows a storage tank blanketed with nitrogen. This configuration resulted in an explosion and fire due to loss of inert. Can you explain why?

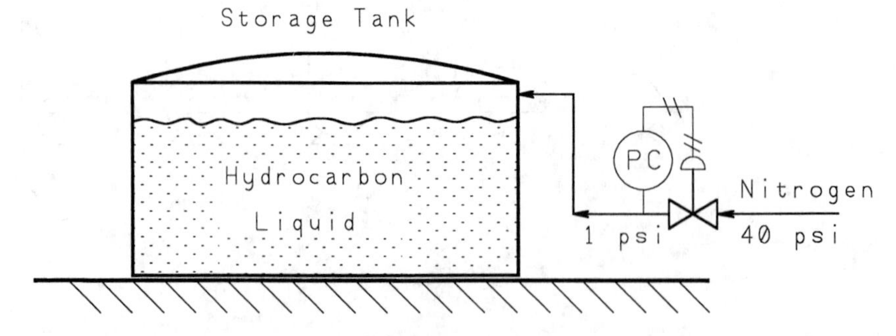

Figure P10-10 Nitrogen padding system for a storage tank.

10-11. Figure P10-11 shows two tanks in series, both with independent level controllers. This configuration will result in the lower tank inevitably overflowing. Can you explain why?

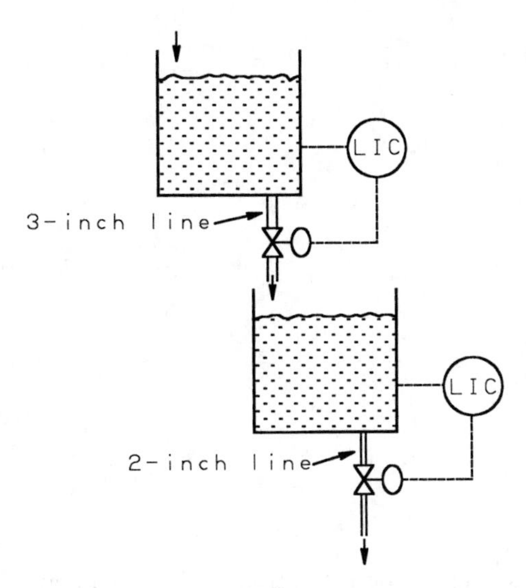

Figure P10-11 Level tanks in series.

10-12. Develop a safety checklist for the system described in Example 10-3 and shown in Figure 10-9. The intention of the checklist is to insure the system is safe prior to operation.

10-13. Prepare a formal safety review memo for the gas-fired furnace described in Problem 10-4 and shown in Figure P10-4. This memo will be given to each committee member prior to the formal safety review committee meeting.

10-14. Describe an informal safety review process for using a cylinder of phosgene to charge gaseous phosgene to a reactor. Review up to the reactor only.

11 Risk Assessment

Risk assessment includes scenario identification and consequence analysis. Scenario identification describes how an accident occurs. It frequently includes an analysis of the probabilities. Consequence analysis describes the damage expected. This includes loss of life, damage to the environment or capital equipment and days outage.

The hazards identification procedures presented in Chapter 10 include some aspects of risk assessment. The Dow Index includes a calculation of the maximum probable property damage (MPPD) and the maximum probable days outage (MPDO). This is a form of consequences analysis. However, these numbers are obtained by some rather simple calculations involving published correlations. Hazard and operability studies (HAZOP) provide information on how a particular accident occurs. This is a form of scenario identification. No probabilities or numbers are used with the typical HAZOP study, although the experience of the review committee is used to decide on an appropriate course of action.

This chapter will

- review probability mathematics, including the mathematics of equipment failure,
- show how the failure probabilities of individual hardware components contribute to the failure of a process, and
- describe two probabalistic methods: event and fault trees.

11-1 REVIEW OF PROBABILITY THEORY

Equipment failures or faults in a process occur as a result of a complex interaction of the individual components. The overall probability of a failure in a process is highly dependent on the nature of this interaction. This section will define the various types of interactions and describe how to perform failure probability computations.

Data is collected on the failure rate of a particular hardware component. With adequate data it can be shown that, on the average, the component fails after a certain period of time. This is called the average failure rate and is represented by μ with units of faults/time. The probability the component will not fail during the time interval $(0, t)$ is given by a Poisson distribution[1]

$$R(t) = e^{-\mu t} \tag{11-1}$$

where R is called the reliability. Equation 11-1 assumes a constant failure rate, μ. As $t \rightarrow \infty$ the reliability goes to 0. The speed at which this occurs is dependent on the value of the failure rate, μ. The higher the failure rate, the faster the reliability decreases. The complement of the reliability is called the failure probability (or sometimes the unreliability), P, and is given by

$$P(t) = 1 - R(t) = 1 - e^{-\mu t} \tag{11-2}$$

The failure density function is defined as the derivative of the failure probability,

$$f(t) = \frac{dP(t)}{dt} = \mu e^{-\mu t} \tag{11-3}$$

The area under the complete failure density function is unity.

The failure density function is used to determine the probability, P, of at least one failure in the time period t_o to t_1,

$$P(t_o \rightarrow t_1) = \int_{t_o}^{t_1} f(t)\, dt = \mu \int_{t_o}^{t_1} e^{-\mu t} dt = e^{-\mu t_o} - e^{-\mu t_1} \tag{11-4}$$

The integral represents the fraction of the total area under the failure density function between time t_o and t_1.

The time interval between two failures of the component is called the mean time between failures (MTBF) and is given by the first moment of the failure density function:

$$E(t) = \text{MTBF} = \int_0^{\infty} t f(t) dt = \frac{1}{\mu} \tag{11-5}$$

Typical plots of the functions, μ, f, P, and R are shown in Figure 11-1.

Equations 11-1 through 11-5 are only valid for a constant failure rate, μ. Many components exhibit a typical "bathtub" failure rate shown in Figure 11-2. The fail-

[1]B. Roffel and J. E. Rijnsdorp, *Process Dynamics, Control and Protection* (Ann Arbor, MI: Ann Arbor Science, 1982), p. 381.

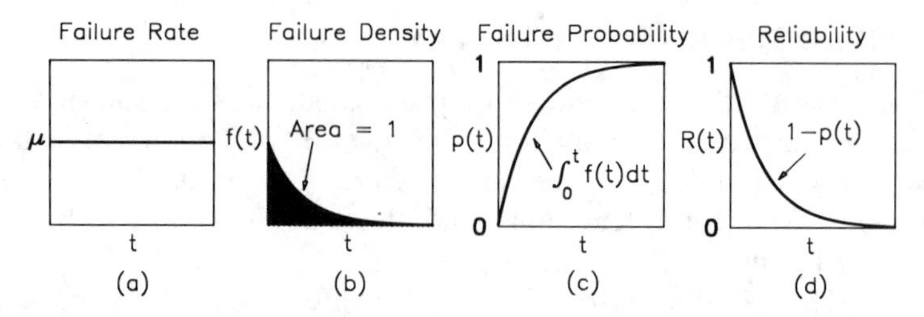

Figure 11-1 Typical plots of (a) the failure rate μ, (b) the failure density $f(t)$, (c) the failure probability $P(t)$, and (d) the reliability $R(t)$.

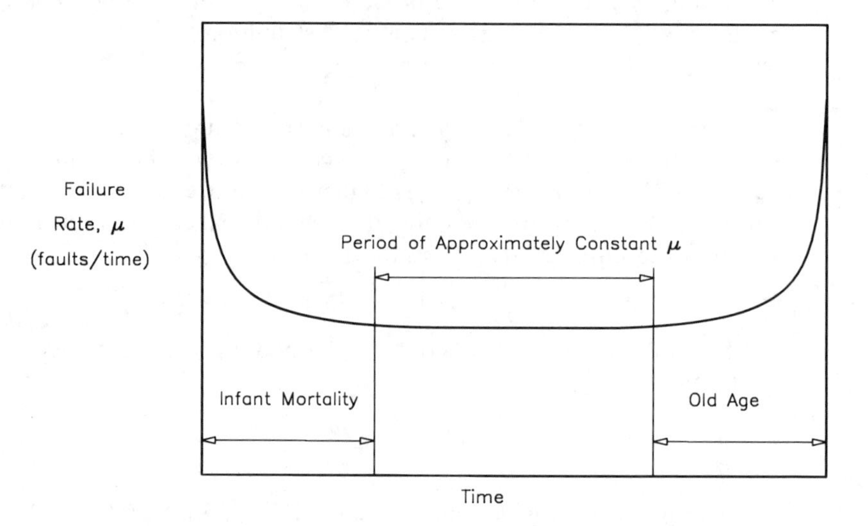

Figure 11-2 A typical "bathtub" failure rate curve for process hardware. The failure rate is approximately constant over the mid-life of the component.

ure rate is highest when the component is new (infant mortality) and when it is old (old age). Between these two periods (denoted by the lines on Figure 11-2), the failure rate is reasonably constant and Equations 11-1 through 11-5 are valid.

Interactions between Process Units

Accidents in chemical plants are usually the result of a complicated interaction of a number of process components. The overall process failure probability is computed from the individual component probabilities.

Process components interact in two different fashions. In some cases, a process failure requires the simultaneous failure of a number of components in parallel. This parallel structure is represented by the logical AND function. This means that the failure probabilities for the individual components must be multiplied:

$$P = \prod_{i=1}^{n} P_i \tag{11-6}$$

where n is the total number of components and P_i is the failure probability of each.

This rule is easily memorized since for *p*arallel components the *p*robabilities are multiplied.

The total reliability for parallel units is given by

$$R = 1 - \prod_{i=1}^{n} (1 - R_i) \tag{11-7}$$

where R_i is the reliability of an individual process component.

Process components also interact in series. This means that a failure of any single component in the series of components will result in failure of the process. The logical OR function represents this case. For series components the overall process reliability is found by multiplying the reliabilities for the individual components:

$$R = \prod_{i=1}^{n} R_i \tag{11-8}$$

The overall failure probability is computed from

$$P = 1 - \prod_{i=1}^{n} (1 - P_i) \tag{11-9}$$

For a system composed of two components A and B, Equation 11-9 is expanded to

$$P(A \text{ or } B) = P(A) + P(B) - P(A)\,P(B) \tag{11-10}$$

The cross-product term $P(A)\,P(B)$ compensates for counting the overlapping cases twice. Consider the example of tossing a single die and determining the probability that the number of points is even *or* divisible by three. In this case,

$$P(\text{even } \textit{or} \text{ divisible by three}) = P(\text{even}) + P(\text{divisible by three}) - P(\text{even } \textit{and} \text{ divisible by three})$$

The last term subtracts the cases where both conditions are satisfied.

If the failure probabilities are small (a common situation), the term $P(A)\,P(B)$ is negligible and Equation 11-10 reduces to

$$P(A \text{ or } B) = P(A) + P(B) \tag{11-11}$$

This result is generalized for any number of components. For this special case, Equation 11-9 reduces to

$$P = \sum_{i=1}^{n} P_i$$

Failure rate data for a number of typical process components are provided in Table 11-1. These are average values determined at a typical chemical process facility. Actual values would depend on manufacturer, materials of construction, design, environment, and other factors.

A summary of computations for parallel and series process components is shown in Figure 11-3.

TABLE 11-1: FAILURE RATE DATA FOR VARIOUS SELECTED PROCESS COMPONENTS[1]

Instrument	Faults/year
Controller	0.29
Control valve	0.60
Flow measurement (fluids)	1.14
Flow measurement (solids)	3.75
Flow switch	1.12
Gas - liquid chromatograph	30.6
Hand valve	0.13
Indicator lamp	0.044
Level measurement (liquids)	1.70
Level measurement (solids)	6.86
Oxygen analyzer	5.65
pH meter	5.88
Pressure measurement	1.41
Pressure relief valve	0.022
Pressure switch	0.14
Solenoid valve	0.42
Stepper motor	0.044
Strip chart recorder	0.22
Thermocouple temperature meas.	0.52
Thermometer temperature meas.	0.027
Valve positioner	0.44

[1]Selected from Frank P. Lees, *Loss Prevention in the Process Industries* (London: Butterworths, 1986), p. 343.

Example 11-1

The water flow to a chemical reactor cooling coil is controlled by the system shown in Figure 11-4. The flow is measured by a differential pressure (DP) device, the controller decides on an appropriate control strategy and the control valve manipulates the flow of coolant. Determine the overall failure rate, the unreliability, the reliability, and the MTBF for this system. Assume a one year period of operation.

Solution These process components are related in series. Thus, if any one of the components fails, the entire system fails. The reliability and failure probability are computed for each component using Equations 11-1 and 11-2. The results are shown in the table below. The failure rates are from Table 11-1.

Component	Failure rate (Faults/yr) μ	Reliability $R = e^{-\mu t}$	Failure probability $P = 1 - R$
Control Valve	0.60	0.55	0.45
Controller	0.29	0.75	0.25
DP Cell	1.41	0.24	0.76

Failure Probability	Reliability	Failure Rate
P_1, P_2 → OR → P	R_1, R_2 → OR → R	μ_1, μ_2 → OR → μ
$P = 1 - (1 - P_1)(1 - P_2)$	$R = R_1 R_2$	$\mu = \mu_1 + \mu_2$
$P = 1 - \prod_{i=1}^{n} (1 - P_i)$	$R = \prod_{i=1}^{n} R_i$	$\mu = \sum_{i=1}^{n} \mu_i$
Series Link of Components:	The failure of either component adds to the total system failure.	
P_1, P_2 → AND → P	R_1, R_2 → AND → R	
$P = P_1 P_2$	$R = 1 - (1 - R_1)(1 - R_2)$	$\mu = (-\text{Ln } R)/t$
$P = \prod_{i=1}^{n} P_i$	$R = 1 - \prod_{i=1}^{n} (1 - R_i)$	
Parallel Link of Components:	The failure of the system requires the failure of both components. Note that there is no convenient way to combine the failure rate.	

Figure 11-3 Computations for various types of component linkages.

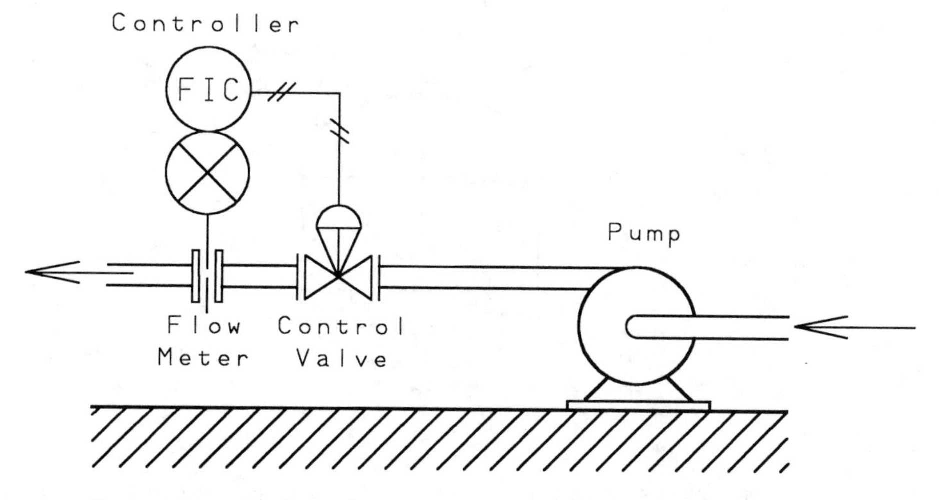

Figure 11-4 Flow control system. The components of the control system are linked in series.

The overall reliability for components in series is computed using Equation 11-8. The result is

$$R = \prod_{i=1}^{3} R_i = (0.55)(0.75)(0.24) = 0.10$$

The failure probability is computed from

$$P = 1 - R = 1 - 0.10 = 0.90/\text{year}$$

The overall failure rate is computed using the definition of the reliability, Equation 11-1,

$$0.10 = e^{-\mu}$$

$$\mu = -\ln(0.10) = 2.30 \text{ failures/year}$$

The mean time between failure (MTBF) is computed using Equation 11-5,

$$\text{MTBF} = \frac{1}{\mu} = 0.43 \text{ years}$$

This system is expected to fail on the average once every 0.43 years.

Example 11-2

A diagram of the safety systems in a certain chemical reactor is shown in Figure 11-5. This reactor contains a high pressure alarm to alert the operator in the event of dangerous reactor pressures. It consists of a pressure switch within the reactor connected to an alarm light indicator. For additional safety, an automatic high pressure reactor shutdown system is installed. This system is activated at a pressure somewhat higher than the alarm system and consists of a pressure switch connected to a solenoid valve in the reactor feed line. The automatic system stops the flow of reactant in the event

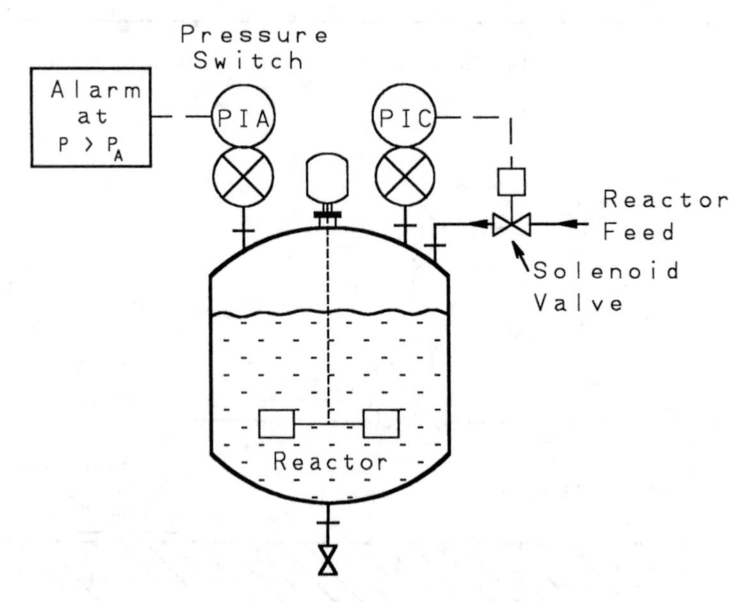

Figure 11-5 A chemical reactor with an alarm and inlet feed solenoid. The alarm and feed shutdown systems are linked in parallel.

of dangerous pressures. Compute the overall failure rate, the failure probability, the reliability and the MTBF for a high pressure condition. Assume a one year period of operation. Also, develop an expression for the overall failure probability based on the component failure probabilities.

Solution Failure rate data is available from Table 11-1. The reliability and failure probabilities of each component are computed using Equations 11-1 and 11-2.

Component	Failure rate (Faults/yr) μ	Reliability $R = e^{-\mu t}$	Failure probability $P = 1 - R$
1. Pressure Switch #1	0.14	0.87	0.13
2. Alarm Indicator	0.044	0.96	0.04
3. Pressure Switch #2	0.14	0.87	0.13
4. Solenoid Valve	0.42	0.66	0.34

A dangerous high pressure reactor situation occurs only when both the alarm system and shutdown system fail. These two components are in parallel.

For the alarm system the components are in series:

$$R = \prod_{i=1}^{2} R_i = (0.87)(0.96) = 0.835$$

$$P = 1 - R = 1 - 0.835 = 0.165$$

$$\mu = -\ln R = -\ln(0.835) = 0.180 \text{ faults/year}$$

$$\text{MTBF} = \frac{1}{\mu} = 5.56 \text{ years}$$

For the shutdown system the components are also in series:

$$R = \prod_{i=1}^{2} R_i = (0.87)(0.66) = 0.574$$

$$P = 1 - R = 1 - 0.574 = 0.426$$

$$\mu = -\ln R = -\ln(0.574) = 0.555 \text{ faults/year}$$

$$\text{MTBF} = \frac{1}{\mu} = 1.80 \text{ years}$$

The two systems are combined using Equation 11-6.

$$P = \prod_{i=1}^{2} P_i = (0.165)(0.426) = 0.070$$

$$R = 1 - P = 0.930$$

$$\mu = -\ln R = -\ln(0.930) = 0.073 \text{ faults/year}$$

$$\text{MTBF} = \frac{1}{\mu} = 13.7 \text{ years}$$

For the alarm system alone, a failure is expected once every 5.5 years. Similarly, for a reactor with a high pressure shutdown system alone, a failure is expected once every

1.80 years. However, with both systems in parallel, the MTBF is significantly improved and a combined failure is expected every 13.7 years.

The overall failure probability is given by

$$P = P(A)P(S)$$

where $P(A)$ is the failure probability of the alarm system and $P(S)$ is the failure probability of the emergency shutdown system. An alternate procedure is to invoke Equation 11-9 directly. For the alarm system,

$$P(A) = P_1 + P_2 - P_1P_2$$

For the shutdown system,

$$P(S) = P_3 + P_4 - P_3P_4$$

The overall failure probability is then

$$P = P(A)P(S) = (P_1 + P_2 - P_1P_2)(P_3 + P_4 - P_3P_4)$$

Substituting the numbers provided in the example

$$P = [0.13 + 0.04 - (0.13)(0.04)][0.34 + 0.13 - (0.34)(0.13)]$$

$$P = (0.165)(0.426) = 0.070$$

This is the same answer as before.

If the products P_1P_2 and P_3P_4 are assumed small,

$$P(A) = P_1 + P_2$$

$$P(S) = P_3 + P_4$$

and

$$P = P(A)P(S) = (P_1 + P_2)(P_3 + P_4)$$

$$P = 0.080$$

The difference between this answer and the answer obtained previously is 14.3%. The component probabilities are not small enough in this example to assume that the cross-products are negligible.

Revealed and Unrevealed Failures

Example 11-2 assumes that all failures in either the alarm or shutdown system are immediately obvious to the operator and are fixed in a negligible amount of time. Emergency alarms and shutdown systems are used only when a dangerous situation occurs. It is possible for the equipment to fail without the operator being aware of the situation. This is called an unrevealed failure. Without regular and reliable equipment testing, alarm and emergency systems can fail without notice. Failures that are immediately obvious are called revealed failures.

A flat tire on a car is immediately obvious to the driver. However, the spare tire in the trunk might also be flat without the driver being aware of the problem until the spare is needed.

Figure 11-6 shows the nomenclature for revealed failures. The time the component is operational is called the period of operation and is denoted by τ_o. After a

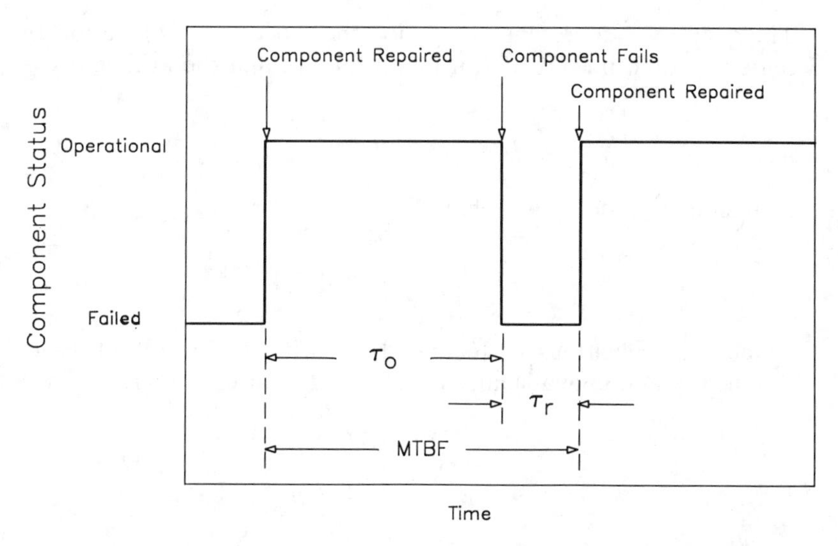

Figure 11-6 Component cycles for revealed failures. A failure requires a period of time for repair.

failure occurs, a period of time, called the period of inactivity or down time (τ_r), is required to repair the component. The mean time between failures (MTBF) is the sum of the period of operation and the down time, as shown.

For revealed failures the period of inactivity or down time for a particular component is computed by averaging the inactive period for a number of failures,

$$\tau_r \cong \frac{1}{n} \sum_{i=1}^{n} \tau_{r_i} \tag{11-12}$$

where n is the number of times the failure or inactivity occurred and τ_{r_i} is the period for repair for a particular failure. Similarly, the time before failure or period of operation is given by

$$\tau_o \cong \frac{1}{n} \sum_{i=1}^{n} \tau_{o_i} \tag{11-13}$$

where τ_{o_i} is the period of operation between a particular set of failures.

The mean time between failures is the sum of the period of operation and the repair period,

$$\text{MTBF} = \frac{1}{\mu} = \tau_r + \tau_o \tag{11-14}$$

It is convenient to define an availability and unavailability. The availability, A, is simply the probability the component or process is found functioning. The unavailability, U, is the probability the component or process is found not functioning. It is obvious that

$$A + U = 1 \tag{11-15}$$

The quantity τ_o represents the period the process is in operation and $\tau_r + \tau_o$ represents the total time. By definition, it follows that the availability is given by

$$A = \frac{\tau_o}{\tau_r + \tau_o} \tag{11-16}$$

and, similarly, the unavailability is

$$U = \frac{\tau_r}{\tau_r + \tau_o} \tag{11-17}$$

Combining Equations 11-16 and 11-17 with the result of Equation 11-14, the availability and unavailability for revealed failures are written as

$$\boxed{\begin{aligned} U &= \mu\tau_r \\ A &= \mu\tau_o \end{aligned}} \tag{11-18}$$

For unrevealed failures, the failure only becomes obvious after regular inspection. This situation is shown in Figure 11-7. If τ_u is the average period of unavailability during the inspection interval and τ_i is the inspection interval, then

$$U = \frac{\tau_u}{\tau_i} \tag{11-19}$$

The average period of unavailability is computed from the failure probability.

$$\tau_u = \int_0^{\tau_i} P(t)\, dt \tag{11-20}$$

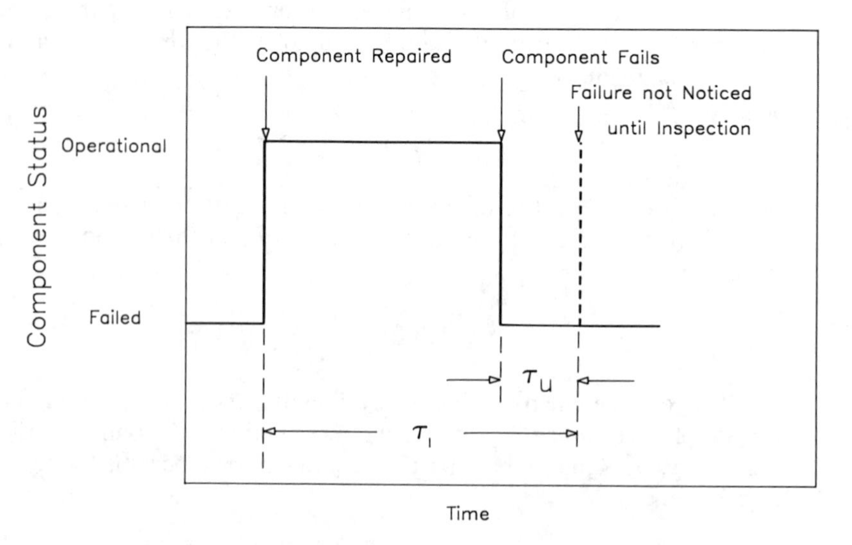

Figure 11-7 Component cycles for unrevealed failures.

Combining with Equation 11-19,

$$U = \frac{1}{\tau_i} \int_0^{\tau_i} P(t)\, dt \tag{11-21}$$

The failure probability $P(t)$ is given by Equation 11-2. This is substituted into Equation 11-21 and integrated. The result is

$$U = 1 - \frac{1}{\mu\tau_i}(1 - e^{-\mu\tau_i}) \tag{11-22}$$

An expression for the availability follows:

$$A = \frac{1}{\mu\tau_i}(1 - e^{\mu\tau_i}) \tag{11-23}$$

If the term $\mu\tau_i << 1$, then the failure probability is approximated by

$$P(t) \approx \mu t \tag{11-24}$$

and Equation 11-21 is integrated to give, for unrevealed failures

$$\boxed{U = \frac{1}{2}\mu\tau_i} \tag{11-25}$$

This is a very useful and convenient result. It demonstrates that, on the average for unrevealed failures, the process or component is unavailable during a period equal to half of the inspection interval. A decrease in the inspection interval is shown to increase the availability of an unrevealed failure.

Equations 11-19 through 11-25 assume a negligible repair time. This is usually a valid assumption since on-line process equipment is generally repaired within hours, while the inspection intervals are usually monthly.

Example 11-3

Compute the availability and the unavailability for both the alarm and shutdown systems of Example 11-2. Assume a maintenance inspection occurs once every month and assume the repair time is negligible.

Solution These systems both demonstrate unrevealed failures. For the alarm system, the failure rate is $\mu = 0.18$ faults/yr. The inspection period is $1/12 = 0.083$ year. The unavailability is computed using Equation 11-25.

$$U = \frac{1}{2}\mu\tau_i = (1/2)(0.18)(0.083) = 0.0075$$

$$A = 1 - U = 0.992$$

The alarm system is available 99.2% of the time. For the shutdown system, $\mu = 0.55$ faults/yr. Thus,

$$U = \frac{1}{2}\mu\tau_i = (1/2)(0.55)(0.083) = 0.023$$

$$A = 1 - 0.023 = 0.977$$

The shutdown system is available 97.7% of the time.

Probability of Coincidence

All process components demonstrate unavailability due to a failure. For alarms and emergency systems it is unlikely that these systems will be unavailable when a dangerous process episode occurs. The danger only results when a process upset occurs and the emergency system is unavailable. This requires a coincidence of events.

Assume that a dangerous process episode occurs p_d times in a time interval T_i. The frequency of this episode is given by

$$\lambda = \frac{p_d}{T_i} \tag{11-26}$$

For an emergency system with unavailability U, a dangerous situation will occur only when the process episode and the emergency system is unavailable. This is every $p_d U$ episodes. The average frequency of dangerous episodes, λ_d, is the number of dangerous coincidences divided by the time period,

$$\lambda_d = \frac{p_d U}{T_i} = \lambda U \tag{11-27}$$

For small failure rates, $U = \frac{1}{2}\mu\tau_i$ and $p_d = \lambda T_i$. Substituting into Equation 11-27 yields

$$\lambda_d = \frac{1}{2}\lambda\mu\tau_i \tag{11-28}$$

The mean time between coincidence (MTBC) is the reciprocal of the average frequency of dangerous coincidences:

$$\text{MTBC} = \frac{1}{\lambda_d} = \frac{2}{\lambda\mu\tau_i} \tag{11-29}$$

Example 11-4

For the reactor of Example 11-3, a high pressure incident is expected once every 14 months. Compute the mean time between coincidence (MTBC) for a high pressure excursion and a failure in the emergency shutdown device. Assume a maintenance inspection occurs every month.

Solution The frequency of process episodes is given by Equation 11-26.

$$\lambda = 1 \text{ episode}/[(14 \text{ months})(1 \text{ yr}/12 \text{ months})] = 0.857\ /\text{year}$$

The unavailability is computed from Equation 11-25.

$$U = \frac{1}{2}\mu\tau_i = (1/2)(0.55)(0.083) = 0.023$$

The average frequency of dangerous coincidences is given by Equation 11-27.

$$\lambda_d = \lambda U = (0.857)(0.023) = 0.020$$

The mean time between coincidence (MTBC) is, from Equation 11-29,

$$\text{MTBC} = \frac{1}{\lambda_d} = \frac{1}{0.020} = 50 \text{ years}$$

It is expected that a simultaneous high pressure incident and failure of the emergency shutdown device will occur once every 50 years.

If the inspection interval, τ_i is halved, then $U = 0.023$, $\lambda_d = 0.010$ and the resulting MTBC is 100 years. This is a significant improvement and shows why a proper and timely maintenance program is important.

11-2 EVENT TREES

Event trees begin with an initiating event and work towards a final result. This approach is inductive. The method provides information on how a failure can occur and the probability of occurrance.

When an accident occurs in a plant, various safety systems come into play to prevent the accident from propagating. These safety systems either fail or succeed. The event tree approach includes the effects of an event initiation followed by the impact of the safety systems.

The typical steps in an event tree analysis are[2]

1. Identify an initiating event of interest,
2. Identify the safety functions designed to deal with the initiating event,
3. Construct the event tree, and
4. Describe the resulting accident event sequences.

If appropriate data are available the procedure is used to assign numerical values to the various events. This is used effectively to determine the probability of a certain sequence of events and to decide what improvements are required.

Consider the chemical reactor system shown in Figure 11-8. This system is identical to the system shown in Figure 10-6, except that a high temperature alarm has been installed to warn the operator of a high temperature within the reactor. The event tree for a loss of coolant initiating event is shown in Figure 11-9. Four safety functions are identified. These are written across the top of the sheet. The first safety function is the high temperature alarm. The second safety function is the operator noticing the high reactor temperature during normal inspection. The third safety function is the operator re-establishing the coolant flow by correcting the problem within time. The final safety function is invoked by the operator performing an emergency shutdown of the reactor. These safety functions are written across the page in the order in which they logically occur.

The event tree is written from left to right. The initiating event is written first in the center of the page on the left. A line is drawn from the initiating event to the first safety function. At this point the safety function can either succeed or fail. By convention, a successful operation is drawn by a straight line upward and a failure is drawn downward. Horizontal lines are drawn from these two states to the next safety function.

If a safety function does not apply, the horizontal line is continued through the safety function without branching. For this example, the upper branch contin-

[2]*Guidelines for Hazard Evaluation Procedures* (New York: American Institute of Chemical Engineers, 1985).

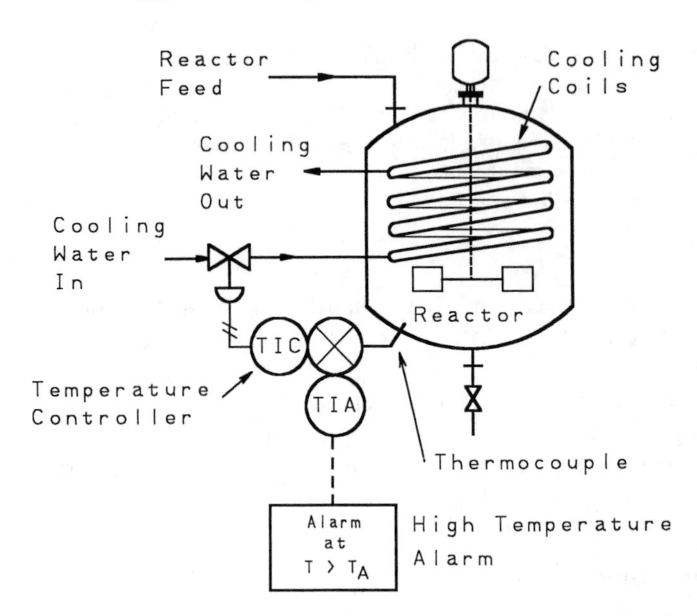

Figure 11-8 Reactor with high temperature alarm and temperature controller.

ues through the second function where the operator notices the high temperature. If the high temperature alarm operates properly, the operator will already be aware of the high temperature condition. The sequence description and consequences are indicated on the extreme right hand side of the event tree. The open circles indicate safe conditions and the circles with the crosses represent unsafe conditions.

The lettering notation in the sequence description column is useful for identifying the particular event. The letters indicate the sequence of failures of the safety systems. The initiating event is always included as the first letter in the notation. An event tree for a different initiating event in this study would use a different letter. For the example here, the lettering sequence ADE represents initiating event A followed by failure of safety functions D and E.

The event tree can be used quantitatively if data are available on the failure rates of the safety functions and the occurrence rate of the initiation event. For this example assume a loss of cooling event occurs once per year. Let us also assume that the hardware safety functions fail 1% of the time they are placed in demand. This is a failure rate of 0.01 failure/demand. Also assume that the operator will notice the high reactor temperature 3 out of 4 times and that 3 out of 4 times the operator will be successful at reestablishing the coolant flow. Both of these cases represent a failure rate of 1 time out of 4, or 0.25 failures/demand. Finally, it is estimated that the operator successfully shuts down the system 9 out of 10 times. This is a failure rate of 0.10 failures/demand.

The failure rates for the safety functions are written below the column headings. The occurrance frequency for the initiating event is written below the line originating from the initiating event.

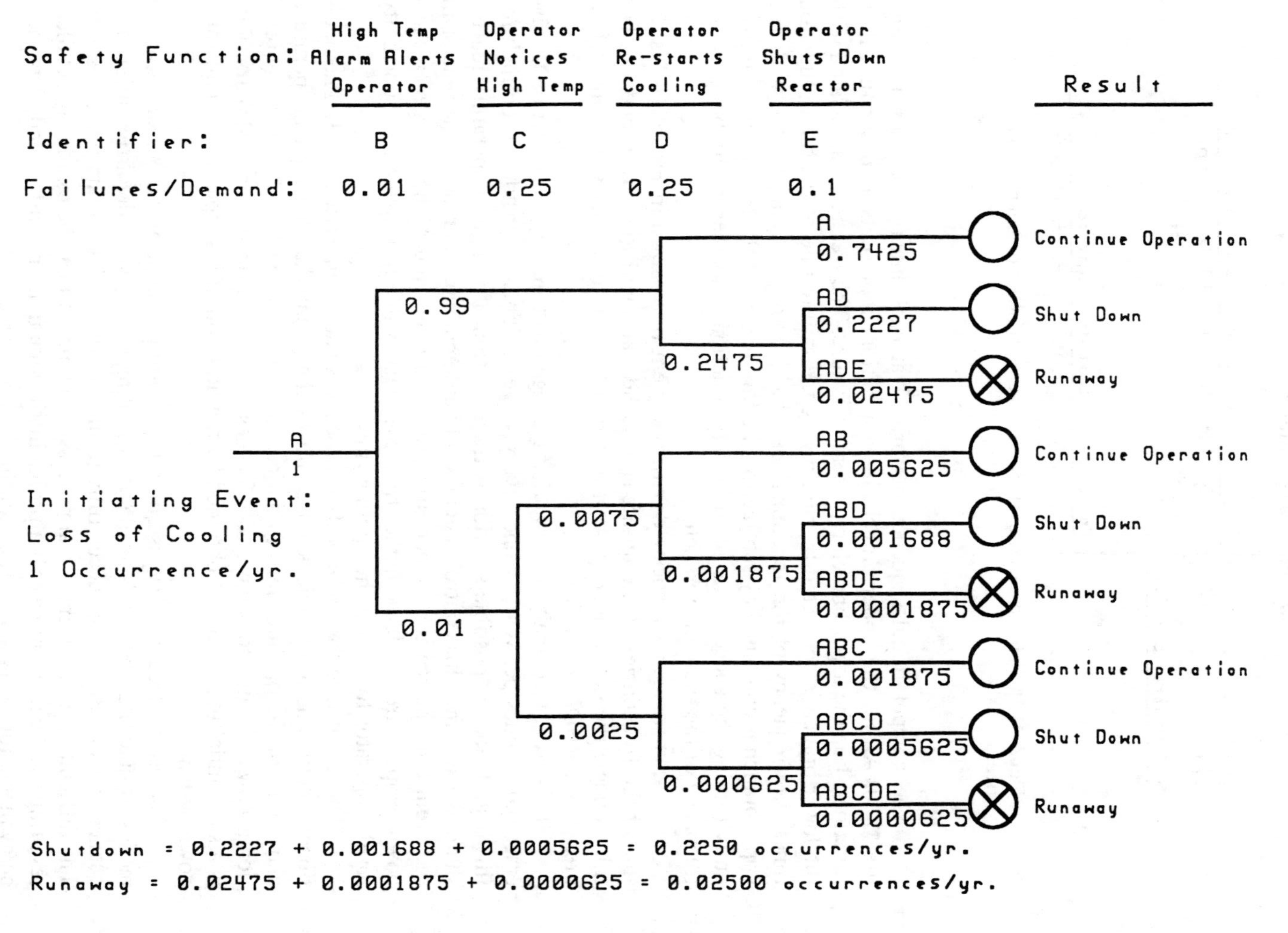

Figure 11-9 Event tree for a loss of coolant accident for the reactor of Figure 11-8.

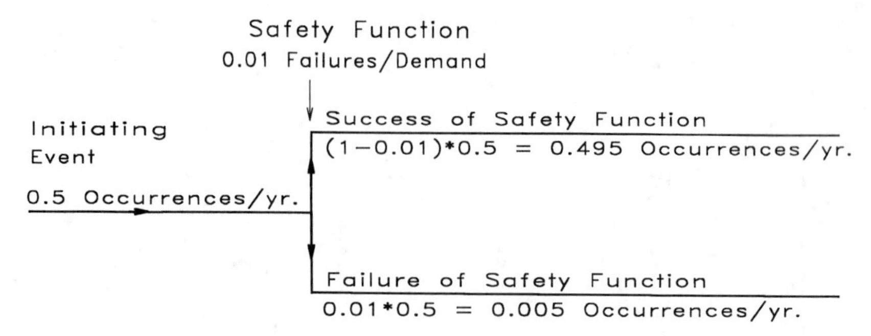

Figure 11-10 The computational sequence across a safety function in an event tree.

The computational sequence performed at each junction is shown in Figure 11-10. Again, the upper branch, by convention, represents a successful safety function while the lower branch represents a failure. The frequency associated with the lower branch is computed by multiplying the failure rate of the safety function times the frequency of the incoming branch. The frequency associated with the upper branch is computed by subtracting the failure rate of the safety function from unity (giving the success rate of the safety function) and then multiplying by the frequency of the incoming branch.

The net frequency associated with the event tree shown in Figure 11-9 is the sum of the frequencies of the unsafe states (the states with the circles and x's). For this example the net frequency is estimated at 0.025 failures per year (sum of failures ADE, ABDE and ABCDE).

This event tree analysis shows that a dangerous runaway will occur on the average 0.025 times per year, or once every 40 years. This is considered too high for this installation. A possible solution is the inclusion of a high temperature reactor shutdown system. This control system would automatically shut down the reactor in the event that the reactor temperature exceeds a fixed value. The emergency shutdown temperature would be higher than the alarm value to provide an opportunity for the operator to restore the cooling flow.

The event tree for the modified process is shown in Figure 11-11. The additional safety function provides a backup in the event that the high temperature alarm fails or the operator fails to notice the high temperature. The runaway reaction is now estimated to occur 0.00025 times per year, or once every 400 years. This is a substantial improvement obtained by the addition of a simple, redundant shutdown system.

The event tree is very useful for providing scenarios on possible failure modes. If quantitative data is available, an estimate can be made of the failure frequency. This is used most successfully to modify the design to improve the safety. The difficulty is that for most real processes the method can be extremely detailed, resulting in a huge event tree. If a probabilistic computation is attempted, data must be available for every safety function in the event tree.

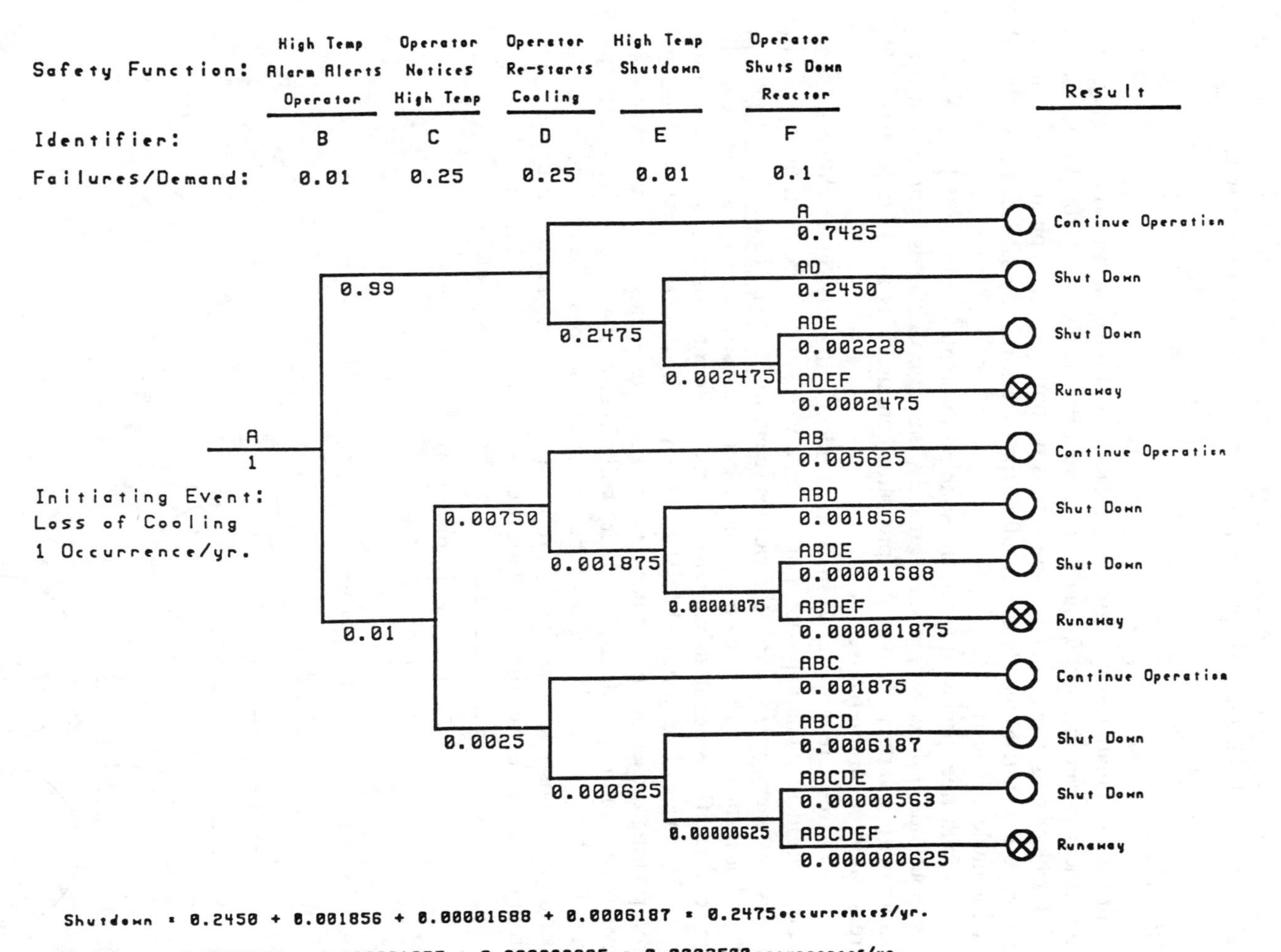

Figure 11-11 Event tree for the reactor of Figure 11-8. This includes a high temperature shutdown system.

An event tree begins with a specified failure and terminates with a number of resulting consequences. If an engineer is concerned about a particular consequence, there is no certainty that the consequence of interest will actually result from the failure selected. This is perhaps the major disadvantage with event trees.

11-3 FAULT TREES

Fault trees originated in the aerospace industry and have been used extensively by the nuclear power industry to qualify and quantify the hazards and risks associated with nuclear power plants. This approach is becoming more popular in the chemical process industries mostly as a result of the successful experiences demonstrated by the nuclear industry.

A fault tree for anything but the simplest of plants can be quite large, involving thousands of process events. Fortunately, this approach lends itself to computerization with a variety of computer programs commercially available to draw fault trees based on an interactive session.

Fault trees are a deductive method for identifying ways in which hazards can lead to accidents. The approach starts with a well defined accident, or top event, and works backwards towards the various scenarios that can cause the accident.

For instance, a flat tire on an automobile is caused by two possible events. In one case the flat is due to driving over debris on the road, such as a nail. The other possible cause is tire failure. The flat tire is identified as the top event. The two contributing causes are either basic or intermediate events. The basic events are events that cannot be defined further, while intermediate events are events that can. For this example, driving over the road debris is a basic event since no further definition is possible. The tire failure is an intermediate event because it results from either a defective tire or a worn tire.

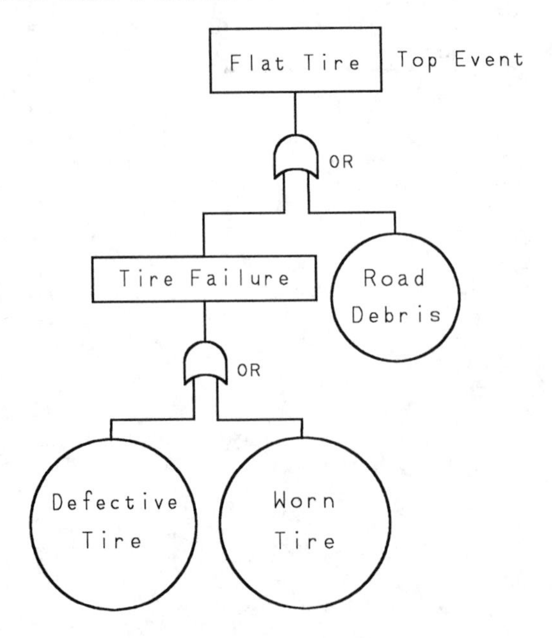

Figure 11-12 A fault tree describing the various events contributing to a flat tire.

The flat tire example is pictured using a fault tree logic diagram shown on Figure 11-12. The circles denote basic events and the rectangles denote intermediate events. The fish-like symbol represents the OR logic function. It means that either of the input events will cause the output state to occur. As shown on Figure 11-12, the flat tire is caused by either debris on the road or tire failure. Similarly, the tire failure is caused by either a defective tire or a worn tire.

Events in a fault tree are not restricted to hardware failures. It can also include software, human, and environmental factors.

For reasonably complex chemical processes, a number of additional logic functions are needed to construct a fault tree. A detailed list is given in Figure 11-13. The AND logic function is very important for describing processes that interact

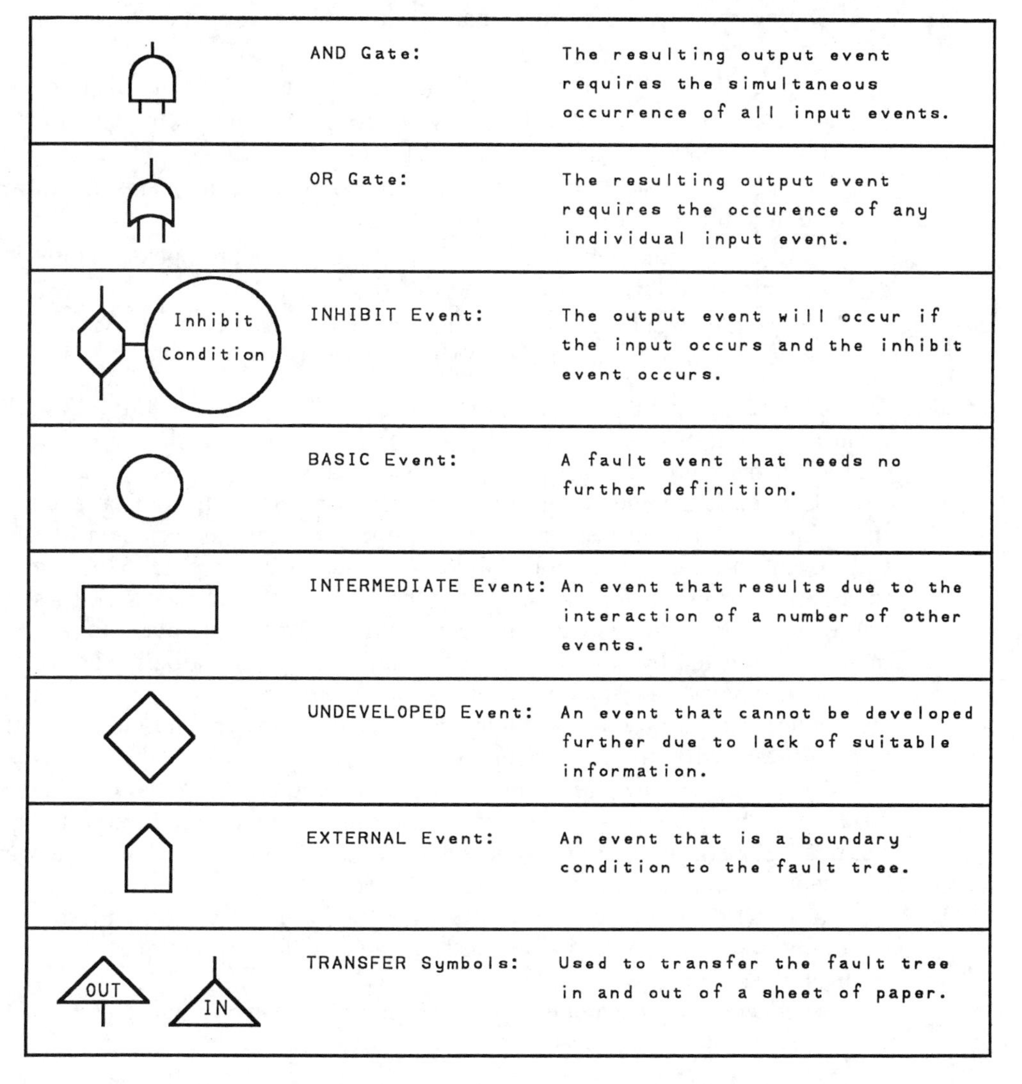

Figure 11-13 The logic transfer components used in a fault tree.

in parallel. This means that the output state of the AND logic function is active only when both of the input states are active. The inhibit function is useful for events that lead to a failure only part of the time. For instance, driving over debris in the road does not always lead to a flat tire. The inhibit gate could be used in the fault tree of Figure 11-12 to represent this situation.

Prior to the drawing of the actual fault tree, a number of preliminary steps must be taken.

1. Define precisely the top event. Events such as HIGH REACTOR TEMPERATURE or LIQUID LEVEL TOO HIGH are precise and appropriate. Events such as EXPLOSION OF REACTOR or FIRE IN PROCESS are too vague, while an event like LEAK IN VALVE is too specific.
2. Define the existing event. What conditions are sure to be present when the top event occurs?
3. Define the unallowed events. These are events that are unlikely or are not under consideration at the present. This could include wiring failures, lightning, tornadoes, hurricanes, and so forth.
4. Define the physical bounds of the process. What components are to be considered in the fault tree?
5. Define the equipment configuration. What valves are open or closed? What are the liquid levels? Is this a normal operation state?
6. Define the level of resolution. Will the analysis consider just a valve, or will it be necessary to consider the valve components?

The next step in the procedure is to draw the fault tree. First draw the top event at the top of the page. Label it as the top event in order to avoid confusion later when the fault tree has spread out to several sheets of paper.

Second, determine the major events that contribute to the top event. Write these down as intermediate, basic, undeveloped, or external events on the sheet. If these events are related in parallel (all events must occur in order for the top event to occur), they must be connected to the top event by an AND gate. If these events are related in series (any event can occur in order for the top event to occur), they must be connected by an OR gate. If the new events cannot be related to the top event by a single logic function, the new events are probably improperly specified. Remember, the purpose of the fault tree is to determine the individual event steps that must occur to produce the top event.

Now consider any one of the new intermediate events. What events must occur to contribute to this single event? Write these down as either intermediate, basic, undeveloped, or external events on the tree. Then decide which logic function represents the interaction of these newest events.

Continue developing the fault tree until all branches have been terminated by basic, undeveloped or external events. All intermediate events must be expanded.

Example 11-5

Consider again the alarm indicator and emergency shutdown system of Example 11-2. Draw a fault tree for this system.

Solution The first step is to define the problem.

1. Top Event: Damage to reactor due to overpressuring
2. Existing event: High process pressure.
3. Unallowed events: Failure of mixer, electrical failures, wiring failures, tornadoes, hurricanes, electrical storms.
4. Physical bounds: The equipment shown in Figure 11-5.
5. Equipment configuration: Solenoid valve open, reactor feed flowing.
6. Level of resolution: Equipment as shown in Figure 11-5.

The top event is written at the top of the fault tree and is indicated as the top event. See Figure 11-14. Two events must occur for overpressuring: failure of the alarm indicator and failure of the emergency shutdown system. These events must occur together so they must be connected by an AND function. The alarm indicator can fail by either a failure of pressure switch #1 or the alarm indicator light. These must be connected by OR functions. The emergency shutdown system can fail by either a failure of pressure switch #2 or the solenoid valve. These must also be connected by an OR function. The complete fault tree is shown in Figure 11-14.

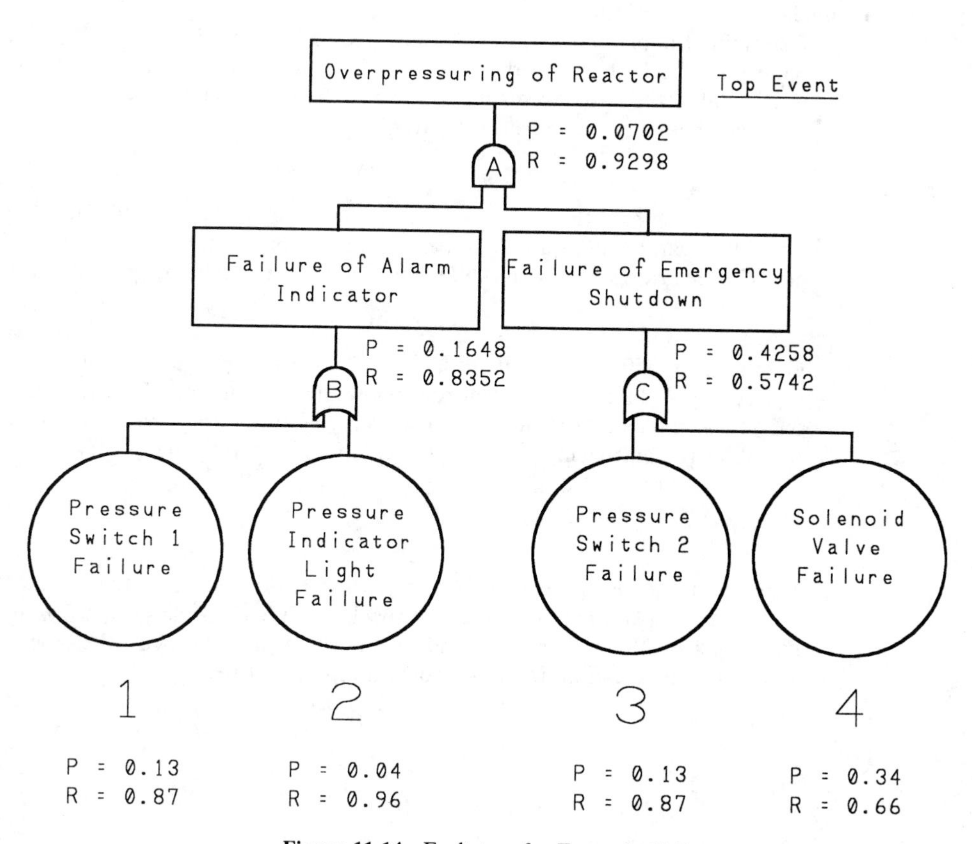

Figure 11-14 Fault tree for Example 11-5.

Determining the Minimal Cut Sets

Once the fault tree has been fully drawn, a number of computations can be performed. The first computation involves determination of the minimal cut sets (or min cut sets). The minimal cut sets are the various sets of events that could lead to the top event. In general, the top event could occur by a variety of different combinations of events. The different unique sets of events leading to the top event are the minimal cut sets.

The minimal cut sets are very useful for determining the various ways in which a top event could occur. Some of the mimimal cut sets will have a higher probability than others. For instance, a set involving just two events will be more likely than a set involving three. Similarly, a set involving human interaction is more likely to fail than one involving hardware alone. Based on these simple rules the minimal cut sets are ordered with respect to failure probability. The higher probability sets are examined carefully to determine if additional safety systems are required.

The minimal cut sets are determined using a procedure developed by Fussell and Vesely.[3] The procedure is best described using an example.

Example 11-6

Determine the min cut sets for the fault tree of Example 11-5.

Solution The first step in the procedure is to label all of the gates using letters and to label all of the basic events using numbers. This is shown in Figure 11-14. The first logic gate below the top event is written.

A

AND gates increase the number of events in the cut sets while OR gates lead to more sets. Logic gate A in Figure 11-14 has two inputs, one from gate B and the other from gate C. Because gate A is an AND gate, gate A is replaced by gates B and C.

A̸ B C

Gate B has inputs from event 1 and event 2. Because gate B is an OR gate, gate B is replaced by adding an additional row below the present row. First, replace gate B by one of the inputs and then create a second row below the first. Copy into this new row all of the entries in the remaining column of the first row.

A̸ B̸ 1 C
2 C

Note that the C in the second column of the first row is copied to the new row.

Next, replace gate C in the first row by its inputs. Because gate C is also an OR gate, replace C by basic event 3 and then create a third row with the other event. Be sure to copy the 1 from the other column of the first row.

A̸ B̸ 1 C̸ 3
2 C
1 4

[3] J. B. Fussell and W. E. Vesely, "A New Methodology for Obtaining Cut Sets for Fault Trees," *Transactions of the American Nuclear Society,* 15, 1972.

Finally, replace gate C in the second row by its inputs. This generates a fourth row.

$\not{A}\ \not{B}\ 1 \quad \not{C}\ 3$

$2 \quad \not{C}\ 3$

$1 \quad 4$

$2 \quad 4$

The cut sets are then:

1, 3

2, 3

1, 4

2, 4

This means that the top event occurs as a result of any one of these sets of basic events.

The procedure does not always deliver the minimal cut sets. Sometimes a set might be of the following form:

1, 2, 2

This is reduced to simply 1, 2. On other occasions the sets might include supersets. For instance, consider

1, 2

1, 2, 4

1, 2, 3

The second and third sets are supersets of the first basic set because events 1 and 2 are in common. The supersets are eliminated to produce the minimal cut sets.

For this example there are no supersets.

Quantitative Calculations Using the Fault Tree

The fault tree can be used to perform quantitative calculations to determine the probability of the top event. This is accomplished in two ways.

With the first approach, the computations are performed using the fault tree diagram itself. The failure probabilities of all of the basic, external, and undeveloped events are written on the fault tree. Then the necessary computations are performed across the various logic gates. Remember that probabilities are multiplied across an AND gate and reliabilities are multiplied across an OR gate. The computations are continued in this fashion until the top event is reached. Inhibit gates are considered a special case of an AND gate.

The results of this procedure are shown on Figure 11-14. The symbol P represents the probability and R represents the reliability. The failure probabilities for the basic events were obtained from Example 11-2.

The other procedure is to use the min cut sets. This procedure will approach the exact result only if the probabilities of all of the events are small. In general, this result will provide a number that will be larger than the actual probability. This approach assumes that the probability cross terms shown in Equation 11-10 are negligible.

The min cut sets represent the various failure modes. For the above example, either events 1, 3 or 2, 3 or 1, 4 or 2, 4 could cause the top event. To estimate the overall failure probability the probabilities from the cut sets are added together. For this case

$$P(1 \text{ AND } 3) = (0.13)(0.13) = 0.0169$$

$$P(2 \text{ AND } 3) = (0.04)(0.13) = 0.0052$$

$$P(1 \text{ AND } 4) = (0.13)(0.34) = 0.0442$$

$$P(2 \text{ AND } 4) = (0.04)(0.34) = 0.0136$$

$$\text{TOTAL:} \quad 0.0799$$

This compares to the exact result of 0.0702 obtained using the actual fault tree. The cut sets are related to each other by the OR function. For the example above all of the cut set probabilities were added. This is an approximate result as shown in Equation 11-10, since the cross product terms were neglected. For small probabilities the cross terms are negligible and the addition will approach the true result.

Advantages and Disadvantages to Fault Trees

The main disadvantage to using fault trees is that for any reasonably complicated process the fault tree will be enormous. Fault trees involving thousands of gates and intermediate events are not unusual. Fault trees of this size require a considerable amount of time, measured in years, to complete.

Furthermore, the developer of a fault tree can never be certain that all of the failure modes have been considered. More complete fault trees are usually developed by more experienced engineers.

Fault trees also assume that failures are "hard," that a particular item of hardware does not fail partially. A leaking valve is a good example of a partial failure. Also, the approach assumes that a failure of one component does not stress the other components resulting in a change in the component failure probabilities.

Fault trees developed by different individuals are usually different in structure. The different trees will generally predict different failure probabilities. This inexact nature of the fault tree is a considerable problem.

If the fault tree is used to compute a failure probability for the top event then failure probabilities are needed for all of the events in the fault tree. These probabilities are not usually known or are not known accurately.

A major advantage to the fault tree approach is that it begins with a top event. This top event is selected by the user to be specific towards the failure of interest. This is opposed to the event tree approach where the events resulting from a single failure might not be the events of specific interest to the user.

Fault trees are also used to determine the minimal cut sets. The min cut sets provide enormous insight into the various ways for top events to occur.

Finally, the entire fault tree procedure enables the application of computers. Software is available for graphically constructing fault trees, determining the min cut sets, and calculating failure probabilities. Reference libraries containing failure probabilities for various types of process equipment can also be included.

SUGGESTED READING

Guidelines for Hazard Evaluation Procedures (New York: American Institute of Chemical Engineers, 1985).

FUSSELL, J. B. and VESELY, W. E., "A New Methodology for Obtaining Cut Sets for Fault Trees," *Transactions of the American Nuclear Society,* Vol. 15 (1972).

LEES, F. P., *Loss Prevention in the Process Industries* (London: Butterworths, 1986), Chapter 7: Reliability Engineering; Chapter 9: Hazard Assessment.

ROFFEL, B. and RIJNSDORP, J. E., *Process Dynamics, Control and Protection* (Ann Arbor, MI: Ann Arbor Science, 1982), Chapter 19: Protection.

PROBLEMS

11-1. Given the fault tree gates shown in Figure P11-1 and the following set of failure probabilities

Component	Failure probability
1	0.1
2	0.2
3	0.3
4	0.4

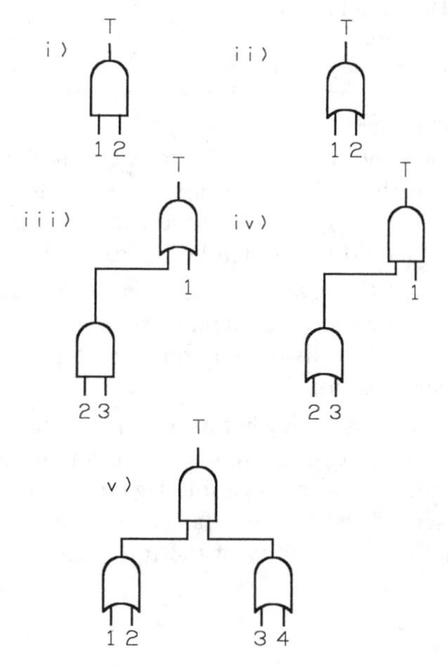

Figure P11-1

a. Determine an expression for the probability of the top event in terms of the component failure probabilities.
b. Determine the minimum cut sets.
c. Compute a value for the failure probability of the top event. Use both the expression of part a and the fault tree itself.

11-2. The storage tank system shown in Figure P11-2 is used to store process feedstock. Overfilling of storage tanks is a common problem in the process industries. To prevent overfilling, the storage tank is equipped with a high level alarm and high level shutdown system. The high level shutdown system is connected to a solenoid valve that stops the flow of input stock.

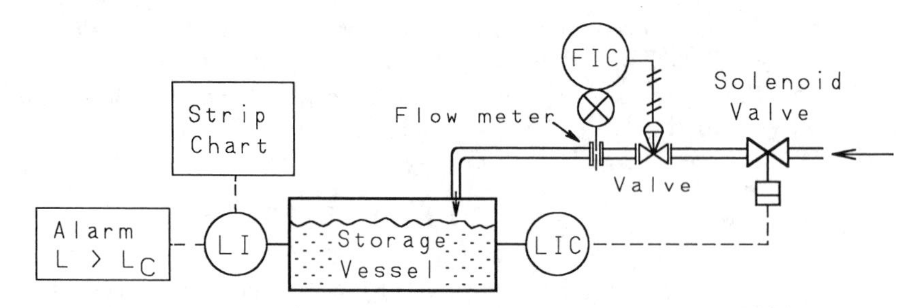

Figure P11-2 Level control system with alarm.

a. Develop an event tree for this system using the "failure of level indicator" as the initiating event. Given that the level indicator fails 4 times/year, estimate the number of overflows expected per year. Use the following data.

System	Failures/demand
High level alarm	0.01
Operator stops flow	0.1
High level switch system	0.01

b. Develop a fault tree for the top event of "storage tank overflows." Use the data in Table 11-1 to estimate the failure probability of the top event and the expected number of occurrences per year. Determine the min cut sets. What are the most likely failure modes? Should the design be improved?

11-3. Compute the availability of the level indicator system and flow shutdown system for Problem 11-2. Assume a one month maintenance schedule. Compute the MTBC for a high level episode and a failure in the shutdown system assuming that a high level episode occurs once every six months.

11-4. The problem of Example 11-5 is somewhat unrealistic in that it is highly likely that the operator will notice the high pressure even if the alarm and shutdown systems are not functioning. Draw a fault tree using an inhibit gate to include this situation. Determine the minimal cut sets. If the operator fails to notice the high pressure in 1 out of 4 occasions, what is the new probability of the top event?

11-5. Derive Equation 11-22.

11-6. Show that for a process protected by two independent protection systems the frequency of dangerous coincidences is given by,

$$\lambda_d = \frac{1}{4}\lambda\mu^2\tau_i^2$$

11-7. A starter is connected to a motor which is connected to a pump. The starter fails once in 50 years and requires 2 hours to repair. The motor fails once in 20 years and requires 36 hours to repair. The pump fails once per 10 years and requires 4 hours to repair. Determine the overall failure frequency, the probability the system will fail in the coming two years, the reliability, and the unavailability for this system.

11-8. A reactor experiences trouble once every 16 months. The protection device fails once every 25 years. Inspection takes place once every month. Calculate the unavailability, the frequency of dangerous coincidences, and the mean time between coincidence (MTBC).

Accident Investigations

12

The investigation of accidents and near misses provides opportunities to learn how to prevent similar events in the future. Accident investigations, including detailed descriptions and recommendations, are commonly shared within the chemical industry. Many professionals believe that this sharing of information about accidents has been a major contributor to the steady improvement in safety performance.

In recent years, important techniques have been developed for improving the effectiveness of investigations. This chapter covers the more important techniques, including

- Learning from accidents
- Layered investigations
- Investigation process
- Investigation summary
- Aids for diagnosis
- Aids for recommendations

An important principle in safety states that the "causes of accidents are visible the day before the accident." These causes are visible to professionals who "see" deficiencies. This vision (knowledge or awareness) is developed by the study and development of accident and near miss investigations.

12-1 LEARNING FROM ACCIDENTS

Every member of an investigation team learns about problems which precipitate accidents. This new knowledge helps every team member avoid similar situations in the future. If the investigation is appropriately reported, many others will also benefit.

This concept is also important for reporting minor accidents or near misses. Minor accidents and near misses are excellent opportunities to obtain "free chances" to prevent larger accidents from occurring in the future. It is much easier to correct minor problems before serious accidents occur than to correct them after they are manifested in major losses.

Accident investigations are designed to enhance learning. The fundamental steps in an investigation include (1) developing a detailed description of the accident, (2) accumulating relevant facts, (3) analyzing the facts and developing potential causes of the accident, (4) studying the system and operating methods relevant to the potential causes of the accident, (5) developing the most likely causes, (6) developing recommendations to eliminate recurrence of this *type* of accident, and (7) using an investigation style which is fact-finding and not faultfinding; faultfinding creates an environment which is not conducive to learning.

Good investigations help organizations use every accident as an opportunity to learn how to prevent future accidents. Investigation results are used to change hazardous practices and procedures and to develop management systems to use this new knowledge on a long term and continuous basis.

12-2 LAYERED INVESTIGATIONS

The important concept of layered investigations is emphasized by T. Kletz.[1] It is a technique which significantly improves the commonly used older methods. Older investigation methods identified only the relatively obvious causes of an accident. Their evidence supported their conclusions, and one or two technical recommendations resulted. According to Kletz this older method developed recommendations which were relatively superficial. Unfortunately, the majority of accidents are investigated in this style.

The newer and better method includes a deeper analysis of the facts and additional levels or "layers" of recommendations. This recommended deeper analysis identifies underlying causes of the accident which are analyzed to develop a multilayered solution to the problem, layered recommendations.

The number of relevant facts accumulated in an accident investigation are usually limited in number. Further investigation usually cannot uncover additional facts. A deeper analysis of the facts, however, often leads to new conclusions and recommendations. This deeper analysis is, for example, similar to a brainstorming session to develop new applications for a common house brick. New and interesting applications will continue to surface.

Kletz emphasizes an extra effort to generate three levels of recommendations for preventing and mitigating accidents:

First layer:	Immediate technical recommendations
Second layer:	Recommendations to avoid the hazards
Third layer:	Recommendations to improve the management system

[1] T. Kletz, "Layered Accident Investigations," *Hydrocarbon Processing,* Nov. 1979, pp. 373-382, and T. Kletz, *Learning from Accidents in Industry* (Boston: Butterworths, 1988).

To fully utilize this layered technique, the investigation process is conducted with a very open mind. Facts about the accident are accumulated which support conclusions at all three levels.

Example 12-1

Illustrate the layered investigation process to develop underlying causes of a municipal pool accident.

A drowning accident occurred during an open swim period. Approximately 100 children, ranging between 5 and 16 years old, were in and around a pool (3 ft to 9 ft deep). An older child unknowingly pushed a 5 year old into the deep water. The pool was relatively crowded, and the 5 year old slipped under the water without being noticed by others, including the lifeguard.

Solution The facts uncovered by an investigation team are

1. Pool did not have deep and shallow markings,
2. The older child was engaged in horseplay,
3. The younger child did not know how to swim,
4. The lifeguard had many blind fields of vision,
5. The pool was overly crowded,
6. The pool did not have an orientation program, and
7. The pool did not offer swimming lessons.

An old style accident investigation report would include only one or two recommendations like paint pool depths at the edges of the pool and add more lifeguards.

Layered recommendations are the result of uncovering the underlying causes of the accident:

First layer recommendations:	Immediate technical recommendations.
1. Paint pool depths at the pool edges. 2. Add more lifeguards. 3. Reduce the number of swimmers.	
Second layer recommendations:	**Avoiding the hazard.**
1. Prohibit horseplay. 2. Zone pool to keep smaller children at shallow end of pool. 3. Add swimming lessons for all age groups. 4. Give all new swimmers (especially young children) a pool orientation. 5. Add a roving lifeguard to monitor and control pool behavior.	
Third layer recommendations:	**Improving the management system.**
1. Train lifeguards to alert supervision of observed potential problems. 2. Assign supervisor to make formal (documented) audits on a regular basis.	

In this particular example, almost all recommendations can be implemented without difficulty. These technical improvements and new management systems will prevent future drownings and also prevent other types of accidents in this pool environment. This example also illustrates the value of having an open mind during the investigation which is a requirement for uncovering underlying causes.

TABLE 12-1 QUESTIONS FOR LAYERED ACCIDENT INVESTIGATIONS[1]

(1) What equipment failed?
How to prevent failure or make it less likely?
How can failure or approaching failure be detected?
How can failure be controlled or consequences minimized?
What does the equipment do?
What other equipment can be used instead?
What could we do instead?
(2) What material leaked (exploded, decomposed, etc.)?
How can leak (etc.) be prevented?
How can leak or approaching leak be detected?
What does material do?
Can we reduce volume of material?
What material can be substituted?
What could we do instead?
(3) Which people could have performed better?
What could they have done better?
How can we help them to perform better?
(4) What is the purpose of the operation involved in the accident?
Why do we do this?
What could we do instead?
How else could we do it?
Who else could do it?
When else could we do it?

[1]Trevor Kletz, *Learning from Accidents in Industry* (Boston: Butterworths, 1988), p. 153.

A set of questions designed to help accident investigators find less obvious ways to prevent accidents is shown in Table 12-1. The team approach of questioning and answering is especially important because the supportive, synergistic, and feedback approach by team members give results which are always greater than the sum of the parts.

12-3 INVESTIGATION PROCESS

Different investigators use different approaches to accident investigations. One approach which can be used for most accidents is described below and shown in Table 12-2; it is an adaptation of a process recommended by A. D. Craven.[2]

The accident investigation report is the major result of the investigation. In general, the format should be flexible and designed specifically to best explain the accident. The format may include the following sections:

a. Introduction
b. Process description (equipment and chemistry)
c. Incident description

[2]Howard H. Fawcett and William S. Wood, eds., *Safety and Accident Prevention in Chemical Operations* (New York: John Wiley and Sons, 1982), pp. 659-680.

TABLE 12-2 ACCIDENT INVESTIGATION PROCESS

Step	Purpose
(1) Investigation team	A team is chosen as quickly as possible. Experience and affiliation is proportional to the magnitude of the accident.
(2) Brief survey	Make overview survey (maximum of 1 hour) to understand the type and value of the information needed to derive causes of the accident.
(3) Set objectives and delegate responsibilities	Based on (1) and (2) the objectives and subobjectives of the investigation are defined by the investigation team. Responsibilities are delegated to team members with suggested completion times.
(4) Preincident facts	Preincident facts are gathered and organized. Flow sheets, procedures, photographs, and data (interviews or recorded data) are used.
(5) Accident facts	Make detailed examinations with photos, inspections, and interviews. Establish origin of accident, and facts relevant to layered causes and recommendations. Record extent of damage and hypothesize the sequence of events. Resist development of potential causes to maintain momentum and objectivity while collecting data.
(6) Research and analyses	Initiate research type experiments and analyze facts to clarify perplexing evidence.
(7) Discussion, conclusions, and recommendations	Study (2) to (6) to develop conclusions and layered recommendations.
(8) Report	Develop accident investigation report: keep report clear, concise, accurate, and technical. Do not smother key results.

d. Investigation results
e. Discussion
f. Conclusions
g. Layered recommendations

The accident investigation report is written using the principles of technical documentation. Sections a to d are objective and should not include the authors' opinions. Section e to g appropriately contain the opinions of the authors (investigation team). This technical style allows any reader to develop their own independent conclusions and recommendations. As a result of this criteria, the accident investigation report is a learning tool which is the major purpose of the investigation.

12-4 INVESTIGATION SUMMARY

The previously described accident investigation report is a logical and necessary result of an investigation. It includes comprehensive details which are of particular interest to specialists. These details, however, are too focused for an average inquirer.

Kletz[3] uses a report format which summarizes the events and recommendations in a diagram. This type of summary is shown in Figure 12-1. It emphasizes

[3]T. Kletz, *Learning from Accidents in Industry,* p. 22.

```
                         Accident Summary

   Accident title:
   Major damage:
   Date:
   Location:

            Recommendations for prevention/mitigation

            1st layer: Immediate technical recommendation
            2nd layer: Recommendations for avoiding hazard
            3rd layer: Recommendations for improving the
                       management systems

Events                      Recommendations

Major accident                          *
                                        *
Events which precipitate                *
   accident                             *
                                        *
                            Recommendations which will break
                            the chain of events (layers 1,2 &
                            3)
Preaccident conditions                  *

Sequence of operations or               *
   steps to derive accident             *
   conditions                           *
                            Recommendations to train, inspect,
                            or change methods (layers 2 & 3)
                                        *
Management influences                   *
  (Decisions and practices)             *
           *                            *
           *                            *
           *                            *
           *                            *
           *                            *
Physical and management                 *
   events to give                       *
   preaccident conditions

                            Recommendations (and events) may
                            address historical events; for
                            example, plant design procedures
```

Figure 12-1 Accident Report Summary.

underlying causes and layered recommendations. These concepts are described in Example 12-2. The illustrated format is similar to the one used by Kletz.

The third layer recommendations shown in Figure 12-1 are highlighted to emphasize the importance of management systems for preventing accidents. Management systems are designed to continuously, and on a long term basis, either prevent the accident or eliminate the hazardous conditions; that is, break the link in the chain of events which led to the accident. Examples may be (a) a quarterly audit program to insure recommendations are understood and used, (b) a semiannual orientation program to review and study accident reports, or (c) a checklist which is initiated by management and checked by operations on a daily basis.

Layered events and recommendations are developed primarily by experienced personnel. For this reason, some experienced personnel are always assigned to investigation teams. Inexperienced team members learn from the experienced personnel, and often they also make significant contributions through an open and probing discussion.

Example 12-2

Use the investigation described in Example 12-1 to develop an investigation summary.

Solution

Accident title:	Drowning in municipal pool.
Major damage:	5 year old fatality.
Date:	xx/xx/xx
Location:	Detroit municipal pool, Park Z.

Recommendations for prevention/mitigation

1st layer: Immediate technical recommendation
2nd layer: Recommendations for avoiding hazard
3rd layer: Recommendations for improving the Management Systems

Events	Recommendations
Drowning(5 yr old)	Add more lifeguards. Add roving lifeguard.
Child knocked in pool	Initiate rule to keep small children away from the deep region of the pool.
Crowded pool	Limit number of swimmers to improve visibility and control.
Horseplay	Give periodic pool orientation and prohibit running and horseplay.
Pool depth not marked	Paint pool depths at edges of pool.
No swimming lessons	Initiate swimming lessons for all levels.
	Train lifeguards to watch for safety problems.
	Initiate periodic audit program to monitor adherence to rules and regulations.

12-5 AIDS FOR DIAGNOSIS[4]

The data collected during an investigation is studied and analyzed to establish the causes of the accident and to develop recommendations to prevent a recurrence. In most cases the evidence clearly supports one or more causes. Sometimes, however, the evidence needs added analysis to uncover explanations. This phase of the investigation may require special techniques or aids to diagnosis to relate the evidence to specific causes.

Fires

The identification of the primary source of ignition is one of the major objectives of investigations. In this regard, observations around the charred remains are helpful. For example, the depth of wood charring is proportional to the duration of burning

[4]Howard H. Fawcett and William S. Wood, eds., *Safety and Accident Prevention in Chemical Operations,* p. 668.

and most woods burn at a rate of 1.5 inches per hour. Therefore, if the time of extinguishment is known, and the depth of char at various locations is known, the region of the origin can be approximated.

Further searching in this region may reveal possible causes of the fire as shown in the following discussion.

The fire temperature for various materials, such as, wood, plastic, and solvents, is approximately 1000°C. Since pure copper melts at 1080°C, copper wire usually survives fires. If copper beads are found around electrical equipment, it may indicate that electrical arcs created temperatures greater than those observed in fires. Sometimes pits at the ends of conductors indicate high temperatures and vaporization of copper while arcing. Although this type of evidence indicates a source of ignition, it may not be the primary source of the fire.

The integrity of steelwork is not very useful evidence. Iron and steel have high melting points (1300-1500°C) compared to fire temperatures. However, steel weakens at approximately 575°C; therefore, steelwork may be completely distorted.

Aluminum and alloys of aluminum have very low melting points (660°C). All aluminum products will, therefore, meltdown during fires. This evidence together with steelwork distortions is not useful and deeper analysis in this regard should be avoided.

Explosions

The classifications of the explosion as either a deflagration or detonation, and the magnitude of the explosion, may be useful for developing causes and recommendations during accident investigations.

Deflagrations. Breaks in pipes or vessels due to deflagrations or simple overpressurizations are usually tears with lengths no longer then a few pipe diameters.

The pressure increases during deflagrations are approximately[5]

$$\frac{p_2}{p_1} \approx 8 \text{ for hydrocarbon-air mixtures} \tag{12-1}$$

$$\frac{p_2}{p_1} \approx 16 \text{ for hydrocarbon-oxygen mixtures} \tag{12-2}$$

For pipe networks, the pressure will increase in front of the flame front as the flame travels through the network. The downstream pressure may be 8 to 16 times greater than the original upstream pressure. This concept is called pressure piling. With pressure piling, therefore, $p_2 = p_1 \times 8 \times 8$ and $p_2 = p_1 \times 16 \times 16$ for Equations 12-1 and 12-2 respectively.

During deflagrations in vessels, the pressure is uniform throughout the vessel; therefore, the failure occurs at the vessel's weakest point. The damage is manifested as tears (detonations give shearing failures) and the point of ignition has no relationship to the ultimate point of failure.

Hydraulic and pneumatic failures. Hydraulic high pressure failures also give relatively small tears, compared to pneumatic failures which are very destructive.

[5]Frank P. Lees, *Loss Prevention in the Process Industries* (Boston: Butterworths, 1983), p. 567.

Rapidly expanding gases give large tears and can propel missiles, drums, and vessels, great distances.

Detonations. As described in Chapter 6, detonations have a rapidly moving flame and/or pressure front. Detonation failures usually occur in pipelines or vessels with large length to diameter ratios.

In a single vessel, detonations increase pressures significantly.[6]

$$\frac{p_2}{p_1} \approx 20 \tag{12-3}$$

When a pipe network is involved, the downstream p_1 increases due to pressure piling; therefore p_2 may increase as much as another factor of 20.

Detonation failures in pipe networks are always downstream from the ignition source. They usually occur at pipe elbows or other pipe constrictions, like valves. Blast pressures may shatter an elbow into many small fragments. A detonation in light gauge ductwork may tear the duct along seams, and also produce a large amount of structural distortion in the torn ducts.

In pipe systems, explosions can initiate as deflagrations and the flame front may accelerate to detonation speeds.

Sources of Ignition in Vessels

When a vessel ruptures due to a deflagration the source of ignition is usually coincident with the point of maximum vessel thinning due to expansion. Therefore, if the vessel parts are reconstructed the source of ignition is at the point with the thinnest walls.

Pressure Effects

When investigating pipe or vessel ruptures it is important to know the pressures required to create the damage and ultimately to determine the magnitude and source of energy.

The pressure necessary to produce a specific stress in a vessel is dependent on the thickness of the vessel, the vessel diameter, and the mechanical properties of the vessel wall.[7] For cylindrical vessels with the pressure, p, not exceeding 0.385 times the mechanical strength of the material, S_M,

$$\boxed{p = \frac{S_M t_v}{r + 0.6 t_v}} \tag{12-4}$$

where

p is the internal gauge pressure,
S_M is the strength of the material,
t_v is the wall thickness of the vessel, and
r is the inside radius of the vessel.

[6] Lees, *Loss Prevention*, p. 569.

[7] Samuel Strelzoff and L. C. Pan, "Designing Pressure Vessels," *Chemical Engineering*, Nov. 4, 1968, p. 191.

For cylindrical vessels and pressures exceeding 0.385 S_M, the following equation applies.

$$p = \frac{S_M\left(\frac{t_v}{r} + 1\right)^2 - S_M}{\left(\frac{t_v}{r} + 1\right)^2 + 1} \tag{12-5}$$

For spherical vessels with pressures not exceeding 0.665 S_M.

$$p = \frac{2\, t_v\, S_M}{r + 0.2\, t_v} \tag{12-6}$$

For spherical vessels and pressures exceeding 0.665 S_M,

$$p = \frac{2\, S_M\left(\frac{t_v}{r} + 1\right)^2 - 2\, S_M}{\left(\frac{t_v}{r} + 1\right)^2 + 2} \tag{12-7}$$

These formulas are used to determine the pressure required to produce elastic deformations by using yield strengths for S_M. They are also used to determine the pressures required to produce failures by using tensile strengths for S_M. Strength of material data is provided in Table 12-3.

High pressure failures are as likely to occur in a pipe or pipe system as they are in vessels. The maximum internal pressure for pipes is calculated using Equations 12-4 and 12-5.

After the maximum internal pressure is computed, the explosive energy is computed using Equation 6-17. The ultimate source of this explosion energy is

TABLE 12-3 STRENGTH OF MATERIALS[1]

Material	Tensile strength (psi)	Yield point (psi)
Borosilicate Glass	10,000	
Carbon	660	
Duriron	60,000	30,000
Hastelloy C	72,000	48,000
Nickel	65,000	48,000
Stainless 304	80,000	35,000
Stainless 316	85,000	40,000
Stainless 420	105,000	55,000

[1]Robert H. Perry and Cecil H. Chilton, eds., *Chemical Engineers' Handbook* (New York: McGraw-Hill Book Company, 1973), p. 6-96, 6-97.

found by developing various reaction or mechanical hypotheses and comparing the reaction energy to the explosion energy, until the most likely hypothesis is identified. After the energy and ignition sources are identified attention is placed on developing conditions to prevent the source of failure.

Medical Evidence

Medical examinations of the accident victims result in evidence which may be useful for identifying the source of the accident or for identifying some circumstances which may help to uncover underlying causes.

The types of medical data which help accident investigations include (a) type and level of toxic or abusive substances in the blood, (b) location and magnitude of injuries, (c) type of poisoning (carbon monoxide, toluene, etc.), (d) signs of suffocation (e) signs of heat exposure or heat exhaustion, and (f) signs of eye irritation.

Miscellaneous Aids to Diagnosis

Other aids for identifying underlying causes of accidents are found throughout this text. During an accident investigation the investigation team must watch for visible evidence, and they must also make supporting calculations to evaluate various hypotheses. A brief review of safety fundamentals prior to the investigation is helpful. This includes, for example, (a) toxicity of chemicals or combinations of chemicals, (b) explosion limits, (c) magnitude of leaks depending on the source, (d) dispersion of vapors outside or inside plants, (e) principles of grounding and bounding, (f) principles of static electricity, (g) design concepts for handling flammables, and (h) methods for performing accident investigations. This knowledge and information will be very useful during an investigation.

Example 12-3

Determine the pressure required to rupture a cylindrical vessel, if the vessel is stainless 316, has a radius of 3 ft, and has a wall thickness of 0.5 in.

Solution Since the pressure is unknown, Equation 12-4 or 12-5 is used by trial and error until the correct equation is identified. Equation 12-4 is applicable for pressures below 0.385 S_M. Since S_M (from Table 12-3) is 85,000 psi, 0.385 S_M = 32,700 psi.

$$r = 3 \text{ ft} = 36 \text{ in}$$

$$t_v = 0.5 \text{ in}$$

Substituting into Equation 12-4 for cylindrical vessels,

$$p = \frac{S_M t_v}{r + 0.6 t_v} = \frac{(85{,}000 \text{ psi})(0.5 \text{ in})}{(36 \text{ in}) + 0.6(0.5 \text{ in})} = 1{,}170 \text{ psi}.$$

Therefore Equation 12-4 is applicable, and a pressure of 1,170 psi is required to rupture this vessel.

Example 12-4

Determine the pressure required to rupture a spherical vessel, if the vessel is stainless 304, has a radius of 5 ft, and has a wall thickness of 0.75 in.

Solution This problem is similar to Example 12-3; Equation 12-6 is applicable if the pressure is less than 0.665 S_M or 0.665 (80,000) = 53,200 psi. Using Equation 12-6 for spherical vessels,

$$p = \frac{2t_v S_M}{r + 0.2t_v} = \frac{2(0.75 \text{ in})(80{,}000 \text{ psi})}{(5 \text{ ft})(12 \text{ in/ft}) + 0.2(0.75 \text{ in})} = 1{,}990 \text{ psi}$$

The pressure criteria is met for this equation. The pressure required to rupture this vessel is 1,990 psi.

Example 12-5

During an accident investigation, it is found that the source of the accident was an explosion which ruptured a 4-in diameter stainless 316, schedule 40 pipe. It is hypothesized that a hydrogen and oxygen deflagration or a detonation was the cause of the accident. Deflagration tests in a small spherical vessel indicate a deflagration pressure of 500 psi. What pressure ruptured the pipe, and was it a deflagration or a detonation that caused this rupture?

Solution A 4-in schedule 40 pipe has an outside diameter of 4.5 in, a wall thickness of 0.237 in, and an inside diameter of 4.026 in. From Table 12-3, the tensile strength, S_M, for stainless 316 is 85,000 psi. Equation 12-4 for cylinders is used to compute the pressure necessary to rupture this pipe.

$$p = \frac{S_M t_v}{r + 0.6t_v} = \frac{(85{,}000 \text{ psi})(0.237 \text{ in})}{(2.013 \text{ in}) + 0.6(0.237 \text{ in})} = 9{,}348 \text{ psi}$$

Equation 12-4 is applicable because the pressure is less than (0.385) S_M = 32,700 psi. The pressure required to rupture this pipe, therefore, is 9,348 psi. Using the deflagration test data which gave a p_2 of 500 psi and assuming pressure piling, the deflagration pressure in the pipe is estimated using Equation 12-2,

$$p_2 = 500 \times 16 = 8{,}000 \text{ psi}$$

For estimating pressures due to a detonation and pressure piling, the original deflagration test pressure, p_1, is estimated using Equation 12-2.

$$p_1 = 500/16 = 31.3 \text{ psi}$$

A detonation with pressure piling is now computed using Equation 12-3.

$$p_2 = 31.3 \times 20 \times 20 = 12{,}500 \text{ psi}$$

This pipe rupture was, therefore, due to a detonation. The next step in the investigation would include the search for a chemical reaction which would give a detonation. A small vessel could be used as a test.

Example 12-6

An explosion rips through a chemical plant. A 1000 cubic ft tank containing compressed air at 100 atm is suspected. Site damage indicates that the windows in a structure 100 yards away are shattered. Is the mechanical explosion of this compressed air tank consistent with the damage reported, or is the explosion the result of some other process?

Solution From Equation 6-17, representing the energy contained in a compressed gas,

$$W_e = \left(\frac{P_1 V_1}{\gamma - 1}\right)\left[1 - \left(\frac{P_2}{P_1}\right)^{(\gamma - 1)/\gamma}\right]$$

For air, $\gamma = 1.4$. Substituting the known quantities,

$$= \frac{(101\text{ atm})\left(14.7\dfrac{\text{lb}_f/\text{in}^2}{\text{atm}}\right)\left(144\dfrac{\text{in}^2}{\text{ft}^2}\right)(1{,}000\text{ ft}^3)}{1.4 - 1} \times \left[1 - \left(\frac{1\text{ atm}}{101\text{ atm}}\right)^{(1.4-1)/1.4}\right]$$

$$= 3.91 \times 10^8 \text{ ft-lb}_f = 1.27 \times 10^8 \text{ cal}$$

The equivalent amount of TNT is

$$m_{\text{TNT}} = 1.27 \times 10^8 \text{ cal}/(1120 \text{ cal/gm TNT}) = 1.13 \times 10^5 \text{ gm TNT}$$
$$= 249 \text{ lb of TNT}$$

From Equation 6-15 the scaling factor is

$$z_e = \frac{r}{(m_{\text{TNT}})^{1/3}}$$

Substituting

$$z_e = \frac{300\text{ ft}}{(249\text{ lb})^{1/3}} = 47.66\text{ ft/lb}^{1/3}$$

From Figure 6-12, the overpressure is estimated at 0.62 psia. From the data provided in Table 6-7 the damage is consistent with the damage observed.

12-6 AIDS FOR RECOMMENDATIONS

Recommendations are the most important result of an accident investigation. They are made to prevent a recurrence of the specific accident, but they are also made to prevent similar accidents within the company and within the industry. The ultimate result of accident investigations is the elimination of the underlying causes of entire families of accidents. One good accident investigation can prevent hundreds of accidents.

There are three overriding principles which are used to influence accident investigation recommendations.

1. Make safety investments on a basis of cost and performance. Evaluate each investment (money and time) to insure there is a true safety improvement proportional to the investment. If the designer is not careful, changes to the system or new procedures may add complexities which result in a more hazardous situation versus an improvement.
2. Develop recommendations to improve the management system to prevent the existence of safety hazards, including training, checklists, inspections, safety reviews, audits, and the like.
3. Develop recommendations to improve the management and staff support to safety with the same enthusiasm, attention, quality, plans, and organization as used in production programs.
4. Develop layered recommendations with an appropriate emphasis on recommendations to eliminate underlying causes of accidents.

All the fundamentals described in this text are commonly used to develop recommendations. Some aids to recommendations are covered in the following sections.

Control Plant Modifications

Modifications to plants are often not given the same attention and concern as a new plant design. Additionally, they are sometimes the result of mechanical problems which shut the plant down, and in these situations all efforts are directed towards a quick restart. Many accidents are the result of plant modifications.

Recommendations are especially designed to prevent this kind of problem.

1. Authorization: All modifications must be authorized by several levels of management.
2. Design: The modification designs must be mechanically constructed with the same quality of equipment and pipes as the original design. Original designs should be studied so that the consequences of any change is understood. The designers must appreciate that for every problem there are many interesting, economically sound, plausible, and *wrong* solutions.
3. Safety Reviews: A safety review (HAZOPS or equivalent for very hazardous operations) must be conducted by engineers, operators, and design specialists while the modification project is in the design phase. This allows (and encourages) safety changes to be made with minimum effort. Once the system is constructed, changes are difficult and costly to make.
4. Training: Engineers and operators need sufficient training to understand and appreciate the modified operation.
5. Audit: Every plant modification needs periodic audits to be sure that the modifications are made and maintained as designed.

User Friendly Designs[8]

New plants or modifications to existing plants must be designed to be friendly, to tolerate departures from the norm without creating hazardous conditions. Examples of friendly designs include using nontoxic and nonflammable solvents when possible, keeping temperatures below the flash point and below the boiling point at atmospheric conditions, keeping inventories low, and designing for safe shutdowns during emergency situations (expect the unexpected).

Block Valves

Block valves are installed throughout plants to return a process to a safe condition under unusual circumstances. For example, the process shown in Figure 12-2 detects a hose leak by comparing flow rates at both ends of the hose. If the hose breaks, the leak is detected and the block valves on the reactor and sewer are immediately closed.

[8]Trevor Kletz, *Learning from Accidents in Industry* (Boston: Butterworths, 1988), p. 148.

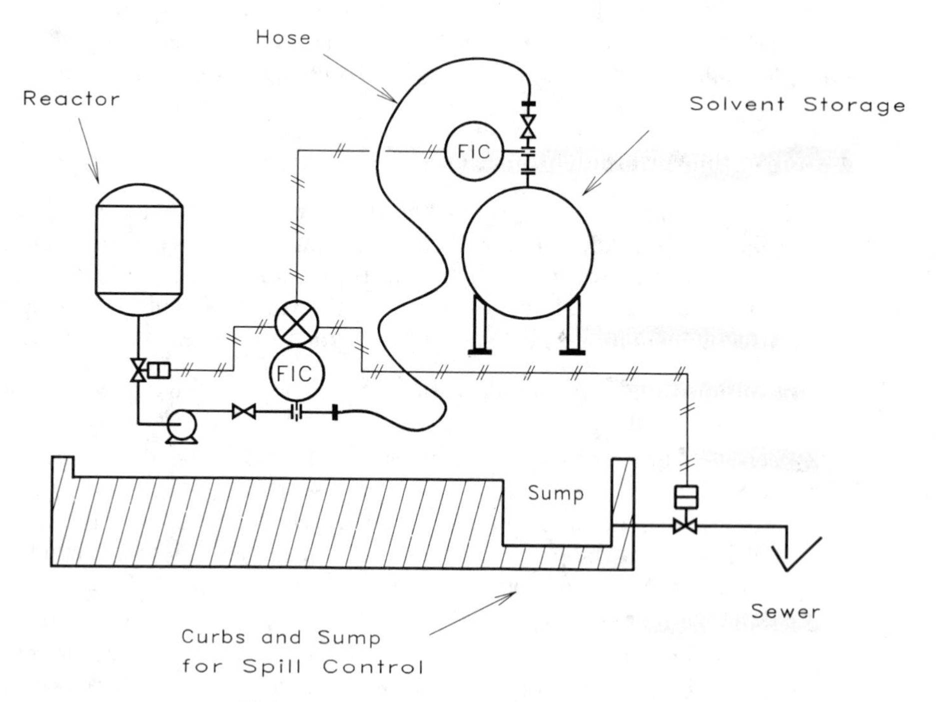

Figure 12-2 A block valve arrangement used to prevent leakage from the connecting hose. If the flow at both ends of the hose is not identical, the block valves are closed.

Block valves are often controlled on a basis of analyzer results like area monitors for detecting solvent leaks, reactor analyzers for detecting runaway reactions (a block valve can be opened to add a reaction inhibitor or to turn on a deluge system), sewer analyzer for detecting high concentrations of contaminants, and vent analyzer to detect high levels of contaminants.

Double Block and Bleed

A special double block and bleed system shown in Figure 12-3, is added to every feed line to a reactor. During normal operating conditions the block valves are open and the bleed line is closed. When the feed pump is shut off, the block valves are closed and the bleed line is open.

This system prevents the reactor contents from siphoning back into the monomer storage vessel, even if the block valves leak. This prevents an unexpected chemical reaction in the storage tank.

This is a relatively simple example of a particularly important application of double block and bleed systems. They are also commonly used for reactive intermediates, and analyzer systems; anywhere a positive break in a line is desired.

Preventative Maintenance

Most engineers are aware of the importance of preventative maintenance programs, especially those owning automobiles or homes. A little neglect can cause serious

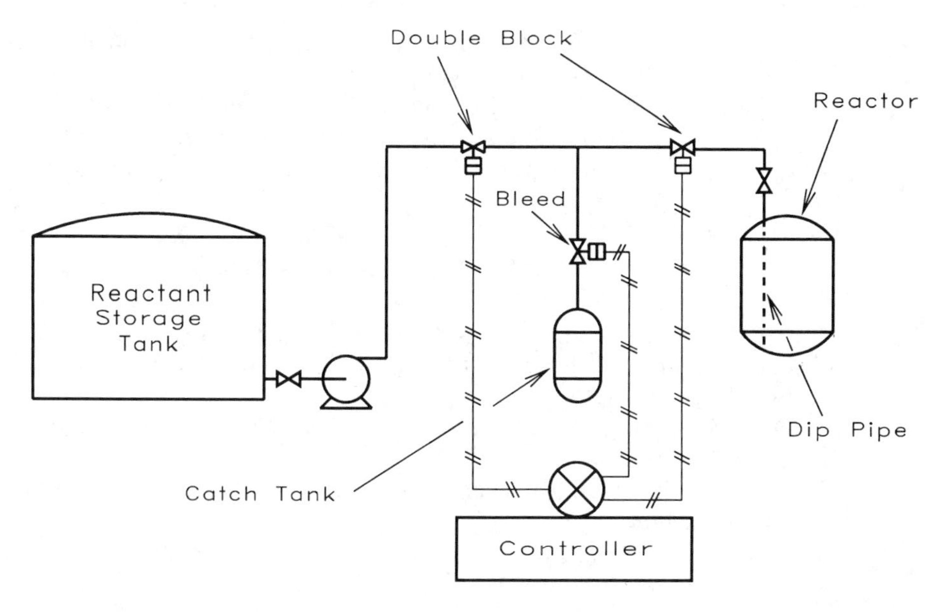

Figure 12-3 A double block and bleed arrangement used to prevent reactant from entering reactor vessel.

property damage and may be the genesis of serious accidents; for example, poorly maintained brake systems have inevitable consequences.

In plants, one major cause of accidents is the failure of emergency protection equipment like cooling water pumps, instruments, and deluge systems. Many times, when evaluating underlying causes of accidents, it is found that protective equipment failed because it was neglected; there was no preventative maintenance program. In this case new procedures or new equipment is not needed; adding more protective equipment or procedures might increase the likelihood of accidents. The only improvement needed is to upgrade the existing maintenance program.

Preventative maintenance programs must be organized, managed, and fully supported by management. Good results may not be immediately apparent, but bad results are very apparent when plants are not appropriately maintained.

Good maintenance programs include scheduled maintenance and a system to keep an inventory of critical maintenance parts. Every maintenance job requires a feedback mechanism based on the inspection of parts while conducting the maintenance. The maintenance schedule is subsequently changed if more frequent maintenance is required.

Analyzers

Chemical analysis of reactor contents and of the surrounding environment is an important way to understand the status of a plant and to identify problems at the incipient state of development. When problems are identified at an early stage action can be taken to return the system to safe operating regions with no adverse consequences.

In recent years, new and better analyzers are being developed. Design engineers should always search for new opportunities to use process analyzers to improve operations and safety within plants. As the reliability and applicability of analyzers are improved, they will become the key control elements in chemical plants, particularly in regard to safety, quality, and yield improvements.

SUGGESTED READING

A Thirty Year Review of One Hundred of the Largest Property Damage Losses in the Hydrocarbon-Chemical Industry, 11th ed. (Chicago: Marsh and McLennan Protection Consultants, 1988).

TREVOR A. KLETZ, "Layered Accident Investigations," *Hydrocarbon Processing,* Nov. 1979, pp. 373-382.

TREVOR A. KLETZ, *What Went Wrong? Case Histories of Process Plant Disasters* (Houston: Gulf Publishing Company, 1985).

TREVOR A. KLETZ, *Learning from Accidents in Industry* (Boston: Butterworths, 1988).

PROBLEMS

12-1. Use the Flixborough Works accident described in Chapter 1, to develop an investigation similar to Example 12-1.

12-2. Use the Flixborough Works accident and the investigation developed in Problem 12-1, to develop an investigation summary similar to Example 12-2. Include layered recommendations to cover the accident causes and underlying causes.

12-3. Use the Bhopal, India, accident described in Chapter 1 to develop an investigation similar to Example 12-1.

12-4. Use the Bhopal accident and the investigation developed in Problem 12-3, to develop an investigation summary similar to Example 12-2. Include layered recommendations.

12-5. Use the Seveso, Italy, accident described in Chapter 1 to develop an investigation similar to Example 12-1.

12-6. Use the Seveso, Italy, accident and the investigation developed in Problem 12-5 to develop an investigation summary similar to Example 12-2. Include layered recommendations.

12-7. Develop an investigation similar to Example 12-1, and an investigation summary for an automobile accident that occurred due to a brake failure. Create your own brief accident scenario for this problem.

12-8. Determine the pressure required for a pipe to swell and the pressure required for a pipe failure. The pipe is 3-in, 316 stainless, schedule 40 pipeline for transporting a gas mixture which is sometimes within the explosive composition range.

12-9. Determine the required thickness of a reactor with cylindrical walls which must be designed to safely contain a deflagration (hydrocarbon plus air). The vessel has a diameter of 4 ft and is constructed with 304 stainless steel. The normal operating pressure is 2 atmospheres.

12-10. An accident occurs which ruptures a high pressure spherical vessel. The vessel is 1.5 ft in diameter, is made of 304 stainless, and the walls are 0.25 in thick. Determine the pressure required to cause this failure. Develop some hypotheses regarding the causes of this accident.

12-11. Compute the theoretical maximum pressure obtained after igniting a stoichiometric quantity of methane and oxygen in a spherical vessel which is 1.5 ft in diameter. Assume an initial pressure of 1-atm.

12-12. Compute the theoretical maximum pressure obtained after igniting a stoichiometric quantity of methane and air in a spherical vessel which is 1.5 ft in diameter. Assume the initial pressure is 1-atm.

12-13. Using the results of Problem 12-11, determine the required vessel wall thickness to contain this explosion if the vessel is stainless 316.

12-14. Using the results of Problem 12-13 determine the vessel wall thickness required to contain an explosion in another vessel which is physically connected to the first vessel with a 1-in pipe. Describe why the second vessel requires a greater wall thickness.

12-15. Describe why accident investigation recommendations must include recommendations to improve the management system.

12-16. Describe a preventative maintenance program which is designed to prevent automobile accidents.

12-17. Describe the concept of using block valves to prevent detonation accidents in a system handling flammable gases. The system has two vessels which are connected with a 4-in vapor line.

12-18. Using the data and results of Example 12-6, determine the wall thickness required to eliminate future failures. Assume the vessel's cylindrical wall height is equal to the vessel's diameter.

12-19. Determine the vessel wall thickness required to contain an explosion of 2 lb of TNT. The spherical vessel is 1.5 ft in diameter and is constructed with 316 stainless steel.

12-20. In the 1930s there were many accidents in homes due to the explosion of hot water heaters. Describe what features are added to water heaters to eliminate accidents.

12-21. A cloud of hydrogen gas is released and subsequently explodes. Glass is shattered 500 feet away. Estimate the quantity of hydrogen gas initially released, assuming stoichiometric quantities of hydrogen and air explode.

Case Histories

13

Case histories are written descriptions of accidents, including the causes, consequences, and methods required to prevent similar events. They are descriptions written by plant foreman and operating personnel. These are the people with the hands-on experience; the ones who know and appreciate the accident and accident prevention methods.

The study of case histories is important in the area of safety. To paraphrase G. Santayana, one learns from history or is doomed to repeat it. This is especially true for safety; anyone working in the chemical industry can learn from case histories and avoid hazardous situations or ignore history and be involved in potentially life threatening accidents.

This chapter covers case histories as reported in the literature. References are provided for more thorough studies. The objective of this chapter is to illustrate, through actual case histories, the importance of applying the fundamentals of chemical process safety.

These case histories are categorized into four sections:

- static electricity
- chemical reactivity
- system design
- procedures

The cause for a specific accident frequently places it into more than one category. Each of these sections will include descriptions of several accidents and a summary of the lessons learned.

The following statements will place these case histories into perspective.

1. These accidents actually occurred. Anyone familiar with the specific equipment or procedures will appreciate the lessons learned.
2. Accidents occur very rapidly and unexpectedly. There is usually inadequate time to manually return a situation back into control after a significant deviation from the norm is observed. Those who believe they can successfully control accident deviations manually, are doomed to repeat history.

13-1 STATIC ELECTRICITY

A very large proportion of the reported fires and explosions are the result of a flammable mixture being ignited by a spark caused by static electricity. Many of these accidents are repeats of previously recorded accidents; engineers are missing some of the important aspects of this subject. The following series of case histories is given to illustrate the complexity of this topic, and to give some important design requirements for preventing future accidents involving static electricity.

Tank Car Loading Explosion[1]

Two plant operators were filling a tank car with vinyl acetate. One operator was on the ground while the other was on top of the car with the nozzle end of a loading hose. A few seconds after the loading operation started, the contents of the tank exploded. The operator on top of the tank was thrown to the ground, sustaining a fractured skull and multiple body burns. He died from these injuries.

The accident investigation indicated that the explosion was caused by a static spark which jumped from the steel nozzle to the tank car. The nozzle was not bonded to the tank car to prevent static accumulation. The use of a nonmetallic hose probably also contributed.

Explosion in Centrifuge[2]

A slurry, containing a solvent mixture of 90 percent methylcyclohexane and 10 percent toluene, was being fed into a basket centrifuge. A foreman was about to look into the centrifuge when it exploded. The lid was lifted and a flame was released between the centrifuge and the lid. The foreman's hand was burned.

The fill line from the reactor to the centrifuge was Teflon-lined steel, running to a point three feet from the centrifuge where there was a rubber sleeve connector. The short line from the sleeve to the centrifuge was steel. The centrifuge was lined.

The accident investigation indicated that a flammable atmosphere was developed due to an air leak. The lined centrifuge was the source of ignition due to static accumulation and discharge.

[1]*Case Histories of Accidents in the Chemical Industry,* Vol. 1 (Washington, DC: Manufacturing Chemists' Association, July, 1962), p. 106.

[2]*Case Histories of Accidents in the Chemical Industry,* Vol. 2 (Washington, DC: Manufacturing Chemists' Association, January, 1966), p. 231.

Later (and successful) processing was conducted in a grounded stainless steel centrifuge which was inerted with nitrogen.

Duct System Explosion[3]

Two duct systems in the same vicinity contained dust transport lines, dryers, and hoppers. One system was recently repaired and left open. The open system emitted some methanol vapors. The other system was being charged through a funnel with a dry organic intermediate. The charge line consisted of a new glass pipe and a six-foot section of plastic pipe. The duct system which was being charged exploded violently and the explosion initiated other fires. Fortunately, no one was seriously injured.

The accident investigation indicated that methanol vapors entered the second charging system. The transportation of the intermediate dust through the glass and plastic line generated a static charge and spark. The ignition source created violent explosions in both systems. Several explosion vents were ruptured and a building blowout panel also ruptured.

This accident points out the need for carefully reviewing systems before, during, and after modifications are made. Open lines should be blanked off when the discharge of flammable vapors is possible. Also, proper grounding and bonding techniques must be used to prevent static buildup.

Lessons Learned

Case histories involving static electricity emphasize the importance of understanding and using the fundamentals described in Chapter 7. In reviewing approximately 30 additional case histories regarding static electricity, some important lessons were identified. (1) A built-in ground line is rendered nonconductive by the use of a nonconductive pipe-dope. (2) A potential is generated between two vessels which are not bonded. (3) Leather arch supporters make shoes ineffective against static. (4) Free fall filling generates static charge and discharge. (5) The use of nonmetallic hoses is a source of static buildup. (6) Large voltages are generated when crumpling and shaking an empty polyethylene bag. (7) A weak grounding clamp may not penetrate the paint on a drum adequately to provide a good electrical contact.

A number of recommendations are also developed. (a) Operators must be cautioned against drawing pipes or tubing through their rubber gloves, resulting in static buildup. (b) Clothing which generates static-electricity must be prohibited. (c) Recirculation lines must be extended into the liquid to prevent static buildup. (d) Shoes with conductive soles are required when handling flammables. (e) Bonding, grounding, humidification, ionization or combinations are recommended when static is a fire hazard. (f) A small water spray will rapidly drain electrical charges during chopping operations. (g) Inert gas blankets must be used when handling flammables. (h) Drums, scoops, and bags should be physically bonded and grounded. (i) Ground connections must be verified with a resistance tester. (j) Spring loaded grounding or bonding clips should be replaced with screw type C

[3] *Case Histories of Accidents in the Chemical Industry,* Vol. 3 (Washington, DC: Manufacturing Chemists' Association, April, 1970), p. 95.

clamps. (k) Conductive grease should be used in bearing seals which need to conduct static charges. (l) Sodium hydride must be handled in static-proof bags. (m) Stainless steel centrifuges must be used when handling flammables. (n) Flanges in piping and duct systems must be bonded.

Example 13-1

Using the layered accident investigation process discussed in Chapter 12, develop the underlying causes of the first accident described in Section 13-1, Tank Car Loading Explosion.

Solution The facts uncovered by the investigation are

a. Contents at the top of vessel were flammable.

b. The charging line was a nonconductive hose.

c. A spark probably jumped between the charging nozzle and the tank car.

d. The explosion knocked the man off the tank car. The fatal injury was probably the fractured skull sustained in the fall.

e. No inspection or safety review procedure was in place to identify problems of this kind.

Layered recommendations are the result of uncovering the underlying causes of the accident.

FIRST LAYER RECOMMENDATIONS: IMMEDIATE TECHNICAL RECOMMENDATIONS.

a. Use a conductive metal hose for transferring flammable fluids.

b. Bond hose to tank car and ground tank car and hose.

c. Provide dip pipe design for charging tank cars.

d. Provide a means to nitrogen pad the tank car during the filling operation.

e. Add guard rails to charging platforms to prevent accidental falls from the top of the tank car to the ground.

SECOND LAYER RECOMMENDATIONS: AVOIDING THE HAZARD.

a. Develop tank car loading procedures.

b. Develop and give operators special training so the hazards are understood for every loading and unloading operation.

THIRD LAYER RECOMMENDATIONS: IMPROVING THE MANAGEMENT SYSTEM.

a. Initiate an immediate inspection of all loading and unloading operations.

b. Initiate, as a standard practice, a policy to give all new loading and unloading applications a safety review. Include engineers and operators in this review.

c. Initiate a periodic (every six months) audit to insure all standards and procedures are effectively utilized.

13-2 CHEMICAL REACTIVITY

Although accidents attributable to chemical reactivity are less frequent compared to fires and explosions, the consequences are dramatic, destructive, and often very injurious to personnel. When working with chemicals the potential for unwanted, unexpected, and hazardous reactions must always be recognized. The following case histories illustrate the importance of understanding the complete chemistry of a reaction system including potential side reactions, decomposition reactions, and reactions due to the accidental and wrong combination of chemicals or reaction conditions (wrong type, wrong concentrations, or the wrong temperature).

Bottle of Isopropyl Ether[4]

A chemist needed isopropyl ether. He found a pint glass bottle. He unsuccessfully tried to open the bottle over a sink. The cap appeared to be stuck tightly so he grasped the bottle in one hand, pressed it to his stomach and twisted the cap with his other hand. Just as the cap broke loose the bottle exploded, practically disemboweling the man and tearing off several fingers. The victim remained conscious and, in fact, coherently described how the accident happened. The man was taken to a hospital and died within two hours of the accident of massive internal hemorrhage.

An accident investigation identified the cause of the accident to be the rapid decomposition of peroxides which were formed in the ether while sitting in storage. It is hypothesized that some of the peroxides may have crystallized in the threads of the cap and exploded when the cap was turned.

As ethers age, especially isopropyl ether, they form peroxides. The peroxides react further to form additional hazardous by-products, such as triacetone peroxide. These materials are very unstable. Light, air, and heat accelerate the formation of peroxides.

Ethers should be stored in metal containers. Only small quantities should be purchased. Ethers should not be kept over six months. Containers should be labeled and dated upon receipt and opened containers should be discarded after three months. All work with ethers should be done behind safety shields. Inhibitors should be used whenever possible.

Nitrobenzene Sulfonic Acid Decomposition[5]

A 300 gallon reactor experienced a violent reaction resulting in the tank being driven through the floor, out the wall of the building and through the roof of an adjoining building. The reactor was designed to contain 60 gallons of sulfuric acid and nitrobenzene sulfonic acid, which was known to decompose at 200°C.

The investigation indicated that the vessel contents were held for eleven hours. A steam leak into the jacket brought the temperature to about 150°C. Although previous tests indicated decomposition at 200°C, subsequent tests showed exothermic decomposition above 145°C.

The underlying cause of this accident was the lack of precise reaction decomposition data. With good data, engineers can design safeguards to absolutely prevent accidental heat-up.

[4]*Case Histories,* Vol. 2, p. 6.

[5]*Case Histories,* Vol. 3, p. 111.

Organic Oxidation[6]

Chemical operators were preparing for an organic oxidation. Steam was applied to the reactor jacket to heat the sulfuric acid and an organic to a temperature of 70°C. The rate of heating was slower than normal. The two operators turned the agitator off and also shut off the steam. One operator went to find a thermometer. Approximately one hour later, the operator was ready to take a temperature reading through the manhole. He turned on the agitator. At this point the material in the kettle erupted through the manhole. The two operators were drenched and both died due to these injuries.

The accident investigation stated that the agitator should never be turned off for this type reaction. Without agitation, cooling is no longer efficient; so heat-up occurs. Without agitation, segregation of chemicals also occurs. When the agitator is subsequently activated, the hotter chemicals mix and react violently.

This type problem is currently preventable through better operator training and installation of electronic safeguards to prevent operators from making this mistake. This is achieved by adding redundant and remote temperature sensors and adding electronic interlocks to prevent the agitator from being turned off while the reaction is still exothermic.

Lessons Learned

Case histories regarding reactive chemicals teach the importance of understanding the reactive properties of chemicals before working with them. The best source of data is the open literature. If data is not available, experimental testing is necessary. Data of special interest includes decomposition temperatures, rate of reaction or activation energy, impact shock sensitivity, and flash point.

Functional Groups. A preliminary indication of the potential hazards can be estimated by knowing something about the chemical structure. Specific functional groups which contribute to the explosive properties of a chemical through rapid combustion or detonation are illustrated in Table 13-1.

Peroxides. Peroxides *and* peroxidizable compounds are dangerous sources of explosions. Structures of peroxidizable compounds are shown in Table 13-2. Some examples of peroxidizable compounds are given in Table 13-3.

When peroxide concentrations increase to 20 ppm or greater, the solution is hazardous. Methods for detecting and controlling peroxides are outlined by H. L. Jackson, et al.[7]

Reaction Hazard Index. D. R. Stull[8] developed a rating system to establish the relative potential hazards of specific chemicals; the rating is called Reaction Hazard

[6] *Case Histories,* Vol. 3, p. 121.

[7] H. L. Jackson, et al., "Control of Peroxidizable Compounds," *Safety in the Chemical Industry,* Vol. 3, Norman V. Steere, ed. (Easton, PA: Division of Chemical Education of the American Chemical Society, 1974), pp. 114-117.

[8] D. R. Stull, "Linking Thermodynamic and Kinetics to Predict Real Chemical Hazards," *Safety in the Chemical Industry,* pp. 106-110.

TABLE 13-1 REACTIVE FUNCTIONAL GROUPS[1]

Azide	N_3
Diazo	$—N{=}N—$
Diazonium	$—N_2^+ X^-$
Nitro	$—NO_2$
Nitroso	$—NO$
Nitrite	$—ONO$
Nitrate	$—ONO_2$
Fulminate	$—ONC$
Peroxide	$—O—O—$
Peracid	$—CO_3H$
Hydroperoxide	$—O—O—H$
Ozonide	O_3
N-haloamine	$—N(X)—Cl$
Amine oxide	$\equiv NO$
Hypohalites	$—OX$
Chlorates	ClO_3
Acetylides of heavy metals	$—C{\equiv}CM$

[1]Conrad Schuerch, "Safe Practice in the Chemistry Laboratory: A Safety Manual," *Safety in the Chemical Laboratory,* Vol. 3, ed. Norman V. Steere (Easton, PA: Division of Chemical Education of the American Chemical Society, 1974), pp. 22-25.

Index (RHI). The RHI is related to the maximum adiabatic temperature reached by the products of a decomposition reaction. It is defined as

$$\boxed{\text{RHI} = \frac{10\,T_d}{T_d + 30\,E_a}} \tag{13-1}$$

where

T_d is the decomposition temperature (K), and
E_a is the Arrhenius activation energy (kcal/mole).

The RHI relationship, Equation 13-1, has a low value (1 to 3) for relatively low reactivities and higher values (5 to 8) for high reactivities. Some RHI data for various chemicals is provided in Table 13-4.

Example 13-2

Compute the Reaction Hazard Index (RHI) for isopropyl ether and compare the result to that shown in Table 13-4. Explain why the RHI is relatively low.

Solution The RHI is computed using Equation 13-1,

$$\text{RHI} = \frac{10\,T_d}{T_d + 30\,E_a}$$

TABLE 13-2 PEROXIDIZABLE COMPOUNDS[1]

Organics

---

1. Ethers, acetals: $-\overset{\displaystyle H}{\underset{|}{C}}-O-$
2. Olefins with allylic hydrogen, chloro and fluoroolefins, terpens, tetrahydronaphthalene:
   $>C=C<$
3. Dienes, vinyl acetylenes: $>C=\overset{|}{C}-\overset{|}{C}=C<$
   and: $>C=\underset{|}{C}-C\equiv CH$
4. Parafins and alkyl-aromatic hydrocarbons, particularly those with tertiary hydrogen:
   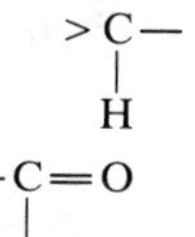
5. Aldehydes: $-\underset{\displaystyle H}{\underset{|}{C}}=O$
6. Ureas, amides, lactones: $-\overset{\displaystyle O}{\overset{\|}{C}}-\overset{|}{N}-\overset{\displaystyle H}{\overset{|}{C}}$
7. Vinyl monomers including vinyl halides, acrylates, methacrylates, vinyl esters: $>C=\overset{|}{C}-$
8. Ketones having an alpha-hydrogen: $-\underset{\displaystyle O}{\underset{\|}{C}}-\underset{\displaystyle H}{\underset{|}{C}}<$

Inorganics

---

1. Alkali metals, particularly potassium
2. Alkali metal alkoxides and amides
3. Organometallics

---

[1]H. L. Jackson, W. B. McCormack, C. S. Rondestvedt, K. C. Smeltz, and I. E. Viele, "Control of Peroxidizable Compounds," *Safety in the Chemical Industry,* Vol. 3, ed. Norman V. Steere (Easton, PA: Division of Chemical Education of the American Chemical Society, 1974), pp. 114-117.

where, from Table 13-4, T_d is 712°K and E_a is 63.5 kcal/mole. The units are compatible with Equation 13-1. Substituting,

$$\text{RHI} = \frac{(10)(712)}{(712) + (30)(63.5)}$$

$$= 2.72$$

which is the same as the value given in Table 13-4. This RHI indicates a chemical with low reactivity. However, isopropyl ether is a peroxidizable compound as indi-

TABLE 13-3 EXAMPLES OF PEROXIDIZABLE COMPOUNDS[1]

Peroxidizable Hazard on Storage:
Isopropyl ether
Divinyl acetylene
Vinylidene chloride
Potassium metal
Sodium amide
Peroxidizable Hazard on Concentration:
Diethyl ether
Tetrahydrofuran
Dioxane
Acetal
Methyl i-butyl ketone
Ethylene glycol dimethyl ether (glyme)
Vinyl ethers
Dicyctapentadiene
Diacetylene
Methyl acetylene
Cumene
Tetrahydronaphthalene
Cyclohexane
Methylcyclopentane
Hazardous when Exposed to Oxygen due to Peroxide Formation and Subsequent Peroxide Initiation of Polymerization:
Styrene
Butadiene
Tetrafluoroethylene
Chlorotrifluoroethylene
Vinyl acetylene
Vinyl acetate
Vinyl chloride
Vinyl pyridine
Chloroprene

[1]Jackson, et. al., "Control of Peroxidizable Compounds," pp. 114-117.

cated in Table 13-3. If we assume an RHI equivalent to diethyl peroxide (RHI = 4.64), the hazards of handling isopropyl ether are high even at concentrations as low as 20 ppm. This example illustrates the importance of understanding the chemistry of the entire system.

13-3 SYSTEM DESIGNS

When new plants are constructed or modifications needed in existing plants, detailed process designs are required. These designs must include special safety features to protect the system and operating personnel. The following case histories emphasize the importance of these special safety design features.

TABLE 13-4 REACTION HAZARD INDEX DATA[1]

No.	Formula	Compound	Decomp. temp. (°K)	Activation energy (kcal/mole)	RHI
1	$CHCl_3$	chloroform	683	47	3.26
2	C_2H_6	ethane	597	89.5	1.82
3	C_7H_8	toluene	859	85	2.52
4	$C_2H_4O_2$	acetic acid	634	67.5	2.38
5	C_3H_6	propylene	866	78	2.70
6	$C_6H_{14}O$	isopropyl ether	712	63.5	2.72
7	C_2H_4	ethylene	1005	46.5	4.19
8	C_4H_6	1,3-butadiene	991	79.4	2.94
9	C_4H_8O	vinyl ethyl ether	880	44.4	3.98
10	C_8H_8	styrene	993	19.2	6.33
11	N_2H_4	hydrazine	1338	60.5	4.25
12	C_2H_4O	ethylene oxide	1062	57.4	3.81
13	C_4H_4	vinylacetylene	2317	28.0	7.33
14	$C_{12}H_{16}N_4O_{18}$	cellulose nitrate	2213	46.7	6.12
15	C_2H_2	acetylene	2898	40.5	7.05
16	$C_3H_5N_3O_9$	nitroglycerine	2895	40.3	7.05
17	$C_4H_{10}O_2$	diethyl peroxide	968	37.3	4.64

[1]D. R. Stull, "Linking Thermodynamics and Kinetics to Predict Real Chemical Hazards," *Safety in the Chemical Industry,* Vol. 3, ed. Norman V. Steere (Easton, PA: Division of Chemical Education of the American Chemical Society, 1974), pp. 106-110.

Ethylene Oxide Explosion[9]

A process storage tank contained 6500 gallons of ethylene oxide. It was accidentally contaminated with ammonia. The tank ruptured and dispersed ethylene oxide into the air. A vapor cloud was formed and almost immediately exploded. It created an explosive force equivalent to 18 tons of TNT as evidenced by the damage. The events happened so rapidly that personnel could not take appropriate cover. One person was killed, nine injured, and property losses exceeded $16.5 million.

This accident was attributed to the lack of design protection to prevent the back-up of ammonia into this storage tank. It also appears that mitigation techniques were not part of the system (deluge systems, dikes, and the like).

Ethylene Explosion[10]

Failure of a 3/8-in compression fitting on a 1000 to 2500 psi ethylene line in a pipe trench resulted in a spill of 200 to 500 lb of ethylene. A cloud was formed and ignited giving an explosion equivalent to 0.12 to 0.30 tons of TNT. This accident took place in a courtyard giving a partially confined vapor cloud explosion. Two people were killed, seventeen injured, and property loss was $6.5 million.

[9]J. A. Davenport, "A Survey of Vapor Cloud Incidents," *Chemical Engineering Progress,* Sept., 1977, pp. 54-63.

[10]Davenport, "A Survey of Vapor Cloud Incidents," pp. 54-63.

The probable causes of this accident include (a) use of nonwelded pipe, (b) installation of pipe in trenches resulting in an accumulation of flammable vapors, and (c) lack of automated vapor detection analyzers and alarms.

Butadiene Explosion[11]

A valve on the bottom of a reactor accidentally opened due to an air failure. The spill generated a vapor cloud which was ignited 50 ft from the source. About 200 gal of butadiene spilled prior to ignition. Overpressures of 0.5 to 1 psi were estimated. Three people were killed and two were injured.

Probable causes of this accident include (a) installation of a fail open valve instead of fail closed, (b) lack of vapor detectors, (c) lack of a block installed as a mitigating device, and (d) failure to eliminate ignition sources in this operating region.

Light Hydrocarbon Explosion[12]

A pipe failed and resulted in a spill of 16,800 lb of light hydrocarbons. A vapor cloud developed and ignited. The explosion knocked out the deluge systems and electrical supplies to the fire pumps. Significant damage resulted from the subsequent fires. The maximum overpressure was estimated from the damage to be 3.5 psi at 120 ft. An equivalent of 1 ton of TNT was estimated giving an explosion yield of approximately 1 per cent of the total energy source. This accident had two fatalities and nine injuries. The total damage was estimated to be $15.6 million.

The magnitude of this accident could have been reduced with (a) improved pipe design, (b) improved deluge system design, (c) backup or more secure electrical supply, and (d) installation of detection analyzers and block valves.

Pump Vibration[13]

Vibration from a bad pump bearing caused a pump seal to fail in a cumene section of a phenol acetone unit. The released flammable liquids and vapors ignited. An explosion ruptured other process pipes adding fuel to the original fire. Damage to the plant exceeded $23 million.

This accident could have been prevented by a good inspection and maintenance program. Potential design improvements include vibration detectors, gas analyzers, block valves, and deluge systems.

Pump Failure[14]

Numerous accidents are unfortunate duplicates of previous accidents, as the following shows.

A pump roller bearing failure in a crude oil refinery initiated the fracture of the motor shaft and the pump bearing bracket. The pump casing then broke releas-

[11]Davenport, "A Survey of Vapor Cloud Incidents," pp. 54-63.

[12]Davenport, "A Survey of Vapor Cloud Incidents," pp. 54-63.

[13]William G. Garrison, *One Hundred Largest Losses, A Thirty-Year Review of Property Damage Losses in the Hydrocarbon Chemical Industries,* 9th ed. (Chicago: Marsh & McLennan Protection Consultants, 1986), p. 7.

[14]Garrison, *One Hundred Largest Losses,* p. 7.

ing hot oil which autoignited. Secondary pipe and flange failures contributed fuel to the fire. Plant damage totaled over $15 million.

Since the pump was only equipped with manually operated suction-side valves, the valves could not be reached during the fire.

Automated block valves would have minimized damage in this fire. A good inspection and maintenance program would have prevented the accident.

Ethylene Explosion[15]

A drain fitting in a high pressure (40 kpsi) compressor line broke allowing ethylene to escape. The ethylene cloud drifted and entered the intake system of an engine which was driving one of the compressors. The ethylene detonated in the engine and this explosion ignited the rest of the vapors.

The explosions were felt six miles away. Twelve buildings were destroyed and fire and explosion damage occurred throughout the polyethylene plant. The damage was estimated at over $15 million.

Automatic equipment promptly detected the hazardous vapor and operated the automatic high-density water-spray system which was designed to wash the ethylene from the atmosphere. The leak was too large for the spray system to handle.

This accident could have been mitigated if the gas detection analyzers alarmed at lower concentrations. Also in the layout design, it should have been noticed that the compressor needed special consideration to eliminate this ignition source.

Ethylene Explosion[16]

Ethylene was accidentally released from a 1/8-in stainless steel instrument tubing line leading to a gauge from a main line on a compressor system. The tubing failed as a result of transverse fatigue caused by vibration from the reciprocating compressor. Ignition may have been by static electricity. This accident caused $21.8 million in damage.

The unmanned compressor building was equipped with a combustible gas detection system. However, it failed to sound an alarm because of a faulty relay in the control room. Automatic fail-safe valves functioned properly blocking in the flow of ethylene but not before 450 to 11,000 lb of gas had already escaped.

This accident emphasizes the importance of adding gas detectors which measure flammable gases at low concentrations so that alarms and block valves can be actuated before large quantities of gas are released.

Ethylene Oxide Explosion[17]

Ethylene oxide is produced by adding ethylene, oxygen, a methane diluent, and recycled carbon dioxide to a continuous reactor. Gaseous compositions are controlled very carefully to keep the concentrations outside of the explosion limits.

[15]Garrison, *One Hundred Largest Losses,* p. 3.

[16]Garrison, *One Hundred Largest Losses,* p. 8.

[17]W. H. Doyle, "Instrument-connected Losses in the CPI," *Instrument Technology,* Oct., 1972, pp. 38-42.

One plant experienced an emergency situation. The emergency procedures specified: Close the oxygen feed valve. The oxygen control valve was normally closed by bleeding air out of the valve bonnet diaphragm (air to open). The bleed line was opened and was noted on the control panel. The air, however, did not bleed off through the bonnet vent because a mud dauber wasp constructed mud cells over the vent hole. Although the vent valve was open, as indicated on the control panel, the air could not escape.

The gases in the ethylene oxide reactor moved into the explosive region while being above the autoignition temperature. A violent explosion occurred, resulting in several injuries and significant plant damage.

It is now an industrial standard to use positive identification of the valve position on all important safety valves; limit switches which are tripped when the valve is open or shut. Additionally, all valve vent lines are now covered with bug screens to prevent blockage.

In this particular case, the accident could also have been prevented with appropriate inspection and maintenance procedures.

Lessons Learned

The case histories related to system design emphasize (a) accidents occur very rapidly, usually with inadequate time to manually return the system back into control once the accident scenario is in progress, (b) the system designs required for preventing accidents or mitigating the consequences of accidents are frequently subtle, requiring only minor process changes, and (c) the time and effort required to develop a safe system design is justified: An engineer is hired for a fraction of the cost of most accidents.

Trevor Kletz[18] and Walter B. Howard[19] have emphasized the special design features for safer plants. The following recommendations also include design features from our own experiences.

- Use the appropriate materials of construction, especially when using old systems for new applications.
- Do not install pipes underground.
- Be sure the quality of construction (like welds) meet the required specifications.
- Check all purchased instruments and equipment for integrity and functionality.
- Do not secure pipes too rigidly. Pipes must be free to expand so that they will not damage other parts of the system.
- Do not install liquid filled flanges above electrical cables. A flange leak will douse the cables with liquid.
- Provide adequate supports for equipment and pipes. Do not allow spring supports to be completely compressed.

[18]Trevor Kletz, *Learning from Accidents in Industry* (Boston: Butterworths, 1988), p. 143.

[19]Walter B. Howard, "Process Safety Technology and the Responsibilities of Industry," *Chemical Engineering Progress,* Sept., 1988, pp. 25-33.

- Design doors and lids so they cannot be opened under pressure. Add interlocks to decrease pressure before the doors can be opened. Also add visible pressure gauges at the doors.
- Do not let pipes touch the ground.
- Remove all temporary supports after construction is completed.
- Remove all temporary start-up or check-out branches, nipples, and plugs, and replace with properly designed welded plugs.
- Do not use screwed joints and fittings when handling hazardous chemicals.
- Be sure all tracing is covered.
- Check to insure all equipment is assembled correctly.
- Do not install pipes in pits, trenches, or depressions, where water can accumulate.
- Do not install relief tail pipes too close to the ground where ice blockage may make them inoperable.
- Be sure all lines which can catch water can be appropriately drained.
- When welding reinforcement pads to pipes or vessels, insure trapped air can escape through a vent during heating.
- Do not install traps in lines where water can collect and develop a corrosion problem.
- Install bellows carefully and according to manufacturers' specifications. Bellows should be used cautiously. If required, inspect frequently and replace when necessary before they fail.
- Make static and dynamic analyses of pipe systems to avoid excessive stresses or excessive vibrations.
- Design systems for easy operation and easy maintenance; for example, install manual valves within easy reach of the operators, and design pipe networks for easy maintenance or with easy access to equipment requiring maintenance.
- Install bug screens on vent lines.
- Make structural analyses of relief systems to avoid structural damage during emergency reliefs.
- Safety technology must work right the first time. Usually, there is no opportunity to adjust or improve its operation.
- Critical safety instruments must have backups.
- Provide hand operated or automatic block valves, or equivalent, for emergency shutdowns.
- Use electronic or mechanical level gauges, not glass sight glasses.
- Add fail safe block valves with a positive indication of the valve position (limit switches).

Example 13-3

Analyze the first ethylene explosion example (3/8-in fitting failure) to determine the percent fuel which actually exploded compared to the quantity of ethylene released in a vapor cloud.

Solution The total energy contained in the vapor cloud is estimated by assuming the Gibbs free energy is a good approximation for the Helmholtz free energy. The combustion reaction is

$$C_2H_4 + 3\ O_2 \rightarrow 2\ CO_2 + 2\ H_2O$$

Therefore the total energy is

$$\Delta G = \sum G_P - \sum G_R$$

Substituting the data from Table 6-8,

$$\Delta G = 2\ (-94.26) + 2\ (-54.636) - 1\ (16.282)$$
$$= -314 \text{ kcal/mole}$$

Assuming 500 lb of ethylene (molecular weight = 28) formed the cloud,

$$\Delta G = -(314{,}000 \text{ cal/gm-mole})\,(8{,}100 \text{ gm-mole})$$
$$= -2.54 \times 10^9 \text{ cal.}$$

Based on the accident investigation, the explosive energy was equivalent to 0.3 ton of TNT, or

$$\text{Energy} = (0.3 \text{ ton TNT})\left(2{,}000\frac{\text{lb}_\text{m}}{\text{ton}}\right)\left(454\frac{\text{gm}}{\text{lb}_\text{m}}\right)\left(1120\frac{\text{cal}}{\text{gm}}\right)$$
$$= 3.05 \times 10^8 \text{ cal.}$$

Therefore the fraction of energy manifested in the explosion is 3.05(100)/25.4 = 12%. This 12% is considerably higher than the 2% normally observed (see Section 6-12) for unconfined vapor cloud explosions. The higher energy conversion is a result of the explosion occurring in a partially confined area.

13-4 PROCEDURES

An organization can develop a good safety program if they have personnel who can identify and eliminate safety problems. An even better safety program, however, is developed by implementing management systems to prevent the existence of safety problems in the first place. The management systems commonly used in industry include safety reviews, operating procedures, and maintenance procedures.

The causes of all accidents can ultimately be attributed to a lack of management systems. Case histories which especially demonstrate this problem are illustrated in this section. In the study of these case histories, one must recognize that the existence of procedures is not enough. There must also be a system of checks in place to insure the procedures are actually used, and used effectively.

Leak Testing a Vessel[20]

A two foot diameter float was fabricated using stainless steel and welded seam construction. Pipefitters were given the job of checking the welds for leaks. They were instructed to use 5 psi of air pressure and a soap solution to identify the leaks.

[20] *Case Histories,* Vol. 2, p. 186.

They clamped a 100 psi air hose to a nipple on the tank. A busy instrument man gave them a gauge. The gauge was incorrectly chosen for vacuum service and not pressure because the vacuum identifier was very small.

A short time later as the fitters were carrying out the tests, the float ruptured violently. Fortunately, there was no fragmentation of the metal, and the two fitters escaped injury.

The accident investigation found that the leak test should have been conducted with a hydraulic procedure and not air, and the vessel should have been protected with a relief device. Additionally, the fitters should have taken more time to check out the gauge to insure it was correct for this application.

Man Working in Vessel[21]

Two maintenance men were replacing part of a ribbon in a large ribbon mixer. The main switch was left energized; the mixer was stopped with one of three start-stop buttons.

As one mechanic was completing his work inside the mixer, another operator on an adjoining floor pushed, by mistake, one of the other start-stop buttons. The mixer started, killing the mechanic between the ribbon flight and the shell of the vessel.

Lock-tag-and-try procedures were developed to prevent accidents of this kind. A padlocked switch at the starter box disconnect, with the key in the mechanics pocket, prevents this type of accident. After the switch gear lock-out, the mechanic should also verify the dead circuit by testing the push-button at all switches; this is the "try" part of the lock-tag-and-try procedure.

Vinyl Chloride Explosion[22]

Two vinyl chloride polymerization reactors were being operated by the same team of operators. Number 3 reactor was in the cool down and dump phase of the process while Number 4 was nearly full of monomer and in the polymerization phase. The foreman and three employees set to work to discharge the contents of Number 3, but in error they opened vessel Number 4 instead. The gaseous vinyl chloride monomer just in the process of polymerization burst out of the vessel, filled the room, and shortly afterwards exploded violently, presumably ignited by a spark from an electric motor or by static electricity generated by the escaping gas. This accident resulted in four fatalities and ten injuries in or around the plant.

The accident could have been prevented with better operating procedures and better training to make the operators appreciate the consequences of mistakes. Modern plants use interlocks or sequence controllers and other special safeguards to prevent this type of error.

Dangerous Water Expansion[23]

A hot oil distillation system was being prepared for operation. The temperature was gradually raised to 500°F. A valve at the bottom of the tower was opened to initiate the transfer of heavy hot oil to a process pump.

[21]*Case Histories,* Vol. 2, p. 225.

[22]*Case Histories,* Vol. 2, p. 113.

[23]*Hazards of Water,* Booklet One (Chicago: Amoco Oil Company, 1984), p. 20.

Before this particular start-up, a double block valve arrangement was installed in the bottom discharge line. It was not realized, however, that the second valve created a dead space between the two block valves and that water was trapped between them.

When the bottom valve was opened, the pocket of water came in contact with the very hot oil. Flashing steam surged upward through the tower. The steam created excessive pressures at the bottom of the tower and all the trays dropped within the tower. In this case the pressure luckily did not exceed the vessel rupture pressure. Although no injuries were sustained, the tower was destroyed by this accident.

Problems similar to this are usually identified in safety reviews. This accident, for example, could have been prevented if the plant used a safety review procedure during the design phase of this plant modification. A bleed line and possibly a nitrogen blow-out line would have prevented the accumulation of this water.

Consequences of contaminating hot and high boiling liquids with low boilers can be estimated using thermodynamics. If these scenarios are possible, relief valves should also be installed to mitigate these events, or adequate safeguards should be added to the system to prevent the specific hazard scenario.

Lessons Learned

Procedures are sometimes incorrectly perceived as bureaucratic regulations which impede progress. When reviewing case histories it is apparent that safety procedures are needed to help the chemical industry (a) eliminate injury to personnel, (b) minimize incapacitating damage to facilities, and (c) maintain steady progress.

In the review of case histories relevant to procedures, additional lessons are identified.[24]

- Use a permit procedure for opening vessels which are normally under pressure.
- Never use gas to open plugged lines.
- Communicate operating changes to other operations which may be affected by the change.
- Train operators and maintenance personnel to understand the consequences of deviations from the norm.
- Make periodic and precise audits of procedures and equipment.
- Use procedures effectively (lock-tag-and try, hot work, vessel entry, emergency, and the like).
- Use safety review procedures during the design phases of projects, including new installations or modifications to existing systems.

13-5 IN CONCLUSION

This chapter on case histories is very brief and does not include all the lessons relevant to accidents. The references provide excellent information for more

[24]T. A. Kletz, *What Went Wrong? Case Histories of Process Plant Disasters* (Houston: Gulf Publishing Company, 1985), pp. 182-188.

studies. There is significant information in the open literature. However, case histories and safety literature is of no value unless it is studied, understood, and used appropriately.

Example 13-4

Using the dangerous water expansion example, compute the approximate pressures which were developed in the bottom of this column. Assume a column diameter of 2 ft, a water slug of 1 gallon, and the column pressure of 10 psia.

Solution The areas of the column trays are 3.14 sq ft. If the tray vapor paths are small openings, the worst case scenario assumes that all the water vapor collects beneath the bottom tray. Assuming a tray spacing of 1 ft, the volume under the first tray is 3.14 cubic ft. Using an equation of state,

$$PV = \left(10.73 \frac{\text{psia ft}^3}{\text{lb-mole °R}}\right) \text{n T}$$

$$P = \left(10.73 \frac{\text{psia ft}^3}{\text{lb-mole °R}}\right) \frac{(0.464 \text{ lb-mole})(500 + 460)\text{°R}}{3.14 \text{ ft}^3}$$

$$P = 1{,}522 \text{ psia if all the water vaporized.}$$

At 500°F the vapor pressure of water is 680 psia. Therefore, the maximum pressure is 680 psi if some water remains as liquid water. The force on the bottom tray is

$$F = (680 \text{ lb}_f/\text{in}^2)(3.14 \text{ ft}^2)(144 \text{ in}^2/\text{ft}^2)$$

$$= 307{,}500 \text{ lb}_f$$

If the tray is bolted to the column with six 1/2-in bolts, the stress on each bolt is

$$S = \left(\frac{307{,}500 \text{ lb}_f}{6 \text{ bolts}}\right)\left[\frac{1 \text{ bolt}}{(3.14)(0.25 \text{ in})^2}\right]$$

$$= 261{,}000 \text{ lb}_f/\text{in}^2$$

Assuming a tensile strength of 85,000 psi for stainless 316, it is clear that the trays are stressed beyond the point of failure. Evidently the vessel could handle 680 psia, otherwise it would have also ruptured.

This example explains why all the column trays were torn away from the supports, and also illustrates the hazards of contaminating a hot oil with a low boiling component.

SUGGESTED READING

Case Histories of Accidents in the Chemical Industry, Vol. 1 (Washington, DC: Manufacturing Chemists' Association, July, 1962).

Case Histories of Accidents, Vol. 2, January, 1966.

Case Histories of Accidents, Vol. 3, April, 1970.

TREVOR A. KLETZ, *What Went Wrong? Case Histories of Process Plant Disasters* (Houston: Gulf Publishing Company, 1985).

FRANK P. LEES, *Loss Prevention in the Process Industries* (London: Butterworths, 1986), Appendix A1: Flixborough; Appendix A2: Seveso; Appendix A3: Case Histories.

Loss Prevention, Vol. 1 through 5 (New York: American Institute of Chemical Engineers).

PROBLEMS

13-1. Illustrate the layered accident investigation process, using Example 13-1 as a guide, to develop the underlying causes of the accident described in Section 13-1, Duct System Explosion.

13-2. Repeat Problem 13-1 for the accident described in Section 13-2, Bottle of Isopropyl Ether.

13-3. Repeat Problem 13-1 for the accident described in Section 13-2, Nitrobenzene Sulfonic Acid Decomposition.

13-4. Repeat Problem 13-1 for the accident described in Section 13-3, Butadiene Explosion.

13-5. Repeat Problem 13-1 for the accident described in Section 13-4, Vinyl Chloride Explosion.

13-6. A square stainless pad (5in × 5in × 1/8 in) is welded to a vessel which is used for high temperature service (1200°C). The welder welds continuously around the pad, forgetting to leave an opening for a vent. Compute the pressure change between the pad and the vessel if the temperature changes from 0°C to 1200°C.

13-7. Vessels normally have a relief device to prevent damage during thermal expansion. A stainless, cylindrical vessel has 1/4-in thick walls and is 4 ft in diameter. It is filled with 400 gallons of water and 0.2 cubic ft of air is trapped at a pressure gauge. Start at 0 psig and 50°F and then heat the vessel. At what temperature will this vessel rupture if it does not have a relief?

13-8. Compute the Reaction Hazard Index for nitroglycerine.

13-9. Compute the Reaction Hazard Index for acetylene.

13-10. A hydrogen peroxide still is used to concentrate peroxide by removing water. The still is of high purity aluminum, a material which is noncatalytic to the decomposition of peroxide vapor. The still is designed to produce 78 percent hydrogen peroxide. It will explode spontaneously at about 90 percent. Illustrate some recommended design features for this still.

13-11. A 1000 gal cylindrical vessel (4 ft in diameter) is nearly filled with water. It has a 10% pad of air at 0 psig and 70°F. If this air is completely soluble at 360°F and 154 psia, what will the vessel pressure be at 380°F? Assume a wall thickness of 1/4 in of stainless 316 and flat cylindrical heads.

13-12. An operation requires the transfer of 50 gal of toluene from a vessel to a 55 gal drum. Develop a set of operator instructions for this operation.

13-13. A reactor system is charged accidentally with benzene and chlorosulfonic acid with the agitator off. At this condition, the two highly reactive reactants form two layers in the reactor. Develop a set of operating instructions for safely handling this situation.

13-14. Develop design features to prevent the situation as described in Problem 13-13.

13-15. Why are bug screens installed on control valve vents?

13-16. Read the article by W. B. Howard (*Chemical Engineering Progress,* Sept., 1988, p. 25). Describe the correct and incorrect designs for installing flame arrestors.

13-17. From W. B. Howard's article (Problem 13-16), describe his concepts concerning combustion venting and thrust forces.

APPENDIX I
Formal Safety Review Report

RESEARCH MEMORANDUM

CHEMICAL ENGINEERING

SAFETY REVIEW FOR PILOT PODBIELNIAK
LIQUID-LIQUID EXTRACTION SYSTEM

AUTHOR: J. Doe SUPERVISOR: W. Smith
November 8, 1988

SUMMARY

Chemical Engineering's Podbielniak (POD) liquid-liquid extraction pilot system has been reassembled. It will be used to evaluate the water-washability of toluene. A formal safety review was held 10/10/88. Main action items from that review included (1) padding all vessels containing solvent with nitrogen, (2) grounding and bonding all tanks containing solvent, (3) adding dip legs to all vessels, (4) using elephant trunks at drum openings, (5) adding heat exchangers equipped with temperature gauges to cool hot solvent, (6) purging all vessels containing solvent with nitrogen prior to start-up, (7) changing the emergency procedure to activate the spill alarm in the event of a spill and to trip the sewer isolation valve, and (8) adding receiving drums for all output streams containing solvent.

Subsequently, a few equipment changes were made during initial system check-out and test runs. These changes were made to enhance operability, not safety; e.g., (1) the pump (P1) generated insufficient head and a stronger spring was installed; and (2) a light-liquid in sample point, a few check valves, and additional temperature and pressure gauges were installed.

ABC chemical	ABC chemical	ABC chemical

DISTRIBUTION:

All

REPORT NUMBER 88-5
SECURITY CLASS None
PROJECT NUMBER 6280
SUPERVISOR(S) APPROVAL(S)

____

____

____

I. INTRODUCTION

A. Process Summary

The following procedure is used to wash toluene in the equipment provided:

1) An appropriate amount of solvent is transferred from the solvent storage tank to the emulsion tank.
2) Water is added to the solvent to form an emulsion.
3) The emulsion is heated to 190°F.
4) The emulsion is separated in the centrifugal contactor (POD) which produces a stream containing water soluble impurities and a stream of washed solvent.

B. Reactions and Stoichiometry

No reaction takes place. As far as stoichiometry is concerned, typically one part water is added to one part solvent. Flow rates for the above are based on a maximum of 1000 cc/min solvent to the POD.

C. Engineering Data

Toluene has a vapor pressure of 7.7 psi at 190°F. System operating pressures will normally be 40-50 psig around the POD, with pumps capable of delivering 140 psig. System temperatures will be maintained between 190 and 200°F. Typical viscosities will be under 10 cp at this temperature.

II. RAW MATERIALS AND PRODUCTS

Solvents:

The most frequently used solvent is toluene. Toluene boils at 231°F, but forms an azeotrope with water boiling at 183°F. Since this is below the system operating temperature, hazards are present due to flammability and volatility. In addition, toluene presents special problems from a personnel exposure viewpoint as a suspected teratagen.

To minimize hazards, the following precautions will be taken:

1) All vessels containing solvent are N_2 padded and grounded.
2) All potential solvent exposure points will be in close proximity to "elephant trunk" exhaust ducts for ventilation.
3) All product streams are cooled before discharge or sampling.
4) Colorimetric sampling tubes will be available for ambient air monitoring.

The possibility exists for using other solvents in the system. Safety reviews for each will be conducted as needed.

III. EQUIPMENT SETUP

A. Equipment Description (Sketches attached)

1) *Emulsion Tank:* The emulsion tank consists of a jacketed, agitated, 50-gal glass lined Pfaudler reactor with N_2 pad and relief valve. Emulsion is heated in the vessel by applying steam to the jacket. Temperature is controlled by means of a temperature indicating controller which measures the temperature in the vessel. The controller modulates a control valve in the steam line to the jacket. Emulsion is circulated from the bottom of the reactor to the POD system and back to the reactor top by means of a Viking pump driven by a 2-HP 1745 RPM motor.

   A slipstream is fed from this loop to the POD system. Pressure in this circulating loop is controlled by means of a back pressure controller located in the return line to the top of the reactor.

2) *Solvent System:* The solvent storage tank is a 75-gal stainless steel pressure vessel (112 psi @ 70°F) with an integral sight glass, N_2 pad and relief valve. Solvent is

pumped from the bottom of the storage tank to the emulsion tank. The pump is a Burks turbine pump driven by an XP rated, 3/4 HP 3450 rpm motor. A dip pipe is used to vacuum charge solvent through a dip leg in the vessel where grounding and bonding is secured.

3) *POD System:* The POD system consists of a Baker-Perkins Model A-1 Contactor (i.e., a Podbielniak centrifugal contactor) fabricated in 316 stainless. A variable speed drive is capable of rotating the unit at speeds up to 10,000 rpm. The normal operating speed is 8100 RPM.

   The solvent/water emulsion is heated in its subsystem, and flows through a Micro Motion mass flow meter. The emulsion is fed to the POD where the water and organic phases are separated. Through this contact and separation, the impurities are extracted into the aqueous phase. This results in a relatively clean solvent.

4) *Washed Solvent System:* The washed solvent tank is a grounded 55-gal drum. An elephant trunk positioned over the bung vents the drum to the exhaust system. Material fed to the drum is cooled from the POD operating temperature of approximately 190°F to 80-110°F by a stainless steel heat exchanger.

5) *Waste Water System:* The waste water tank is also a grounded, 55-gal drum vented to the exhaust system. The heavy liquid out (HLO) stream from the POD system is cooled before discharge into the drum by a stainless steel heat exchanger. Disposal will depend on the solvent used, its solubility in water and environmental constraints.

B. Equipment Specifications

1) Emulsion System

   Reactor: 50-gal, glass-lined, jacketed Pfaudler;
   Operating pressures: reactor — 150 psi @ 450°F
   jacket — 130 psi
   Safety relief valves: reactor — 60 psi
   jacket — 125 psi

   Agitator: Turbine, 3.6 HP, 1750 RPM, XP rated motor, variable speed drive.

   Circulating pump: Viking series HL124, motor — 2 HP, 1745 RPM, XP rated.

   Micro Motion Mass Flowmeter: 316L stainless steel, 0-80 lb/min mass flow range, accuracy of 0.4% of range, XP rated with electronics unit mounted separately in non-hazardous area.

2) Solvent System

   Tank: 75 gal., stainless steel, rupture disk set at 112 psi.

   Pump: Burks turbine, model ET6MYSS; motor: 3/4 HP, 3450 RPM, XP rated.

3) POD System

   POD: Baker-Perkins A-1 centrifugal contactor, 316 SS, max. temp: 250°F, max. pressure: 250 psig, max. speed: 10,000 RPM.

   Drive: Variable speed Reeves Motodrive, 935-3950 RPM, 3 HP, 1745 RPM motor, XP rated.

4) Washed Solvent System

   Tank: 55-gal drum.

   Light liquid out (LLO) Cooler: American Standard, single pass, SS, model 5-160-03-024-001, max. temperature: 450°F, max. working pressure: 225 psig shell, 150 psig tube.

5) Waste Water System

   Tank: 55 gal drum.

   HLO Cooler: Same as LLO cooler.

IV. PROCEDURES

A. Normal Operating Procedures

1. Purge solvent and emulsion tanks with nitrogen by valves 1(a) and 1(b).
2. If necessary, solvent and emulsion tanks are vented to "elephant trunks" and into the exhaust system through valves 2(a) and 2(b).
3. Pull a vacuum (15 in Hg) on the solvent storage tank and charge with solvent by sucking it from the appropriate drum. Check the tank level using the level glass. Periodically check the air for toluene by using colorimetric tubes.
4. Break the vacuum and pad with nitrogen through valve 1(a).
5. Make sure valve 3 is closed from the water head tank to the emulsion tank.
6. Charge the proper amount of softened water through valve 4 to the water head tank located above the emulsion tank.
7. Close valve 4 and pad the water head tank with nitrogen through valve 5.
8. Turn on the emulsion tank agitator.
9. Pump solvent from the solvent storage tank to the emulsion tank:
   a. Line up valves from solvent storage tank through pump P2, to the top of the emulsion tank.
   b. Start pump P2.
   c. Stop pump and close valves when addition is complete.
10. Open valve 3 and add water in the head tank to the emulsion tank. Close valve 3 when addition is complete.
11. Establish circulation in the emulsion system:
    a. Close valve 6 on the feed stream to the Micro Motion.
    b. Line up valves from the bottom of the tank to pump P1 and from the return line back to the top of the vessel.
    c. Start pump P1.
    d. Open steam flow to the jacket of the feed tank.
    e. Bring emulsion up to temperature (190°F).
12. Turn on cooling water to the solvent (LLO) discharge cooler and to the aqueous (HLO) discharge cooler.
13. Line up valves on the HLO and LLO streams from the POD, to the coolers, and to their respective waste tanks.
14. Open valve 10 to fill the POD.
15. Start the motor for the POD and slowly bring up to the desired rpm.
16. Open valve 6 to begin emulsion flow.
17. Adjust flow to obtain desired rate on Micro Motion meter.
18. Control back pressure on the POD LLO and HLO streams by adjusting valves 11(a) and 11(b) respectively.
19. Samples can be obtained from the LLO stream via valve 12(a) and from the HLO stream via valve 12(b).
20. To shut down the POD after a run has been completed:
    a. Close valve 6.
    b. Reduce pressure on the LLO stream (valve 11a) and slowly reduce rotor speed.
    c. Turn off POD motor.
    d. Close valve 10 after the rotor has stopped.
    e. Shut down emulsion system.
    f. Shut off steam and cooling water.

B. Safety Procedures

The safety concerns unique to this operation are due to:

a. The solvent used is volatile and flammable and is also being used at a temperature above its normal atmospheric boiling point.
b. The materials are all hot (190°F or greater) and capable of producing thermal burns.
c. Toluene presents a special handling problem due to potential health hazards.

The specific procedures to be followed to minimize the risks associated with the above are:

a. Flammable Solvents
   (1) Solvents are exposed to atmosphere only with adequate ventilation.
   (2) Solvents are only transferred into and out of the system when cold. Do not operate if coolers are not functioning properly.
   (3) All solvent containing process vessels are N_2 purged and maintained under N_2 pad or blanket.
   (4) Vapors containing solvent are vented only to the exhaust ducts, never into the worker area.
   (5) Initial opening of sample and product valves to atmosphere is done slowly to avoid flashing.
   (6) All transfers of solvent-containing streams to or from drums are done in accordance with accepted bonding and grounding procedures.
   (7) All equipment is electrically grounded.
b. Hot Material
   (1) Avoid contact with hot process lines and vessels. Most lines are insulated for personnel protection.
   (2) Wear gloves when working on potentially hot equipment.
   (3) Periodically check stream temperatures and cooling water flow to insure coolers are working properly.
c. Health Hazards (Toxicity, etc.)
   (1) Handle potentially hazardous material only when material is cool and adequate ventilation is present.
   (2) Periodically check operating area for leaks with colorimetric tubes.
   (3) Repair any leaks immediately.

1. Emergency Shutdown
   a. Close solvent valve at bottom of solvent storage tank (if open).
   b. Shut off solvent pump P2 (if operating).
   c. Close valve at bottom of emulsion tank.
   d. Shut off emulsion pump P1.
   e. Shut off steam to the emulsion tank jacket.
   f. Shut down POD drive system.
2. Fail-Safe Procedures
   a. Steam failure — no negative consequences.
   b. Cooling water failure — shut down system.
      (1) LLO to washed solvent drum will flash and be sucked into vent system.
      (2) HLO to waste drum — some solvent may flash off and be sucked into vent system.
   c. Electrical failure — close HLO and LLO valves to protect the unit while it coasts to a stop.

d. N_2 failure — stop any operational procedures.
e. Exhaust system failure — shut down system.
f. Pump failure — shut down system.
g. Air failure — all steam control valves fail closed. All cooling water control valves fail open.

3. Spill and Release Procedures
   a. Solvent spill — follow hazardous spill response as outlined in Safety Manual.
      (1) Sound alarm and evacuate if warranted (e.g., large drum quantity spill or hot solvent spill).
      (2) Vent system on high speed.
      (3) Trip sewer isolation valves.
      (4) If safe to do so, isolate equipment and ignition sources, and absorb or dike the spill.
      (5) Allow excess to evaporate. Check area with explosimeter and colorimetric tubes. Do *not* enter explosive atmosphere.
      (6) When safe to do so, sweep up any absorbant into waste drums for proper disposal.
      (7) Consult with Environmental department if material is trapped in sewer system.

C. Waste Disposal

The washed solvent is collected in drums for disposal. The aqueous stream, after analysis, can be sent directly to the publicly owned treatment works (POTW). Limits have not yet been set for dumping *vs* waste disposal in drums. If the solvent being used is a regulated substance (such as toluene), drum disposal of the HLO may be the only acceptable way.

D. Clean-up Procedure

Minor spills are soaked up with absorbant material and disposed of in drums. Equipment is washed with hot and/or cold water as necessary.

V. SAFETY CHECKLIST

____ Purge emulsion tank with nitrogen, fill, establish nitrogen pad.
____ Purge solvent storage tank with nitrogen, fill, establish nitrogen pad.
____ Purge washed solvent tank with nitrogen, establish nitrogen pad.
____ Check cooling water flow in two coolers.
____ Vent system operational.
____ Availability of absorbant material and disposal drum.
____ Availability of impervious gloves, goggles/face shield.
____ Sniff area with colorimetric tubes for hazardous solvents.
____ Availability of air line hood.
____ Check all drums for proper grounding.

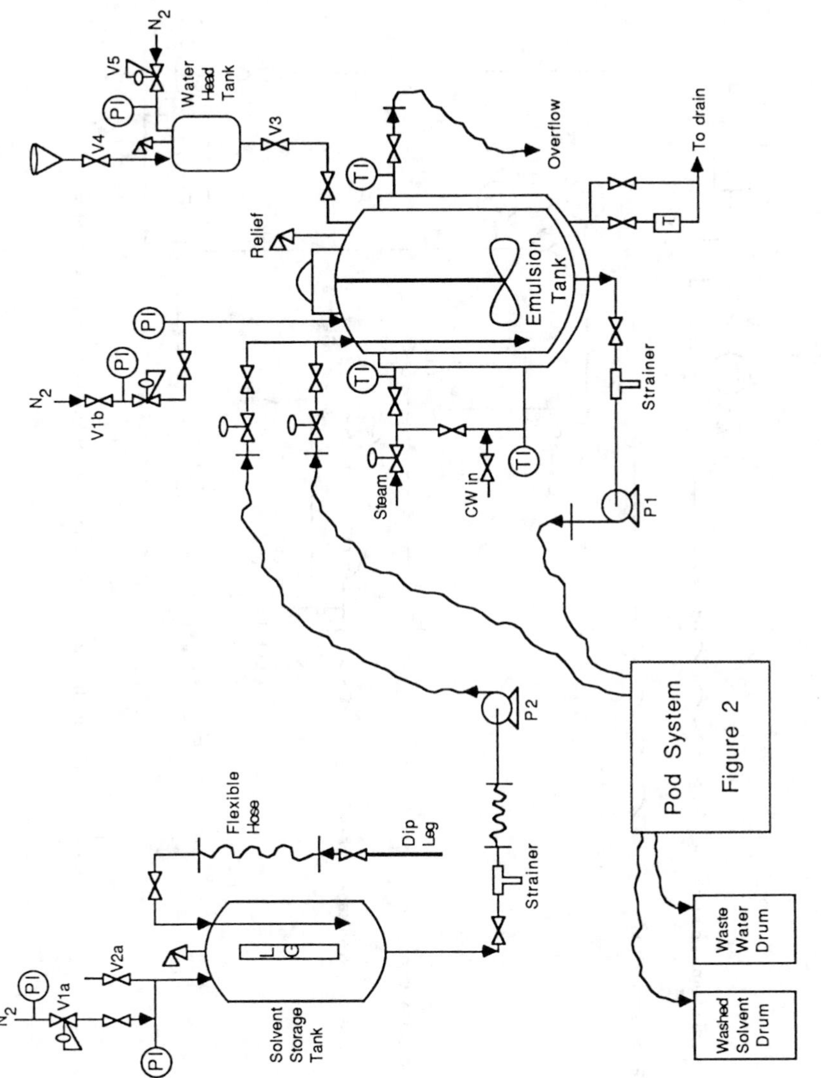

Figure A-1 Podbielniak extraction system.

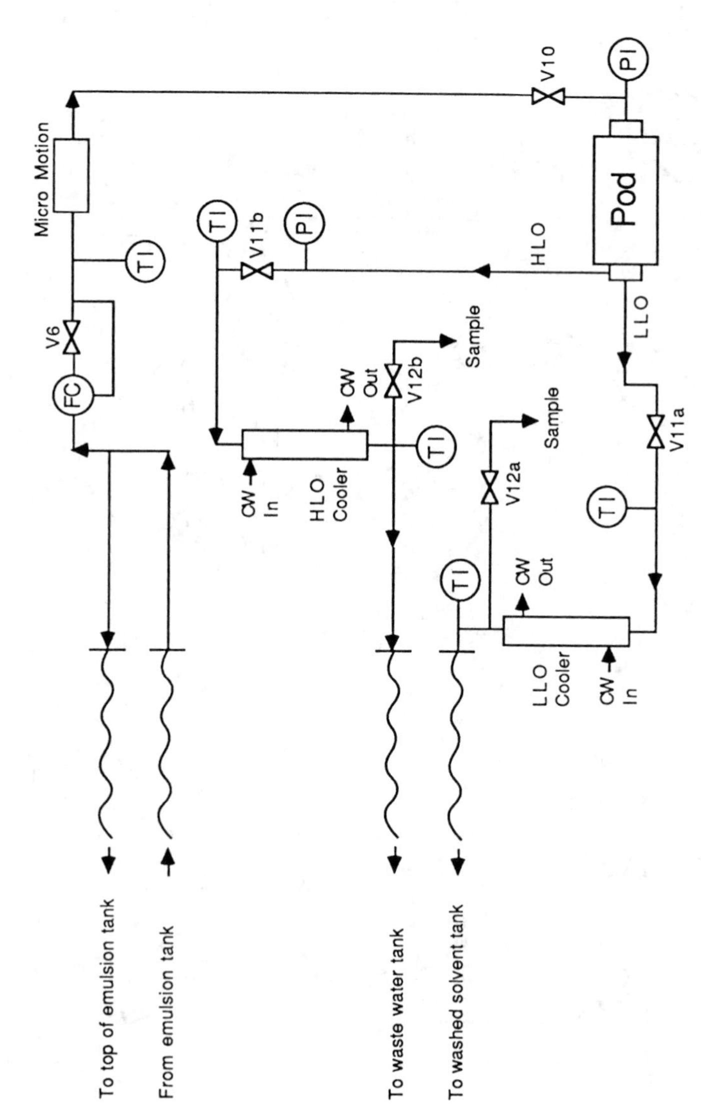

Figure A-2 Piping diagram for Podbielniak solvent water wash system.

MATERIAL SAFETY DATA SHEET

Common chemical name: Toluene	**Physical state:** Colorless liquid	**Odor:** sweet, pungent	
Synonym: Methylbenzene	**Molecular weight:** 92.13	**Odor threshold:** 2-4 ppm	**CAS No:** 108-88-3
Chemical formula: C_7H_8	**Explosive limits:** 1.27-7.0%	**Vapor pressure:** 36.7 mm Hg @ 30°C	**PEL:** 100 ppm-skin

Toxic Properties:

Eyes:	Moderately irritating
Skin:	Moderately irritating
Inhalation:	CNS effects
Ingestion:	Moderately toxic

Vapors may cause eye irritation. Eye contact with liquid may result in corneal damage and conjunctival irritation which lasts for 48 hours. Inhalation may be irritating and result in fatigue, headaches, CNS effects, and narcosis at high concentrations. Toluene is skin absorbed. Repeated or prolonged skin contact may result in irritation, defatting, and dermatitis.

Occasionally, chronic poisoning may result in anemia, leukopenia, and enlarged liver.

Some commercial grades of toluene contain small amounts of benzene as an impurity. Benzene is an OSHA regulated material.

Personal Protection:

Goggles, impervious gloves, protective clothes and shoes are recommended. Chemical cartridge respirators are sufficient for routine handling. Air-line respirators or self-contained breathing apparatus are recommended for high concentrations.

First Aid:

Eyes:	Flush thoroughly with water. Consult a physician.
Skin:	Wash affected areas with plenty of water. If irritation persists, get medical attention.
Inhalation:	Remove to fresh air. Aid in breathing if necessary. Consult a physician.
Ingestion:	If swallowed, do *not* induce vomiting. Call a physician immediately.

Special Precautions/Considerations:

This is a flammable liquid. The flash point is 40°F and should be handled accordingly. During transport and storage, protect against physical damage. Outside or detached storage is preferable. Separate from oxidizing materials.

APPENDIX II
Saturation Vapor Pressure Data[1]

$$\ln(P^{sat}) = A - \frac{B}{C + T}$$

where

P^{sat} is the saturation vapor pressure (mm Hg),
T is the temperature (K), and
A, B, C are constants given below.

Species	Formula	Range(K)	A	B	C
Acetone	C_3H_6O	241-350	16.6513	2940.46	−35.93
Benzene	C_6H_6	280-377	15.9008	2788.51	−52.36
Carbon tetrachloride	CCl_4	253-374	15.8742	2808.19	−45.99
Chloroform	$CHCl_3$	260-370	15.9732	2696.79	−46.19
Cyclohexane	C_6H_{12}	280-380	15.7527	2766.63	−50.50
Ethyl acetate	$C_4H_8O_2$	260-385	16.1516	2790.50	−57.15
Ethyl alcohol	C_2H_6O	270-369	18.9119	3803.98	−41.68
n-Heptane	C_7H_{16}	270-400	15.8737	2911.32	−56.51
n-Hexane	C_6H_{14}	245-370	15.8366	2697.55	−48.78
Methyl alcohol	CH_4O	257-364	18.5875	3626.55	−34.29
n-Pentane	C_5H_{12}	220-330	15.8333	2477.07	−39.94
Toluene	$C_6H_5CH_3$	280-410	16.0137	3096.52	−53.67
Water	H_2O	284-441	18.3036	3816.44	−46.11

[1]Selected from David Himmelblau, *Basic Principles and Calculations in Chemical Engineering* (New Jersey: Prentice Hall, 1982), p. 591.

APPENDIX III
Unit Conversion Constants[1]

READ ACROSS

Volume equivalents				
in^3	ft^3	U.S. gal	liters	m^3
1	5.787×10^{-4}	4.329×10^{-3}	1.639×10^{-2}	1.639×10^{-5}
1728	1	7.481	28.32	2.832×10^{-2}
231	0.1337	1	3.785	3.785×10^{-3}
61.03	3.531×10^{-2}	0.2642	1	1.000×10^{-3}
6.102×10^4	35.31	264.2	1000	1

Mass equivalents			
avoir oz	lb_m	grains	grams
1	6.25×10^{-2}	437.5	28.35
16	1	7000	453.6
2.286×10^{-3}	1.429×10^{-4}	1	6.48×10^{-2}
3.527×10^{-2}	2.20×10^{-3}	15.432	1

[1]Selected from David M. Himmelblau, *Basic Principles and Calculations in Chemical Engineering,* 4th ed. (New Jersey: Prentice Hall, 1982).

Linear measure			
m	in	ft	mi
1	39.37	3.2808	6.214×10^{-4}
2.54×10^{-2}	1	8.333×10^{-2}	1.58×10^{-5}
0.3048	12	1	1.8939×10^{-4}
1609	6.336×10^{4}	5280	1

Power equivalents				
HP	kw	ft-lb_f/sec	BTU/s	J/s
1	0.7457	550	0.7068	745.7
1.341	1	737.56	0.9478	1000
1.818×10^{-3}	1.356×10^{-3}	1	1.285×10^{-3}	1.356
1.415	1.055	778.16	1	1055
1.341×10^{-3}	1.000×10^{-3}	0.7376	9.478×10^{-4}	1

Heat, energy, or work equivalents					
ft-lb_f	kWh	HP hr	BTU	Calorie	Joule
1	3.766×10^{-7}	5.0505×10^{-7}	1.285×10^{-3}	0.3241	1.356
2.655×10^{6}	1	1.341	3412.8	8.6057×10^{5}	3.6×10^{6}
1.98×10^{6}	0.7455	1	2545	6.4162×10^{5}	2.6845×10^{6}
778.16	2.930×10^{-4}	3.930×10^{-4}	1	252	1055
3.086	1.162×10^{-6}	1.558×10^{-6}	3.97×10^{-3}	1	4.184
0.7376	2.773×10^{-7}	3.725×10^{-7}	9.484×10^{-4}	0.2390	1

Pressure equivalents				
mm Hg	in Hg	bar	atm	kPa
1	3.937×10^{-2}	1.333×10^{-3}	1.316×10^{-3}	0.1333
25.40	1	33.87	3.342×10^{-2}	3.387
750.06	29.53	1	0.9869	100.0
760.0	29.92	1.013	1	101.3
75.02	0.2954	0.01000	9.872×10^{-3}	1

Ideal gas constant, R_g
1.9872 cal/gm-mole K
1.9872 BTU/lb-mole °R
10.731 psia ft^3/lb-mole °R
8.3143 kPa m^3/kg-mole K = 8.314 J/gm-mole K
82.057 cm^3 atm/gm-mole K
0.082057 (1) atm/gm-mole K
21.9 (in Hg) ft^3/lb-mole °R
0.7302 ft^3 atm/lb-mole °R
1,545.3 ft lb$_f$/lb-mole °R

Gravitational constant, g_c
32.174 ft-lb_m/lb_f-s^2
1 (kg m/s^2)/N
1 (gm cm/s^2)/dyne

Miscellaneous
1 Poise = 100 centipoise = 0.1 kg/m s = 0.1 Pa s = 0.1 N s/m^2
1 N = 1 kg m/s^2
1 J = 1 N m = 1 kg m^2/s^2

Index